COMPLEX VARIABLES
AND THE
LAPLACE TRANSFORM
FOR
ENGINEERS

COMPLEX VARIABLES
AND THE
LAPLACE TRANSFORM
FOR
ENGINEERS

Wilbur R. LePage

Department of Electrical and Computer Engineering
Syracuse University

Dover Publications, Inc.
New York

Published in Canada by General Publishing Com-
pany, Ltd., 30 Lesmill Road, Don Mills, Toronto,
Ontario.
Published in the United Kingdom by Constable
and Company, Ltd., 10 Orange Street, London
WC2H 7EG.

This Dover edition, first published in 1980, is an
unabridged and corrected republication of the work
originally published in 1961 by McGraw-Hill, Inc.

International Standard Book Number: 0-486-63926-6
Library of Congress Catalog Card Number: 79-055908

Manufactured in the United States of America
Dover Publications, Inc.
180 Varick Street
New York, N.Y. 10014

To THOSE WHO *find satisfaction in reflective thought and who regard the scholarly quest for understanding as inherently valuable to the individual and to society, this book is dedicated*

PREFACE

This book is written for the serious student, probably at the graduate level, who is interested in obtaining an understanding of the theory of Fourier and Laplace transforms, together with the basic theory of functions of a complex variable, without which the transform theory cannot be understood. No prior knowledge other than a good grounding in the calculus is necessary, although undoubtedly the material will have more meaning in the initial stages for the student who has the motivation provided by some understanding of the simpler applications of the Laplace transform. Such prior knowledge will usually be at an introductory level, having to do with the mechanical manipulation of formulas. It is reasonable to begin a subject by the manipulative approach, but to do so should leave the serious student in a state of unrest and perhaps mild confusion. If he is alert, many of the manipulative procedures will not really make sense. If you have experienced this kind of confusion and if it bothers you, you are ready to profit from a study of this book, which occupies a position between the usual engineering treatments and the abstract treatments of the mathematicians. The book is intended to prepare you for creative work, not merely to solve stereotyped problems. The approach is intended for workers in an age of mature technology, in which the scientific method occupies a position of dominance. Because of the heavy emphasis on interpretation and because of the lack of generality in the proofs, this should be regarded as an engineering book, in spite of the extensive use of mathematics.

The highly personal aspect of the learning process makes it impossible for an author to write a book that is ideal for anyone except himself. Recognition of this reality provides the key to how you can benefit most from a book such as this. Probably you will want first to search for the main pattern of ideas, with the details to be filled in at such time as your interest is aroused. Learning is essentially a random process, and an author cannot insist that events in your program of learning will occur in any predictable order. Therefore, it is recommended that you remain alert to points of interest, and particularly to points of confusion. To acknowledge that a concept is not fully understood is to recognize it as a point of interest. It is suggested that you give due respect to such a point, at the time it cries for attention, without regard for whether it is the next topic in the book—searching for related ideas, referring to other texts, and, above all, experimenting with your own ideas. There is such a wealth of interrelatedness of topics that, if you do this with complete intellectual honesty, you will eventually find that you have more than covered the text, without ever having read it in continuous fashion from cover to cover.

The text is roughly in two parts. The first part, on functions of a complex variable, begins at a relatively low level. Experience with graduate students in electrical engineering at Syracuse University, over a period of five years during which the first eleven chapters of this material were used in note form, indicates that the approach is acceptable to most beginning graduate students. The level of difficulty gradually increases throughout the book, and the material beyond Chapter 7 attains a relatively high degree of sophistication. However, it is anticipated that with a gradual increase in your knowledge the material will present an aspect of approximately constant difficulty.

The material on functions of a complex variable is quite similar to many of the standard beginning engineering-oriented texts on the subject, except perhaps for the amount of interpretation and illustrative material. One other difference will be noticed immediately: the use of $s = \sigma + j\omega$ instead of the usual $z = x + iy$. To use j in place of i is established practice in engineering literature and probably is not controversial. The choice of s, σ, and ω in place of z, x, and y was a calculated risk in terms of reader reaction. It provides a unity in this one book, but it will necessitate a symbol translation when comparing with other books on function theory. My apologies are offered to anyone for whom this is a nuisance.

A few suggestions are offered here, to both student and teacher, as to what material might be considered superfluous in an initial course of study. Chapter 1 provides motivation and a perspective viewpoint. It will not serve all students equally well, possibly being too concise for some and too elementary for others. It has no essential position in the stream of continuity and therefore can be omitted. Chapter 2 gives the main introductory concepts and is essential. Chapter 3 should be covered to the extent of firmly establishing the geometrical interpretation of a function of a complex variable as a point transformation between two planes, and the basic ideas of conformality of the transformation as related to analyticity of the function. However, on first reading it may be advisable not to go into details of all the examples given. Much of this is reference material.

Chapters 4 and 5 are very basic and should not be omitted. Chapter 6 bears a relation to the general text material similar to that of Chapter 3.

Some knowledge of multivalued functions is certainly essential, but the student should adjust to his own taste how much detail and how many practical illustrations are appropriate. Much of this chapter may be regarded as reference material. Chapter 7, the last of the chapters devoted to function theory, consists almost completely of reference material pertinent to network theory, and can be omitted without loss of continuity. In fact the encyclopedic nature of this chapter causes some of the topics to appear out of logical order.

Chapter 8 contains background on certain properties of integrals, particularly improper integrals, in anticipation of applications in the later chapters. This chapter deals with difficult mathematical concepts and, compared with the standards of rigor set in the other chapters, is largely intuitive. The main purpose is to alert the reader to the major problems arising when an improper integral is used to represent a function. The chapter can be skipped without loss of continuity, but at least a cursory reading is recommended, followed by deeper consideration of appropriate parts while studying the later chapters.

Chapters 9 and 10 form the core of the second part of the book— the Fourier and Laplace transform theory. In the transition from the Fourier integral to the one-sided Laplace integral, the two-sided Laplace integral is introduced, on the argument that conceptually the two-sided Laplace integral lies midway between the other two. This seems to smooth the way for the student to negotiate the subtle conceptual bridge between the Fourier and Laplace transform theories. If the Laplace transform theory is to be understood at the level intended, there seems to be no alternative but to include the two-sided Laplace transform.

The theory of convolution integrals presented in Chapter 11 is certainly fundamental, although not wholly a part of Laplace transform theory, and therefore should be included in a comprehensive course of study. The remaining chapters deal with special topics, and each has its roots in the all-important Chapter 10. Chapter 12 is essentially a continuation of Chapter 10 and is primarily a reference work. The practical applications treated in Chapter 13 provide a brief summary of the theory of linear systems and preferably should be studied together with, or following, a course in network theory. Otherwise the treatment may be too abstract. However, taken at the proper time, it can be helpful in unifying the ideas about this important field of application.

In regard to Chapter 14, on impulse functions, a critical response from some readers is anticipated by saying that this chapter represents one particular viewpoint. Many will say that 'it labors the point and that all the useful ideas contained therein can be reduced to one page. This is a matter of opinion, and it is thought that a significant number of students can benefit from this analysis. Everyone knows that impulse functions are not functions in the true sense of the word, and this chapter has something to say about this question, casting the usual results in such a form that no doubts can arise as to the meaning. A knowledge of the customary formalisms associated with impulse functions and their symbolic transforms represents a bare minimum of accomplishment in this chapter.

Finally, Chapters 15 and 16 on periodic functions and the Z transform are related and provide background material for many of the practical applications which you will probably study elsewhere. The justification for including these chapters is to be found in the desire to present this fundamental applied material with the same degree of completeness as the Laplace transform itself.

No particular claim is made for originality in the basic theory, other than in organization and details of presentation. Nor is it claimed that the proofs are always as short or as elegant as possible. The general criterion used was to select proofs that are realistically straightforward, with the hope that this would ensure a high degree of intellectual honesty, while always keeping in touch with simple concepts.

In its preliminary versions, this material has been taught by about twenty different colleagues. All of these persons have made helpful suggestions, and a list of their names would be too long to give in its entirety. However, I would like to single out Professors Norman Balabanian, David Cheng, Harry Gruenberg, Richard McFee, Fazlollah Reza, and Sundaram Seshu as having been especially helpful. Also, Professor Rajendra Nanavati and Messrs. Joseph Cornacchio and Robert Richardson deserve special acknowledgment for reading and constructively criticizing the entire manuscript. Similar acknowledgment is made to Professors Erik Hemmingsen and Jerome Blackman, of the Syracuse Mathematics Department, for their careful reviews of Chapters 8 through 12. Finally, my sincere thanks go to Miss Anne L. Woods for her skill and untiring efforts in typing the various versions of the notes and the final manuscript.

Wilbur R. LePage

CONTENTS

Preface ix

Chapter 1. Conceptual Structure of System Analysis 1
1-1   Introduction 1
1-2   Classical Steady-state Response of a Linear System 1
1-3   Characterization of the System Function as a Function of a Complex Variable 2
1-4   Fourier Series 5
1-5   Fourier Integral 6
1-6   The Laplace Integral 8
1-7   Frequency, and the Generalized Frequency Variable 10
1-8   Stability 12
1-9   Convolution-type Integrals 12
1-10 Idealized Systems 13
1-11 Linear Systems with Time-varying Parameters 14
1-12 Other Systems 14
Problems 14

Chapter 2. Introduction to Function Theory 19
2-1   Introduction 19
2-2   Definition of a Function 24
2-3   Limit, Continuity 26
2-4   Derivative of a Function 29
2-5   Definition of Regularity, Singular Points, and Analyticity . . 31
2-6   The Cauchy-Riemann Equations 33
2-7   Transcendental Functions 35
2-8   Harmonic Functions 41
Problems 42

Chapter 3. Conformal Mapping 46
3-1   Introduction 46
3-2   Some Simple Examples of Transformations 46
3-3   Practical Applications 52
3-4   The Function $w = 1/s$ 56
3-5   The Function $w = \frac{1}{2}(s + 1/s)$ 57
3-6   The Exponential Function 61
3-7   Hyperbolic and Trigonometric Functions 62
3-8   The Point at Infinity; The Riemann Sphere 64
3-9   Further Properties of the Reciprocal Function 66
3-10 The Bilinear Transformation 70
3-11 Conformal Mapping 73

3-12　Solution of Two-dimensional-field Problems 77
Problems 81

Chapter 4.　Integration 85

4-1　Introduction 85
4-2　Some Definitions 85
4-3　Integration 88
4-4　Upper Bound of a Contour Integral 94
4-5　Cauchy Integral Theorem 94
4-6　Independence of Integration Path 98
4-7　Significance of Connectivity 99
4-8　Primitive Function (Antiderivative) 100
4-9　The Logarithm 102
4-10　Cauchy Integral Formulas 105
4-11　Implications of the Cauchy Integral Formulas 108
4-12　Morera's Theorem 109
4-13　Use of Primitive Function to Evaluate a Contour Integral . . . 109
Problems 110

Chapter 5.　Infinite Series 116

5-1　Introduction 116
5-2　Series of Constants 116
5-3　Series of Functions 120
5-4　Integration of Series 124
5-5　Convergence of Power Series 125
5-6　Properties of Power Series 128
5-7　Taylor Series 129
5-8　Laurent Series 134
5-9　Comparison of Taylor and Laurent Series 136
5-10　Laurent Expansions about a Singular Point 139
5-11　Poles and Essential Singularities; Residues 142
5-12　Residue Theorem 145
5-13　Analytic Continuation 147
5-14　Classification of Single-valued Functions 152
5-15　Partial-fraction Expansion 153
5-16　Partial-fraction Expansion of Meromorphic Functions (Mittag-Leffler Theorem) 157
Problems 162

Chapter 6.　Multivalued Functions 169

6-1　Introduction 169
6-2　Examples of Inverse Functions Which Are Multivalued 170
6-3　The Logarithmic Function 176
6-4　Differentiability of Multivalued Functions 177
6-5　Integration around a Branch Point 180
6-6　Position of Branch Cut 185
6-7　The Function $w = s + (s^2 - 1)^{1/2}$ 185
6-8　Locating Branch Points 186
6-9　Expansion of Multivalued Functions in Series 188
6-10　Application to Root Locus 190
Problems 197

Chapter 7. Some Useful Theorems 201

7-1 Introduction 201
7-2 Properties of Real Functions 201
7-3 Gauss Mean-value Theorem (and Related Theorems). 205
7-4 Principle of the Maximum and Minimum 207
7-5 An Application to Network Theory 208
7-6 The Index Principle 211
7-7 Applications of the Index Principle, Nyquist Criterion 213
7-8 Poisson's Integrals 215
7-9 Poisson's Integrals Transformed to the Imaginary Axis 220
7-10 Relationships between Real and Imaginary Parts, for Real Frequencies 223
7-11 Gain and Angle Functions 229
Problems 231

Chapter 8. Theorems on Real Integrals 234

8-1 Introduction 234
8-2 Piecewise Continuous Functions of a Real Variable 234
8-3 Theorems and Definitions for Real Integrals 236
8-4 Improper Integrals. 237
8-5 Almost Piecewise Continuous Functions 240
8-6 Iterated Integrals of Functions of Two Variables (Finite Limits) . . 242
8-7 Iterated Integrals of Functions of Two Variables (Infinite Limits) . . 247
8-8 Limit under the Integral for Improper Integrals 250
8-9 M Test for Uniform Convergence of an Improper Integral of the First Kind 251
8-10 A Theorem for Trigonometric Integrals. 252
8-11 Two Theorems on Integration over Large Semicircles. 254
8-12 Evaluation of Improper Real Integrals by Contour Integration . . . 259
Problems 263

Chapter 9. The Fourier Integral. 268

9-1 Introduction 268
9-2 Derivation of the Fourier Integral Theorem. 268
9-3 Some Properties of the Fourier Transform 273
9-4 Remarks about Uniqueness and Symmetry 273
9-5 Parseval's Theorem 279
Problems 282

Chapter 10. The Laplace Transform 285

10-1 Introduction 285
10-2 The Two-sided Laplace Transform 285
10-3 Functions of Exponential Order 287
10-4 The Laplace Integral for Functions of Exponential Order 288
10-5 Convergence of the Laplace Integral for the General Case 289
10-6 Further Ideas about Uniform Convergence. 293
10-7 Convergence of the Two-sided Laplace Integral 295
10-8 The One- and Two-sided Laplace Transforms 297
10-9 Significance of Analytic Continuation in Evaluating the Laplace Integral 298
10-10 Linear Combinations of Laplace Transforms 299
10-11 Laplace Transforms of Some Typical Functions 300
10-12 Elementary Properties of $F(s)$ 306

10-13   The Shifting Theorems 309
10-14   Laplace Transform of the Derivative of $f(t)$ 311
10-15   Laplace Transform of the Integral of a Function 312
10-16   Initial- and Final-value Theorems 314
10-17   Nonuniqueness of Function Pairs for the Two-sided Laplace Transform 315
10-18   The Inversion Formula 318
10-19   Evaluation of the Inversion Formula 322
10-20   Evaluating the Residues (The Heaviside Expansion Theorem) . . . 324
10-21   Evaluating the Inversion Integral When $F(s)$ Is Multivalued . . . 326
Problems 328

Chapter 11.   Convolution Theorems 336

11-1   Introduction 336
11-2   Convolution in the t Plane (Fourier Transform) 337
11-3   Convolution in the t Plane (Two-sided Laplace Transform) 338
11-4   Convolution in the t Plane (One-sided Transform) 342
11-5   Convolution in the s Plane (One-sided Transform) 343
11-6   Application of Convolution in the s Plane to Amplitude Modulation . 347
11-7   Convolution in the s Plane (Two-sided Transform) 349
Problems 350

Chapter 12.   Further Properties of the Laplace Transform 353

12-1   Introduction 353
12-2   Behavior of $F(s)$ at Infinity 353
12-3   Functions of Exponential Type 357
12-4   A Special Class of Piecewise Continuous Functions 362
12-5   Laplace Transform of the Derivative of a Piecewise Continuous Function
        of Exponential Order 367
12-6   Approximation of $f(t)$ by Polynomials 370
12-7   Initial- and Final-value Theorems 372
12-8   Conditions Sufficient to Make $F(s)$ a Laplace Transform 374
12-9   Relationships between Properties of $f(t)$ and $F(s)$ 376
Problems 378

Chapter 13.   Solution of Ordinary Linear Equations with Constant Coefficients 381

13-1   Introduction 381
13-2   Existence of a Laplace Transform Solution for a Second-order Equation 381
13-3   Solution of Simultaneous Equations 384
13-4   The Natural Response 388
13-5   Stability 390
13-6   The Forced Response 390
13-7   Illustrative Examples 391
13-8   Solution for the Integral Function 395
13-9   Sinusoidal Steady-state Response 397
13-10   Immittance Functions 398
13-11   Which Is the Driving Function? 400
13-12   Combination of Immittance Functions 400
13-13   Helmholtz Theorem 403
13-14   Appraisal of the Immittance Concept and the Helmholtz Theorem . . 405
13-15   The System Function 406
Problems 407

Chapter 14.   Impulse Functions 410

14-1   Introduction 410
14-2   Examples of an Impulse Response 410
14-3   Impulse Response for the General Case. 412
14-4   Impulsive Response 415
14-5   Impulse Excitation Occurring at $t = T_1$ 418
14-6   Generalization of the "Laplace Transform" of the Derivative . . . 419
14-7   Response to the Derivative and Integral of an Excitation 422
14-8   The Singularity Functions 424
14-9   Interchangeability of Order of Differentiation and Integration . . . 425
14-10   Integrands with Impulsive Factors 426
14-11   Convolution Extended to Impulse Functions 428
14-12   Superposition 430
14-13   Summary 431
Problems 433

Chapter 15.   Periodic Functions 435

15-1   Introduction 435
15-2   Laplace Transform of a Periodic Function 436
15-3   Application to the Response of a Physical Lumped-parameter System . 438
15-4   Proof That $\mathcal{L}^{-1}[P(s)]$ Is Periodic 440
15-5   The Case Where $H(s)$ Has a Pole at Infinity 441
15-6   Illustrative Example 442
Problems 444

Chapter 16.   The Z Transform 445

16-1   Introduction 445
16-2   The Laplace Transform of $f^*(t)$ 446
16-3   Z Transform of Powers of t 448
16-4   Z Transform of a Function Multiplied by e^{-at} 449
16-5   The Shifting Theorem 450
16-6   Initial- and Final-value Theorems 450
16-7   The Inversion Formula 451
16-8   Periodic Properties of $F^*(s)$, and Relationship to $F(s)$ 453
16-9   Transmission of a System with Synchronized Sampling of Input and
        Output 456
16-10   Convolution. 457
16-11   The Two-sided Z Transform 458
16-12   Systems with Sampled Input and Continuous Output 459
16-13   Discontinuous Functions 462
Problems 462

Appendix A 465

Appendix B 468

Bibliography 469

Index 471

COMPLEX VARIABLES
AND THE
LAPLACE TRANSFORM
FOR
ENGINEERS

CHAPTER 1

CONCEPTUAL STRUCTURE OF SYSTEM ANALYSIS

1-1. Introduction. It is worthwhile for a serious student of the analytical approach to engineering to recognize that one important facet of his education consists in a transition from preoccupation with techniques of problem solving, with which he is usually initially concerned, to the more sophisticated levels of understanding which make it possible for him to approach a subject more creatively than at the purely manipulative level. Lack of adequate motivation to carry out this transition can be a serious deterrent to learning. This chapter is directed at dealing with this matter. Although it is assumed that you are familiar with the Laplace transform techniques of solving a problem, at least to the extent covered in a typical undergraduate curriculum, it cannot be assumed that you are fully aware of the importance of functions of a complex variable or of the wide applicability of the Laplace transform theory.

Since motivation is the primary purpose of this chapter, for the most part we shall make little effort to attain a precision of logic. Our aim is to form a bridge between your present knowledge, which is assumed to be at the level described above, and the more sophisticated level of the relatively carefully constructed logical developments of the succeeding chapters. In this first chapter we briefly use several concepts which are reintroduced in succeeding chapters. For example, we make free use of complex numbers in Chap. 1, although they are not defined until Chap. 2. Presumably a student with no background in electric-circuit theory or other applications of the algebra of complex numbers could study from this book; but he would probably be well advised to start with Chap. 2.

Most of Chap. 1 is devoted to a review of the roles played by complex numbers, the Fourier series and integral, and the Laplace transform in the analysis of linear systems. However, the theory ultimately to be developed in this book has applicability beyond the purely linear system, particularly through the various convolution theorems of Chap. 11 and the stability considerations in Chaps. 6, 7, and 13.

1-2. Classical Steady-state Response of a Linear System. A brief summary of the essence of the sinusoidal steady-state analysis of the response of a linear system requires a prediction of the relationship between the magnitudes A and B and initial angles α and β for two functions such as

$$v_a = A \cos (\omega t + \alpha)$$
$$v_b = B \cos (\omega t + \beta)$$
(1-1)

where v_a, for example, is a driving function* and v_b is a response function. From a steady-state analysis we learn that it is convenient to define two complex quantities

$$V_a = A e^{j\alpha} \qquad V_b = B e^{j\beta}$$
(1-2)

which are related to each other through a system function $H(j\omega)$ by the equation

$$V_b = H(j\omega)V_a$$
(1-3)

$H(j\omega)$, a complex function of the real variable ω, provides all the information required to determine the magnitude relationship and the phase difference between input and output sinusoidal functions. Presently we shall point out that $H(j\omega)$ also completely determines the nonperiodic response of the system to a sudden disturbance.

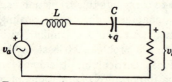

FIG. 1-1. A physical system described by the function in Eq. (1-4).

In the example of Fig. 1-1, the $H(j\omega)$ function is

$$H(j\omega) = \frac{j\omega RC}{1 - \omega^2 LC + j\omega RC}$$
(1-3a)

$$H(j\omega) = \frac{\omega RC}{\sqrt{(1 - \omega^2 LC)^2 + \omega^2 R^2 C^2}} \, e^{j[\pi/2 - \tan^{-1} \omega RC/(1-\omega^2 LC)]}$$
(1-3b)

Equation (1-3a) emphasizes the fact that $H(j\omega)$ is a rational function (ratio of polynomials) of the variable $j\omega$, and Eq. (1-3b) places in evidence the factors of $H(j\omega)$ which are responsible for changing the magnitude and angle of V_a, to give V_b. Evaluation of the steady-state properties of a system is usually in terms of magnitude and angle functions given in Eq. (1-3b), but the rational form is more convenient for analysis.

This brief summary leaves out the details of the procedure for finding $H(j\omega)$ from the differential equations of a system. It should be recognized that $H(j\omega)$ is a rational function only for systems which are described by ordinary linear differential equations with constant coefficients.

1-3. Characterization of the System Function as a Function of a Complex Variable. The material of the preceding section provides our first

---

* The terms driving function, forcing function, and excitation function are used interchangeably in this text.

point of motivation for a study of functions of a complex variable. In the first place, purely for convenience of writing, it is simpler to write

$$H(s) = \frac{RCs}{1 + RCs + LCs^2} \tag{1-4}$$

which reduces to Eq. (1-3a) if we make the substitution $s = j\omega$. However, wherever we write an expression like this, with s indicated as the variable, we understand that s is a *complex variable*, not necessarily $j\omega$. In fact, throughout the text we shall use the notation $s = \sigma + j\omega$. Another advantage of Eq. (1-4) is recognized when it appears in the factored form

$$H(s) = \frac{(R/L)s}{(s - s_1)(s - s_2)} \tag{1-5}$$

Carrying these ideas a bit further, we observe that the general steady-state-system response function can be characterized as a rational function

$$H(s) = K \frac{(s - s_1)(s - s_3) \cdots (s - s_n)}{(s - s_2)(s - s_4) \cdots (s - s_m)} \tag{1-6}$$

Various systems differ with respect to the degree of numerator and denominator of Eq. (1-6), in the factor K, and in the locations of the critical values s_1, s_2, s_3, etc. The quantities s_1, s_3, etc., in the numerator are called zeros of the function, and the corresponding s_2, s_4, etc., in the denominator are poles of the function. In general, these critical values of s, where $H(s)$ becomes either zero or infinite, are complex numbers, emphasizing the need to deal with complex numbers in the analysis of a linear system.

Equation (1-6) provides an example of the importance of becoming accustomed to thinking in terms of a *function* of a complex variable, since, with $s = j\omega$ and ω variable, this function represents the variation of system response as a function of frequency. In particular, the variation of response magnitude with frequency is often important, as in filter design; and Eq. (1-6) provides a convenient vehicle for obtaining this functional variation. Geometrically, each factor in the numerator or denominator of Eq. (1-6) has a magnitude represented graphically by line AB in Fig. 1-2a, shown for the particular case where s_k is a negative real number. Except for the real multiplying factor K, for any complex value of s the complex number $H(s)$ has a magnitude which can be calculated as a product and quotient of line lengths like AB in the figure and an angle which is made up of sums and differences of angles like α_k. Thus, a plot in the complex s plane provides a pictorial aid in understanding the properties of the function $H(s)$. In particular, steady-state response for variable frequency is characterized by allowing point s to move along the vertical axis.

This formulation is also helpful when we are concerned with variation of the magnitude $|H(j\omega)|$ as a function of ω. The function $H(s)H(-s)$ plays a central role in this question. $H(-s)$ is made up of products and quotients of factors like $-s - s_k$, one of which is portrayed by magnitude

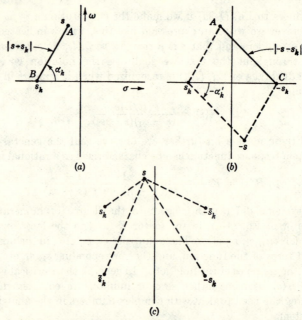

FIG. 1-2. Geometrical interpretation of factors in the numerator and denominator of $H(s)$, $H(-s)$, and $H(s)H(-s)$. (a) A factor of $H(s)$; (b) a factor of $H(-s)$; (c) factors of $H(s)H(-s)$ due to a pair of conjugate zeros or poles of $H(s)$.

$|-s - s_k|$ and angle α'_k in Fig. 1-2b. Thus, if each s_k is real, $H(s)H(-s)$ is formed from the product and quotient of factors like

$$|s - s_k| \, |-s - s_k| e^{j(\alpha_k + \alpha_k')}$$

The geometry of Fig. 1-2a makes it evident that when $s = j\omega$ (placing s on the vertical axis) the sum of angles $\alpha_k + \alpha'_k$ is zero and therefore $H(j\omega)H(-j\omega)$ is real. Furthermore, $AB = AC$ when $s = j\omega$, and therefore $H(j\omega)H(-j\omega)$ is the square of the magnitude of $H(j\omega)$. It can be shown, from physical considerations, that, if s_k is complex, the factor $s - s_k$ is accompanied by a companion factor $s - \bar{s}_k$, where $\bar{s}_k$ is the complex conjugate of s_k, as illustrated in Fig. 1-2c.* In that case, both factors

* A bar above a symbol designates its complex conjugate. Conjugates bear to each other the relation shown in the figure, having the same real components and imaginary components of opposite signs. The conjugate is defined and discussed in Chap. 2.

are considered together, with the conclusion that the product of four factors $(s - s_k)(s - \bar{s}_k)(-s - s_k)(-s - \bar{s}_k)$ is real when $s = j\omega$. Thus, since K is real, we find generally that the function $H(s)H(-s)$ is a function of a *complex variable* which has the peculiar property of *being real* when $s = j\omega$ and furthermore of being the square of the magnitude of $H(j\omega)$. We summarize by writing

$$|H(j\omega)|^2 = H(s)H(-s)\Big|_{s=j\omega} \tag{1-7}$$

Equation (1-4) can be used for an illustration, where

$$H(-s) = \frac{-RCs}{1 - RCs + LCs^2} \tag{1-8}$$

giving $\qquad H(s)H(-s) = \dfrac{-R^2C^2s^2}{(1 + LCs^2)^2 - R^2C^2s^2} \tag{1-9}$

When $s = j\omega$, since $(j\omega)^2 = -\omega^2$, we obtain

$$H(j\omega)H(-j\omega) = \frac{R^2C^2\omega^2}{(1 - LC\omega^2)^2 + R^2C^2\omega^2} \tag{1-10}$$

which is the square of the magnitude factor in Eq. (1-3b).

The function $H(s)H(-s)$ is particularly important in the design of filter and corrective networks because of the property just demonstrated. Again we can say that analytical work is easier if we deal with the complex function $H(s)H(-s)$ than if we deal only with the real function $|H(j\omega)|$.

1-4. Fourier Series. The sinusoidal function described in Sec. 1-3 plays a vital role beyond the sinusoidal case for which it is defined. The reason is provided by the Fourier series, whereby a periodic function $v_a(t)$, of angular frequency ω_a, can be described as a sum of sinusoidal components. One way to write the Fourier series for the driving function is

$$v_a(t) = \sum_{n=0}^{\infty} A_n \cos(n\omega_a t + \alpha_n) \tag{1-11}$$

each term of which is like Eqs. (1-1). Assuming that the principle of superposition is applicable, the response is

$$v_b(t) = \sum_{n=0}^{\infty} B_n \cos(n\omega_a t + \beta_n) \tag{1-12}$$

Upon comparing with Eqs. (1-2) and (1-3), it is evident that A_n and B_n and α_n and β_n are related by

$$B_n e^{j\beta_n} = H(jn\omega_a)A_n e^{j\alpha_n} \tag{1-13}$$

The important concept here is the emergence of the idea of a signal spectrum (a line spectrum in the case of a periodic function) represented by the sequence of complex numbers $A_1e^{j\alpha_1}$, $A_2e^{j\alpha_2}$, etc., and of the modification of this signal spectrum by the system function $H(jn\omega_a)$ to give the spectrum of the output signal. Emphasis is on the importance of $H(j\omega)$ as a function of ω, where in this case the function is used only at discrete values of ω.

1-5. Fourier Integral. It is a short intuitive step from the Fourier series to the Fourier integral.* In this case, $v_a(t)$ is not periodic, but we can define a periodic function, of period $\pi/\Delta\omega$,

$$w(t) = v_a(t) \qquad |t| < \frac{\pi}{\Delta\omega} \tag{1-14}$$

and for all t,

$$w(t) = w\left(t \pm \frac{2\pi}{\Delta\omega}\right) \tag{1-15}$$

as illustrated in Fig. 1-3. The period of $w(t)$ can be made as large as

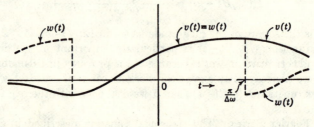

Fig. 1-3. Range of validity of Fourier series for a nonperiodic function.

desired by making $\Delta\omega$ arbitrarily small. As another indication of the usefulness of complex quantities, it is also known that we can write

$$A\cos(\omega t + \alpha) = \frac{A}{2}\left(e^{j\alpha}e^{j\omega t} + e^{-j\alpha}e^{-j\omega t}\right) \tag{1-16}$$

thereby putting into evidence the fact that $w(t)$ can be represented as a complex Fourier series involving a summation over negative as well as positive n, as follows:

$$w(t) = \frac{\Delta\omega}{2\pi}\sum_{n=-\infty}^{\infty} A_n e^{jn\,\Delta\omega t} \tag{1-17}†$$

* The procedure outlined here is good as a learning aid, although a rigorous development of the basic Fourier integral theorem requires a more complicated treatment, as given in Chap. 9.

† The factor $\Delta\omega/2\pi$ is the reciprocal of the period and normally is incorporated as a factor of A_n. The notation used here is adopted to blend with standard practice in Laplace transform theory.

where each A_n is *complex*, rather than real and positive as in Eq. (1-11). The function $w(t)$ is identical with $v_a(t)$ over the arbitrarily large interval $-\pi/\Delta\omega < t < \pi/\Delta\omega$ and will have a response $r(t)$ given by

$$r(t) = \frac{\Delta\omega}{2\pi} \sum_{n=-\infty}^{\infty} B_n e^{jn\,\Delta\omega t} \tag{1-18}$$

where
$$B_n = H(jn\,\Delta\omega)A_n \tag{1-19}$$

The last two equations are similar to Eqs. (1-12) and (1-13), with the exception that A_n and B_n are complex in Eq. (1-19). As defined, $r(t)$ is the response to a fictitious periodic function which approximates $v_a(t)$ over the interval $-\pi/\Delta\omega < t < \pi/\Delta\omega$. If $v_a(t)$ reduces essentially to zero for t outside this interval, it is intuitively reasonable to expect $r(t)$ to be an approximation for $v_b(t)$, valid over the same interval. Assuming this to be the case, one might expect to be able to take the limit as $\Delta\omega$ goes to zero, thereby obtaining exact expressions which we formally write for the two functions:

$$v_a(t) = \lim_{\Delta\omega\to 0} w(t) = \lim_{\Delta\omega\to 0} \frac{\Delta\omega}{2\pi} \sum_{n=-\infty}^{\infty} A_n e^{jn\,\Delta\omega t} \tag{1-20a}$$

and
$$v_b(t) = \lim_{\Delta\omega\to 0} r(t) = \lim_{\Delta\omega\to 0} \frac{\Delta\omega}{2\pi} \sum_{n=-\infty}^{\infty} H(jn\,\Delta\omega)A_n e^{jn\,\Delta\omega t} \tag{1-20b}$$

Of course, according to the well-known theory of Fourier series, the coefficients A_n are given by

$$A_n = \int_{-\pi/\Delta\omega}^{\pi/\Delta\omega} v_a(t)e^{-jn\,\Delta\omega t}\,dt \tag{1-21}$$

We now allow $\Delta\omega$ to approach zero, with no concern for the question of whether the limits exist. Equation (1-21), which is a function of the discrete variable $n\,\Delta\omega$, becomes a function of a continuous variable $j\omega$, and we formally write

$$\mathcal{U}_a(j\omega) = \int_{-\infty}^{\infty} v_a(t)e^{-j\omega t}\,dt \tag{1-22}$$

The summations in Eqs. (1-20) become integrals, as follows:

$$v_a(t) = \frac{1}{2\pi} \int_{-\infty}^{\infty} \mathcal{U}_a(j\omega)e^{j\omega t}\,d\omega \tag{1-23a}$$

$$v_b(t) = \frac{1}{2\pi} \int_{-\infty}^{\infty} \mathcal{U}_b(j\omega)e^{j\omega t}\,d\omega \tag{1-23b}$$

where
$$\mathcal{U}_b(j\omega) = H(j\omega)\mathcal{U}_a(j\omega) \tag{1-24}$$

takes the place of Eq. (1-19). The system response function now "acts on" the continuous-spectrum function $\mathcal{U}_a(j\omega)$ of the excitation, to produce the continuous spectrum of the response $\mathcal{U}_b(j\omega)$.

The above brief outline gives the essential ideas of the Fourier integral treatment of a linear system, showing how the response to a general nonperiodic excitation is determined by the system function $H(s)$, again emphasizing the importance of this function.

1-6. The Laplace Integral. The Fourier integral approach, although powerful, is not satisfying for solving certain practical problems and does not provide as general a basis for theoretical analysis as we should like. Its shortcomings are two in number:

1. An integral like Eq. (1-22) does not exist for most $v_a(t)$ functions of practical interest. For example, it will not handle such a simple case as the unit-step function.*

2. The formulation does not conveniently take into consideration the transient effects when energy stored in system components is suddenly released. That is, arbitrary *initial conditions* cannot be handled.

Both these problems are dealt with by making two simple modifications. The excitation function is replaced by a function which is defined to be zero for negative t; and $v_a(t)$ and $v_b(t)$ are multiplied by a "converging factor" $e^{-\sigma t}$. For positive values of the real number σ, this function approaches zero fast enough to allow many integrals like

$$\int_0^\infty v_a(t)e^{-\sigma t}e^{-j\omega t}\,dt \quad\text{and}\quad \int_0^\infty v_b(t)e^{-\sigma t}e^{-j\omega t}\,dt$$

to converge when they do not converge in the absence of the $e^{-\sigma t}$ factor. For example, if $v_a(t)$ is the unit step,

$$\int_0^\infty e^{-\sigma t}(\cos\omega t - j\sin\omega t)\,dt = \int_0^\infty e^{-\sigma t}\cos\omega t\,dt - j\int_0^\infty e^{-\sigma t}\sin\omega t\,dt$$

and each integral on the right exists if $\sigma > 0$. This permits definition of a function of $\sigma + j\omega$, which bears the same relationship to $v_a(t)e^{-\sigma t}$ as $\mathcal{U}_a(j\omega)$ bears to $v_a(t)$, with the additional stipulation that $v_a(t)$ is now zero for $t < 0$. Thus, we define

$$V_a(\sigma + j\omega) = \int_0^\infty v_a(t)e^{-(\sigma+j\omega)t}\,dt \tag{1-25}$$

and a formula corresponding to Eq. (1-23a) can be derived, giving

$$v_a(t)e^{-\sigma t} = \frac{1}{2\pi}\int_{-\infty}^\infty V_a(\sigma + j\omega)e^{j\omega t}\,d\omega$$

---

* The Fourier integral for the unit step, the function which is zero for $t < 0$ and 1 for $t > 0$, is

$$\int_0^\infty e^{j\omega t}\,dt = \int_0^\infty (\cos\omega t + \sin\omega t)\,dt$$

Neither cosine nor sine can be integrated from zero to infinity.

which is more conveniently written

$$v_a(t) = \frac{1}{2\pi} \int_{-\infty}^{\infty} V_a(\sigma + j\omega)e^{(\sigma+j\omega)t} \, d\omega \qquad (1\text{-}26)$$

Similar expressions apply for $v_b(t)$, for which we have

$$V_b(\sigma + j\omega) = \int_0^{\infty} v_b(t)e^{-(\sigma+j\omega)t} \, dt \qquad (1\text{-}27)$$

as the Fourier integral of $v_b(t)e^{-\sigma t}$, where $v_b(t) = 0$ when $t < 0$; and also, in similarity with Eq. (1-26),

$$v_b(t) = \frac{1}{2\pi} \int_{-\infty}^{\infty} V_b(\sigma + j\omega)e^{(\sigma+j\omega)t} \, d\omega \qquad (1\text{-}28)$$

It is beyond the scope of this chapter to show that $V_a(\sigma + j\omega)$ and $V_b(\sigma + j\omega)$ bear a relationship similar to Eq. (1-24), namely,

$$V_b(\sigma + j\omega) = H(\sigma + j\omega)V_a(\sigma + j\omega) \qquad (1\text{-}29)$$

It now becomes apparent that Eqs. (1-25) through (1-29) are materially simplified by regarding σ and ω as the components of the complex variable s, in which case Eqs. (1-25) and (1-26) become

$$V_a(s) = \int_0^{\infty} v_a(t)e^{-st} \, dt \qquad (1\text{-}30)$$

$$v_a(t) = \frac{1}{2\pi j} \int_{\text{Br}} V_a(s)e^{st} \, ds \qquad (1\text{-}31)$$

where the last integral is a contour integral of the complex function $V_a(s)e^{st}$ taken over a vertical line for which the real component σ is constant. That is, on the contour of integration, $s = \sigma + j\omega$ and $ds = j \, d\omega$. (Contour integration is the topic of Chap. 4.) A similar pair of equations applies to $v_b(t)$, giving

$$V_b(s) = \int_0^{\infty} v_b(t)e^{-st} \, dt \qquad (1\text{-}32)$$

$$v_b(t) = \frac{1}{2\pi j} \int_{\text{Br}} V_b(s)e^{st} \, ds \qquad (1\text{-}33)$$

and Eq. (1-29) becomes

$$V_b(s) = H(s)V_a(s) \qquad (1\text{-}34)$$

The functions $V_a(s)$ and $V_b(s)$ are called *Laplace transforms*. The possibility of having prescribed initial conditions at $t = 0$ is not admitted by Eq. (1-34); but it is a relatively simple matter to show how this can be handled by adding another term, giving

$$V_b(s) = H(s)V_a(s) + G(s) \qquad (1\text{-}35)$$

as the general expression $V_b(s)$. $G(s)$ is a function of initial-energy terms. For details, you are referred to Chap. 13.

In view of the implications of Eqs. (1-30) and (1-31), $H(s)$ takes on added significance when the Fourier integral is extended to the Laplace integral formulation. Until this introduction of the Laplace integral, we have been interested primarily in $H(j\omega)$, although the observation has been made that considerable simplification ensues if $H(j\omega)$ is regarded as a special case of $H(s)$, thereby making it possible for general properties of s to be used in interpretation and design problems in which $H(j\omega)$ is the primary function. Now, with the Laplace integral formulation we find $H(s)$ appearing explicitly as a function of s rather than of $j\omega$.

In the developments of the last two sections we have made free use of *improper integrals*. This fact points to another of the topics which must be considered, the question of properties of the integrand functions that will make the integrals exist. Perhaps more important is the fact that a formula like Eq. (1-33) is useful only if it can be evaluated and if it can be interpreted to determine its properties as a function. Therefore, techniques of evaluating and manipulating improper integrals provide one of the later objectives of this study.

1-7. Frequency, and the Generalized Frequency Variable. The Fourier integral carries the limits $-\infty$ and ∞, where integration is with respect to the variable ω, implying that we are interested in functions of ω for negative as well as positive values of ω. In the analysis of time response of systems it is customary to call ω the angular frequency, recognizing it as related to the actual frequency f by the simple formula

$$\omega = 2\pi f$$

What, then, is the physical meaning of ω and f when they become negative numbers? No physical interpretation seems possible, since frequency is by definition a count of number of cycles or radians per second. The error is in calling f and ω frequency; the proper terminology is

$$\text{Frequency} = |f|$$
and $$\text{Angular frequency} = |\omega|$$

The alternative is to refer to f as the frequency variable, rather than frequency. However, the quantity ω is so much more prevalent than f in analytical work that we shall consider ω to be the *frequency variable*.

Sometimes in analysis it is convenient to regard the Laplace generalization of the Fourier integral as equivalent to a generalization of a sinusoidal driving function. For example, in Eq. (1-23a) we may think of $v_a(t)$ as due to a superposition (via the integral) of sinusoidal components like

$$[\mathcal{V}_a(j\omega)e^{j\omega t} + \mathcal{V}_a(-j\omega)e^{-j\omega t}] \, d\omega$$

which is essentially the cosine function. To clarify this statement, it can be shown that for a practical system, having a real response to a real excitation, $\mathcal{V}_a(-j\omega)$ is the conjugate of $\mathcal{V}_a(j\omega)$, and so if $A(\omega)$ is the magnitude function and $\alpha(\omega)$ is the angle function,

$$\mathcal{V}_a(j\omega) = A(\omega)e^{j\alpha(\omega)} \qquad \text{and} \qquad \mathcal{V}_a(-j\omega) = A(\omega)e^{-j\alpha(\omega)}$$

in terms of which the above becomes

$$2A(\omega)\cos\left[\omega t + \alpha(\omega)\right] d\omega$$

In the Laplace case, the corresponding formula is

$$v_a(t) = \frac{1}{2\pi j}\int_{\text{Br}} V_a(s)e^{st}\,ds = \frac{1}{2\pi}\int_{-\infty}^{\infty} V_a(\sigma + j\omega)e^{(\sigma+j\omega)t}\,d\omega$$

which implies a summation of components like

$$[V_a(\sigma + j\omega)e^{(\sigma+j\omega)t} + V_a(\sigma - j\omega)e^{(\sigma-j\omega)t}]\,d\omega = [V_a(s)e^{st} + V_a(\bar{s})e^{\bar{s}t}]\,d\omega$$

This focuses attention on e^{st} instead of $e^{j\omega t}$ as the basic building block.

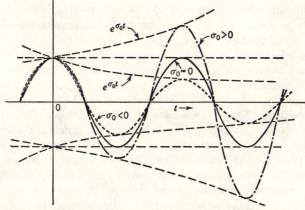

FIG. 1-4. Plots of the function $\frac{1}{2}(e^{s_0 t} + e^{\bar{s}_0 t})$, where $s_0 = \sigma_0 + j\omega_0$, for three values of σ_0 and with ω_0 constant.

The above can be made to look more like the previous case by writing it in the form

$$e^{\sigma t}[V_a(s)e^{j\omega t} + V_a(\bar{s})e^{-j\omega t}]$$

showing it to be a sinusoidal function multiplied by an exponential.* Examples of the special case

$$\frac{1}{2}e^{\sigma_0 t}(e^{j\omega_0 t} + e^{-j\omega_0 t})$$

are plotted in Fig. 1-4.

* This is true because in practical problems $V_a(\bar{s})$ is the conjugate of $V_a(s)$.

In view of the fact that e^{st} is a generalization of $e^{j\omega t}$, it is customary to call s the *generalized frequency variable*. This is of course in complete consonance with the previously observed fact that $H(j\omega)$ can be generalized by replacing it by $H(s)$. In that case also, the variable s should be thought of as the generalized frequency variable, an idea which is implied by Eq. (1-34). There is one unfortunate consequence of this terminology; the frequency variable ω is the *imaginary* component of the generalized frequency variable. It would be conceptually more satisfying if ω were the real part of s. However, as a consequence of certain factors which lead to simplifications elsewhere in the theory, the subject has developed with this apparently anomalous situation.

1-8. Stability. Stability is one of the important considerations in all problems of system design. This comment applies whether the system is linear or nonlinear. In fact, one way to determine whether or not a nonlinear system is stable is to consider that initial disturbances are small and to consider the system momentarily linear. In that case, there is no difference in the consideration of stability between a system that remains linear and one that is basically nonlinear. In fact, every unstable, physically realizable system must eventually become nonlinear as the response continues to build up.

A detailed analysis of system response, such as is given in Chap. 13, shows that the values of s at which $H(s)$ becomes infinite carry the essential information as to whether or not the system is stable. Referring to Eq. (1-6), the system is stable if the real parts of the numbers s_1, s_2, etc., are nonpositive. Thus, the question of stability provides further reason to study the various properties of $H(s)$. Two important engineering techniques for dealing with the question of stability are taken up in Chap. 6 (the root locus) and in Chap. 7 (the Nyquist criterion).

These methods of studying stability can be used, purely as techniques, with only superficial knowledge; but their justifications are grounded in quite subtle properties of functions of a complex variable. Therefore, if a satisfying degree of understanding of the question of stability of both linear and nonlinear systems is to be acquired, there is no recourse but to become acquainted with the theory of functions of a complex variable.

1-9. Convolution-type Integrals. Integrals of the form

$$\int_0^t f(\tau)g(t - \tau)\, d\tau \qquad \text{and} \qquad \int_{-\infty}^{\infty} f(\tau)g(t + \tau)\, d\tau$$

arise from situations which are essentially divorced from the complex-function viewpoint of network response or Laplace transform theory. For instance, the first of these is the result we get by applying the super-position principle to obtain the response of a linear system. The second type of integral occurs in the theory of correlation.   Similar integrals also occur in the theory of the response of nonlinear systems.

Integrals of this type bear a relationship to Laplace and Fourier transforms by virtue of their Laplace and Fourier transforms being *products* of the transforms of the functions appearing in the integrand. This property provides a vehicle whereby the Laplace transform can be brought into play in situations other than the basic one described in the bulk of this chapter.

One example, given in Chap. 11, makes use of a convolution integral in transform functions, showing how Laplace transform theory is applicable in the essentially nonlinear processes of modulation and demodulation.

1-10. Idealized Systems.   In many practical situations, linear systems of the types considered here are parts of larger systems.   In the design of these larger systems it is often convenient to idealize the component subsystems.   When this is done, the component parts are described by idealized magnitude and angle (phase) response functions of real frequency.   In this discussion, it is not possible to generalize the aspects of all design problems in one sentence.   However, it is generally true that an idealized response is chosen to give an adequate (and possibly optimum) time response to a desired signal, while rejecting unwanted signals, and to provide a system that is stable.   The following two examples can be given: In communication systems filtering is used to provide an intelligible signal in the presence of noise; and in control systems an accurate reproduction of a control signal is required.   Filters and corrective systems (electrical networks, and sometimes mechanical or other systems) are encountered in all cases.   Because the time response, or an estimate thereof, is the usual end result, if we attempt to think in terms of idealized frequency-response functions, a link between time- and frequency-response functions is essential.   This link is provided by the Fourier integral theorem.

One then naturally asks why the Fourier integral theorem, which is basically a theorem relating real functions of real variables, is not sufficient.   As a partial answer to this question we submit the following ideas: In the first place, once the idealized characteristics of a system component (filter, for example) have been decided upon, the designer is faced with the problem of creating a physically realizable device which will approximate the ideal.   This is the synthesis problem.   We have seen that electric networks, for example, are characterized by functions of a complex variable, and hence a translation of idealized response functions into realizable functions inevitably involves functions of a complex variable. Also, once a realizable system has been designed, an analysis of its specific time-response characteristics requires the solution of integrodifferential equations. Then the Laplace integral and transform become important and are more closely related to the Fourier integral theorem than the various other methods available for solving these same equations.

1-11. Linear Systems with Time-varying Parameters. The emphasis in this chapter on linear equations with constant coefficients should not be construed to imply neglect of systems with time-varying parameters. Such systems are important, and many of them fall within the realm of linear systems. They are omitted from detailed consideration here because the treatment of this chapter is basically superficial, and to add this further complication would magnify the appearance of superficiality while contributing little to the main objective.

1-12. Other Systems. Systems in which time is the independent variable are certainly important and provide the main vehicle for the examples in this text. However, they do not exhaust the practical applications of the material presented. Many field problems yield linear equations, and it is shown in Chap. 3 that the theory of functions of a complex variable is directly applicable to certain field problems in two dimensions. Also, the linear antenna is another important application. When applied to antennas, the Fourier integral plays a role very similar to that played in the theory of the time response of linear systems. Thus, the material presented in this text is applicable in several areas not illustrated in this introductory chapter.

PROBLEMS

1-1. Obtain the function $H(s)$ for Fig. P 1-1, assuming that displacement x is the driving function and y is the response.

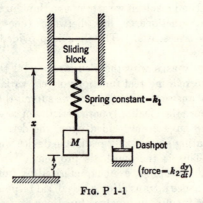

Fig. P 1-1

1-2. Referring to Fig. P 1-2, let v_1 be the driving function and v_2 the response. Obtain an expression for $H(s)$ for this system.

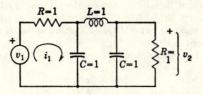

FIG. P 1-2

1-3. An inductor of L has a saw-tooth current flowing, of the form shown in Fig. P 1-3.

(a) Write the Fourier series for this current.

(b) From this, using $H(j\omega)$ on each term, obtain the Fourier series for the voltage across the inductor.

(c) From the differential equation $v = L\, di/dt$, determine the waveshape of the voltage, and find its Fourier series, using the formula for the Fourier coefficients. Compare the result with part b.

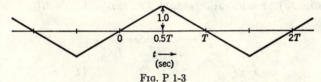

FIG. P 1-3

1-4. A periodic function can be represented by a Fourier series in either of the following equivalent forms:

$$\sum_{n=0}^{\infty} A_n \cos(n\omega_0 t + \alpha_n) = \sum_{-\infty}^{\infty} C_n e^{jn\omega_0 t}$$

where A_n and α_n are real and C_n is complex. Obtain a formula for C_n in terms of A_n and α_n.

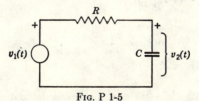

FIG. P 1-5

1-5. Consider the circuit of Fig. P 1-5, for which the excitation is the voltage pulse

$$v_1(t) = \begin{cases} 0 & t < -T \\ 1 & -T < t < T \\ 0 & T < t \end{cases}$$

The capacitor is uncharged at $t = -\infty$. Voltage $v_2(t)$ is the output.

(a) Using any method of solution for transient response known to you, obtain an expression for the output $v_2(t)$, valid for all t.

(b) Obtain the function $\mathcal{U}_1(j\omega)$ from the Fourier integral of $v_1(t)$.

(c) Obtain the function $\mathcal{U}_2(j\omega)$ from the Fourier integral of $v_2(t)$.

(d) Check whether or not $\mathcal{U}_2(j\omega) = H(j\omega)\,\mathcal{U}_1(j\omega)$, using the $H(j\omega)$ you would obtain from steady state-circuit analysis.

1-6. Do Prob. 1-5, but using a driving voltage

$$v_1(t) = \begin{cases} 0 & t < 0 \\ 1 & 0 < t < T \\ 0 & T < t \end{cases}$$

1-7. In Fig. P 1-7 the voltage source is of the form

$$v_1(t) = e^{-b|t|}$$

where b is real and positive and $b \neq R/L$. The current is zero at $t = -\infty$.

(a) Using any method you know, obtain a formula for i which is valid for all t.

(b) Obtain $\mathcal{U}_1(j\omega)$ from the Fourier integral.

(c) Obtain $\mathcal{I}(j\omega)$ from the Fourier integral.

(d) Obtain $H(j\omega)$ for this circuit, and check the relation $\mathcal{I}(j\omega) = H(j\omega)\mathcal{U}_1(j\omega)$.

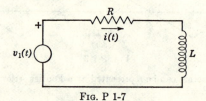

FIG. P 1-7

1-8. Obtain the Fourier integral for a pulse of sinusoidal waves consisting of an even number (N) of cycles, for the following two cases:

(a) $v(t) = \begin{cases} \sin \omega_0 t & |t| < \dfrac{N\pi}{\omega_0} \\[2mm] 0 & \dfrac{N\pi}{\omega_0} < |t| \end{cases}$

(b) $v(t) = \begin{cases} 0 & t < 0 \\[2mm] \sin \omega_0 t & 0 < t < \dfrac{2N\pi}{\omega_0} \\[2mm] 0 & \dfrac{2N\pi}{\omega_0} < t \end{cases}$

1-9. For which of the following does the Fourier integral exist? Justify your answer.

(a) $f(t) = e^t$ (b) $f(t) = e^{-|t|}$ (c) $f(t) = e^{-t^2}$

(d) $f(t) = \sin \omega_0 t$ (e) $f(t) = \dfrac{1 - \cos t}{t^2}$

1-10. Let $f(t)$ be a real function of the real variable t, and assume that its Fourier integral yields a function $\mathfrak{F}(j\omega)$. Prove the following:

(a) $\mathfrak{F}(j\omega)$ is real, and $\mathfrak{F}(j\omega) = \mathfrak{F}(-j\omega)$, if $f(t) = f(-t)$.

(b) $\mathfrak{F}(j\omega)$ is imaginary, and $\mathfrak{F}(j\omega) = -\mathfrak{F}(-j\omega)$, if $f(t) = -f(-t)$.

1-11. Let $f(t)$ and $g(t)$ be two given functions for which the integral

$$\int_{-\infty}^{\infty} f(\tau)g(t - \tau)\, d\tau$$

exists. Prove that, if $f(t)$ and $g(t)$ are both identically zero for negative t, then the above integral becomes

$$\int_{0}^{t} f(\tau)g(t - \tau)\, d\tau$$

1-12. Using the formulas

$$\mathcal{F}(j\omega) = \int_{-\infty}^{\infty} f(t)e^{-j\omega t}\, dt \qquad \mathcal{G}(j\omega) = \int_{-\infty}^{\infty} g(t)e^{-j\omega t}\, dt$$

(a) Show that $\mathcal{F}(j\omega)\mathcal{G}(-j\omega)$ is the function obtained from the Fourier integral of

$$\int_{-\infty}^{\infty} f(\tau)g(t + \tau)\, d\tau$$

(b) Show that $\mathcal{F}(j\omega)\mathcal{G}(j\omega)$ is the function obtained from the Fourier integral of

$$\int_{-\infty}^{\infty} f(\tau)g(t - \tau)\, d\tau$$

HINT: Write the Fourier integral for the function in question, make appropriate changes in the order of integration (a process which will be assumed to be justified), and make suitable changes in the variable of integration. Note that a definite integral is independent of the variable of integration, so that, for example,

$$\int_{-\infty}^{\infty} f(t)e^{-j\omega t}\, dt = \int_{-\infty}^{\infty} f(x)e^{-j\omega x}\, dx$$

1-13. Begin with the function

$$f(t) = \begin{cases} 1 & |t| < 1 \\ 0 & 1 < |t| \end{cases}$$

and find the function $\mathcal{F}(j\omega)$ from the Fourier integral. Next, evaluate the integral

$$\frac{1}{2\pi} \int_{-\infty}^{\infty} \mathcal{F}(j\omega)e^{j\omega t}\, d\omega$$

to establish that this yields $f(t)$ as originally defined. HINT: Use the known integral

$$\int_{0}^{\infty} \frac{\sin x}{x}\, dx = \frac{\pi}{2}$$

1-14. The functions

$$f(t) = \begin{cases} 0 & |t| > 2 \\ 1 & |t| < 2 \end{cases} \qquad g(t) = \begin{cases} 0 & |t| > 1 \\ 1 & |t| < 1 \end{cases}$$

are given, and let $\mathcal{F}(j\omega)$ and $\mathcal{G}(j\omega)$ be the corresponding functions obtained from the Fourier integral.

(a) Find $\mathcal{F}(j\omega)$ and $\mathcal{G}(j\omega)$.

(b) Find $\int_{-\infty}^{\infty} f(\tau)g(t - \tau)\,d\tau$ and $\int_{-\infty}^{\infty} g(\tau)f(t - \tau)\,d\tau$, showing they are the same function.

(c) Evaluate the Fourier integral for the function obtained in part b, and check whether or not this result is identical with $\mathcal{F}(j\omega)\mathcal{G}(j\omega)$.

1-15. The functions $f(t)$ and $g(t)$ are zero for negative t, and for $t \geqq 0$ are given by

$$f(t) = e^{-t} \qquad g(t) = e^{t}$$

Let $F(s)$ and $G(s)$ be the respective functions obtained from the Laplace integral.

(a) Find $F(s)$ and $G(s)$.

(b) Find $\int_{0}^{t} f(\tau)g(t - \tau)\,d\tau$ and $\int_{0}^{t} g(\tau)f(t - \tau)\,d\tau$, showing that they are the same.

(c) Find the function obtained from the Laplace integral of the function obtained in part b, and compare the result with $F(s)G(s)$.

CHAPTER 2

INTRODUCTION TO FUNCTION THEORY

2-1. Introduction. The theory of linear systems, particularly when cast in terms of the Laplace transform, relies heavily on the theory of functions of a complex variable. A brief insight into this dependence was given in Chap. 1. In the next few chapters we shall develop the theory of functions of a complex variable to provide the background for further study of linear systems and related subjects, particularly the Laplace transform and convolution integrals.

Before continuing, a word about how we shall approach the subject is in order. We shall not proceed as would a mathematician, who would place emphasis on rigor and generality of the theorems. However, it will be the generality more than the rigor that we shall give up. In mathematics one of the objectives is always to prove theorems for the most general cases possible. For us to do this would be a waste of time, looking as we are toward the utilitarian value of the subject, because the most general conditions are not needed. By this we mean that you will encounter most of the standard theorems, but applied to relatively simple cases. There will be no significant loss of rigor, and therefore the work should be satisfying to the thoughtful reader. However, because of the reduction in generality, you should not regard this work as a mathematics course in functions of a complex variable.

At the beginning we shall assume that you are familiar with algebraic manipulation of complex numbers, but the subject will be reviewed. You should understand that a complex number A is an *ordered pair* of *real* numbers A_1 and A_2 which can be written symbolically

$$A = (A_1, A_2) \tag{2-1}$$

A second complex number may be designated by

$$B = (B_1, B_2)$$

Using these as examples, the algebraic operations are defined as follows:
1. Identity

$$A = B \tag{2-2}$$

if and only if $A_1 = B_1$ and $A_2 = B_2$.

19

2. Addition

$$A + B = (A_1 + B_1, A_2 + B_2) \qquad (2\text{-}3)$$

3. Multiplication

$$AB = (A_1B_1 - A_2B_2, A_1B_2 + A_2B_1) \qquad (2\text{-}4)$$

It is left as an exercise for you to show from these definitions that addition and multiplication obey the commutative, associative, and distributive laws of algebra.

4. Division. In a system consistent with real numbers we cannot define division independently. We shall want C, where

$$C = \frac{A}{B}$$

to be the number such that when multiplied by B it will give A. It is then possible to prove

$$\frac{A}{B} = \left(\frac{A_1B_1 + A_2B_2}{B_1{}^2 + B_2{}^2}, \frac{A_2B_1 - A_1B_2}{B_1{}^2 + B_2{}^2} \right) \qquad (2\text{-}5)$$

If a complex number has the special form

$$(R,0)$$

it is said to be *real* and we can write

$$R = (R,0)$$

Thus, we make a distinction between a real number R and a complex number which has the real value R.

Another frequently occurring form is

$$(0,I)$$

This is said to be an *imaginary* number, but as yet we have introduced no symbol for it.

As a result of the above terminology, it has become the custom, given $A = (A_1,A_2)$, to call A_1 the *real part* (or the *real component*) and to call A_2 the *imaginary component*. Also, the number $(0,A_2)$ is called the *imaginary part* of A.

It is convenient to have a notation to denote real and imaginary components. For this we use

$$A_1 = \operatorname{Re} A$$
$$A_2 = \operatorname{Im} A$$

Here are three complex numbers of great importance:

$$0 = (0,0)$$
$$1 = (1,0)$$
$$j = (0,1)$$

The complex numbers 0 and 1 play the same roles in the operations with complex numbers as do their counterparts in real numbers. A number added to 0 is unchanged, and a number multiplied by 1 is unchanged.

The special imaginary number j (written i in mathematics literature) has no counterpart in real numbers. From the rule of multiplication note that

$$jA = (0,1)(A_1,A_2) = (-A_2,A_1)$$

Thus, multiplying a complex number by j interchanges its real and imaginary components with a subsequent sign change of the new real component. In particular note that

$$jj = (-1,0) \qquad (2\text{-}6)$$

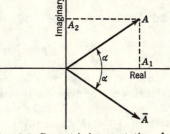

The number j is important because it provides a handy way to write a complex number. By applying the rules of algebra we get

$$
\begin{aligned}
A &= (A_1,A_2) \\
&= (A_1,0) + (0,A_2) \\
&= (A_1,0) + (0,1)(A_2,0) \\
&= A_1 + jA_2 \qquad (2\text{-}7)
\end{aligned}
$$

Fig. 2-1. Geometric interpretation of a complex number.

Because a complex number is an ordered pair of real numbers, it can be represented pictorially as a point in a plane, as shown in Fig. 2-1. This portrayal suggests defining the *magnitude* and *angle* of a complex number as follows:

$$|A| = \sqrt{A_1{}^2 + A_2{}^2} \qquad \text{magnitude} \qquad (2\text{-}8a)$$

$$\alpha = \tan^{-1}\frac{A_2}{A_1} \qquad \text{angle} \qquad (2\text{-}8b)*$$

It will sometimes be convenient to designate these respectively by the notation

$$|A| = \text{mag } A \qquad \alpha = \text{ang } A$$

These two quantities (magnitude and angle) can be interpreted geometrically as the polar coordinates of a point in a plane.

In writing a complex number it is sometimes convenient to draw on this geometrical interpretation and to write

$$A = |A|\underline{/\alpha} \qquad (2\text{-}9)$$

However, $|A|$ and α do not have quite the fundamental significance of A_1 and A_2. It would be inconvenient to use $|A|$ and α to define a complex

* In much of the literature $|A|$ is called the *modulus* and α the *argument*, in which case the abbreviations are mod A and arg A.

number because of the multivaluedness of α; $|A|\underline{/\alpha} = |A|\underline{/\alpha + 2\pi}$, for example. Thus, for a given complex number, the angle is not a unique number.

It is apparent from the geometrical interpretation that

$$
\begin{align}
A_1 &= |A| \cos \alpha \\
A_2 &= |A| \sin \alpha
\end{align}
\tag{2-10}
$$

Also, it takes only a little trigonometry to show that

$$
\begin{align}
|AB| &= |A|\,|B| \\
\text{ang } AB &= \text{ang } A + \text{ang } B
\end{align}
\tag{2-11}
$$

and

$$
\begin{align}
\left|\frac{A}{B}\right| &= \frac{|A|}{|B|} \\
\text{ang } \frac{A}{B} &= \text{ang } A - \text{ang } B
\end{align}
\tag{2-12}
$$

In view of these rules for multiplication and division, a consistent definition can be given for the root of a complex number. We shall write $(A)^{1/2}$ as the symbol for the number which multiplied by itself gives A. Thus, if $A^{1/2} = |B|\underline{/\beta}$, and $A = |A|\underline{/\alpha}$, it follows that

$$
|B|^2 \underline{/2\beta} = |A|\underline{/\alpha}
$$

and therefore

$$
B = \sqrt{|A|}
$$

$$
\beta = \frac{\alpha}{2}
$$

However, we note that $|A|\underline{/\alpha} = |A|\underline{/\alpha + 2\pi}$, and therefore a second value of angle

$$
\beta = \frac{\alpha}{2} + \pi
$$

is possible. Since

$$
|B|\underline{\Big/\frac{\alpha}{2}} = -|B|\underline{\Big/\frac{\alpha}{2} + \pi}
$$

this possibility of adding π corresponds to the usual sign ambiguity of the square root. A third value corresponding to $|A|\underline{/\alpha} = |A|\underline{/\alpha + 4\pi}$ is not obtained, because $\alpha/2 + 2\pi$ is geometrically the same as $\alpha/2$. Thus, in similarity with real numbers, $A^{1/2}$ has two roots. Likewise, $A^{1/3}$ has three roots:

$$
\sqrt[3]{|A|}\underline{\Big/\frac{\alpha}{3}} \qquad \sqrt[3]{|A|}\underline{\Big/\frac{\alpha + 2\pi}{3}} \qquad \sqrt[3]{|A|}\underline{\Big/\frac{\alpha + 4\pi}{3}}
$$

which correspond to the three geometrically equivalent values

$$
|A|\underline{/\alpha} = |A|\underline{/\alpha + 2\pi} = |A|\underline{/\alpha + 4\pi}
$$

Finally, for the general case, $A^{1/n}$ has n distinct roots:

$$A^{1/n} = \sqrt[n]{|A|} \bigg/ \frac{\alpha + 2k\pi}{n} \qquad k = 0, 1, 2, \ldots, n-1 \qquad (2\text{-}13)$$

We conclude this introduction by mentioning the *complex conjugate* of A, which is written $\bar{A}$ and defined by

$$\begin{aligned} \bar{A} &= (A_1, -A_2) \\ &= A_1 - jA_2 \end{aligned} \qquad (2\text{-}14)$$

The complex conjugate (or conjugate) of a number is obtained by changing the sign of the imaginary part.

From the rules of algebra it follows that

$$\overline{A + B} = \bar{A} + \bar{B} \qquad (2\text{-}15a)$$
$$\overline{AB} = \bar{A}\bar{B} \qquad (2\text{-}15b)$$
$$\overline{\left(\frac{A}{B}\right)} = \frac{\bar{A}}{\bar{B}} \qquad (2\text{-}15c)$$

and also

$$|A|^2 = A\bar{A} \qquad (2\text{-}16a)$$
$$\text{Re } A = \text{Re } \bar{A} = \tfrac{1}{2}(A + \bar{A}) \qquad (2\text{-}16b)$$
$$\text{Im } A = -\text{Im } \bar{A} = \frac{1}{2j}(A - \bar{A}) \qquad (2\text{-}16c)$$

Certain inequality relationships are important in the subsequent work. From the geometry of Fig. 2-1 it is evident that

$$|\text{Re } A| \leqq |A| \qquad (2\text{-}17a)$$
$$|\text{Im } A| \leqq |A| \qquad (2\text{-}17b)$$

Now consider $|A + B|^2$, which, in accordance with Eqs. (2-15a) and (2-16a), can be written

$$\begin{aligned} |A + B|^2 &= (A + B)(\bar{A} + \bar{B}) \\ &= A\bar{A} + B\bar{B} + A\bar{B} + \bar{A}B \\ &= |A|^2 + |B|^2 + 2\text{ Re }(A\bar{B}) \end{aligned}$$

In the last line above, $\text{Re }(A\bar{B})$ may be negative. The right-hand side will be increased or unchanged if we write $|\text{Re }(A\bar{B})|$ or decreased or unchanged if we write $-|\text{Re }(A\bar{B})|$. Therefore, it follows that

$$|A|^2 + |B|^2 - 2|\text{Re }(A\bar{B})| \leqq |A + B|^2 \leqq |A|^2 + |B|^2 + 2|\text{Re }(A\bar{B})|$$

Also, from Eq. (2-17a)

$$|\text{Re }(A\bar{B})| \leqq |A\bar{B}| = |A|\,|\bar{B}| = |A|\,|B|$$

and therefore the previous inequality simplifies to

$$|A|^2 + |B|^2 - 2|A|\,|B| \leqq |A + B|^2 \leqq |A|^2 + |B|^2 + 2|A|\,|B|$$

Finally, by taking square roots we get

$$||A| - |B|| \leqq |A + B| \leqq |A| + |B| \tag{2-18}$$

This result is an analytical statement of the fact that the length of one side of a triangle is less than the sum of the lengths of the other two sides but greater than their difference.

2-2. Definition of a Function. One of the most important concepts to be established is the idea of a complex number being a *function* of another complex number. Let the symbol

$$s = \sigma + j\omega \tag{2-19}$$

represent a complex number, where σ and ω each may have any real value between negative and positive infinity. (A complex number designated

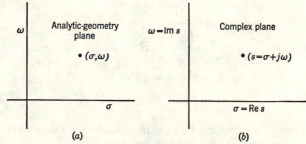

Fig. 2-2. Comparison of analytic-geometry plane and complex plane.

in this way is commonly called a complex *variable*, although in reality it is no more "variable" than any other complex number designated by a letter symbol.) You are familiar with the use of an "analytic-geometry" plane for plotting the relationship between two numbers (variables) such as σ and ω. Such a plane is shown in Fig. 2-2a, in which a representative point has coordinates (σ,ω) and the axes are labeled accordingly.

The complex number s, as defined by Eq. (2-19), provides a slightly different way to represent a point in the plane. Figure 2-2b shows what we shall call the *complex s plane*. Geometrically it is the same as the *analytic-geometry plane*, but philosophically it is quite different. In the s plane the axes are labeled "real" and "imaginary," and a typical point is labeled with a *single* symbol, namely, s. As you read this, you should begin to acquire a feeling for the idea of using *one* symbol to represent *two* real variables.

In the subsequent developments we shall have much use for the idea of a complex plane, but occasionally we shall relate it back to the analytic-geometry plane for interpretations. Meanwhile, even though the axes may be labeled σ and ω, these symbols will mean Re s and Im s.

Now we are ready to introduce the notion of a function of a complex variable. In addition to the s plane, imagine a second complex plane, which we shall call the w plane. Let w have the form

$$w = u + jv \tag{2-20}$$

and suppose that a rule is stated whereby for each point in the s plane (or portion thereof) a unique point is specified in the w plane. We can say that w is a function of s, and we may indicate that fact symbolically by writing

$$w = f(s)$$

In this definition of a function we understand that for each point in the s plane there is only one point to correspond to it in the w plane. In other words, when we say *function* we shall understand the word to mean a *single-valued* function. At a later time we shall be interested in multivalued "functions," but for the present they will be avoided.

You should understand thoroughly that the w plane is geometrically similar to the s plane, differing only in the symbol used to designate it. By this we mean that the w plane is also a geometric idea for portraying a complex variable and that it also is quite similar to the analytic-geometry plane of the *pair of real* variables u and v.

To pursue further the idea of a function of a complex variable, consider the particular case

$$w = s^2 \tag{2-21}$$

Note that there is no question about the meaning of this, since s^2 has the meaning $(s)(s)$, and multiplication has been defined. Thus, for each point in the s plane

$$w = (\sigma + j\omega)(\sigma + j\omega)$$
$$= \sigma^2 - \omega^2 + j2\sigma\omega \tag{2-22}$$

or

$$u = \sigma^2 - \omega^2$$
$$v = 2\sigma\omega \tag{2-23}$$

The first idea you should get from this example is that a formula such as (2-21) does give a rule for determining points in the w plane to correspond to points in the s plane. Equations (2-23) actually give the rectangular coordinates of the w-plane points in terms of the rectangular coordinates of the s-plane points. Equation (2-21) is simpler to write than Eqs. (2-23), but they are geometrically equivalent.

Note that Eq. (2-21) is only one of many functions that can be defined through nothing more than the rules for addition, multiplication, and division. Additional functions like

$$w = 1 + s \qquad w = s^3 \qquad w = \frac{1}{s} \qquad w = s + s^2 \qquad \text{etc.}$$

can be constructed. All that we require at this point is that the formula tell us how, given a value of s, the corresponding value of w should be determined. Several of these functions are illustrated in Fig. 2-3. A generalization of functions consisting of linear combinations of powers of s can be written

$$P(s) = a_0 + a_1 s + a_2 s^2 + \cdots + a_n s^n \qquad (2\text{-}24)$$

where the a's are complex constants and n can be any positive integer. The notation $P(s)$ implies that the function in this case is a polynomial in s. A further generalization of the functions we are prepared to deal with now is obtained if we have a second polynomial

$$Q(s) = b_0 + b_1 s + b_2 s^2 + \cdots + b_m s^m$$

and then let w be the ratio

$$w = \frac{P(s)}{Q(s)} = \frac{a_0 + a_1 s + a_2 s^2 + \cdots + a_n s^n}{b_0 + b_1 s + b_2 s^2 + \cdots + b_m s^m} \qquad (2\text{-}25)$$

You should have no difficulty in understanding that when the a's and b's are all known each value of s gives a value of w which can be calculated.

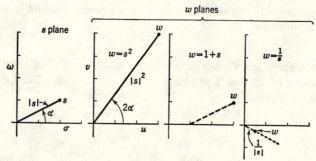

FIG. 2-3. Some examples of functions of s.

We shall seldom actually make such calculations numerically; in fact one of our objectives is to get interpretations and meanings out of such functions without making calculations, or with a minimum of calculations.

2-3. Limit, Continuity. So far we have reviewed the algebra of complex numbers, and introduced the concepts of a complex variable and functions thereof. We found that relatively simple functions could be defined wholly on the basis of algebraic operations of addition and multiplication. Eventually we shall define many other functions, but at this point we have done about all we can without going into the ideas of calculus.

To continue, we next examine the concepts of limit and continuity. Consider a function

$$w = f(s)$$

and allow s to approach a number s_0 along a line such as a in Fig. 2-4. In the function plane w will in general approach a point labeled w_0', along a line a'. Now suppose that w always approaches w_0', regardless of the direction in which s approaches s_0. If this is the case, we say

$$w_0' = \lim_{s \to s_0} f(s) \tag{2-26}$$

In the above we use the symbol w_0', rather than w_0, because the limit can exist even when $f(s_0)$ does not exist; and we shall reserve w_0 as a symbol for $f(s_0)$.

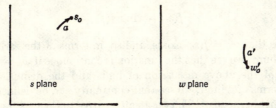

FIG. 2-4. Geometric interpretation of a function approaching a limit.

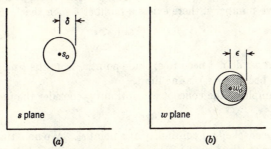

FIG. 2-5. Definition of limit in terms of ϵ and δ neighborhoods.

In precise mathematical language, we define the limit w_0' as existing if, when given a small arbitrary positive number ϵ, it is possible to find a number δ such that

$$|f(s) - w_0'| < \epsilon$$

when $$0 < |s - s_0| < \delta$$

The zero on the left of the above inequality is necessary in order to admit the cases where $f(s_0)$ does not exist, or where $f(s_0) \neq w_0'$. The two neighborhoods specified above are shown in Fig. 2-5, with the shaded area being an example of the region actually covered when s lies in the circular neighborhood specified. It is emphasized that, for the limit to exist, it must be possible to find the number δ no matter how small we choose to make ϵ.

The concept of a neighborhood introduced in the above paragraph is used occasionally throughout the text, and so a definition is in order. A *neighborhood* of a given point consists of the portion of the plane within a circle centered at the point in question. The radius of the circle (ϵ and δ, respectively, in the above examples) is usually arbitrarily small, but not necessarily so. Sometimes it is convenient to specify the radius (say ϵ) by describing the neighborhood as an ϵ neighborhood. If the center point is omitted, as in the neighborhood defined above for s_0, the result is called a *deleted neighborhood*.

With the concept of limit firmly established, it is a simple matter to define continuity. The function $f(s)$ is defined as being continuous at a point s_0 if

$$f(s_0) = \lim_{s \to s_0} f(s) \tag{2-27}$$

This implies that $w_0 = f(s_0)$ exists, and so, in terms of the earlier discussion of the limit, we see that the function is continuous if $w_0 = w_0'$.

In view of the above discussion of limit and the definition of continuity in terms of a limit, we see that continuity can be defined in terms of ϵ and δ neighborhoods. Thus as an alternative to Eq. (2-27) we say that $f(s)$ is defined as being continuous at s_0 if, corresponding to a small arbitrary positive number ϵ, there exists a number δ such that

$$|f(s) - f(s_0)| < \epsilon$$
when
$$|s - s_0| < \delta$$

We note that we do not need to exclude point s_0 from the δ neighborhood in this case, because $f(s_0)$ is the limit.

As an example of the concept of continuity, consider the function

$$w = s^2$$

which will be checked for continuity at a point s_0. We note that $w_0 = s_0^2$ exists and write

$$|w - w_0| = |s^2 - s_0^2| = |s - s_0|\,|s + s_0|$$
Now, if
$$|s - s_0| < \delta$$

we also have

$$|s + s_0| = |s - s_0 + 2s_0| < \delta + 2|s_0|$$
and so
$$|w - w_0| < \delta^2 + 2|s_0|\delta$$

Accordingly, if we now think of ϵ being given, we find that

$$|w - w_0| < \epsilon$$
when
$$|s - s_0| < \delta$$
if
$$\delta^2 + 2|s_0|\delta = \epsilon$$
or
$$\delta = \sqrt{|s_0|^2 + \epsilon} - |s_0| \tag{2-28}$$

Thus, with ϵ given, an appropriate value of δ has been found, and so continuity of the function is established.

In most cases the limit process indicated in Eq. (2-27) can be applied directly. Thus, continuity of $w = s^2$ can be checked by writing

$$s - s_0 = |r|\underline{/\theta}$$

Then
$$w = |r|^2\underline{/2\theta} + 2(|r|\underline{/\theta})(s_0) + s_0{}^2$$

Clearly, w approaches $s_0{}^2$ as $|r|$ approaches zero for *all* values of θ. The *definition* of continuity is needed for use in some of the analytical proofs to follow.

It is a simple matter to prove that products and sums of two functions are continuous at points where each of them is continuous.

2-4. Derivative of a Function. Let a function $f(s)$ be continuous at the point s, which means that

$$\lim_{s' \to s} [f(s') - f(s)] = 0$$

This being the case, it is possible, but not certain, that the quotient $[f(s') - f(s)]/(s' - s)$ can approach a limit as s' approaches s. When such a limit exists, it is a function of s, designated by

$$f'(s) = \lim_{s' \to s} \frac{f(s') - f(s)}{s' - s} \tag{2-29}$$

The term on the right is called the *differential quotient*, and the limit is called the *derivative* of $f(s)$. An alternative definition is

$$f'(s) = \lim_{\Delta s \to 0} \frac{f(s + \Delta s) - f(s)}{\Delta s}$$

in which the term $s' - s$ of Eq. (2-29) has been replaced by Δs. Other symbols are used for the derivative, just as with real variables. Thus, if $w = f(s)$, the derivative may also be designated by

$$\frac{dw}{ds} \quad \text{or} \quad \frac{df(s)}{ds}$$

It was observed above that continuity is necessary, but not sufficient, for existence of the derivative. The function $f(s) = |s|$ serves to illustrate that continuity is not sufficient. This function is continuous at the origin, but the differential quotient

$$\frac{|s'| - 0}{s'} = \underline{/-\text{ang } s'}$$

does not approach a limit when s' approaches 0, as evidenced by its dependence on the angle of s'.

As an important example we now consider the derivative of the function

$$f(s) = s^n \tag{2-30}$$

where n is a positive integer. An expression for $(s + \Delta s)^n$ is provided by the binomial theorem, as follows:

$$(s + \Delta s)^n = s^n + n s^{n-1} \Delta s + \frac{n(n-1)}{2} s^{n-2}(\Delta s)^2 + \cdots + (\Delta s)^n$$

Note that since the complex numbers obey the rules of algebra we need have no qualms about using the binomial theorem. The differential quotient is

$$\lim_{\Delta s \to 0} \frac{s^n + n s^{n-1} \Delta s + \dfrac{n(n-1)}{2} s^{n-2}(\Delta s)^2 + \cdots + (\Delta s)^n - s^n}{\Delta s}$$

$$= \lim_{\Delta s \to 0} n s^{n-1} + \frac{n(n-1)}{2} s^{n-2} \Delta s + \cdots + (\Delta s)^{n-1} = n s^{n-1}$$

Thus,
$$\frac{ds^n}{ds} = n s^{n-1} \tag{2-31}$$

which is the same formula as for functions of a real variable.

Since the differential quotient has the same form for complex variables as for real variables, we should expect the rules for differentiating sums, products, and quotients to be the same as for real variables. In case you do not recall the proofs of these rules for real variables, they are repeated here, in the notation of complex variables. Let $f(s)$ and $g(s)$ be two functions of the complex variables s, each of which possesses a derivative at a point s. First consider the derivative of the sum $f(s) + g(s)$. Forming the differential quotient gives

$$\frac{d[f(s) + g(s)]}{ds} = \lim_{\Delta s \to 0} \frac{f(s + \Delta s) + g(s + \Delta s) - f(s) - g(s)}{\Delta s}$$

$$= \lim_{\Delta s \to 0} \frac{f(s + \Delta s) - f(s)}{\Delta s} + \lim_{\Delta s \to 0} \frac{g(s + \Delta s) - g(s)}{\Delta s}$$

$$= \frac{df}{ds} + \frac{dg}{ds} \tag{2-32}$$

Next consider the derivative of the product $f(s)g(s)$, obtained as follows:

$$\lim_{\Delta s \to 0} \frac{f(s + \Delta s)g(s + \Delta s) - f(s)g(s)}{\Delta s}$$

$$= \lim_{\Delta s \to 0} \frac{f(s + \Delta s)g(s + \Delta s) - f(s + \Delta s)g(s) + f(s + \Delta s)g(s) - f(s)g(s)}{\Delta s}$$

$$= \lim_{\Delta s \to 0} \left[f(s + \Delta s) \frac{g(s + \Delta s) - g(s)}{\Delta s} + g(s) \frac{f(s + \Delta s) - f(s)}{\Delta s} \right]$$

Thus, we have the result

$$\frac{d[f(s)g(s)]}{ds} = f(s)\frac{dg(s)}{ds} + g(s)\frac{df(s)}{ds} \qquad (2\text{-}33)$$

The derivative of the reciprocal of $f(s)$ is formally given by

$$\lim_{\Delta s \to 0} \frac{1/f(s+\Delta s) - 1/f(s)}{\Delta s} = \lim_{\Delta s \to 0}\left[\frac{1}{f(s+\Delta s)f(s)}\frac{f(s) - f(s+\Delta s)}{\Delta s}\right]$$

In the limit this becomes

$$\frac{d[1/f(s)]}{ds} = -\frac{1}{[f(s)]^2}\frac{df(s)}{ds} \qquad (2\text{-}34)$$

This formula suggests that the derivative of $1/f(s)$ becomes infinite at any point where $f(s)$ is zero. However, one may question whether the derivative may still exist at such a point if $f'(s)$ is also zero, so that Eq. (2-34) would then be indeterminate. This is not possible, as we see by recalling that, if s_0 is a point where $f(s)$ is zero, then $1/f(s)$ must be discontinuous at s_0 because $1/f(s_0)$ is infinite. Therefore, the necessary condition of continuity is not satisfied, and the differential quotient cannot have a limit. Then the sequence of steps leading to Eq. (2-34) is not valid, and $1/f(s)$ can in no case have a derivative at a point where $f(s)$ is zero.

By combining Eqs. (2-33) and (2-34) we get the derivative of a quotient of two functions, as follows:

$$\frac{d[f(s)/g(s)]}{ds} = f(s)\frac{d[1/g(s)]}{ds} + \frac{1}{g(s)}\frac{d[f(s)]}{ds}$$

$$= [g(s)]^{-2}\left[g(s)\frac{df(s)}{ds} - f(s)\frac{dg(s)}{ds}\right] \qquad (2\text{-}35)$$

Taking a cue from the previous case, it is expected that this derivative does not exist at any point where $g(s)$ is zero. This conclusion is correct with the exception of certain degenerate cases in which $f(s)$ is also zero at the point where $g(s)$ is zero.

2-5. Definition of Regularity, Singular Points, and Analyticity. If a function possesses a derivative at some point s_0 in the s plane, and if we can draw a small circle around the point such that the derivative exists at all points in this circle (that is, in a neighborhood of s_0) then the function is said to be *regular* at s_0 and s_0 is said to be a *regular point*. A point at which a function is not regular is said to be a *singular point*, or merely a *singularity*.

A point where a function becomes infinite, such as a value of s which reduces the denominator of a function like Eq. (2-25) to zero, is always a singular point. However, there can be singularities of other types, where the function does not become infinite. At a later time these other types of singular points will be introduced, and functions will be classified in accordance with the types of singularities they possess.

As an example, consider the function

$$f(s) = \frac{s^2}{s^2 - 1} = \frac{s^2}{(s + 1)(s - 1)}$$

The points $s = 1, -1$ are recognized as points where the denominator is zero, and therefore they are singular points. At all other points the formula for differentiation of a quotient gives

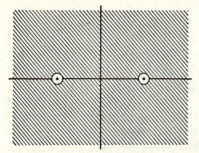

$$\frac{df(s)}{ds} = -\frac{2s}{(s^2 - 1)^2}$$

This function is regular throughout the shaded region of Fig. 2-6. The circles surrounding the singular points can be arbitrarily small.

A regular point is defined in such a way that every point in a neighborhood of that point is also a regular point. As the subject de-

FIG. 2-6. Singular points and region of regularity of $f(s) = s^2/(s^2 - 1)$.

velops you will find that the property of having a point which is regular places a powerful restriction on a function. Accordingly, those functions which have at least one regular point are put into a special category and classified as *analytic functions*.

In this definition of analyticity we must exclude "composite" functions which have artificially created boundaries. For example, the function

$$f(s) = \begin{cases} s & |s| \leqq 1 \\ |s| & |s| > 1 \end{cases}$$

is not classed as an analytic function, although it is regular at each point where $|s| < 1$. An explanation of this comes under the topic of Analytic Continuation in Chap. 5.

Many authors use the word analytic in two ways, as a synonym for regular, and also in the sense defined above. In the terminology adopted here, regularity implies existence of a derivative at a point and in a neighborhood of the point. It is a point-by-point characteristic of a function. On the other hand, an analytic function can have points where *it is not regular*, as in the above example. The function $s^2/(s^2 - 1)$ is an analytic function which is singular at 1 and -1 and regular at all other points. As defined here, analyticity does not refer to a property at a point; rather, it describes certain over-all properties of a function.

A simple example can be given of a function which has a derivative at a point but which is not an analytic function. This is the function

$$f(s) = |s|^2$$

At $s = 0$, this function is differentiable, as can be seen by writing

$$\lim_{\Delta s \to 0} \frac{f(0 + \Delta s) - f(0)}{\Delta s} = \lim_{\Delta s \to 0} \frac{|\Delta s|^2}{\Delta s}$$
$$= \lim_{\Delta s \to 0} \frac{(\Delta s)\overline{(\Delta s)}}{\Delta s} = \lim_{\Delta s \to 0} \overline{\Delta s} = 0$$

However, if we check this function at any other point, we find that the derivative does not exist, and therefore all points are singular, including the origin. Although the function is differentiable at $s = 0$, this point is not a regular point because every circle around the origin must enclose points at which the function is not differentiable. The function has no regular points and is not analytic.

2-6. The Cauchy-Riemann Equations. We have established the definition of the derivative of a function of a complex variable, and some examples of analytic and nonanalytic functions have been presented. The analytic functions so far considered are the so-called *rational functions,* consisting of algebraic combinations of powers of s. Next we want to consider the transcendental functions. For this purpose it is convenient to have a set of conditions which can be used in place of Eq. (2-29) to determine whether or not a derivative exists.

In this development it is convenient to return to the notation of Eq. (2-20), namely,

$$f(s) = w = u + jv$$

where u and v are each a function of the two variables σ and ω, implied by the notation

$$u = m(\sigma,\omega) \qquad v = n(\sigma,\omega) \tag{2-36}$$

You may refer to Eqs. (2-22) and (2-23) for an example. Now look at Eq. (2-29), which can be written

$$\lim_{\Delta s \to 0} \frac{f(s + \Delta s) - f(s)}{\Delta s}$$
$$= \lim_{\substack{\Delta \sigma \to 0 \\ \Delta \omega \to 0}} \frac{m(\sigma + \Delta\sigma, \omega + \Delta\omega) + jn(\sigma + \Delta\sigma, \omega + \Delta\omega) - m(\sigma,\omega) - jn(\sigma,\omega)}{\Delta\sigma + j\,\Delta\omega}$$

$$\tag{2-37}$$

Now let
$$m(\sigma + \Delta\sigma, \omega + \Delta\omega) - m(\omega,\sigma) = \Delta u$$
$$n(\sigma + \Delta\sigma, \omega + \Delta\omega) - n(\omega,\sigma) = \Delta v$$

so that Eq. (2-37) becomes

$$\lim_{\Delta s \to 0} \frac{f(s + \Delta s) - f(s)}{\Delta s} = \lim_{\substack{\Delta\sigma \to 0 \\ \Delta\omega \to 0}} \frac{\Delta u + j\,\Delta v}{\Delta\sigma + j\,\Delta\omega} \qquad (2\text{-}38)$$

If the derivative exists, this limit must be independent of the manner in which $\Delta\sigma$ and $\Delta\omega$ approach zero. In particular, the limit must be the same if $\Delta\sigma \equiv 0$ and $\Delta\omega$ approaches zero or if $\Delta\omega \equiv 0$ and $\Delta\sigma$ approaches zero. Thus, assuming the derivative exists, it is necessary that

$$\lim_{\Delta\sigma \to 0} \frac{\Delta u + j\,\Delta v}{\Delta\sigma} = \lim_{\Delta\omega \to 0} \frac{\Delta u + j\,\Delta v}{j\,\Delta\omega}$$

or

$$\frac{\partial u}{\partial\sigma} + j\frac{\partial v}{\partial\sigma} = -j\frac{\partial u}{\partial\omega} + \frac{\partial v}{\partial\omega}$$

which yields the pair of equations

$$\frac{\partial u}{\partial\sigma} = \frac{\partial v}{\partial\omega}$$

$$\frac{\partial u}{\partial\omega} = -\frac{\partial v}{\partial\sigma} \qquad (2\text{-}39)$$

Thus, we see that these equations, which are called the *Cauchy-Riemann differential equations*, are necessarily satisfied if the derivative exists.

Now we shall check whether or not Eqs. (2-39) are sufficient to ensure existence of the derivative of $f(s)$. Let us assume that all the partial derivatives in Eq. (2-39) are continuous. In this case, from real-variable theory, it is known that

$$\frac{du}{d\sigma} = \frac{\partial m}{\partial\sigma} + \frac{\partial m}{\partial\omega}\frac{d\omega}{d\sigma}$$

$$\frac{dv}{d\sigma} = \frac{\partial n}{\partial\sigma} + \frac{\partial n}{\partial\omega}\frac{d\omega}{d\sigma} \qquad (2\text{-}40)$$

Under these conditions Eq. (2-38) becomes

$$\lim_{\Delta s \to 0} \frac{\Delta u + j\,\Delta v}{\Delta\sigma + j\,\Delta\omega} = \lim_{\Delta s \to 0} \frac{\Delta u/\Delta\sigma + j(\Delta v/\Delta\sigma)}{1 + j(\Delta\omega/\Delta\sigma)}$$

$$= \frac{du/d\sigma + j(dv/d\sigma)}{1 + j(d\omega/d\sigma)}$$

Equations (2-40) can be substituted in this last result, giving

$$\lim_{\Delta s \to 0} \frac{\Delta u + j\,\Delta v}{\Delta\sigma + j\,\Delta\omega} = \frac{\left(\dfrac{\partial m}{\partial\sigma} + j\dfrac{\partial n}{\partial\sigma}\right) + \left(\dfrac{\partial m}{\partial\omega} + j\dfrac{\partial n}{\partial\omega}\right)\dfrac{d\omega}{d\sigma}}{1 + j\dfrac{d\omega}{d\sigma}} \qquad (2\text{-}41)$$

If Eqs. (2-39) are satisfied, the above quantities in parentheses are related by

$$j\left(\frac{\partial m}{\partial \sigma} + j\frac{\partial n}{\partial \sigma}\right) = \left(\frac{\partial m}{\partial \omega} + j\frac{\partial n}{\partial \omega}\right)$$

and so under the assumed conditions the factor

$$1 + j\frac{d\omega}{d\sigma}$$

cancels in the numerator and the denominator. Referring to Fig. 2-7, it is evident that $d\omega/d\sigma$ is the slope of the line along which Δs approaches zero, and since $d\omega/d\sigma$ has been eliminated from the expression for the limit of the differential quotient, this limit is independent of the direction of approach and can be called the derivative. It is now evident that Eq. (2-41) yields a limit which can be written in either of the following ways:

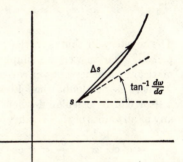

$$\frac{df}{ds} = \frac{\partial u}{\partial \sigma} + j\frac{\partial v}{\partial \sigma} \quad (2\text{-}42a)$$

$$\frac{df}{ds} = \frac{\partial v}{\partial \omega} - j\frac{\partial u}{\partial \omega} \quad (2\text{-}42b)$$

To summarize: first, we proved that, if the derivative exists, the real and imaginary parts of a function must satisfy the Cauchy-Riemann

FIG. 2-7. Interpretation of $d\omega/d\sigma$ in relation to the increment Δs.

equations; second, we proved that, if the Cauchy-Riemann equations are satisfied by two functions whose partial derivatives are continuous, then these functions are the real and imaginary parts of a function which has a derivative.

We have arrived at the point of knowing how to test for the existence of the derivative, and also how to get the derivative, while dealing only with the real and imaginary parts of the function. This is in contrast with Sec. 2-4, where no consideration was given to the real and imaginary parts. The usefulness of this new development is brought out in the next section.

2-7. Transcendental Functions. In Sec. 2-2 the functions which can be specified by algebraic combinations were defined. They comprise the so-called rational algebraic functions. They were convenient for introducing the idea of a function, because the operations of addition, multiplication, and division are relatively simple. It is easy to understand how, in a function like $f(s) = s + s^2$, one would proceed to find $f(s)$ when s is given.

From real variables you recognize that the exponential and trigonometric functions are among the most frequently encountered transcendental functions. For this reason we shall begin by defining similar functions of a complex variable. However, a new problem arises. Whereas we know from the definition of multiplication what is meant by s^2, we have no idea what is meant by sin s or cos s.

This state of affairs leads us to do something which you may regard as being slightly mysterious, if not illogical. We pick two functions, in what seems to be an arbitrary fashion, which will serve as the real and imaginary parts of a function $f(s)$. Continuing to use the notation

$$f(s) = u + jv$$

the two functions are

$$u = e^\sigma \cos \omega$$
$$v = e^\sigma \sin \omega \tag{2-43}$$

The mystery in this is in the question: Why were these two functions picked? To answer, we simply admit that we are drawing on mathematical experience not yet revealed to you.

There is no doubt that Eq. (2-43) *defines* a function $f(s)$, because $u + jv$ can be calculated when given any $s = \sigma + j\omega$. Is it regular at any or all values of s? To answer, apply the Cauchy-Riemann equations:

$$\frac{\partial u}{\partial \sigma} = e^\sigma \cos \omega \quad \frac{\partial v}{\partial \omega} = e^\sigma \cos \omega \quad \frac{\partial u}{\partial \omega} = -e^\sigma \sin \omega \quad \frac{\partial v}{\partial \sigma} = e^\sigma \sin \omega$$

These derivatives are obviously continuous, and they satisfy the Cauchy-Riemann equations for all values of s. Therefore, the function

$$f(s) = e^\sigma(\cos \omega + j \sin \omega) \tag{2-44}$$

has a derivative at every point in the s plane. It is an analytic function. Furthermore, by Eqs. (2-42), its derivative is

$$\frac{df}{ds} = \frac{\partial u}{\partial \sigma} + j \frac{\partial v}{\partial \sigma}$$
$$= e^\sigma(\cos \omega + j \sin \omega) = f(s) \tag{2-45}$$

The fact that this function has a derivative equal to itself is reminiscent of the exponential function in *real* variables. Accordingly, it is appropriate to adopt a similar notation and to write e^s. In summary: *by definition*

$$e^s = e^\sigma(\cos \omega + j \sin \omega) \tag{2-46}$$

and, by the proof given above, this function has a derivative equal to itself at every value of s. It is called the *exponential* function and is a *transcendental* function.

Note that if s is real ($s = \sigma$) Eq. (2-46) reduces to e^{σ}. There is consistency in using e^s for the new function; it reduces to the standard notation for real variables when s becomes real.

An important special case occurs when $\sigma = 0$, giving

$$e^{j\omega} = \cos \omega + j \sin \omega$$
$$= 1 \left/ \tan^{-1} \frac{\sin \omega}{\cos \omega} \right.$$
$$= 1 \underline{/\omega} \qquad (2\text{-}47)$$

In addition to the two methods already given for writing a complex number, namely,

$$A = A_1 + jA_2$$
$$A = |A| \underline{/\alpha}$$

a third is now possible. Since $A_1 = |A| \cos \alpha$ and $A_2 = |A| \sin \alpha$, it follows from Eq. (2-47) that we can also write

$$A = |A| e^{j\alpha} \qquad (2\text{-}48)$$

This is called the *exponential form*.

The exponential form can also be used in writing the nth root of a complex number, as follows:

$$A^{1/n} = \sqrt[n]{|A|} e^{j(\alpha + 2k\pi)/n} \qquad k = 0, 1, 2, \ldots, n-1 \qquad (2\text{-}49)$$

Now return to the original stream of thought. It has been shown that the function e^s obeys the usual law for differentiation. This fact is enough to warrant calling it an exponential function; but it is instructive to observe that it also obeys the usual law of multiplication. Observe that a combination of Eqs. (2-46) and (2-47) gives

$$e^s = e^{\sigma} e^{j\omega}$$

but since $s = \sigma + j\omega$, the above becomes

$$e^{\sigma} e^{j\omega} = e^{\sigma + j\omega}$$

Thus, for the special case of the sum $\sigma + j\omega$ we find that the law of exponentials holds. If $s_1 = \sigma_1 + j\omega_1$ and $s_2 = \sigma_2 + j\omega_2$ are two specific values of s, the above equation gives

$$e^{s_1 + s_2} = e^{(\sigma_1 + \sigma_2) + j(\omega_1 + \omega_2)} = e^{\sigma_1 + \sigma_2} e^{j(\omega_1 + \omega_2)}$$

From the known law of real exponents,

$$e^{\sigma_1 + \sigma_2} = e^{\sigma_1} e^{\sigma_2}$$

and from Eq. (2-47)

$$e^{j(\omega_1+\omega_2)} = \cos(\omega_1 + \omega_2) + j\sin(\omega_1 + \omega_2)$$
$$= (\cos\omega_1 + j\sin\omega_1)(\cos\omega_2 + j\sin\omega_2)$$
$$= e^{j\omega_1}e^{j\omega_2}$$

Therefore,

$$e^{s_1+s_2} = e^{\sigma_1}e^{\sigma_2}e^{j\omega_1}e^{j\omega_2} = e^{\sigma_1+j\omega_1}e^{\sigma_2+j\omega_2}$$
$$= e^{s_1}e^{s_2}$$

Having established the validity of the usual law of exponents, it is interesting to observe that, since $|A| = e^{\log|A|}$, Eq. (2-48) can be written

$$A = e^{\log|A|}e^{j\alpha}$$
$$= e^{\log|A|+j\alpha}$$

In other words, a complex number A can be written

$$A = e^C$$

where

$$C = \log|A| + j\alpha$$

In Chap. 5 we shall find that we can also write $C = \log A$, after the logarithm of a complex number has been defined.

With the exponential function in hand, the way is open to many other transcendental functions. The four basic trigonometric and hyperbolic functions will be defined. For the complex variable s, we *define* the following by the expressions given:

$\sinh s = \dfrac{e^s - e^{-s}}{2}$	hyperbolic sine of s	(2-50a)
$\cosh s = \dfrac{e^s + e^{-s}}{2}$	hyperbolic cosine of s	(2-50b)
$\tanh s = \dfrac{\sinh s}{\cosh s}$	hyperbolic tangent of s	(2-50c)
$\coth s = \dfrac{\cosh s}{\sinh s}$	hyperbolic cotangent of s	(2-50d)
$\sin s = \dfrac{e^{js} - e^{-js}}{2j}$	sine of s	(2-50e)
$\cos s = \dfrac{e^{js} + e^{-js}}{2}$	cosine of s	(2-50f)
$\tan s = \dfrac{\sin s}{\cos s}$	tangent of s	(2-50g)
$\cot s = \dfrac{\cos s}{\sin s}$	cotangent of s	(2-50h)

The names are tabulated to emphasize that we use the same words as for real variables, but the meanings are different. The geometrical interpretations are lost. However, in each case, if s becomes real ($s = \sigma$), these definitions reduce to the customary ones for a real variable. An illustration for cos s should be sufficient.   From Eq. (2-50f)

$$\cos \sigma = \frac{e^{j\sigma} + e^{-j\sigma}}{2}$$

But each exponent here is complex (imaginary), and we must go to Eq. (2-46) to get

$$e^{j\sigma} = \cos \sigma + j \sin \sigma$$
$$e^{-j\sigma} = \cos \sigma - j \sin \sigma$$

and thus the previous equation leads to an identity.   Again referring to Eqs. (2-50) we get the important set of identities

$\sin s = -j \sinh js$	$\sinh s = -j \sin js$	(2-51a)
$\cos s = \cosh js$	$\cosh s = \cos js$	(2-51b)
$\tan s = -j \tanh js$	$\tanh s = -j \tan js$	(2-51c)
$\cot s = j \coth js$	$\coth s = j \cot js$	(2-51d)

Differentiation of the trigonometric and hyperbolic functions can be accomplished with the theorems of Sec. 2-4.   Thus

$$\frac{d(\sinh s)}{ds} = \frac{1}{2}\left(\frac{de^s}{ds} - \frac{de^{-s}}{ds}\right)$$
$$= \tfrac{1}{2}(e^s + e^{-s}) = \cosh s \qquad (2\text{-}52a)$$

and similarly

$$\frac{d(\cosh s)}{ds} = \sinh s \qquad (2\text{-}52b)$$

$$\frac{d(\sin s)}{ds} = \cos s \qquad (2\text{-}52c)$$

$$\frac{d(\cos s)}{ds} = -\sin s \qquad (2\text{-}52d)$$

Also, using the procedure given in Sec. 2-4 for differentiating a quotient,

$$\frac{d(\tan s)}{ds} = \frac{1}{\cos^2 s} \qquad (2\text{-}53a)$$

$$\frac{d(\cot s)}{ds} = -\frac{1}{\sin^2 s} \qquad (2\text{-}53b)$$

$$\frac{d(\tanh s)}{ds} = \frac{1}{\cosh^2 s} \qquad (2\text{-}53c)$$

$$\frac{d(\coth s)}{ds} = -\frac{1}{\sinh^2 s} \qquad (2\text{-}53d)$$

Formulas (2-52) and (2-53) provide information about points of regularity and singularity.   We see that sin, cos, sinh, and cosh are regular at all points.   However, tan, cot, tanh, and coth are each singular at points where the derivative (and also the function) becomes infinite.   Thus, each point at which $\cos^2 s$ is zero is a singular point of tan s, and each point where $\cosh^2 s$ is zero is a singular point of tanh s.   Let us find these points.

Consider the zeros of $\cos^2 s$, which are also the zeros of $\cos s$. We have

$$\cos s = \frac{e^{js} + e^{-js}}{2}$$

and

$$e^{js} = e^{-\omega + j\sigma} = e^{-\omega}(\cos \sigma + j \sin \sigma)$$
$$e^{-js} = e^{\omega - j\sigma} = e^{\omega}(\cos \sigma - j \sin \sigma)$$

Thus

$$\cos s = \frac{e^{-\omega}(\cos \sigma + j \sin \sigma) + e^{\omega}(\cos \sigma - j \sin \sigma)}{2}$$

$$= \cos \sigma \frac{e^{\omega} + e^{-\omega}}{2} - j \sin \sigma \frac{e^{\omega} - e^{-\omega}}{2}$$

For this to be zero, each term must be zero:

$$\cos \sigma \frac{e^{\omega} + e^{-\omega}}{2} = 0 \tag{2-54a}$$

$$\sin \sigma \frac{e^{\omega} - e^{-\omega}}{2} = 0 \tag{2-54b}$$

The function $e^{\omega} + e^{-\omega}$ can never be zero (with the understanding that ω is always real), and hence one condition is

$$\cos \sigma = 0$$

Also, when $\cos \sigma = 0$, $|\sin \sigma| = 1$, and hence for Eq. (2-54b) to be satisfied we need

$$\frac{e^{\omega} - e^{-\omega}}{2} = 0 \tag{2-55}$$

which is true only if

$$\omega = 0 \tag{2-56}$$

Equations (2-55) and (2-56) together give the zero points of $\cos s$, namely,

$$\sigma = \pm (2n + 1)\frac{\pi}{2} \qquad n = 0, 1, 2, \ldots$$
$$\omega = 0$$

or, expressed as values of s,

$$s = \pm (2n + 1)\frac{\pi}{2} + j0 \tag{2-57}$$

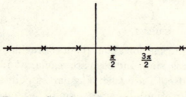

FIG. 2-8. Singular points of $f(s) = \tan s$.

Thus we see that the function $\tan s$ has an infinite number of singular points distributed at uniform intervals along the real axis, as suggested in Fig. 2-8.

As a second example let us find the singular points of $\tanh s$. Equation (2-53c) shows that they occur at points where $\cosh s = 0$. Referring to Eq. (2-51b) we see that these are points where

$$\cos js = 0$$

and from the previous case this means

$$js = \pm(2n + 1)\frac{\pi}{2} \qquad n = 0, 1, 2, \ldots$$

Therefore, $\qquad\qquad s = \pm j(2n + 1)\frac{\pi}{2}$ $\qquad\qquad$ (2-58)

are the singular points of tanh s. Some of these are shown in Fig. 2-9.

We could pursue this sort of development further, considering many more of the transcendental functions such as coth s, $1/\cos s$, etc. As a matter of fact, it should be recognized that there are many transcendental functions other than those given here. Most of the functions we meet in real variables have counterparts in functions of a complex variable, as was the case with the exponential and trigonometric functions. The examples mentioned here are sufficient to present the ideas involved when defining a function, and to help to clarify the similarities and differences between the real-variable and complex-variable cases. Of course, there are many possibilities with the few functions given here, because they themselves can be combined algebraically to give still more transcendental functions.

Note that the transcendental functions considered here have points at which they are regular. Therefore, they are analytic functions.

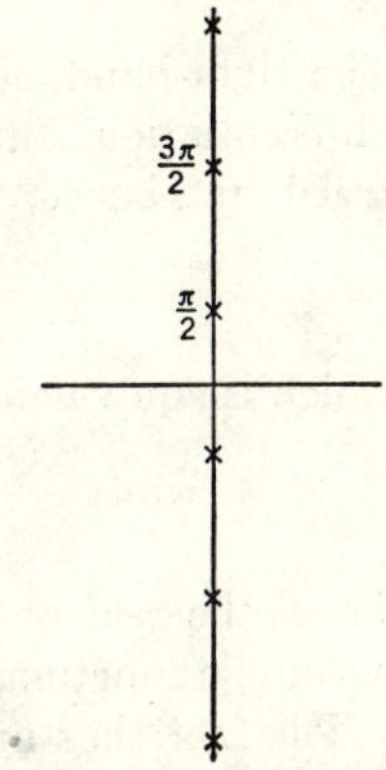

Fig. 2-9. Singular points of $f(s) =$ tanh s.

2-8. Harmonic Functions. A function of two real variables (such as σ and ω) satisfying the two-dimensional Laplace equation

$$\frac{\partial^2 u}{\partial \sigma^2} + \frac{\partial^2 u}{\partial \omega^2} = 0 \qquad\qquad (2\text{-}59)$$

throughout some region of the $\sigma\omega$ plane is said to be a *harmonic function*. We shall show that the real and imaginary components of an analytic function are harmonic functions, satisfying Eq. (2-59) in the region of the $\sigma\omega$ plane corresponding to the region of the s plane where the function is regular.

To prove this, we need merely assume that $f(s)$ is analytic and employ the usual notation

$$s = \sigma + j\omega$$
$$f(s) = w = u + jv$$

At each regular point the Cauchy-Riemann equations are satisfied, and so

$$\frac{\partial u}{\partial \sigma} = \frac{\partial v}{\partial \omega}$$

$$\frac{\partial u}{\partial \omega} = -\frac{\partial v}{\partial \sigma}$$

If we differentiate the first of these with respect to σ and the second with respect to ω, we get*

$$\frac{\partial^2 u}{\partial \sigma^2} = \frac{\partial^2 v}{\partial \sigma\, \partial \omega}$$

$$\frac{\partial^2 u}{\partial \omega^2} = -\frac{\partial^2 v}{\partial \omega\, \partial \sigma}$$

The right-hand sides of these two equations differ only in the order of differentiation, but the order of taking two partial derivatives is immaterial, and hence

$$\frac{\partial^2 u}{\partial \sigma^2} = -\frac{\partial^2 u}{\partial \omega^2}$$

which is equivalent to Eq. (2-59).   A similar proof shows also that

$$\frac{\partial^2 v}{\partial \sigma^2} + \frac{\partial^2 v}{\partial \omega^2} = 0$$

Thus, the real and imaginary components of an analytic function are harmonic functions.

The fact that an analytic function of s has real and imaginary components satisfying the two-dimensional Laplace equation is of far-reaching importance.   It provides the bridge whereby the theory of functions of a complex variable yields techniques for solving two-dimensional-field problems.   This relationship is also used in some methods of network design.

PROBLEMS

2-1. Prove that Eq. (2-5) is consistent with the rule for multiplication.

2-2. Define the two complex numbers

$$P = \left(\frac{1}{\sqrt{2}}, \frac{1}{\sqrt{2}}\right)$$

$$Q = \left(-\frac{1}{\sqrt{2}}, \frac{1}{\sqrt{2}}\right)$$

and write

$$A = xP + yQ$$

where x and y are real.   Obtain formulas for x and y in terms of A_1 and A_2.

---

* Some question could be raised as to whether these second derivatives exist.   A proof can be given that they do, but this proof depends on theory not yet presented.

2-3. Assume that C is a positive real number and that the numbers A and B are complex. Prove that if

$$|A - B| < C$$

then $$|B| - C < |A| < |B| + C$$

and $$|A| - C < |B| < |A| + C$$

Also, show that to have either of the last two relations satisfied is not sufficient to ensure that $|A - B| < C$.

2-4. Using the formulas which define multiplication and division, prove the validity of Eqs. (2-11) and (2-12).

2-5. Prove relations (2-15) and (2-16).

2-6. For the function

$$w = s + \frac{1}{s}$$

let s take on the eight values specified by $s = 2\underline{/n\pi/4}$, where $n = 0, 1, \ldots, 7$, and calculate and plot the corresponding values of w.

2-7. For the function given in Prob. 2-6 show that the same set of w points would be obtained if $s = \frac{1}{2}\underline{/-n\pi/4}$, where $n = 0, 1, \ldots, 7$. Prove this analytically, without making the actual calculations.

2-8. If $f(s)$ is a polynomial in s, with real coefficients, prove that $f(\bar{s}) = \overline{f(s)}$.

2-9. You are given the function

$$f(s) = \begin{cases} 1 & s \text{ on the unit circle} \\ s & s \text{ not on the unit circle} \end{cases}$$

Show that this function is continuous at $s = 1$, but not continuous at other points on the unit circle, by calculating $\lim_{\epsilon \to 0} (\rho + \epsilon)e^{j\phi}$ and $\lim_{\epsilon \to 0} (\rho - \epsilon)e^{j\phi}$.

2-10. Do Prob. 2-9 for the function

$$f(s) = \begin{cases} s & |s| \leqq 1 \\ |s| & |s| > 1 \end{cases}$$

2-11. Show that the function

$$f(s) = |s|^2$$

is continuous at all finite points.

2-12. Show that the function

$$f(s) = \frac{1}{s}$$

is continuous at all points except $s = 0$.

2-13. Prove that the function

$$f(s) = \frac{s + 1}{s^2 + 3s + 2}$$

approaches a limit as s approaches -1.

2-14. Prove that the function

$$f(s) = \frac{\sigma^2 - \sigma + \omega^2}{(\sigma - 1)^2 + \omega^2} + \frac{j}{(\sigma - 1)^2 + \omega^2}$$

is continuous at all finite points except $s = 1$.

2-15. If $f(s)$ is continuous at s_0 and $f(s_0) \neq 0$, prove that $1/f(s)$ is continuous at s_0.

2-16. Show that the following functions are continuous at all finite points:

(a) $f(s) = s\bar{s}$ (b) $f(s) = s + \bar{s}$

2-17. Prove that $f(s) + g(s)$ and $f(s)g(s)$ are continuous at any point where $f(s)$ and $g(s)$ are continuous.

2-18. If

$$z = s^2 + s + 1$$

and

$$w = \frac{z^2}{z + 1}$$

obtain a formula for dw/ds, and indicate what values of s are the singular points of $w = f(s)$.

2-19. Find the singular points of

(a) $f(s) = \dfrac{1}{s^2 + 3s + 2}$ (b) $f(s) = \dfrac{1}{s^4 - 1}$

2-20. Prove that $f(s) = s^2$ satisfies the Cauchy-Riemann differential equations.

2-21. Prove that $f(s) = 1/s$ satisfies the Cauchy-Riemann differential equations at all points except at the origin.

2-22. Show that the function $f(s) = |s|^3$ has a derivative only at the origin.

2-23. Show that $f(s) = s|s|$ has a derivative only at the origin.

2-24. Consider the three functions, all of which are defined to be zero at $s = 0$:

(a) $f(s) = s^2 \sin \dfrac{1}{|s|}$ $s \neq 0$

(b) $f(s) = s \sin \dfrac{1}{|s|}$ $s \neq 0$

(c) $f(s) = s^2 \sin \dfrac{1}{s}$ $s \neq 0$

Show that (a) and (b) are continuous at the origin, that (c) is not continuous at the origin, and that (a) has a derivative at the origin, but that (b) and (c) do not have derivatives at the origin.

2-25. Find formulas for the real and imaginary parts of s^n, where n is any positive integer. (HINT: Use the binomial theorem.) Apply the Cauchy-Riemann differential equations to establish that $f(s) = s^n$ is an analytic function.

2-26. Consider the function $f(s) = u + jv$, where

$$u = e^\sigma(\sigma \cos \omega - \omega \sin \omega)$$
$$v = e^\sigma(\omega \cos \omega + \sigma \sin \omega)$$

(a) Establish that $f(s)$ is an analytic function.

(b) Write $f(s)$ explicitly in terms of the single variable s.

2-27. Do Prob. 2-26 for the pair of functions

$$u = e^{\sin \sigma \cosh \omega} \cos (\cos \sigma \sinh \omega)$$
$$v = e^{\sin \sigma \cosh \omega} \sin (\cos \sigma \sinh \omega)$$

2-28. Do Prob. 2-26 for the pair of functions

$$u = \frac{\sin \sigma \cosh \omega}{\sin^2 \sigma + \sinh^2 \omega}$$
$$v = -\frac{\cos \sigma \sinh \omega}{\sin^2 \sigma + \sinh^2 \omega}$$

2-29. Show that the singular points of $f(s) = 1/(A + s^n)$ are equally spaced on a circle, and obtain a general formula for specifying the positions of the singular points.

2-30. For each of the following functions find the limit indicated, and evaluate the derivative at the point being approached:

(a) $\lim\limits_{s \to 0} \dfrac{\sinh s^2}{s}$

(b) $\lim\limits_{s \to \pi/2} \dfrac{s^2 - \pi s + \pi^2/4}{\cos 2s + 1}$

2-31. Show that

$$f(s) = \frac{1}{\cosh s + e^s}$$

is singular at $s = \log \sqrt{\frac13} + j[(2n + 1)/2]\pi,\ n = 0, 1, 2, \ldots$.

2-32. Prove the identities:

(a) $\cos s = \cos \sigma \cosh \omega - j \sin \sigma \sinh \omega$
(b) $\cosh s = \cosh \sigma \cos \omega + j \sinh \sigma \sin \omega$

2-33. Prove the identities:

(a) $\dfrac{2 \cosh^2 s + \cosh 2s - 1}{\sinh s \cosh s} = 4 \coth 2s$

(b) $\sinh e^s + \sinh e^{-s} = 2 \sinh (\cosh s) \cosh (\sinh s)$

(c) $\tanh s = \dfrac{\sinh 2\sigma + j \sin 2\omega}{\cos 2\omega + \cosh 2\sigma}$

2-34. Obtain the polar-coordinate equivalent of the Cauchy-Riemann differential equations.

CHAPTER 3

CONFORMAL MAPPING

3-1. Introduction. It is customary to use graphical aids in considering functions of a real variable. Graphs of the functions serve to give helpful pictorial representations. You may have noted in Chap. 2 that such graphical interpretations were absent, although the idea of a functional relationship between pairs of points in two planes was introduced. This is an inherent difficulty in the study of functions of a complex variable. It is impossible to plot a function of a complex variable in the sense that we can plot a function of a real variable, because there are four varying quantities: the real and imaginary parts of both the independent variable and of the function of that variable.

Something can be done by way of graphical interpretation, however, by plotting the function as the independent variable moves along some prescribed path in the s plane. Some of the more important ideas related to this method of interpretation are presented in this chapter.

3-2. Some Simple Examples of Transformations. *Linear Function.* We shall begin with the simplest nontrivial case, the function

$$f(s) = A + Bs \tag{3-1}$$

where A and B are complex. An example with numerical coefficients is chosen so that numerical scales can be used. Accordingly, take

$$A = 1 + j$$
$$B = 2 + j$$

and write $f(s) = u + jv$ and $s = \sigma + j\omega$. We get

$$u + jv = 1 + j + (2 + j)(\sigma + j\omega)$$
$$= 1 + 2\sigma - \omega + j(1 + \sigma + 2\omega)$$

and equating real and imaginary parts yields

$$u = 1 + 2\sigma - \omega$$
$$v = 1 + \sigma + 2\omega$$

which give the explicit functional dependence of u and v on the independent variables σ and ω.

The existence of such pairs of functions as analytical equivalents of a single function of a complex variable was illustrated in Chap. 2. We seek two relationships between u and v: one with ω appearing as a parameter, and the other with σ as a parameter. Routine algebra, eliminating first σ and then ω from the above equations, yields, respectively,

$$v = \tfrac{1}{2}u + \tfrac{1}{2}(1 + 5\omega) \qquad (3\text{-}2a)$$
$$v = -2u + (3 + 5\sigma) \qquad (3\text{-}2b)$$

Equation (3-2a) represents the line in the w plane corresponding to points along a horizontal line ($\omega =$ constant) in the s plane. The solid lines in Fig. 3-1 illustrate the case $\omega = 1$. Equation (3-2b) is the equation for the line in the w plane traced out in the w plane as a point moves along

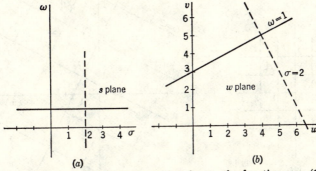

FIG. 3-1. Development of the transformation due to the function $w = (1 + j) + (2 + j)s$.

a vertical line ($\sigma =$ constant) in the s plane. The dashed lines in Fig. 3-1 are examples for $\sigma = 2$. The points of intersection in the two planes are corresponding points, as related by Eq. (3-1).

Note that this graphical interpretation, which is only partially completed in Fig. 3-1, gives some information about the point-transforming properties of the function. A straight line in the s plane is translated and rotated.

Now imagine a coordinate "grid" in the s plane, as shown in Fig. 3-2a. Each straight line of the s plane transforms to a corresponding line in the w plane; and Eqs. (3-2) describe these lines. It may be said that the grid of lines in the w plane is a map of the grid of lines in the s plane. By allowing this grid structure to become finer and finer, an increasingly more accurate picture of the function can be obtained.

The mapping idea can be extended to recognize that any arbitrary geometrical figure in the s plane, such as the curvilinear triangle shown, goes into a corresponding figure in the w plane. The grid development of Fig. 3-2 adds the idea of change of size. The grid spacing is greater in the w plane than in the s plane, and a geometrical figure is magnified.

The process whereby a set of points in one plane is converted into a set of points in another plane is called a *transformation*. The law of a transformation can be expressed either by a *pair* of functions in real variables or by a *single* function of a complex variable.

This example illustrates the transformation properties of the comparatively simple linear function. Geometrical shapes are affected in the manner noted above, but they are undistorted. Other functions cannot be described so simply, because they introduce distortions.

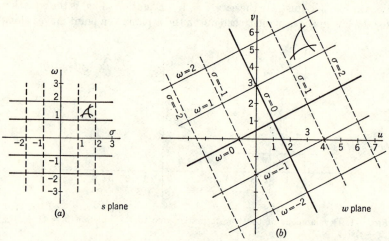

FIG. 3-2. Portrayal of a transformation by the use of coordinate grids.

The linear function is the simplest one we could employ to emphasize the necessity of using two planes to represent a function graphically and to describe the utility of considering the map in the w plane of a coordinate grid in the s plane. In some cases it is more convenient to use the coordinate lines of polar coordinates.

Quadratic Function. The next function to be considered is

$$w = s^2 \tag{3-3}$$

We shall consider this case twice, first using polar coordinates and then using rectangular coordinates.

Let the polar coordinates be $\rho = |s|$ and $\phi = $ ang s, giving

$$w = \rho^2 / 2\phi \tag{3-4}$$

From the right side of the above it is apparent that a coordinate circle in the s plane goes into a coordinate circle in the w plane, of radius equal to the square of the radius in the s plane. A radial line coordinate in the s plane goes into a radial line in the w plane, but making twice the angle with the real axis.

These relationships are shown in Fig. 3-3. If s is restricted to the right half of the s plane, the whole w plane will be covered by the transformation. A similar conclusion follows if s is restricted to the left half plane, as

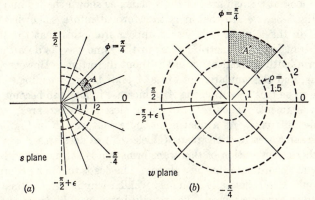

Fig. 3-3. Transformation of the right-half s plane by $w = s^2$.

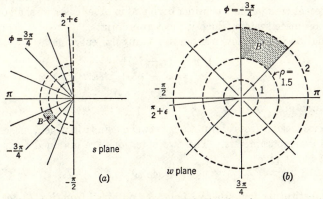

Fig. 3-4. Transformation of the left-half s plane by $w = s^2$.

illustrated in Fig. 3-4. This behavior, whereby each half of the s plane maps onto the entire w plane, arises because $s^2 = (-s)^2$. This is the first of several functions to be considered in which only part of the s plane is needed to cover the entire w plane.

The wedge-shaped "cut" along the negative real axis in each of the portrayals of the w plane needs explaining. If the whole $j\omega$ axis had been kept in each case, it would then have been used twice. To avoid using it twice, while using the remainder of the s plane only once, the negative imaginary axis is deleted in Fig. 3-3a, and the positive imaginary axis is omitted from Fig. 3-4a. (This choice was arbitrary; it could have been the other way.) To make the picture as complete as possible, a radial line is shown just to the right of the negative imaginary axis in the s plane of Fig. 3-3a. This line, taken with the positive $j\omega$ axis, forms the wedge in the w plane. Figure 3-4 is treated in a similar fashion.

If we look at a small area of the s plane, as shown shaded and labeled A in Fig. 3-3a, we see that it is transformed into a somewhat similar area A' in the w plane. It would appear from this that the function $w = s^2$ transforms geometric figures in the same way as does the linear transform, with magnification but without distortion. However, this is not true, and the similarity of areas A and A' is misleading. We can see this to be true by supposing A to increase in angle to become a semicircular segment. Then A' will be a closed annular area, which is certainly different. As a final point, note that area B in Fig. 3-4 goes into the same area in the w plane (labeled B') as does area A.

Further consideration of this transformation is accomplished by using rectangular coordinates in the s plane. This leads to a more complicated situation. Parenthetically, you may wonder why polar coordinates were not used in the linear case. They could have been, but not much more would have been learned, because the transformation is so simple. For the present case we shall consider both and in other cases we shall use only the coordinate system which is most appropriate.

Equation (3-3) is now written in rectangular coordinates:

$$u + jv = (\sigma + j\omega)^2$$
$$= \sigma^2 - \omega^2 + j2\sigma\omega$$

or

$$u = \sigma^2 - \omega^2 \tag{3-5a}$$
$$v = 2\sigma\omega \tag{3-5b}$$

Again we seek two sets of equations between u and v in which, successively, ω and σ are parameters. The two equations are obtained by squaring Eq. (3-5b) to give

$$v^2 = 4\sigma^2\omega^2$$

and substituting for either σ^2 or ω^2, as obtained from Eq. (3-5a). The results are

$$u + \omega^2 = \frac{1}{4\omega^2}v^2 \tag{3-6a}$$

$$u - \sigma^2 = -\frac{1}{4\sigma^2}v^2 \tag{3-6b}$$

Equation (3-6a) is a family of parabolas with vertices at point $-\omega^2$. They are the maps in the w plane of the coordinate lines $\omega = $ constant of the s plane. The solid-line parabolas in Fig. 3-5b show this graphically, being the w-plane transformations of the horizontal solid coordinate lines of Fig. 3-5a. Referring to Eq. (3-6a), you will note that the parabola is a function of ω^2, but not of ω. Therefore, there is no distinction in the w plane between coordinate lines for positive and negative values of ω. This is related to the fact that we got two w planes when considering polar coordinates. Thus, by taking the top half of the s plane we could cover the w plane with lines and would again cover it with the same set of lines by taking the bottom half of the s plane.

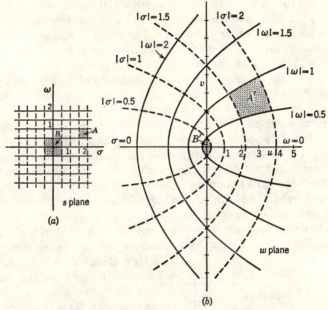

FIG. 3-5. Study of the transformation $w = s^2$, using rectangular coordinates in the s·plane. (Areas marked B and B' are referred to in Sec. 3-11.)

Next we consider the transformation of lines of constant σ. According to Eq. (3-6b) these lines go into parabolas with vertices at point $+\sigma^2$, as shown by the dashed lines of Fig. 3-5b. Again we find the parameter appearing only as a square, and therefore each dashed parabola is the transformation of two vertical lines in the s plane. If we take the right half of the s plane, it would cover the w plane with lines, and similarly for the left half. With rectangular coordinates, it is more difficult to relate the whole s plane to two w planes, because for one set of coordinate lines we would use the top and bottom and for the other the right and left halves of the s plane. The polar-coordinate system shows the relationship so well that it would be unreasonable to undertake it again for this more difficult portrayal.

Note that the real axis of the w plane is in two parts. The right half is a degenerate case of the system of solid-line parabolas, for $\omega = 0$; and the left half is a degenerate case of the dashed-line parabolas, for $\sigma = 0$.

This transformation treats the coordinate grid more violently than did the linear function. The simple idea of a translation, rotation, and change of scale no longer applies. A somewhat satisfying description of what happens is possible, however. Suppose that the right-half s plane in Fig. 3-3a is made of a rubber membrane which can be "fanned out" by drawing the two halves of the $j\omega$ axis together as if they were hinged at the origin. Figure 3-3b would be the result. This picture provides an insight into the kind of distortion produced. The result of this stretching can also be seen in Fig. 3-5, where the vertical coordinate lines in the s plane become bent to form the dashed lines in the w plane.

It is a bit more difficult to picture the distortion of the left-half s plane to form the w plane. To attempt such a description would not add materially to the concept we wish to describe.

The rectangular-coordinate portrayal serves to emphasize the distortion of geometrical shapes when transformed by the function $w = s^2$. An inspection of the labels on the transformed coordinates shows that the square area A in Fig. 3-5a will transform to the curvilinear square labeled A' in Fig. 3-5b. The important conclusion is that with this transformation some straight lines can go into curved lines and geometrical shapes are not preserved exactly.

3-3. Practical Applications. The transformation properties of a number of additional functions will be considered later in this chapter. But before going on, a few practical examples taken from the field of network theory will be presented, showing why these functions, and their graphical portrayals, are useful.

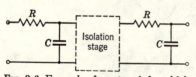

FIG. 3-6. Example of a network for which the response function is $w = z^2$.

The first example is an application of the function treated in the previous section. Consider two RC networks connected in cascade by an isolating stage, such as an amplifier with negligible input admittance, as shown in Fig. 3-6. The output/input function for each network, as a function of real angular frequency ω, is

$$z = \frac{1/j\omega C}{R + 1/j\omega C} = \frac{1}{1 + j\omega RC} \tag{3-7}$$

and for the combined network the transfer function, assuming unity gain, is

$$w = z^2$$

As ω increases from zero to infinity, we find that z traces out the semicircle shown in Fig. 3-7a (proof of this is given in the following section). The transformation properties of the function $w = z^2$ can then be used to determine the locus of the response function of the two networks in cascade. Figure 3-7b shows the transformation of polar coordinates of the z plane and also the transformation of the curve traced out by w as ω varies.

A graph paper with a grid on it like Fig. 3-7b can in this way be used to find loci of steady-state response functions for two identical isolated cascaded amplifiers, when the locus for one stage is known. Other functions (z^3, z^4, etc.) could be plotted to take care of more than two sections in cascade.

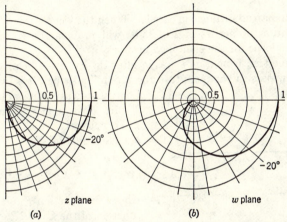

FIG. 3-7. Graphical interpretation of the response function of Fig. 3-6.

Steady-state network theory provides other examples of functions whose transformation properties are worth investigating. In the previous example the transfer function z can also be written

$$z = \frac{1}{RC} \frac{1}{1/RC + j\omega}$$

If we interpret s as $1/RC + j\omega$, the locus of z with variable ω can be obtained from a study of the map of the rectangular coordinates of the s plane due to the function $z = 1/s$. Thus, the reciprocal function is a useful aid in the analysis of a commonly used coupling network. The general problem of determining an admittance locus from an impedance locus, or vice versa, is a more general application of the reciprocal function.

The function

$$w = az + \frac{1}{bz} = \sqrt{\frac{a}{b}} \left(\sqrt{ab}\, z + \frac{1}{\sqrt{ab}\, z} \right) \qquad (3\text{-}8)$$

provides another practical example. This function gives the impedance (or admittance) of the series (or parallel) connection of two reciprocally related impedances (i.e., impedances having a product which is a real constant). Two examples are shown in Fig. 3-8, where w is an impedance

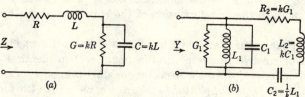

FIG. 3-8. Two network applications of the function $w = az + 1/bz$.

in case a and an admittance in case b. When the locus of z is known, the transformation properties of Eq. (3-8) will provide the corresponding immittance locus for the combined network.

This same transformation arises in filter theory, often being described as the "low-pass to bandpass transformation," although it is really more

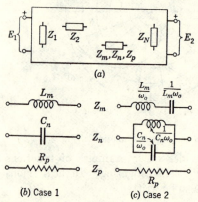

(a)

(b) Case 1   (c) Case 2

FIG. 3-9. Network application of the function $w = az + 1/bz$, relating band-frequency properties of one network with low-frequency properties of another network having a similar structure.

general than this name would imply. Figure 3-9a represents a general filter network, which will have a transfer function

$$\frac{E_2}{E_1} = F(Z_1, Z_2, \ldots, Z_N) ,$$

where the function F is a rational function of the Z's. Now suppose that each Z of this network has one of the forms shown at (b) in the figure, which we shall call case 1; and then let corresponding branches have the forms shown at (c), which will be called case 2. Note that there is a systematic transition from case 1 to case 2; each R is unchanged, each L is replaced by L and C in series, and each C is replaced by L and C in parallel, with specific value relationships as indicated in the figure. The respective impedances are readily shown to be as follows:

Case 1

$$Z_m = L_m s$$
$$Z_n = \frac{1}{C_n s}$$
$$Z_p = R_p$$

Case 2
$$Z_m = L_m w$$
$$Z_n = \frac{1}{C_n w}$$
$$Z_p = R_p$$

where
$$w = \frac{s}{\omega_0} + \frac{\omega_0}{s}$$

and ω_0 is an arbitrary design constant. This last equation is a special form of Eq. (3-8).

The above tabulation places in evidence the fact that the functional dependence of each Z on s in case 1 and on w in case 2 is the same. Also, since the structure is the same in both cases, the response function is given by the same function of the Z's, namely, $F(Z_1, \ldots, Z_N)$ defined above. Therefore, if we define

$$\frac{E_2}{E_1} = T_1(s)$$

for case 1, it follows that

$$\frac{E_2}{E_1} = T_1(w)$$

for case 2. This response function for case 2 is also a function of s, which may be designated $T_2(s)$. Therefore, for case 2 we have

$$T_2(s) = T_1(w)$$

This information can be useful if a solution is available for case 1. From this the transformation properties of the function $T_1(w)$ would be known, giving a map of a coordinate grid in the w plane on the E_2/E_1 plane. Then, with the known mapping properties of Eq. (3-8), the coordinate grid of the s plane can be mapped onto the w plane and thence onto the E_2/E_1 plane.

We see now why the "low-pass to bandpass" designation is not an adequate description. If $T_1(s)$ is a low-pass function, $T_2(s)$ will be a bandpass function. However, $T_1(s)$ is not necessarily low-pass; it could eliminate low frequencies, and then $T_2(s)$ would be a bandstop function.

Earlier in this section it was pointed out that z raised to an integral power arises as the function which describes certain cases of networks in cascade. In many applications it is preferable to write the transmission function of each of the cascaded sections as an exponential, so that multiplication of functions is replaced by addition. Accordingly, in network theory we are sometimes interested in the function

$$w = e^z \tag{3-9}$$

and so part of this chapter is devoted to an investigation of its transformation properties. This function and its related hyperbolic functions also arise in the solution of the wave equations, and so we also consider the function

$$w = \cosh s$$

in detail; and from it we deduce the transformation properties of the other circular and hyperbolic functions.

Finally, we shall consider the bilinear function

$$w = \frac{z - 1}{z + 1} \qquad (3\text{-}10)$$

which is of interest partly because, if z is replaced by $1/z$, w merely changes sign. Therefore, a graphical interpretation of Eq. (3-10) is useful for finding reciprocals of complex numbers. This function is also the reflection coefficient which appears in the scattering matrix formulation of the theory of multiport networks. Specifically, the reflection coefficient of an impedance Z in series with an impedance Z_0 is

$$\frac{Z - Z_0}{Z + Z_0} = \frac{Z/Z_0 - 1}{Z/Z_0 + 1}$$

which is identical with Eq. (3-10) if we let $z = Z/Z_0$. This function also plays an important role in the graphical analysis of transmission lines; it is the function which leads to the "circular" (or Smith) transmission-line chart.

In presenting the functions to be analyzed, z was used as the independent variable, as a reminder that the independent variable will not always be the generalized frequency variable s. However, with this understanding, in the following sections we shall revert to s as the independent variable, recognizing that in the presentation s is an abstract variable which can represent any complex variable derived from a physical situation.

3-4. The Function $w = 1/s$. It would be easy to get the polar-coordinate map of the function

$$w = \frac{1}{s} \qquad (3\text{-}11)$$

but little would be learned by so doing. Accordingly, rectangular coordinates will be used, giving

$$(u + jv)(\sigma + j\omega) = 1$$

from which we get, by equating real and imaginary parts,

$$\begin{aligned} u\sigma - v\omega &= 1 \\ v\sigma + u\omega &= 0 \end{aligned} \qquad (3\text{-}12)$$

By a purely algebraic elimination of first σ and then ω, we get the two families of circles

$$\left(v + \frac{1}{2\omega}\right)^2 + u^2 = \frac{1}{4\omega^2}$$

$$v^2 + \left(u - \frac{1}{2\sigma}\right)^2 = \frac{1}{4\sigma^2}$$

(3-13)

The degenerate cases $\sigma = 0$ and $\omega = 0$ cannot be handled by Eqs. (3-13), but from Eq. (3-11) we find that they, respectively, give the two axes $u = 0$ and $v = 0$.

The transformation is shown in Fig. 3-10. Notice that this justifies using the circle in the previous section in dealing with Eq. (3-7), which is covered by the present example if $s = 1 + j\omega RC$.

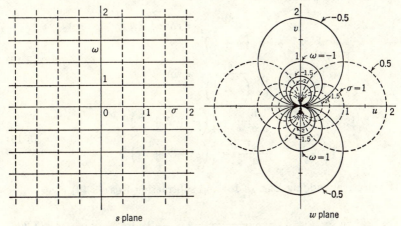

Fig. 3-10. Map of the function $w = 1/s$.

The reciprocal function treats the plane more violently than the other cases so far considered. The "edge of the plane at infinity" is pulled in to the center, while the center is stretched out in all directions to infinity. It is perhaps not too difficult to visualize this process and thus to get some perception as to the reasonableness of the families of curves shown in Fig. 3-10.

3-5. The Function $w = \frac{1}{2}(s + 1/s)$. We turn to the case of Eq. (3-8), which is now written in the slightly different form

$$w = \frac{1}{2}\left(s + \frac{1}{s}\right)$$

(3-14)

where the factor $\frac{1}{2}$ is introduced for later convenience in treating the hyperbolic functions. This case is analyzed by using polar coordinates,

$$s = \rho\underline{/\phi} \tag{3-15}$$

which gives

$$u + jv = \frac{1}{2}\left(\rho + \frac{1}{\rho}\right)\cos\phi + j\frac{1}{2}\left(\rho - \frac{1}{\rho}\right)\sin\phi \tag{3-16}$$

and

$$u = \frac{1}{2}\left(\rho + \frac{1}{\rho}\right)\cos\phi \tag{3-17a}$$

$$v = \frac{1}{2}\left(\rho - \frac{1}{\rho}\right)\sin\phi \tag{3-17b}$$

The required pair of equations relating u and v, with ρ or ϕ as parameters, is obtained by squaring both of Eqs. (3-17) and writing them as either of the following pairs:

$$\frac{4u^2}{\left(\rho + \dfrac{1}{\rho}\right)^2} = \cos^2\phi$$

$$\frac{4v^2}{\left(\rho - \dfrac{1}{\rho}\right)^2} = \sin^2\phi \tag{3-18}$$

or

$$\frac{u^2}{\cos^2\phi} = \frac{1}{4}\left(\rho^2 + 2 + \frac{1}{\rho^2}\right)$$

$$\frac{v^2}{\sin^2\phi} = \frac{1}{4}\left(\rho^2 - 2 + \frac{1}{\rho^2}\right) \tag{3-19}$$

Respectively adding Eqs. (3-18) and subtracting Eqs. (3-19) gives

$$\frac{4u^2}{\left(\rho + \dfrac{1}{\rho}\right)^2} + \frac{4v^2}{\left(\rho - \dfrac{1}{\rho}\right)^2} = 1 \tag{3-20a}$$

$$\frac{u^2}{\cos^2\phi} - \frac{v^2}{\sin^2\phi} = 1 \tag{3-20b}$$

Equation (3-20a) describes a symmetrically located ellipse of major axis $\rho + 1/\rho$ along the u axis and minor axis $\rho - 1/\rho$ along the v axis. Equation (3-20b) represents a pair of symmetrically located hyperbolas with intercepts on the real axis at $u = \pm\cos\phi$. Notice that Eq. (3-20a) is unchanged if ρ is replaced by $1/\rho$ and that Eq. (3-20b) is unchanged if ϕ is replaced by $-\phi$. This, of course, is due to the squaring operation used in arriving at Eqs. (3-18). We can see from Eqs. (3-17) that this ambiguity in ρ and ϕ is spurious. Replacing ρ by $1/\rho$ or replacing ϕ by $-\phi$ (not both) changes the sign of v, a fact which is not indicated by Eqs.

(3-20). We shall use Eqs. (3-20) to give the shapes of the curves but shall need Eqs. (3-17) to help in the process of labeling.

First note from Eqs. (3-17) that if ρ is replaced by $1/\rho$ and the sign of ϕ is changed there is no change in u and v. Thus, if we keep $\rho \geqq 1$ and let ϕ vary from $-\pi$ to π, then u and v will each take on all values from $-\infty$ to ∞ and w will cover the w plane. (We could also cover the w plane by restricting ϕ to the range from 0 to π and letting ρ have all values from 0 to ∞.) The former alternative is chosen arbitrarily; and this means restricting s to points on and outside the unit circle. Clearly to designate this, let

$$
\begin{aligned}
\rho &= 1 & 0 \leqq \phi \leqq \pi \\
\rho &> 1 & -\pi < \phi \leqq \pi
\end{aligned} \tag{3-21}
$$

Figure 3-11a shows the portion of the s plane being considered and several curves of the families for $\rho =$ constant and $\phi =$ constant. We must, however, remove the ambiguity due to the fact that the latter family is unchanged if ϕ is replaced by $-\phi$. This is easily done by looking at Eq. (3-17b), which shows that v is positive when $\rho \geqq 1$ and $0 < \phi \leqq \pi$ and negative when $-\pi < \phi \leqq 0$. Thus, each branch of a hyperbola carries different labels on the upper and lower halves.

The real axis of the s plane ($\rho > 1$ still being retained) is a degenerate case because $\sin \phi = \sin \pi = 0$ in Eq. (3-20b). For the equation to be satisfied, v must be zero; but this leaves us with an indeterminate form; and so we look to Eqs. (3-17). When $\phi = 0$ or π and $\rho > 1$,

$$
u = \begin{cases} \dfrac{1}{2}\left(\rho + \dfrac{1}{\rho}\right) & \phi = 0 \\[2mm] -\dfrac{1}{2}\left(\rho + \dfrac{1}{\rho}\right) & \phi = \pi \end{cases} \tag{3-22}
$$
$$
v = 0
$$

The function $\rho + 1/\rho$ has a minimum value of 2 at $\rho = 1$. Thus, the real axis of the s plane, outside the unit circle, goes into two sections of the u axis, omitting the space between -1 and $+1$.

Mapping onto the remainder of the real axis of the w plane is described by the conditions

$$
\rho = 1 \qquad 0 \leqq \phi \leqq \pi
$$

It is obvious from Eq. (3-17b) that

$$
v = 0
$$
and
$$
u = \cos \phi
$$

is obtained from Eq. (3-17a). If ϕ had been permitted the range

$$-\pi \leqq \phi \leqq \pi$$

each value of u between -1 and $+1$ would have been covered twice. This is undesirable; we wish to use just enough of the s plane to cover the w plane once. Therefore, for the one value $\rho = 1$, ϕ is restricted to the range $0 \leqq \phi \leqq \pi$. The dashed part of the unit circle is not used, and so the line $-1 \leqq u \leqq 1$ is the map of the top half of the unit circle of the s plane. A semicircle in the lower-half s plane just outside the unit circle goes into half an ellipse just below the $-1 \leqq u \leqq 1$ segment, as shown in

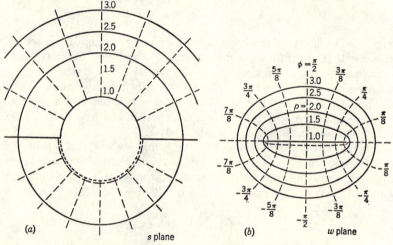

FIG. 3-11. Map of outside unit circle of $w = \frac{1}{2}(s + 1/s)$.

Fig. 3-11b. Deleting part of the unit circle is analogous to deleting part of the $j\omega$ axis in Figs. 3-3 and 3-4.

Up to this point we have shown how the portion of the s plane of Fig. 3-11a maps into the entire w plane. We observe that at points $s = \pm 1$ two curves which *intersect at right angles* in the s plane transform into a straight line, the u axis of the w plane.

Finally, we need to know what happens to points inside the unit circle. The answer was implied when it was noted that u and v are unchanged if ρ is replaced by $1/\rho$ and ϕ by $-\phi$ (that is, if s is replaced by $1/s$). It is necessary only to take Fig. 3-11b and change the labels, changing sign on all ϕ labels and replacing ρ labels by their reciprocals. The result is shown in Fig. 3-12. Note that now the top half of the unit circle is omitted, and accordingly observe that the two portions of the s plane in Figs. 3-11 and 3-12 are complementary.

To complete the description, imagine the portion of the s plane of Fig. 3-11a to be made of rubber. If the top and bottom halves of the unit circle are drawn together, the w plane is produced. Circles in the s plane go into ellipses and radial lines into hyperbolas. The distortion of the inside of the unit circle is harder to visualize because as the two halves of the unit circle are drawn together the center must be picked up and put at infinity. The best visualization for Fig. 3-12a is to take reciprocals of points within the unit circle, which reverts back to Fig. 3-11a.

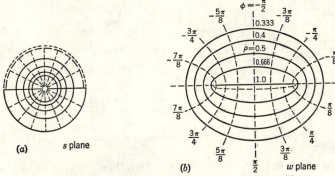

Fig. 3-12. Map of inside unit circle of $w = \frac{1}{2}(s + 1/s)$.

3-6. The Exponential Function. The function

$$w = e^s \tag{3-23}$$

is of particular interest because it takes rectangular coordinates in the s plane into polar coordinates in the w plane. Referring to Chap. 2, we write

$$w = e^{\sigma + j\omega}$$
$$= e^{\sigma} e^{j\omega} \tag{3-24}$$

or $\quad w = e^{\sigma}/\underline{\omega}$

and then make the important observation that when ω goes through a range 2π (say from $-\pi$ to π) the entire w plane will be covered. Thus, the strip shown in Fig. 3-13a goes into the entire w plane, and the rectangular-coordinate lines in the s plane go into the polar-coordinate lines shown in Fig. 3-13b. This is another case where less than the entire s plane is needed to cover the whole w plane. The line $\omega = -\pi$ is deleted from the strip in the s plane in order to avoid double coverage of the negative real axis of the w plane.

Of course, the function will transform s points outside the strip shown. A point such as B at $s = 2 + j3\pi/2$ goes into the same point in the w

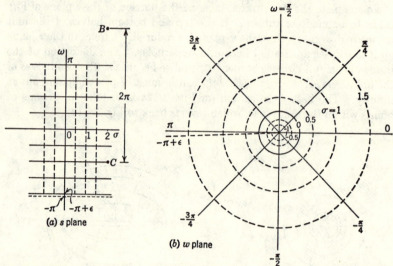

Fig. 3-13. Transformation of a strip of the s plane onto the entire w plane, by the transformation $w = e^s$.

plane as does point C ($s = 2 - j\pi/2$). Thus there are an infinite number of strips in the s plane each of width 2π; and each of them maps onto the

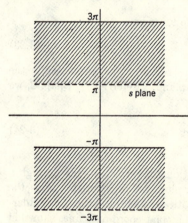

entire w plane. Two more such strips are shown in Fig. 3-14. It is to be emphasized that each strip has an open boundary at the bottom. The decision to close the strip at the top and leave it open at the bottom was arbitrary.

Two interesting points may be noted about Fig. 3-13a. First the $j\omega$ axis transforms to the unit circle in the w plane, and second circles having radii reciprocally related in the w plane correspond to vertical lines equal distances to the right and left of the $j\omega$ axis.

Fig. 3-14. Additional s-plane strips which map onto the entire $w = e^s$ plane.

3-7. Hyperbolic and Trigonometric Functions. In this section the functions $\cosh s$, $\sinh s$, $\cos s$, and $\sin s$ will be dealt with as a group, by relating them to Eq. (3-14), which is now written

$$w = \frac{1}{2}\left(z + \frac{1}{z}\right) \tag{3-25}$$

The change in notation is merely a convenience so that s can be retained as the independent variable. Accordingly, four cases will be considered, as follows:

Case a	$z = e^s$	$w = \cosh s$
Case b	$z = e^{s+j\pi/2}$	$w = j \sinh s$
Case c	$z = e^{js}$	$w = \cos s$
Case d	$z = e^{j(s-\pi/2)}$	$w = \sin s$

$$(3\text{-}26)$$

The correctness of the above formulas for w can be checked by using Eq. (3-25) and referring to Eqs. (2-50) for definitions of these functions.

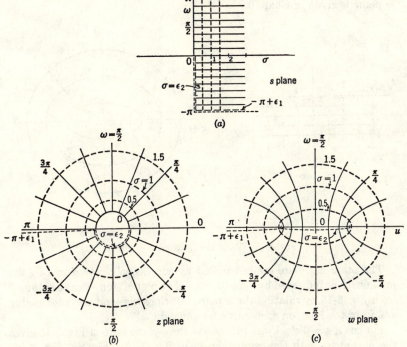

FIG. 3-15. Plots used to describe the function $w = \cosh s$.

Case a will be considered in some detail, with simple modifications taking care of the other cases. The graphical interpretations of Fig. 3-11 and 3-13 provide the needed clues. However, the s plane in Fig. 3-11 and the w plane in Fig. 3-13 will both be called the z plane. In the interest of clarity both figures are combined in Fig. 3-15, with appropriate changes in notation. Part of the $j\omega$ axis is omitted in order to make Fig. 3-15b look like Fig. 3-11a. The transformation from Fig. 3-15b to 3-15c has already been treated in Fig. 3-11. Thus, we see that the function

$$w = \cosh s$$

maps the *semi-infinite* strip of the s plane at (a) in Fig. 3-15 on the whole w plane. This strip is open along the dashed edges, thereby avoiding multiple coverage of points in the w plane. Other similar strips in the right-half s plane again cover the w plane. Similar strips in the left-half s plane go into the inside of the unit circle of the z plane, and from Fig. 3-12 we therefore see that each of these strips also covers the w plane. Most of what we want to portray at present about this function is provided by Fig. 3-15. Further information relevant to the multiple covering of the w plane is given in Chap. 6.

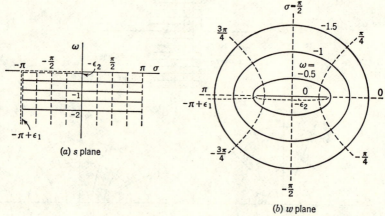

(a) s plane

(b) w plane

Fig. 3-16. Plot of the function $w = \cos s$.

The other cases are easily handled by successively replacing s by the modifications indicated by Eqs. (3-26). The plot of sinh s can be obtained from Fig. 3-15 by rotating the w plane 90° clockwise and also subtracting $\pi/2$ from the label on each curve for constant ω.

For cos s, a $\pi/2$ clockwise rotation of the s-plane strip of Fig. 3-15 gives the new strip, with the results shown in Fig. 3-16. Finally, Fig. 3-16b can be used for sin s by adding $\pi/2$ to each label on the curves for σ = constant. Actual plots are not shown for sinh s and sin s, because they differ only slightly from cosh s and cos s, as noted above.

3-8. The Point at Infinity; The Riemann Sphere. In Chap. 2, reference was made to infinity not being a unique point. In a strict sense this is correct. However, in considering the transformation properties of a function, it is sometimes convenient to use the words *point at infinity*, as if infinity were a geometrical point. Then it is possible, for example, to say that the function

$$w = \frac{1}{s}$$

transforms the point $s = 0$ into the point at infinity in the w plane. However, in this usage, the point at infinity is an abstraction; it cannot be regarded as a geometrical point. All points on the circumference of an expanding circle would have to approach the point at infinity, and this is absurd from a geometrical viewpoint.

The point at infinity is no more a geometrical point than is ∞ a symbol for a number. Nevertheless, both these vaguely inaccurate concepts, the "point at infinity" and the "number infinity," find limited usage in the

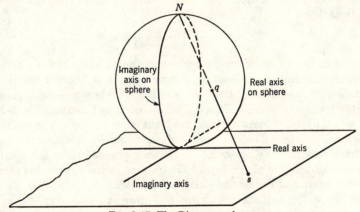

FIG. 3-17. The Riemann sphere.

language of mathematics. The inability to identify infinity as a geometrical point is related to the fact that operations like $\infty - \infty$ and ∞ / ∞ have no meaning. Therefore, the concept of a point at infinity is limited to nonprecise descriptions of mapping operations.

There is an artifice whereby the point at infinity can be interpreted as a finite point. This is accomplished by mapping the infinite plane onto the finite area of a sphere. Such a sphere is called the Riemann sphere and is constructed by establishing a one-to-one correspondence between points on the plane and the sphere, in the manner shown in Fig. 3-17. A sphere sits on the complex plane at the origin, and a ray is drawn from point N to the point s in the plane. The point q where this ray pierces the sphere corresponds to the point s. As point s goes to infinity in any direction, point q moves toward the unique point N on the sphere, which therefore is the "projection" on the sphere of the point at infinity. Note that the real and imaginary axes go into great circles on the sphere.

All the maps so far presented could be placed on a spherical surface, and then the point at infinity would become unique. Of course, the families of curves would become distorted, and the equations describing the loci would be changed. Thus, the Riemann sphere is not very helpful quantitatively, but occasionally it has pictorial and conceptual value.

3-9. Further Properties of the Reciprocal Function. In Sec. 3-4 you were given an introductory analysis of the function $w = 1/s$. Now we shall consider the more general case

$$w = \frac{1}{s+1} \tag{3-27}$$

Study of this function yields further insight into properties of $1/s$, and Eq. (3-27) has the further value that it can be used to construct other functions, like

$$1 - \frac{2}{s+1} = \frac{s-1}{s+1}$$

Turning to Eq. (3-27), if we use rectangular coordinates in the s plane, the analysis is readily obtained by writing $s + 1 = s'$, thereby reverting to the earlier case. Then, the s plane of Fig. 3-10 becomes the s' plane, and the transformation for Eq. (3-27) is obtained by repeating Fig. 3-10, but with the label reduced by 1 on each locus of constant σ. The result is shown in Fig. 3-18. This simple shift of origin has an interesting effect on the transformation. Whereas Eq. (3-11) transforms the right-half s plane into the right-half w plane, Eq. (3-27) transforms the right-half s plane into the interior of the shaded circle in Fig. 3-18.

All loci in the w plane are circles, if a straight line is regarded as a degenerate circle. Equations for these circles can be obtained readily by calculating the intercepts from the original equation or from Eqs. (3-13) by replacing σ by $\sigma + 1$. The new equations are

$$\left(v + \frac{1}{2\omega}\right)^2 + u^2 = \frac{1}{4\omega^2}$$
$$v^2 + \left(u - \frac{1}{2\sigma + 2}\right)^2 = \frac{1}{4(\sigma + 1)^2} \tag{3-28}$$

The situation becomes more challenging when we go to polar coordinates in the s plane. In the usual notation for polar coordinates, Eq. (3-27) is conveniently written

$$(u + jv)(1 + \rho \cos \phi + j\rho \sin \phi) = 1$$

When the multiplication is carried out and real and imaginary parts are equated, we get

$$\rho u \cos \phi - \rho v \sin \phi = 1 - u$$
$$\rho v \cos \phi + \rho u \sin \phi = -v \tag{3-29}$$

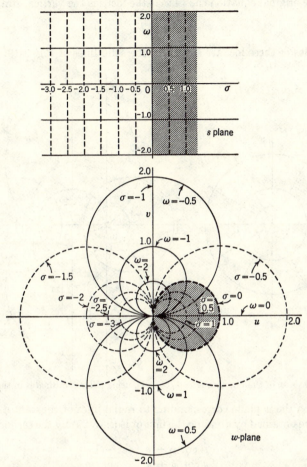

Fig. 3-18. Transformation $w = 1/(s + 1)$ for rectangular coordinates in the s plane.

Squaring each of these equations and adding the results gives

$$(u^2 + v^2)(1 - \rho^2) - 2u + 1 = 0$$

and completing the square in u gives

$$\left(u - \frac{1}{1 - \rho^2}\right)^2 + v^2 = \left(\frac{\rho}{1 - \rho^2}\right)^2 \qquad (3\text{-}30)$$

This equation describes the family of loci in the w plane for circles of constant ρ in the s plane, for all but the degenerate case $\rho = 1$. For that case the original equation shows that the locus is the vertical straight line

$$u = \tfrac{1}{2}$$

Examples of these loci are shown by the solid lines of Fig. 3-19.

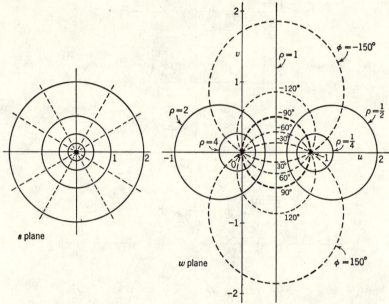

FIG. 3-19. Map of the function $w = 1/(s + 1)$, using polar coordinates in the s plane.

Loci in the w plane corresponding to radial lines of constant ϕ in the s plane are obtained by dividing the first of Eqs. (3-29) by the second, giving

$$\frac{u \cot \phi - v}{v \cot \phi + u} = \frac{u - 1}{v}$$

and this in turn becomes

$$u^2 - u + v^2 - v \cot \phi = 0$$

Completing the square in both u and v then gives

$$\left(u - \frac{1}{2}\right)^2 + \left(v - \frac{\cot \phi}{2}\right)^2 = \left(\frac{1}{2 \sin \phi}\right)^2 \qquad (3\text{-}31)$$

Thus it is found that the radial straight lines (ϕ = constant) in the s plane go into circles in the w plane. When $\sin \phi = 0$, we get a degenerate case; but this is the real axis in the s plane, and it has been shown that this goes into the real axis of the w plane. Thus, when $\sin \phi = 0$, the locus is

$$v = 0$$

Loci for constant ϕ are shown by the dashed circles in Fig. 3-19. Here we make the interesting observation that each of these circles is interpreted in two parts, an upper part and a lower part, for which the angle designations differ by 180°. This is a natural consequence of the fact that each circle is the trace of a radial line passing *through* the origin of the s plane; and of course the angular designation for such a line changes as the origin is passed.

In the w plane we again find the circle into which is mapped the right half of the s plane. This is the circle carrying the designation $\pm 90°$, which of course is the same as labeling it $\sigma = 0$, as in Fig. 3-18.

This treatment of Eq. (3-27) leads to an interesting general conclusion about the function

$$w = \frac{1}{s}$$

Let s describe a circle defined by

$$s = A + Re^{j\theta}$$

where the complex A and real R are constants and θ is variable. Then we can write

$$w = \frac{1}{A + Re^{j\theta}}$$

$$= \frac{1}{A} \frac{1}{1 + (R/|A|)e^{j(\theta-\alpha)}}$$

where $\alpha = \text{ang } A$. This is of the form

$$w = \frac{1}{A} \frac{1}{s' + 1} \tag{3-32}$$

where

$$s' = \frac{R}{|A|} e^{j(\theta-\alpha)}$$

We recognize Eq. (3-32) as being similar to Eq. (3-27), and it is recalled from the analysis of the latter equation that circles centered at the origin in the s plane go into circles in the w plane. Therefore, Eq. (3-32) yields a circle in the w plane when s' describes a circle centered at the origin. Furthermore, we have defined s' in such a way that it is centered at the origin when s describes the prescribed circle. It is concluded that $w = 1/s$ traces out a circle whenever s follows a circular path.

3-10. The Bilinear Transformation. In Sec. 3-3 it is mentioned that the function

$$w = \frac{s - 1}{s + 1} \tag{3-33}$$

has important applications in the analysis of linear systems. Because of this importance, we now give it a brief consideration. Equation (3-33) is a special case of the general bilinear function

$$w = \frac{as + b}{cs + d}$$

It is preferable to study Eq. (3-33), and from it we can learn all we need to know about the general case. The treatment of the previous section serves as the point of departure because, as was pointed out there, Eq. (3-33) can be written

$$w = 1 - \frac{2}{s + 1}$$

We shall consider the map of rectangular coordinates in the right half of the s plane. Figure 3-18 shows the transformation of the right half plane by the function $1/(s + 1)$. We merely take the mirror image of this transformation, scaled by a factor 2, and translate it one unit to the right. The result is shown in Fig. 3-20a. From Eqs. (3-28) we get the equations of the loci by changing sign on u and v and multiplying them by the factor $\frac{1}{2}$ and by subtracting 1 from u to perform the required translation to the right. The results are

$$\left(v - \frac{1}{\omega}\right)^2 + (u - 1)^2 = \left(\frac{1}{\omega}\right)^2$$
$$v^2 + \left(u - \frac{\sigma}{\sigma + 1}\right)^2 = \left(\frac{1}{\sigma + 1}\right)^2 \tag{3-34}$$

A particularly important portrayal of this transformation is obtained by using polar coordinates in the w plane. In order to obtain the corresponding loci in the s plane, write Eq. (3-33) in the inverse form

$$s = \frac{1 + w}{1 - w} = \frac{2}{1 + (-w)} - 1$$

The term

$$\frac{2}{1 + (-w)}$$

is like Eq. (3-27), but with $-w$ in place of s. Thus, to get the loci of the *present s plane*, we take the loci of the *w plane of Fig.* 3-19, with a scale factor of 2, and with all angle labels changed by $\pm\pi$ (to account for s

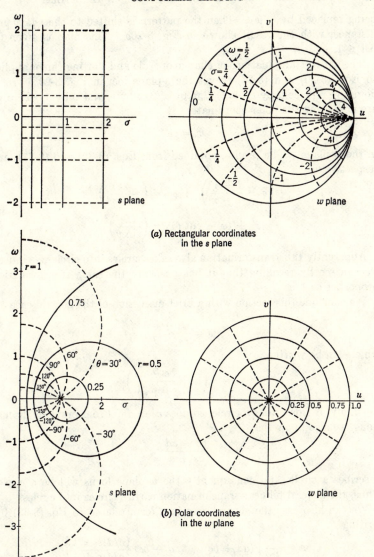

(a) Rectangular coordinates
in the s plane

(b) Polar coordinates
in the w plane

Fig. 3-20. Transformation due to the bilinear function $w = (s - 1)/(s + 1)$.

being replaced by $-w$). Then the pattern is shifted to the left a unit distance, with the result shown in Fig. 3-20b. This map is given for $|w| \leqq 1$.

The above manipulations of change of scale and shifting can be applied to Eqs. (3-30) and (3-31) to get the s-plane loci in Fig. 3-20b. In so doing we replace u by $(\sigma + 1)/2$, v by $\omega/2$, ρ by r, and ϕ by θ, where we are now using the polar coordinates

$$w = re^{j\theta}$$

in the w plane. The results, obtained from Eqs. (3-30) and (3-31), are, respectively,

$$\left(\sigma - \frac{1 + r^2}{1 - r^2}\right)^2 + \omega^2 = \left(\frac{2r}{1 - r^2}\right)^2$$

$$\sigma^2 + (\omega - \cot \theta)^2 = \left(\frac{1}{\sin \theta}\right)^2 \qquad (3\text{-}35)$$

Apparently this transformation also takes circles into circles, as indeed we can see by recalling that it has basically the same transformation properties as $1/s$.

We conclude this section with a brief discussion of the general case

$$w = \frac{as + b}{cs + d}$$

which can be written

$$w = \frac{1}{c}\left(a + \frac{bc - ad}{cs + d}\right)$$

Certainly, if s describes a circle, so does $cs + d$. From previous discussions, we also know that

$$\frac{bc - ad}{cs + d}$$

describes a circle, and consequently the w-plane locus will be a circle. Thus, the general bilinear transformation carries circles into circles.

We can also cast the general case in a form related to Eq. (3-33) by writing

$$w = \frac{1}{2cd}\left[(ad + bc) + (ad - bc)\frac{(c/d)s - 1}{(c/d)s + 1}\right]$$

The second term is a constant multiplied by $(s' - 1)/(s' + 1)$, where $s' = (c/d)s$, and this transformation has been treated. The other term in the above equation is merely a constant. Thus we see how with suitable change of variable the general case can be obtained from the specific one.

3-11. Conformal Mapping. In all the foregoing examples it is observed that in most cases mutually perpendicular lines in the s plane transform into mutually perpendicular lines in the w plane. However, we find there are some points in these illustrations where this is not true, the point $s = 0$ for the function $w = s^2$ being one example. Here the mutually perpendicular real and imaginary axes in the s plane transform into lines in the w plane intersecting at 180°, as shown in Figs. 3-3 and 3-4.

Another property exhibited in most instances is that a small geometric figure, like a curvilinear rectangle formed by four coordinate lines, transforms into a similar figure in the w plane. The corresponding areas labeled A and A' in Figs. 3-3 and 3-5 are illustrations. However, again using $w = s^2$ as the example, we find that this preservation of general shape is not true at the origin. This can be seen by referring to Fig. 3-5 and observing that the rectangle B formed around the origin by the lines $\sigma = \pm 0.5$, $\omega = \pm 0.5$ transforms into a figure B' which has no resemblance to a rectangle.

Thus, experience gained from these examples implies a certain geometrical regularity of the transformations, as embodied in preservation of angles of intersection and approximate geometrical shapes, in going from one plane to the other. However, we also find certain exceptions. In the present section we shall explain this behavior of transformation maps, showing why the indicated relationships are usually found and under what conditions exceptions are to be expected.

The derivative serves as the point of departure. Recall that the derivative of $w = f(s)$ at a point s_1 is defined as

$$\lim_{\Delta s \to 0} \frac{\Delta w}{\Delta s} = f'(s_1)$$

where Δs is an increment from point s_1 and Δw is the corresponding increment in the w plane. Existence of this limit implies that corresponding to an arbitrary small $\epsilon > 0$ there can be found a δ such that

$$\left| \frac{\Delta w}{\Delta s} - f'(s_1) \right| < \epsilon$$

when $$|\Delta s| < \delta$$

and therefore $$|\Delta w - f'(s_1)\, \Delta s| < \epsilon |\Delta s|$$

This inequality is portrayed in Fig. 3-21, at (a) for $|f'(s_1)| \neq 0$ and at (b) for $f'(s_1) = 0$. If we use the notation $f'(s_1) = A e^{j\alpha}$, it is apparent from the geometry of Fig. 3-21a that

$$\big| |\Delta w| - A |\Delta s| \big| < \epsilon |\Delta s| \tag{3-36a}$$

and $$|\text{ang } \Delta w - \alpha - \text{ang } \Delta s| < \sin^{-1} \frac{\epsilon}{A} \tag{3-36b}$$

It is emphasized that relation (3-36b) is true only if

$$|f'(s_1)| = A > \epsilon$$

This can be understood by referring to Fig. 3-21b. In this case it is certain that Δw lies within the circle shown, but no estimate of the angle of Δw is possible merely from knowing the radius and location of the circle because w_1 is at the center of the circle.

We now allow ϵ to approach zero, while recognizing that ϵ determines δ and that $|\Delta s| < \delta$. Clearly, we are dealing with increments which may be required in certain cases to be quite small, depending on the size of A. However, so long as $f'(s_1)$ is not zero, as ϵ approaches zero the

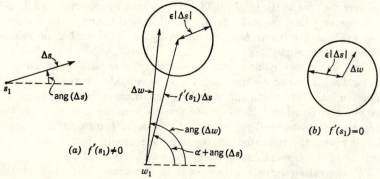

FIG. 3-21. Transformation of an increment, at a point where the derivative exists.

condition $\epsilon < A$ will ultimately be attained and finally the right-hand sides of relations (3-36) will be considered substantially zero, giving the pair of approximate equalities

$$|\Delta w| \doteq A|\Delta s| \tag{3-37}$$
$$\text{ang } \Delta w \doteq \alpha + \text{ang } \Delta s \tag{3-38}$$

In words, these equations say that if an increment Δs is formed at a point s_1, where $w = f(s)$ has a nonzero derivative, then the corresponding increment Δw is obtained from Δs by scaling its magnitude by the factor $|f'(s_1)|$ and by increasing its angle by the amount ang $f'(s_1)$.

Equations (3-37) and (3-38) lead to two important conclusions about the properties of transformation maps in small regions. In Fig. 3-22a two curves are shown intersecting at point s_1, where ϕ' is the angle of intersection. Each of these curves is considered to be generated by increment Δs_1 or Δs_2 as it approaches zero. Two corresponding increments Δw_1 and Δw_2 trace out a pair of intersecting curves in the w plane, as indicated in Fig. 3-22b. These curves intersect at an angle θ'. From

Eq. (3-38) we have two approximate equations

$$\text{ang } \Delta w_1 \doteq \alpha + \text{ang } \Delta s_1 \qquad \text{ang } \Delta w_2 \doteq \alpha + \text{ang } \Delta s_2$$

and therefore $\quad \text{ang } \Delta w_1 - \text{ang } \Delta w_2 \doteq \text{ang } \Delta s_1 - \text{ang } \Delta s_2$

This approximate inequality becomes increasingly accurate as each increment approaches zero. Furthermore, in the limit, as the increments approach zero,

$$\phi' = \text{ang } \Delta s_1 - \text{ang } \Delta s_2 \qquad \theta' = \text{ang } \Delta w_1 - \text{ang } \Delta w_2$$

Thus, we conclude that

$$\phi' = \theta'$$

Emphasis is placed on the fact that this equation is exact. We have shown that, if two curves intersect in the s plane at a point where a function has a nonzero derivative, the traces of these curves in the function plane will intersect at the same angle.

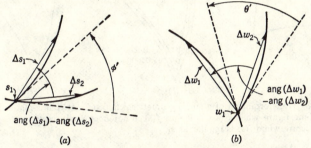

Fig. 3-22. Preservation of angles at points where $f'(s) \neq 0$.

Now consider a small geometrical figure in the s plane, such as the curvilinear triangle T in Fig. 3-23a, surrounding a point s_1 where the derivative exists and is not zero. Assume that the triangle is small enough to allow Eqs. (3-37) and (3-38) to apply with reasonable accuracy for any increment Δs drawn from s_1 to a point on the periphery of T. Equations (3-37) and (3-38) tell us that each corresponding Δw increment is obtained from the Δs increment by scaling its magnitude by the same multiplying factor and by rotation through the same angle α. Therefore, the figure T' generated in the w plane will be similar, but with changed size and orientation. However, Eq. (3-37) is only approximate, and so there is a certain indefiniteness for each Δw, which depends upon the size of T, and therefore we draw a double line in the w plane in such a way that the true boundary of T' is known only to lie between this pair of lines.

This development is based only on the fact that $f'(s_1)$ exists and is not zero. As a result, we get no conclusion as to the behavior of the boundary of T' as it wavers between the two bounding lines shown in Fig. 3-23b. For example, we have no assurance that the boundary of T' will have the same angles as T at corresponding vertices, nor are we sure whether or not the boundary of T' will be as smooth as the boundary of T. We can say nothing more about the boundary of T' without placing greater restriction on $f(s)$, which we now do by stipulating that $f(s)$ shall have a nonzero derivative at every point in a neighborhood large enough to include figure T. Now we can say that angles are preserved at every point on T, and therefore angles at the vertices of T' will be the same as for corresponding vertices of T. Furthermore, if we take some point b, not at a vertex, we can interpret this as the intersection of two curves at 180°. It follows that the curves must intersect at 180° at the transformed point b'

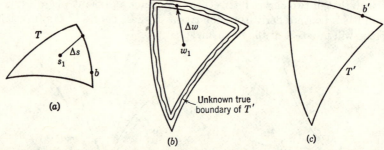

FIG. 3-23. Transformation of a small geometrical figure. (a) The s-plane figure; (b) the w-plane figure when $f'(s_1) \neq 0$, but $f(s)$ is not necessarily regular at s_1; (c) the w-plane figure when $f(s)$ is regular at s_1.

and therefore that T' must be smooth where T is smooth. The result will be as shown in Fig. 3-23c. However, it is emphasized that T' is not an exact scaled and rotated replica of T and that preservation of angles at the vertices is dependent upon $f(s)$ having a nonzero derivative in a neighborhood of s_1 large enough to include T.

Transformations which have the property of preserving angles of intersection and of approximately preserving small shapes are said to be *conformal*. An analytic function is thus seen to provide a conformal transformation at any point where the function is regular and has a nonzero derivative.

In conclusion, your attention is drawn to a comparison of the statements about angles and shapes. Angles of intersection are preserved *exactly;* but shapes are preserved only *approximately*. This difference arises because the angle between Δs_1 and Δs_2 retains its meaning down to zero size for the increments, but, in order to have figures with describable shapes, the increments must remain finite.

In the case of $w = s^2$ we find that angles are doubled at point $s = 0$, and for $w = s^3$ it is found that angles are tripled. Furthermore, we note that $w = s^2$ has a zero first derivative at $s = 0$, and $w = s^3$ has first and second derivatives which are zero at $s = 0$. These observations seem to indicate a general relation between the behavior of angles at a point of zero derivative and the number of derivatives which are zero. If we can take $w = s^n$ as typical, it has $n - 1$ derivatives which are zero at $s = 0$, and the angle is increased by the factor n (one greater than the number of zero derivatives). A more general discussion of this topic is found in Chap. 6.

3-12. Solution of Two-dimensional-field Problems. The conformal transformation property of analytic functions is related to their usefulness

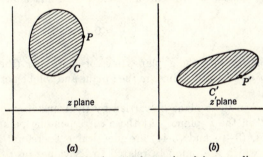

z plane   *z′* plane

(a)   (b)

Fig. 3-24. General transformation $z' = f(z)$ for use in solving two-dimensional boundary-value problems.

in arriving at solutions of the two-dimensional Laplace equation

$$\frac{\partial^2 u}{\partial x^2} + \frac{\partial^2 u}{\partial y^2} = 0 \tag{3-39}$$

subject to certain types of boundary conditions. In this equation u stands for a scalar potential function, like electric potential or temperature.

Referring to Fig. 3-24, suppose that an analytic function

$$w = g(z') \tag{3-40}$$

is prescribed. Its real component

$$u = \mathrm{Re}\,[g(z')] \tag{3-41}$$

will be a solution of

$$\frac{\partial^2 u}{\partial (x')^2} + \frac{\partial^2 u}{\partial (y')^2} = 0 \tag{3-42}$$

and will have a unique set of values at points on curve C'. The imaginary component v will also be a solution of the Laplace equation. Lines of constant u and lines of constant v form a curvilinear grid of mutually perpendicular lines in the z' plane. If u is the potential, then lines of constant v are flow lines of the field. Now suppose that there is a second analytic function

$$z' = f(z) \tag{3-43}$$

which carries curve C of the z plane into curve C', with the interior of C going into the interior of C'. It is assumed that C and C' are both simple closed curves. The function

$$w = g[f(z)] \tag{3-44}$$

is an analytic function of an analytic function. It can easily be shown that this in turn is an analytic function. Thus, again using u to designate the real component of w, we conclude that u is a solution of

$$\frac{\partial^2 u}{\partial x^2} + \frac{\partial^2 u}{\partial y^2} = 0$$

Also, in similarity with the z' plane, lines of constant v (the imaginary component of w) will be flow lines in the z plane and will be normal to lines of constant u.

Let point P in the z plane be carried by the transformation $z' = f(z)$ into point P' in the z' plane. At these corresponding points the arguments of Eqs. (3-40) and (3-44) are identical. Therefore, u will have values on C which are uniquely related to its values on C'. The gist of these comments is that u is simultaneously the solution of *two* different boundary-value problems, depending on whether z' or z is considered to be the independent variable.

This is a useful idea, because the two boundaries are different, making it possible by using analytic functions to get additional solutions from one known solution. In the practical application of this principle, the right half of the z' plane is usually taken for the region "inside" C', so that C' becomes the imaginary axis.* (In this case C' is not a simple closed curve, but this fact is normally unimportant.) It is also possible to use the unit circle as C'. The significant point is that C' is chosen to have a simple geometrical shape. The appropriate pair of planes for this case is shown in Fig. 3-25.

Now assume that we are to solve Laplace's equation for the region inside C, in the z plane. It is assumed that u is the electric potential, subject to the conditions that u is identically zero on C, and that there is a line charge of q electrostatic units (esu) per unit length piercing the z plane

---

* In much of the literature it is the top half of the z' plane that is used. The right half plane is used here because electrical applications frequently involve analysis of functions in the right half plane.

at point P_0. The transformation

$$z' = f(z)$$

which carries the interior of C into the right half of the z' plane is assumed to be known at this time. Point P_0 will be carried into P'_0, and in the z' plane the imaginary axis will be at zero potential. The geometry of the z' plane is simple, and the problem of finding u as a function of x' and y' is an elementary one in field theory. The potential is equivalent to the free-space potential of a line charge q at P'_0 and a line charge $-q$ located

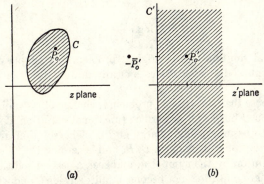

(a)   (b)

FIG. 3-25. Special case of Fig. 3-24, in which curve C' is the imaginary axis.

at the image in the j axis of point P'_0. For the moment, by use of real variables, this potential is known to be

$$u = \frac{q}{2} \log \frac{(x' - x'_0)^2 + (y' - y'_0)^2}{(x' + x'_0)^2 + (y' - y'_0)^2} \tag{3-45}$$

In order to cast this into the notation of complex variables, it is necessary to use the logarithmic function, which is defined in the next chapter. However, we shall use this function, with the suggestion that you look ahead to Eq. (4-21) for information about its fundamental properties. From this we find that

$$\log (z' - z'_0) = \frac{1}{2} \log [(x' - x'_0)^2 + (y' - y'_0)^2] + j \tan^{-1} \frac{y' - y'_0}{x' - x'_0}$$

It is shown in Chap. 4 that this can be considered an analytic function, provided that we make the restriction

$$-\pi < \tan^{-1} \frac{y' - y'_0}{x' - x'_0} \leqq \pi \tag{3-46}$$

Thus, Eq. (3-45) can be written in the equivalent form

$$u = q \operatorname{Re} \left(\log \frac{z' - z'_0}{z' + \bar{z}'_0} \right)$$

or
$$u = q \log \left| \frac{z' - z'_0}{z' + \bar{z}'_0} \right| \tag{3-47}$$

In the above, the symbol $-\bar{z}'_0$ has been used for the image of z'_0. Note that when $z' = jy'$ the quantity between the absolute-value signs in Eq. (3-47) becomes unity, correctly giving zero u on the imaginary axis.

Continuing to assume that $f(z)$ is a known analytic function, Eq. (3-47) provides

$$u = q \log \left| \frac{f(z) - z'_0}{f(z) + \bar{z}'_0} \right| \tag{3-48}$$

as the formal solution to the original problem. A simple extension of this treatment would cover the case of any finite number of line charges inside C.

Now suppose that the original problem is changed by stipulating that u is not zero on C but has a prescribed functional variation along C, while continuing to assume a line charge at P_0. The solution given by Eq. (3-48) ensures proper behavior near P_0 and is zero on C. We now designate this solution by the symbol u_1 and seek a second solution u_2 which has no singular point (does not become infinite) inside C and which has the prescribed values on C. Then, by superposition,

$$u = u_1 + u_2 \tag{3-49}$$

is the required solution.

The next objective is to determine u_2. Continuing to assume that the transformation $z' = f(z)$ is available, it is possible to determine how u_2 varies with y', on the jy' axis. Thus, $u_2(0,y')$ can be considered a known function. It is shown in Chap. 7 that, if $u_2(0,y')$ has continuous derivatives with respect to y', it is possible to determine an analytic function which is regular in the right half plane and such that $u_2(0,y')$ is the real component of this function on the j axis. This function is expressed in terms of the *Poisson integrals* for the real and imaginary components. In its entirety, this function would be written

$$\begin{aligned} w_2 &= g_2(z') \\ &= u_2(x',y') + jv_2(x',y') \end{aligned} \tag{3-50}$$

but we are presently interested only in the real component.

Equation (7-48a) is pertinent, and in the present notation it yields

$$u_2(x',y') = \frac{x'}{\pi} \int_{-\infty}^{\infty} \frac{u_2(0,y'')}{(x')^2 + (y' - y'')^2} \, dy'' \tag{3-51}$$

This is a solution in the z' plane. To get it into the z plane, we recall that

$$x' = \text{Re}\,[f(z)]$$
$$y' = \text{Im}\,[f(z)]$$

which converts Eq. (3-51) into a function of x and y. In claiming that u_2 is a function of Laplace's equation in the z plane we are relying on theory presented in Chap. 7 to the effect that Eq. (3-51) is the real component of the analytic function designated by Eq. (3-50).

By the process just described, two-dimensional boundary-value problems can be solved for a region inside a simple closed curve C when the scalar function is prescribed, and not necessarily constant, on C. However, it has been assumed that the analytic function $z' = f(z)$ is known. We now briefly address our attention to methods of finding this function for specific problems. You will recognize that certain of the functions described in this chapter transform a simple closed curve into a coordinate axis. An extension of this catalogue of such functions would therefore be of help. However, there is a general function, known as the *Schwarz-Christoffel transformation*, which will transform the inside of a polygon into the right half plane.* This function can be used in the general case to get an approximate solution, by choosing a polygon which will approximate the given curve C.

This section is concluded with the observation that we have provided here only a glimpse of an extensive branch of applied mathematics. For further information you are referred to the literature.

PROBLEMS

3-1. For the transformation shown in Fig. 3-2 determine the locus in the w plane of the curve $\omega = \sigma^2$.

3-2. Show a set of plots like Fig. 3-3 for the function $w = s^2$, but let the range of angle of w be from 0 to 2π.

3-3. Make sketches of Fig. 3-5, labeling w-plane loci with appropriate values of σ and ω such that the w plane is the transformation of

(a) The top half of the s plane
(b) The bottom half of the s plane
(c) The right half of the s plane
(d) The left half of the s plane

* See, for example, Z. Nehari, "Conformal Mapping," p. 189, McGraw-Hill Book Company, Inc., New York, 1952; P. Morse and H. Feshbach, "Methods of Theoretical Physics," vol. 1, p. 445, McGraw-Hill Book Company, Inc., New York, 1953; and E. A. Guillemin, 'The Mathematics of Circuit Analysis," p. 380, John Wiley & Sons, Inc., New York, 1949; R. V. Churchill, "Complex Variables and Applications," McGraw-Hill Book Company, Inc., New York, 1960.

3-4. In the notation $s = \sigma + j\omega$ and $w = u + jv$, determine the equations in σ and ω for curves of constant u and constant v, for the function

$$w = s^2$$

Draw a sketch of the s plane, showing two areas and their boundaries into which are transformed the w-plane square with sides $u = 0, 1; v = 0, 1$.

3-5. Derive Eqs. (3-13) of the text.

3-6. Obtain a function $w = f(s)$ that will transform the half plane Im $s > 2$ into the interior of the unit circle in the w plane. Sketch and label orthogonal curves in the s and w planes showing properties of the transformation.

3-7. In the notation of Sec. 3-3 analyze each of the circuits of Fig. P 3-7 for $T = E_2/E_1$, and determine values for the unlabeled elements such that $T(s)$ in Fig. P 3-7a will be the same as $T(w)$ for Fig. P 3-7b.

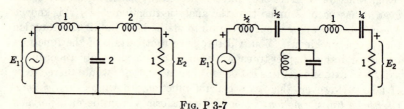

FIG. P 3-7

3-8. Sketch plots of the function

$$w = \frac{1}{2}\left(s + \frac{1}{s}\right)$$

similar to those given in the text, but for the following conditions:

$$(a) \quad 0 \leqq \phi < \pi \qquad \rho \neq 1 \qquad\qquad 0 \leqq \phi \leqq \pi \qquad \rho = 1$$
$$(b) \quad -\pi \leqq \phi < 0 \qquad \rho \neq 1 \qquad\qquad -\pi \leqq \phi \leqq 0 \qquad \rho = 1$$

3-9. Use the known properties of $w = \frac{1}{2}(s + 1/s)$, as described in the text, to obtain the transformation properties of

$$w = \frac{1}{2}\left(s - \frac{1}{s}\right)$$

3-10. Use the known properties of $w = \frac{1}{2}(s + 1/s)$, as described in the text, to obtain information about the transformation properties of

$$w = \frac{(s + 1)^2}{s + 2}$$

In this problem it is not necessary to use either of the standard coordinate systems in the s plane.

3-11. From the results given in the text, obtain the lines in the w plane which are the traces of the polar-coordinate lines in the s plane for the function

$$w = \frac{s^4 + 1}{s^2}$$

3-12. Obtain a function $w = f(s)$ that will transform a general ellipse of major axis a and minor axis b of the w plane (with axes coinciding with the u and v axes) into a circle centered at the origin in the s plane.

3-13. Derive Eqs. (3-28).

3-14. Obtain a function $w = f(s)$ which will transform the $s =$ plane strip $\sigma \geqq 1$, $0 \leqq \omega \leqq 1$, into the right half $u \geqq 0$ of the w plane. Obtain equations in u and v of lines of constant σ and of constant ω. Use the notation $s = \sigma + j\omega$, $w = u + jv$.

3-15. For the function

$$w = e^{s^2}$$

specify the region in the s plane that maps into the inside of the unit circle of the w plane. Specify the equation between σ and ω (where $s = \sigma + j\omega$) for a circle $|w| = R < 1$.

3-16. Investigate the mapping properties of

(a) $w = \sin s$ \hfill (b) $w = \sinh s^2$

3-17. Obtain the mapping properties of the function

$$w = \tanh s$$

3-18. Derive Eqs. (3-35) directly from the function

$$w = \frac{s - 1}{s + 1}$$

3-19. Determine how rectangular coordinates in the left half of the s plane are transformed by the function

$$w = \frac{s - 1}{s + 1}$$

3-20. Determine how polar coordinates in the s plane are transformed by the function

$$w = \frac{s - 1}{s + 1}$$

3-21. In the s plane let a circle C_1 have intercepts at points 5 and 20 on the j axis, and let a circle C_2 have intercepts at points 6 and 12 on the j axis. Circles C_1 and C_2 have centers on the j axis. Find a function $w = f(s)$ that will transform C_1 into a circle C_1' of radius 2, centered at the origin in the w plane, and such that C_2 is transformed to a circle C_2' which is concentric with C_1' and of radius less than 2. Find this radius.

3-22. Referring to the data given in Prob. 3-21 for circles C_1 and C_2 in the s plane, obtain a transformation that will transform C_1 into a circle with intercepts 1 and 5 and C_2 into a circle with intercepts 2 and 3, all intercepts being on the real axis. These transformed circles have centers on the real axis.

3-23. The function

$$w = \frac{s - 1}{s + 1}$$

transforms the right-half s plane inside the unit circle in the w plane. Find a function

$$z = g(s)$$

such that the top half of the s plane goes into a circle of radius unity with center at $z = 1 + j$.

3-24. For the general function

$$w = \frac{s^2 + bs + c}{s + a} \qquad \text{where } a, b, c \text{ are real}$$

show that there is a circle of radius R centered at A in the s plane, such that the inside of this circle and the outside each map onto the entire w plane.

(a) Find A and R.

(b) Show that, if $a^2 + c > ab$, the j axis of the s plane maps twice on to the j axis of the w plane and that, if $a^2 + c < ab$, the real axis of the s plane maps twice onto the real axis of the w plane.

3-25. A polynomial function $w = f(s)$ is to have the following properties: Point $s = 1$ shall transform to $w = -1$, and $s = -1$ shall transform to $w = 7$; and lines radiating from point $s = -1$ shall have their angles tripled in the transformation, and angles radiating from $s = 1$ shall have their angles doubled.

(a) Determine the coefficients of the polynomial.

(b) Determine how a small figure near $s = 0$ will be transformed.

(c) Sketch w-plane traces of the following s-plane lines: the real axis, the imaginary axis, and the unit circle.

3-26. Investigate the preservation of angles and small shapes for the function

$$w = s + |s|^2$$

(a) At $s = 0$ (b) At $s = 1$ (c) At $s = j$

3-27. Investigate the mapping properties of the function

$$w = \frac{2s + 3}{4s + 2}$$

3-28. Obtain the solution for the potential due to an infinite line charge of q units charge per unit length parallel to the axis of a zero-potential infinite circular cylinder of radius R. The distance from the center to the line charge is $b < R$. Show that the potential inside the cylinder is the same as the potential in free space due to the given charge plus an image line charge $-q$ located at a distance R^2/b from the center, along a radial line through the given line charge.

CHAPTER 4

INTEGRATION

4-1. Introduction. In the usual undergraduate course in calculus the student often falls into the practice of relating the definite integral most closely to differentiation, frequently losing sight of the real meaning of the definite integral. In functions of a complex variable it is necessary to be more careful about this; and in the following treatment we shall therefore go back to first principles in defining the definite integral. Because of the two-dimensional character of complex variables we shall find some interesting extensions of the concepts of integration which make it imperative that the definition of the definite integral be thoroughly understood. There are significant differences between real and complex variables as far as theorems on integration are concerned.

4-2. Some Definitions. The primary purpose of Chaps. 2 and 3 is to introduce the concept of a complex variable and a function thereof. Now we are ready to start some relatively precise mathematics, and for that purpose several definitions will be needed. You will recognize that some of the words to be defined here have already been used, with their meanings left to your intuitive interpretation. With the background of Chaps. 2 and 3 you should now be ready for more precise definitions.

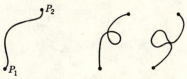

(a) Simple arc   (b) Not simple arcs

Fig. 4-1. Example of simple and nonsimple arcs.

Simple Arc, Differentiable Arc, Simple Closed Curve. Imagine a continuous path in the complex plane starting at a point P_1 and ending at a point P_2. If there are no multiple points in the path (points through which the path goes more than once), such a path is called a *simple arc.* A simple arc is to be distinguished from paths in general in that the simple arc has no multiple points. If a simple arc has a tangent at every point it is said to be a *differentiable arc.* Figure 4-1 illustrates the concept of a simple arc.

Consider the definition of a simple arc, and then allow points P_1 and P_2 to coincide. The result is called a *simple closed curve* (often called a *simple closed Jordan curve*).   Figure 4-2 gives an example of two closed paths, one of which is a simple closed curve.   At (*b*) in that figure each loop can be identified as a simple closed curve, but the complete path is not a simple closed curve.   The point at infinity will not be admitted as a possible point on a simple closed curve.

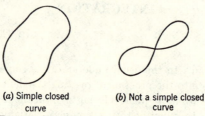

(*a*) Simple closed
curve

(*b*) Not a simple closed
curve

Fig. 4-2. Simple and nonsimple closed curves.

Point Set.   A *point set* (or *set*) is a collection of complex numbers.   For example, all the integers from 1 to 10 form a point set.   Another example would be all the points on the unit circle.

Connected Set.   A *connected set* is a set for which any two points in the set can be joined by a line such that each point in the line will belong to the set. The set of integers from 1 to 10 is not connected, whereas the set of points

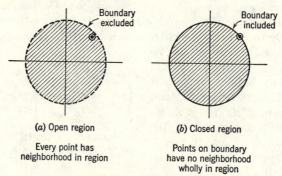

(*a*) Open region

Every point has
neighborhood in region

(*b*) Closed region

Points on boundary
have no neighborhood
wholly in region

Fig. 4-3. Open and closed regions.

on the unit circle and the set of points inside the unit circle are both connected sets.

Region.   Let us imagine a connected set having the property that for each point in the set there is a neighborhood such that each point of the neighborhood is also a member of the set.   Such a set is called an *open region* (or, more briefly, a *region*).   The set of points inside, but not on the perimeter of, a unit circle is an example of an open region.

The significance of requiring that every point shall have a neighborhood in the set is illustrated in Fig. 4-3.   As long as a point is inside the circle,

a neighborhood can be found which also is inside the circle.   This is not
true of a point on the circle itself, as can be seen in Fig. 4-3*b*.

If the point set used to define a region includes points on the boundary
curve as well as points inside, the result is called a *closed region*.   A closed
region is more complicated than an open region because there are two
kinds of points in the closed region, points "inside,"
which have neighborhoods in the region, and points
on the boundary, having neighborhoods not wholly
in the region.

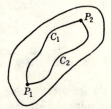

Order of Connectivity.   Regions are classified in
still another way, according to their *connectivity*.
A *simply connected region* is shown in Fig. 4-4.
Take *any* two points P_1 and P_2 lying in the region
and two arbitrarily chosen simple arcs joining the
points, as indicated by C_1 and C_2.   If it is possible

Fig. 4-4. Test for sim-
ple connectivity.

to distort one of these arcs to coincide with the other one, without having
it pass out of the region at any time, then the region is *simply connected*.
Emphasis is placed on the fact that this must be possible for any two
points and any pair of arcs joining them.   The full significance of this
definition is perhaps best understood by considering the doubly connected
region of Fig. 4-5*a*.   In this case arcs C_1 and C_2 cannot be distorted, one
into the other, without passing over the area in the center, which is not
part of the region.

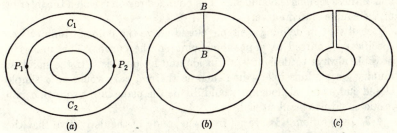

Fig. 4-5. Reducing a doubly connected region to a simply connected one.

A doubly connected region can be converted to a simply connected
region by excluding from the region a set of points which forms a *barrier*
line, such as BB in Fig. 4-5*b*.   When this is done, an arc such as C_1 in
Fig. 4-5*a* is inadmissible, because it will have one point not in the modified
region.   Thus, any two arcs between a pair of points in the modified
region can be brought into coalescence, showing that the modified region
is simply connected.   Figure 4-5*c* shows an alternative viewpoint.

Regions can have higher orders of connectivity.   For example, Fig. 4-6
shows a triply connected region, which can be reduced to a simply con-

nected region by introducing two barrier lines. These can be any two of the three dashed lines shown. The order of connectivity is one greater than the number of barrier lines required to make the region simply connected.

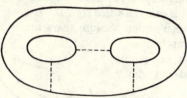

FIG. 4-6. A triply connected region.

Finite Plane, Infinite Plane. In Chap. 3 the point at infinity was presented as a concept worth using occasionally. The *finite plane* (sometimes called the *entire finite plane*) is the set of all the points of the complex plane except the point at infinity. When the point at infinity is included, the result is called the *extended plane.*

Bounded Region. If a region satisfies the condition

$$|s| < M$$

for all points in the region, where M is any fixed number, then the region is said to be *bounded.*

Jordan-curve Theorem. We shall now state an important theorem without proof. For a text at this level the proof is scarcely needed, because the truth of the theorem seems quite obvious. The *Jordan-curve theorem* states that every simple closed curve divides the complex plane into two regions having the curve as their common boundary and that one of these regions is bounded. The bounded region is called the interior of the simple closed curve.

Recall that in defining a simple closed curve the point at infinity was specifically omitted as a possible point on the curve. This makes it possible always to designate an inside and an outside. If the point at infinity were admissible, a line such as the real axis would be a simple closed curve and, although it would divide the plane into two regions, an inside could not be identified.

4-3. Integration. We recall from the study of the derivative that the limit of the differential quotient might depend on the direction from which the increment of the variable approaches zero. We run up against a similar situation in defining the definite integral of a function of a complex variable. In real integration, the variable of integration has a range on the real axis between the limits of integration. For complex variables the integration limits are replaced by two points in the complex plane, and a *path of integration* * is specified between these points. How-

---

* A path (or contour) of integration is an oriented continuous line, which is not necessarily a simple arc or a simple closed curve. By *continuous* we mean that the points of the line form a connected set.

ever, an infinite number of paths lead from one point to the other, and it is possible that the value of the integral may depend on the path.

You are expected to recall the following rudiments of the process of defining a real definite integral: The interval between the limits is broken up into a number of small increments, the length of each increment is multiplied by the value of the function at some point within that interval, and a summation of these products is taken. Then the length of each interval is allowed to approach zero, while their number approaches infinity. If the sum approaches a limit, this limit is defined as the definite integral. If the limit does not exist, the integral does not exist. The real definite integral can be interpreted geometrically as an area.

In the case of complex integration no such simple geometrical interpretation is possible. Figure 4-7 shows a contour C connecting two points s_a and s_b. The *contour integral* of $f(s)$ between the two points s_a and s_b in the direction indicated along C is defined as

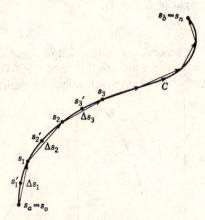

FIG. 4-7. Approximation of an integral along an arc C.

$$\int_C f(s)\ ds = \lim_{\substack{n \to \infty \\ \max|\Delta s_k| \to 0}} \sum_{k=1}^{n} f(s'_k)\ \Delta s_k \tag{4-1}$$

where $\max|\Delta s_k|$ is the maximum value of all the numbers $|\Delta s_k|$ and s'_k is a value of s on the path somewhere between s_{k-1} and s_k. Here $|\Delta s_k| = |s_k - s_{k-1}|$ is the length of the chord joining two adjacent points of subdivision on the path. Now write

$$f(s'_k) = u'_k + jv'_k \tag{4-2a}$$
$$\Delta s_k = \Delta \sigma_k + j\ \Delta \omega_k \tag{4-2b}$$

so that

$$\sum_{k=1}^{n} f(s'_k)\ \Delta s_k = \sum_{k=1}^{n} (u'_k\ \Delta \sigma_k - v'_k\ \Delta \omega_k) + j \sum_{k=1}^{n} (u'_k\ \Delta \omega_k + v'_k\ \Delta \sigma_k) \tag{4-3}$$

Each of the summations involves only real quantities. In the limit these summations become real line integrals, giving

$$\int_C f(s)\ ds = \int_C u\ d\sigma - \int_C v\ d\omega + j \left(\int_C u\ d\omega + \int_C v\ d\sigma \right) \tag{4-4}$$

If the integrals on the right of Eq. (4-4) exist, then the integral in Eq. (4-1) exists.

In order to understand Eq. (4-4), it is necessary to know the meaning of a *line integral* of a real variable. Consider the first integral on the right of Eq. (4-4),

$$\int_C u \, d\sigma$$

In general, u is a function of σ and ω, but the notation C implies that σ and ω are related by path C. With this restriction, u is actually a function of σ, which might be designated $u_c(\sigma)$, and the integral then

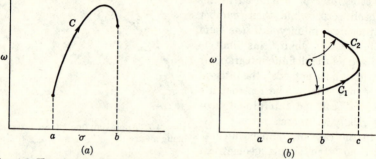

FIG. 4-8. How multivalued integrands can appear. (a) For integrals with respect to ω; (b) for integrals with respect to σ.

becomes an ordinary real integral. For example, for the path C in Fig. 4-8a we could write

$$\int_C u \, d\sigma = \int_a^b u_c(\sigma) \, d\sigma$$

A slight complexity arises in a case like Fig. 4-8b, because $u_c(\sigma)$ would be a double-valued function of σ. To avoid this situation, the path C is broken into two parts, labeled C_1 and C_2. Over each part a single-valued u_c function can be defined, designated, respectively, $u_{c1}(\sigma)$ and $u_{c2}(\sigma)$. Then

$$\int_C u \, d\sigma = \int_{C_1} u \, d\sigma + \int_{C_2} u \, d\sigma$$

and

$$\int_C u \, d\sigma = \int_a^c u_{c1}(\sigma) \, d\sigma + \int_c^b u_{c2}(\sigma) \, d\sigma$$

$$= \int_a^c u_{c1}(\sigma) \, d\sigma - \int_b^c u_{c2}(\sigma) \, d\sigma$$

In evaluating line integrals, paths may need to be split up into more than two parts. It is always necessary to reduce the u_c function to a sequence of *single-valued* functions.

Equation (4-4) contains one other integral with respect to σ and two integrals with respect to ω. These are treated similarly. The only significant difference, when integration is with respect to ω, is that for a curve like Fig. 4-8a the integrand will be multivalued; and for Fig. 4-8b it will be single-valued.

From this development it is seen that if the real integrals like those in Eq. (4-4) exist, then the integral of Eq. (4-1) exists. Furthermore, from the theory of real integration we know that continuity of the integrand over a finite interval of integration is *sufficient* to ensure existence of a real integral over that interval. It is a simple matter to prove that $u_{c1}(\sigma)$, $u_{c2}(\sigma)$, etc., are continuous when $f(s)$ is continuous on C. Therefore we have the following theorem:

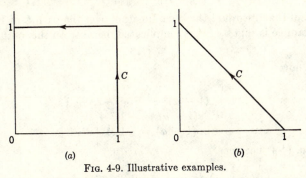

(a)   (b)

FIG. 4-9. Illustrative examples.

Theorem 4-1.   If a function $f(s)$ is continuous at all points on a path C of finite length, then the integral

$$\int_C f(s)\, ds$$

exists.

Much of the subsequent work has to do with theorems which simplify the finding of integrals; and eventually the real integrals of Eq. (4-4) will rarely occur. However, in preparation for this later work it is very important to understand the interpretation of the integral by Eq. (4-4). To that end, the following examples are given.

Illustrative Examples.   1. Find the integral of the function

$$f(s) = s^2$$

over the path shown in Fig. 4-9a.

Over the vertical part of the path $d\sigma$ is zero, and over the horizontal part $d\omega$ is zero. Furthermore,

$$u = \sigma^2 - \omega^2$$
$$v = 2\sigma\omega$$

Then, from Eq. (4-4)

$$\int_C f(s)\, ds = -\int_0^1 2\sigma\omega\, d\omega + j\int_0^1 (\sigma^2 - \omega^2)\, d\omega$$

$$+ \int_1^0 (\sigma^2 - \omega^2)\, d\sigma + j\int_1^0 2\sigma\omega\, d\sigma$$

In the first and second integral, $\sigma = 1$; and in the third and fourth integral, $\omega = 1$. Thus,

$$\int_C f(s)\, ds = -\omega^2 \Big|_0^1 + j\omega \Big|_0^1 - j\frac{\omega^3}{3}\Big|_0^1 + \frac{\sigma^3}{3}\Big|_1^0 - \sigma \Big|_1^0 + j\sigma^2 \Big|_1^0$$
$$= -1 + j - j\tfrac{1}{3} - \tfrac{1}{3} + 1 - j = -\tfrac{1}{3}(1 + j)$$

2. Find the integral of the same function over the path C of Fig. 4-9b. This example is slightly more complicated, because on this path

$$\omega = 1 - \sigma$$

so that now

$$u = \sigma^2 - (1 - \sigma)^2 = 2\sigma - 1$$
or
$$u = (1 - \omega)^2 - \omega^2 = -2\omega + 1$$
and
$$v = 2\sigma(1 - \sigma) = -2\sigma^2 + 2\sigma$$
or
$$v = 2\omega(1 - \omega) = -2\omega^2 + 2\omega$$

Now Eq. (4-4) becomes

$$\int_C f(s)\, ds = \int_1^0 (2\sigma - 1)\, d\sigma - \int_0^1 (-2\omega^2 + 2\omega)\, d\omega$$

$$+ j\left[\int_0^1 (-2\omega + 1)\, d\omega + \int_1^0 (-2\sigma^2 + 2\sigma)\, d\sigma \right]$$

$$= \sigma^2 \Big|_1^0 - \sigma \Big|_1^0 + \frac{2\omega^3}{3}\Big|_0^1 - \omega^2 \Big|_0^1$$

$$+ j\left(-\omega^2 \Big|_0^1 + \omega \Big|_0^1 - \frac{2\sigma^3}{3}\Big|_1^0 + \sigma^2 \Big|_1^0 \right)$$

$$= -1 + 1 + \tfrac{2}{3} - 1 + j(-1 + 1 + \tfrac{2}{3} - 1) = -\tfrac{1}{3}(1 + j)$$

Both examples give the same result; apparently the integral of s^2 from $s = 1 + j0$ to $s = 0 + j$ is independent of the path. At least the integral is the same for the two paths tried. One of the theorems to be developed deals with the question of when an integral between two fixed points is independent of the path.

The contour integral is seen to be closely related to a set of *line integrals*. However,

$$\int_C f(s)\, ds$$

is not itself a line integral although a path C is involved in its evaluation. To make this distinction in this text, integrals of functions of a complex variable are called *contour integrals*.

It has been implied that a direction along a path (or contour) must be assigned in designating a contour integral. Thus, whenever a symbol such as C is appended to an integral sign, a direction must be stipulated, usually on a diagram or by a statement describing C (for example, "C is a circle taken clockwise"). If the direction of integration along a given path is reversed, the integral changes sign. Proof of this is left to you as an exercise.

Another important, and almost obvious, property of contour integrals is that

$$\int_{C_1+C_2} f(s)\, ds = \int_{C_1} f(s)\, ds + \int_{C_2} f(s)\, ds \qquad (4\text{-}5)$$

where C_1 and C_2 are two paths having a point common to the end of one path and the beginning of the other, and where $C_1 + C_2$ is the path consisting of both C_1 and C_2 and having the same direction as C_1 and C_2.

Another property is stated without proof. By going back to the definition, it can be proved that, if $f(s)$ and $g(s)$ are two functions each having an integral over a contour C, then $f(s) + g(s)$ is integrable over C and

$$\int_C [f(s) + g(s)]\, ds = \int_C f(s)\, ds + \int_C g(s)\, ds \qquad (4\text{-}6)$$

The contour integral was interpreted in terms of rectangular coordinates. However, in many situations, particularly when integration paths are circular arcs, a polar-coordinate interpretation is convenient. Since

$$\sigma = \rho \cos \phi$$
$$\omega = \rho \sin \phi$$

we have

$$d\sigma = -\rho \sin \phi \, d\phi + \cos \phi \, d\rho$$
$$d\omega = \rho \cos \phi \, d\phi + \sin \phi \, d\rho$$

and Eq. (4-4) can then be written

$$\int_C f(s)\, ds = \int_C (u \cos \phi - v \sin \phi)\, d\rho - \int_C \rho(u \sin \phi + v \cos \phi)\, d\phi$$
$$+ j \left[\int_C (u \sin \phi + v \cos \phi)\, d\rho + \int_C \rho(u \cos \phi - v \sin \phi)\, d\phi \right] \qquad (4\text{-}7)$$

This looks more complicated than Eq. (4-4), and it is therefore normally used only for integration along radial lines or along origin-centered circular arcs, in which case $d\phi$ or $d\rho$ is respectively zero and the remaining integrals reduce to ordinary real integrals. Thus, if C is a radial path at angle ϕ_0, extending from ρ_1 to ρ_2, we get

$$\int_C f(s)\, ds = \int_{\rho_1}^{\rho_2} (u \cos \phi_0 - v \sin \phi_0)\, d\rho + j \int_{\rho_1}^{\rho_2} (u \sin \phi_0 + v \cos \phi_0)\, d\rho$$

Also, if C is a circular arc of radius ρ_0 centered at the origin and extending from ϕ_1 to ϕ_2, the integral is

$$\int_C f(s)\, ds = -\rho_0 \int_{\phi_1}^{\phi_2} (u \sin\phi + v\cos\phi)\, d\phi + j\rho_0 \int_{\phi_1}^{\phi_2} (u\cos\phi - v\sin\phi)\, d\phi$$

4-4. Upper Bound of a Contour Integral. Suppose, over the contour of integration, that the function $f(s)$ has the property

$$|f(s)| \leqq M$$

where M is a positive constant. By repeated application of inequality (2-18) we can write, in the notation of Eq. (4-1) and Fig. 4-7,

$$\left| \sum_{k=1}^{n} f(s_k')\, \Delta s_k \right| \leqq \sum_{k=1}^{n} |f(s_k')|\, |\Delta s_k| \leqq M \sum_{k=1}^{n} |\Delta s_k|$$

In the limit as $n \to \infty$ the left side becomes the absolute value of the integral, and the term in the middle becomes the integral of the absolute value of $f(s)$, which we write

$$\int_C |f(s)|\, |ds|$$

This is a real integral. Furthermore, the sum on the right approaches the length of C, which we shall designate L_c, giving

$$\left| \int_C f(s)\, ds \right| \leqq \int_C |f(s)|\, |ds| \leqq M L_c \tag{4-8}$$

as a generally valid relationship.

4-5. Cauchy Integral Theorem.* We shall develop the Cauchy integral theorem by first showing it to be true for two simple functions. In doing so we provide illustrations of the theorem and also obtain results which are needed in proving the general theorem later on.

Let R be a simply connected region in which lies a simple closed curve C. Figure 4-10a may be considered typical. Now consider the integral of $f(s) \equiv 1$ around contour C, in a counterclockwise (positive) sense. The integral is broken up into two real integrals, as follows:

$$\int_C ds = \int_C d\sigma + j \int_C d\omega$$

However, in similarity with Eq. (4-5), we have

$$\int_C d\sigma = \int_{\sigma_1}^{\sigma_2} d\sigma - \int_{\sigma_1}^{\sigma_2} d\sigma = 0 \quad \text{and} \quad \int_C d\omega = \int_{\omega_1}^{\omega_2} d\omega - \int_{\omega_1}^{\omega_2} d\omega = 0$$

* This theorem is frequently referred to as the Cauchy-Goursat theorem.

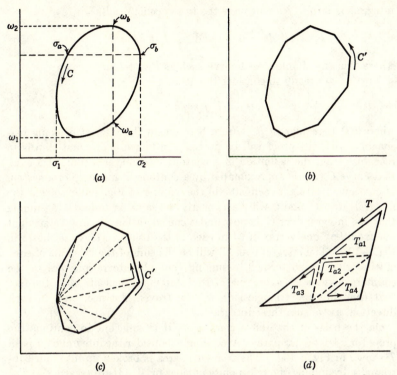

FIG. 4-10. Stages in the approximation of a contour integral over a simple closed curve by a sum of integrations over a set of triangles.

Now consider $f(s) = s$, for which the integral is

$$\int_C f(s) \, ds = \int_C \sigma \, d\sigma - \int_C \omega \, d\omega + j \left(\int_C \omega \, d\sigma + \int_C \sigma \, d\omega \right)$$

The first two integrals on the right are treated like the previous case, again giving zero. Each of the two remaining integrals is nonzero, but we are to show that their sum is zero. First look at

$$\int_C \omega \, d\sigma = \int_{\sigma_1}^{\sigma_2} \omega_a \, d\sigma - \int_{\sigma_1}^{\sigma_2} \omega_b \, d\sigma = - \int_{\sigma_1}^{\sigma_2} (\omega_b - \omega_a) \, d\sigma$$

This is clearly the negative of the area enclosed by C. The other integral gives

$$\int_C \sigma \, d\omega = \int_{\omega_1}^{\omega_2} \sigma_b \, d\omega - \int_{\omega_1}^{\omega_2} \sigma_a \, d\omega = \int_{\omega_1}^{\omega_2} (\sigma_b - \sigma_a) \, d\omega$$

and this is the area enclosed by C. Therefore, the sum of these two integrals is zero. We now have the two specific results

$$\int_C ds = 0 \quad \text{and} \quad \int_C s \, ds = 0$$

where C is any simple closed curve such as Fig. 4-10a.*

Now we are ready to consider the general case

$$\int_C f(s) \, ds$$

where $f(s)$ is regular in R. From here on, in the interest of being reasonably brief, the proof will be partially intuitive. The first step is to recognize from the definition of a contour integral, as a limit of a sum, that curve C can be approximated by a contour C' made up of a system of contiguous straight segments, in the manner of Fig. 4-10b, so that the integrals over C' and C will be as nearly the same as desired. Assuming that the integral over C' is now under consideration, we draw a straight line from any one vertex of C' to each of the other vertices, as shown in Fig. 4-10c. The integral over C' will be the sum of the integrals of each of the triangles so formed, assuming that all integrations are in the counterclockwise sense. Proof of this statement depends upon the fact that each internal triangle side will be traversed twice, once in each direction, as we sum these integrals.

On the basis of the above conclusion, it is sufficient to continue the proof for a triangular path. Accordingly, consider the integral over path T shown in Fig. 4-10d. The triangular area is broken up into four subtriangles by joining the mid-points of sides of T. If we regard T_{a1}, T_{a2}, T_{a3}, T_{a4} as labels for oriented paths of integration around these respective subtriangles, then

$$\int_T = \int_{T_{a1}} + \int_{T_{a2}} + \int_{T_{a3}} + \int_{T_{a4}}$$

and

$$\left| \int_T \right| \leqq \left| \int_{T_{a1}} \right| + \left| \int_{T_{a2}} \right| + \left| \int_{T_{a3}} \right| + \left| \int_{T_{a4}} \right|$$

Let T_1 designate the one of the four paths which gives the largest term on the right, or if there is no largest term, because two or more unexceeded terms are equal, then T_1 can be any one of these paths which yield equal integrals. Then we can say

$$\left| \int_T f(s) \, ds \right| \leqq 4 \left| \int_{T_1} f(s) \, ds \right|$$

Now let T_1 serve in place of the original triangle, and repeat the process to get another triangle T_2 which is a similarly defined subtriangle of T_1.

---

* Proof is given for a nonreentrant curve C, one which intersects any straight line only twice. However, any reentrant closed curve can be subdivided into a number of nonreentrant ones such that the contour over the given curve is the sum of the integrals over the others.

This process can be repeated n times, thereby yielding a triangle T_n such that

$$\left| \int_T f(s) \, ds \right| \leq 4^n \left| \int_{T_n} f(s) \, ds \right| \tag{4-9}$$

where triangle T_n is contained in all previous triangles $T_{n-1}, \ldots, T_1, T$. Now let n become infinite. The area of T_n approaches zero, and so it is intuitively to be expected that T_n will shrink down to a single point.* Call this point s_0 and recall that $f(s)$ is regular at s_0. Therefore, corresponding to a small $\epsilon > 0$ there exists a δ such that

$$\left| \frac{f(s) - f(s_0)}{s - s_0} - f'(s_0) \right| < \epsilon$$

when $|s - s_0| < \delta$. This can also be written

$$\frac{f(s) - f(s_0)}{s - s_0} - f'(s_0) = m(s)$$

where $|m(s)| < \epsilon$, and then

$$f(s) = f(s_0) + f'(s_0)(s - s_0) + m(s)(s - s_0)$$

Now choose n large enough so that P_n (the length of the largest side of T_n) is less than δ. The above expression can then be used in the integration over T_n, giving

$$\left| \int_{T_n} f(s) \, ds \right| \leq \left| \int_{T_n} f(s_0) \, ds \right|$$
$$+ \left| \int_{T_n} f'(s_0)(s - s_0) \, ds \right| + \left| \int_{T_n} m(s)(s - s_0) \, ds \right|$$

if each integral on the right exists. We have already proved that a closed-path integral of a constant, or of s, is zero; and therefore the first two integrals on the right are each zero. Thus,

$$\left| \int_{T_n} f(s) \, ds \right| \leq \left| \int_{T_n} m(s)(s - s_0) \, ds \right| < \epsilon \int_{T_n} |s - s_0| \, |ds|$$

But, from the geometry of T_n, when s is on its boundary,

$$|s - s_0| \leq P_n$$
and $\qquad\qquad$ Perimeter of $T_n \leq 3P_n$

It is also true that, if P is the maximum side of T, then

$$P_n = \frac{P}{2^n}$$

---

* For a proof of this see K. Knopp, "Theory of Functions," vol. 1, p. 9, Dover Publications, New York, 1945.

Thus, by relation (4-8), we arrive at the estimate

$$\left| \int_{T_n} f(s) \, ds \right| < \frac{3\epsilon P^2}{4^n}$$

and from relation (4-9)

$$\left| \int_T f(s) \, ds \right| < 3\epsilon P^2$$

But ϵ is arbitrarily small, and so we have the conclusion

$$\int_T f(s) \, ds = 0$$

and, by the earlier argument that any curve can be made up as accurately as we like by segments of straight lines, it follows that

$$\int_C f(s) \, ds = 0 \tag{4-10}$$

for all simple closed paths C in the region of regularity R.

In the proof just completed we were restricted to simple closed curves and found that in a simply connected region of regularity of $f(s)$ the integral is zero around such a curve. However, since a nonsimple closed curve can be made up of a finite number of simple closed curves, with the integral being zero for each of them, we conclude that in a simply connected region of regularity of $f(s)$ the integral is zero for any closed curve. This is a statement of the *Cauchy integral theorem*.

4-6. Independence of Integration Path. Now consider a function $f(s)$ having the property

$$\int_C f(s) \, ds = 0$$

for every closed curve C in a region R. [This will of course be true if $f(s)$ and R meet the conditions of the Cauchy integral theorem; but these conditions are presently not assumed.]

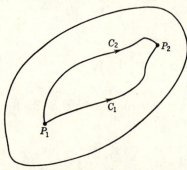

FIG. 4-11. Independence of path of integration in a simply connected region of regularity.

We take any two paths C_1 and C_2 joining a pair of points, as shown in Fig. 4-11, and let C designate the closed curve formed by C_1 and C_2 taken together. The direction of integration associated with C will be the same as the direction defined for C_1 and therefore is opposite to the direction defined for C_2. This choice of directions permits us to write

$$\int_C f(s) \, ds = \int_{C_1} f(s) \, ds - \int_{C_2} f(s) \, ds$$

and therefore

$$\int_{C_1} f(s)\ ds = \int_{C_2} f(s)\ ds \qquad (4\text{-}11)$$

When an integral is independent of path, in a certain region R, the contour need not be designated. Only the beginning and ending points need to be shown, so long as these points are in R. Therefore, for integrals such as Eq. (4-11) the notation

$$\int_{P_1}^{P_2} f(s)\ ds$$

is commonly used. The lower limit is the starting point. Now you see why the two examples of Sec. 4-3 gave the same result.

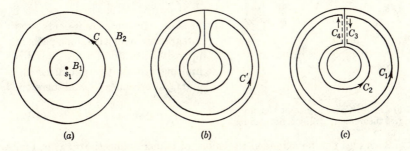

C_1 and C_2 are closed. C_1' and C_2' are open at the cut.

Fig. 4-12. Consequence of the Cauchy integral theorem in a doubly connected region.

4-7. Significance of Connectivity. In Sec. 4-5, R was carefully designated as being simply connected. Suppose that a closed curve C encloses an isolated singular point s_1 in Fig. 4-12a. Note that C lies in a region of regularity, between boundaries B_1 and B_2, but that this is a doubly connected region. The Cauchy integral theorem does not then apply, and there is no reason to believe that the contour integral around C is zero. If a barrier is introduced in Fig. 4-12b, to make the region simply connected, then a closed path like C' is admissible; but not C, because it would go outside the region.

One of the purposes in defining connectivity of regions was to simplify the statement of the Cauchy integral theorem.

Although the contour integral around a path such as C is not necessarily zero, it can be shown to be *independent of path for all paths that encircle the inside boundary* B_1 (that is, for all distortions of C that can be made without leaving the region of regularity). Consider two such paths, C_1 and C_2 in Fig. 4-12c, for which the small gaps shown are assumed closed.

Join C_1 and C_2 by two auxiliary paths C_3 and C_4, which can be as close together as we like, and let C_1' and C_2' be the same as C_1 and C_2 but with the slight gaps shown in the figure. Also, put a barrier in the region between C_3 and C_4 so that the path $C_1' + C_3 - C_2' + C_4$ is a closed path in a simply connected region of regularity. By the Cauchy integral theorem

$$\int_{C_1'} f(s)\ ds + \int_{C_3} f(s)\ ds - \int_{C_2'} f(s)\ ds + \int_{C_4} f(s)\ ds = 0$$

Now let paths C_3 and C_4 come together, and note that they are opposite in direction. In the limit

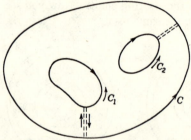

$$\int_{C_3} f(s)\ ds + \int_{C_4} f(s)\ ds = 0$$

because the sum represents two integrations over a common path, in opposite directions. Also, in this limiting process C_1' approaches C_1, and C_2' approaches C_2, and therefore

Fig. 4-13. Consequence of Cauchy integral theorem in a triply connected region.

$$\int_{C_1'} f(s)\ ds = \int_{C_1} f(s)\ ds \qquad (4\text{-}12)$$

This idea can be extended to regions of higher connectivity, as in Fig. 4-13. By cutting the region in much the same way as in Fig. 4-12 it can be shown that

$$\int_C f(s)\ ds = \int_{C_1} f(s)\ ds + \int_{C_2} f(s)\ ds \qquad (4\text{-}13)$$

The proof is left to you as an exercise. Similar equations can be written for a region of any order of connectivity, showing that the integral around a closed path surrounding a finite number of isolated singular points equals the sum of the integrals around individual singular points, with due regard for path orientations.

4-8. Primitive Function (Antiderivative)

Theorem 4-2. Let $f(s)$ be a function which is continuous in an open region R and which has the property

$$\int_C f(s)\ ds = 0$$

for every closed path C in R. Then, the function

$$F(s) = \int_{s_0}^{s} f(z)\ dz \qquad s_0 \text{ and } s \text{ in } R$$

is a single-valued function of s and is regular in R with the derivative

$$\frac{dF(s)}{ds} = f(s)$$

The proof begins by noting that $F(s)$ is single-valued because, by Sec. 4-6, the integral is independent of path from s_0 to s. Next refer to Fig. 4-14, and consider the differential*

$$F(s + \Delta s) - F(s) = \int_{s_0}^{s+\Delta s} f(z)\ dz - \int_{s_0}^{s} f(z)\ dz = \int_{s}^{s+\Delta s} f(z)\ dz \quad (4\text{-}14)$$

If we write

$$\int_{s}^{s+\Delta s} f(z)\ dz = f(s) \int_{s}^{s+\Delta s} dz + \int_{s}^{s+\Delta s} [f(z) - f(s)]\ dz$$

$$= f(s)\ \Delta s + \int_{s}^{s+\Delta s} [f(z) - f(s)]\ dz$$

then Eq. (4-14) becomes

$$F(s + \Delta s) - F(s) - f(s)\ \Delta s = \int_{s}^{s+\Delta s} [f(z) - f(s)]\ dz \quad (4\text{-}15)$$

Since $f(s)$ is continuous, corresponding to an arbitrary small positive number ϵ there is a number δ such that

$$|f(s) - f(z)| < \epsilon$$

when $|s - z| < \delta$. Now choose $|\Delta s| < \delta$, which will ensure $|s - z| < \delta$ for z on the path in question. Therefore, we have

$$\left| \int_{s}^{s+\Delta s} [f(z) - f(s)]\ dz \right|$$

$$\leq \int_{s}^{s+\Delta s} |f(z) - f(s)|\ |dz| < \epsilon |\Delta s|$$

and Eq. (4-15) can now be written

$$\left| \frac{F(s + \Delta s) - F(s)}{\Delta s} - f(s) \right| < \epsilon \quad (4\text{-}16)$$

when $\qquad\qquad |\Delta s| < \delta$

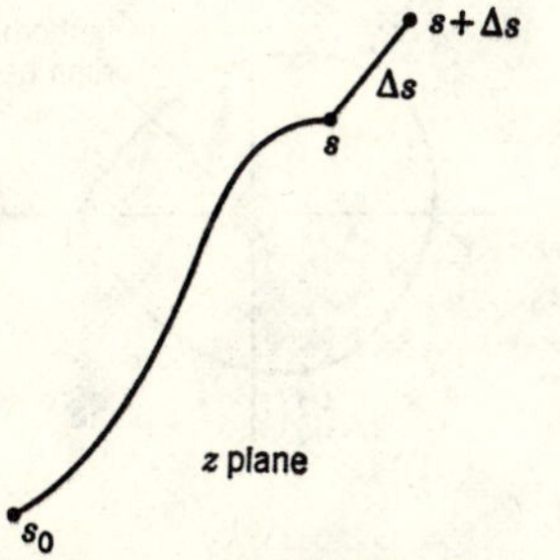

Fig. 4-14. Derivative of an integral as function of upper limit.

Since ϵ can be arbitrarily small, we conclude that

$$\lim_{\Delta s \to 0} \frac{F(s + \Delta s) - F(s)}{\Delta s} = f(s)$$

or $\qquad\qquad\qquad \dfrac{dF(s)}{ds} = f(s) \qquad\qquad\qquad (4\text{-}17)$

* An increment Δs can always be found within R because each point in R must have a neighborhood also in R, and Δs can be in that neighborhood.

This result is obtained for any point in R, and so $F(s)$ is regular in R. The function $F(s)$ is called the primitive (or *antiderivative*) of $f(s)$.

If $f(s)$ is regular in a *simply connected* region R, we know from the Cauchy integral theorem that the integral of $f(s)$ is zero over any closed path in R. Therefore, such a function would meet the conditions of Theorem 4-2. This fact makes it possible to state a corollary to Theorem 4-2, namely, that, if $f(s)$ is regular in a simply connected region R, then the conclusions of Theorem 4-2 are true.

4-9. The Logarithm. The last section leads naturally to the definition of an important new function and in the process emphasizes the importance of connectivity. We begin with the function

$$f(s) = \frac{1}{s}$$

which is regular in the doubly connected region outside a small circle centered at the origin. Now consider the integral

$$\int_1^s \frac{dz}{z}$$

which would be a single-valued function of s if the path of integration from 1 to s should lie in a simply connected region. However, referring

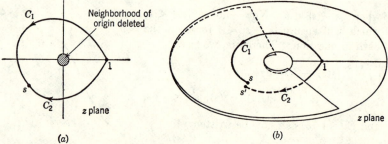

FIG. 4-15. Definition of unique path of integration for defining log s. (a) Original doubly connected region in the z plane; (b) Riemann surface defined to make Eq. (4-18) single-valued, shown in perspective view.

to Fig. 4-15a, you see that paths C_1 and C_2 cannot be brought into coalescence while continuously remaining in the region. Thus, there is doubt whether the above integral is a single-valued function of s.

This is our first encounter with a multivalued function. Figure 4-15b shows a conceptual device for dealing with the doubly connected region. The complex plane is expanded into a sequence of overlapping sheets connected together in helical fashion (like a circular staircase). Now point s becomes two points (or more, if more sheets are included) labeled

s and s'. Inability to distort C_1 into C_2 is now unimportant, because $s \neq s'$. This idea can be extended, so that corresponding to any point

$$s = \rho\underline{/\phi}$$

there is a doubly infinite set of points on other sheets having angles differing from ϕ by integral multiples of $\pm 2\pi$.

The surface described above, consisting of a set of connected sheets, is called a *Riemann surface*. In Chap. 6 additional functions are considered for which Riemann surfaces are invented to make them single-valued.

It is now possible to write

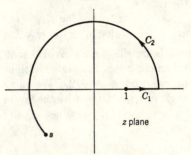

$$F(s) = \int_1^s \frac{dz}{z} \qquad (4\text{-}18)$$

and $F(s)$ will be single-valued on the Riemann surface of Fig. 4-15b. $F(s)$ is independent of path from 1 to s, so long as the path remains in the surface (i.e., winds around enough times to get to the sheet in which s is located). The path shown in Fig.

Fig. 4-16. Specific path of integration for the function log s.

4-16 will be used. C_1 is a radial line, and C_2 is a circular arc. Two special cases of Eq. (4-7) are used to evaluate the integral. Along C_1

$$z = r \qquad dz = dr$$

and along C_2

$$z = \rho e^{j\theta} \qquad dz = j\rho e^{j\theta} \, d\theta$$

where ρ is the magnitude of $s = \rho e^{j\phi}$. Equation (4-18) now becomes

$$F(s) = \int_{C_1+C_2} \frac{dz}{z} = \int_1^\rho \frac{dr}{r} + j \int_0^\phi d\theta$$
$$= \log \rho + j\phi$$

From Theorem 4-2 we know that $F(s)$ is regular in the Riemann surface and that

$$\frac{dF(s)}{ds} = \frac{1}{s} \qquad (4\text{-}19)$$

Recognizing similarity with the logarithm of a real variable, we are prompted to use the same notation, *defining*

$$\log s = \int_1^s \frac{dz}{z} \qquad (4\text{-}20)$$
$$\log s = \log \rho + j\phi \qquad (4\text{-}21)$$

The question arises whether or not log s is the inverse of e^s (that is, whether or not $e^{\log s} = s$). To check, we write

$$e^{\log s} = e^{\log \rho + j\phi} = e^{\log \rho} e^{j\phi} = \rho e^{j\phi} = s \tag{4-22}$$

as expected. It follows that the logarithm of a complex number, defined by Eq. (4-21), can be used in the same way as logarithms of real numbers.

By looking at the logarithm of a product, the importance of the Riemann surface is appreciated. Thus, if $s_1 = \rho_1 \underline{/\phi_1}$ and $s_2 = \rho_2 \underline{/\phi_2}$, then

$$\log s_1 s_2 = \log \rho_1 \rho_2 + j(\phi_1 + \phi_2)$$

and

$$\log s_1 + \log s_2 = \log \rho_1 + j\phi_1 + \log \rho_2 + j\phi_2$$

which are the same provided that $\phi_1 + \phi_2$ is not modified by adding or subtracting an integral multiple of 2π. For example, suppose that $\phi_1 = \pi$ and $\phi_2 = \pi/2$. We must then use $3\pi/2$ (not $-\pi/2$) for the angle of $s_1 s_2$. Otherwise the law of adding logarithms would not be valid.

As a useful by-product of this development we are now ready to consider

$$\int_C \frac{ds}{s}$$

where C is a simple closed curve encircling the origin in a counterclockwise direction. The interpretation given above allows us to write

$$\int_C \frac{ds}{s} = \int_{s_1'}^{s_1} \frac{dz}{z} = \log s_1 - \log s_1'$$

where s_1' and s_1 are two points directly over one another, but on adjacent sheets of the Riemann surface. Since C is counterclockwise, the angle of s_1 is 2π greater than the angle of s_1'. Thus

$$\int_C \frac{ds}{s} = j2\pi \tag{4-23}$$

Another important case is

$$\int_C \frac{ds}{s^n}$$

where C is the same curve we have been considering and n is a positive or negative integer not equal to 1. There is no loss in generality if C is the circle

$$s = \rho e^{j\phi}$$

Then, if we use the specific values $s_1' = \rho \underline{/-\pi}$, $s_1 = \rho \underline{/\pi}$, we get

$$\int_C \frac{ds}{s^n} = \int_{s_1'}^{s_1} \frac{ds}{s^n} = j \int_{-\pi}^{\pi} \rho^{1-n} e^{j(1-n)\phi} \, d\phi$$

$$= j\rho^{1-n} \int_{-\pi}^{\pi} [\cos (1-n)\phi + j \sin (1-n)\phi] \, d\phi$$

which is zero when $n \neq 1$ and confirms Eq. (4-23) when $n = 1$. Thus,

$$\int_C \frac{ds}{s^n} = 0 \qquad n \neq 1 \qquad (4\text{-}24)$$

4-10. Cauchy Integral Formulas. Let a simple closed contour C lie in a simply connected region of regularity of a function $f(s)$, and let s be a point inside C. Also, let C_1 be a circle of radius r and center at s, lying wholly within C.

The function

$$\frac{f(z)}{z - s}$$

is a regular function of z, in a doubly connected region, with a neighborhood of s deleted, as shown in Fig. 4-17. Now, by the principles established in Sec. 4-7, it can be said that

$$\int_C \frac{f(z)}{z - s} \, dz = \int_{C_1} \frac{f(z)}{z - s} \, dz \qquad (4\text{-}25)$$

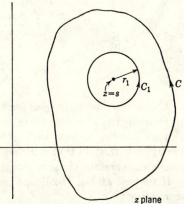

FIG. 4-17. Substitution of path of integration by a circle approaching zero radius.

and then

$$\int_{C_1} \frac{f(z)}{z - s} \, dz = \int_{C_1} \frac{f(s) + f(z) - f(s)}{z - s} \, dz$$

$$= f(s) \int_{C_1} \frac{dz}{z - s} - \int_{C_1} \frac{f(s) - f(z)}{z - s} \, dz$$

In the first integral on the right replace $z - s$ by s' and dz by ds' to give an integral like Eq. (4-23), and thus

$$f(s) \int_{C_1} \frac{dz}{z - s} = j2\pi f(s)$$

To deal with the second integral, note that the only possible point where the integrand might become infinite or discontinuous is at $z = s$; but

$$\lim_{z \to s} \frac{f(s) - f(z)}{s - z} = f'(s)$$

which is noninfinite because $f(z)$ is regular at $z = s$. Thus we can write

$$\left| \frac{f(s) - f(z)}{z - s} \right| < |f'(s)| + \epsilon = M$$

when z is on C_1, and so

$$\left| \int_{C_1} \frac{f(s) - f(z)}{z - s} \, dz \right| < 2\pi r_1 M$$

where r_1 is the radius of C_1. Furthermore, the integral is unchanged if r_1 approaches zero. In so doing M remains constant. The conclusion is that

$$\int_{C_1} \frac{f(s) - f(z)}{z - s} \, dz = 0$$

and so, finally,

$$f(s) = \frac{1}{2\pi j} \int_C \frac{f(z)}{z - s} \, dz \qquad (4\text{-}26)$$

This is the *Cauchy integral formula* for $f(s)$. Equation (4-26) is not to be construed as a formula for calculating $f(s)$. It is useful as a *representation* for $f(s)$, to be used in later analytical work. Its usefulness stems partly from the fact that it includes an integral and therefore can be used to evaluate certain kinds of integrals.

If we look at Eq. (4-26) and formally differentiate under the integral sign, we get

$$f'(s) = \frac{1}{2\pi j} \int_C \frac{f(z)}{(z - s)^2} \, dz \qquad (4\text{-}27a)$$

$$f''(s) = \frac{2}{2\pi j} \int_C \frac{f(z)}{(z - s)^3} \, dz \qquad (4\text{-}27b)$$

$$\cdots \cdots \cdots \cdots \cdots \cdots \cdots$$

$$f^{(n)}(s) = \frac{n!}{2\pi j} \int_C \frac{f(z)}{(z - s)^{n+1}} \, dz \qquad (4\text{-}27c)$$

This process is not justified, since the theorem on differentiating under an integral for real integrals cannot be extended to contour integrals.* The proof for real integrals depends on the mean-value theorems; and there is no such theorem for a contour integral of a function of a complex variable. Therefore, the above process is merely suggestive of the forms we might expect. The formulas are correct, but for proof we use a different approach.

We shall prove that Eq. (4-27a) is correct by a procedure which can be applied repeatedly to verify the formulas for the higher derivatives. Since Eq. (4-27a) is the anticipated formula, we compare it with the limit

---

* A theorem on differentiating under the integral for contour integrals can be proved, but not in the manner of the real-integral case. In fact, the Cauchy integral formulas are used in the proof. See P. Franklin, "Treatise on Advanced Calculus," p. 448, John Wiley & Sons, Inc., New York, 1940; and also Prob. 4-27.

of the differential quotient. In other words, we are to determine whether

$$\lim_{\Delta s \to 0} \left[\frac{f(s + \Delta s) - f(s)}{\Delta s} - \frac{1}{2\pi j} \int_C \frac{f(z)}{(z - s)^2} \, dz \right] = 0$$

Equation (4-26) provides an integral formulation for the differential quotient, namely,

$$\frac{f(s + \Delta s) - f(s)}{\Delta s} = \frac{1}{2\pi j} \int_C \left[\frac{f(z)}{\Delta s(z - s - \Delta s)} - \frac{f(z)}{\Delta s(z - s)} \right] dz \quad (4\text{-}28)$$

Inserting this in the relation to be checked establishes that we are to determine whether or not

$$\lim_{\Delta s \to 0} \left\{ \int_C \left[\frac{1}{\Delta s(z - s - \Delta s)} - \frac{1}{\Delta s(z - s)} - \frac{1}{(z - s)^2} \right] f(z) \, dz \right\} = 0 \quad (4\text{-}29)$$

Adding the three terms in the brackets reduces this to the simpler question of whether or not

$$\lim_{\Delta s \to 0} \left[\int_C \frac{\Delta s \, f(z)}{(z - s)^2(z - s - \Delta s)} \, dz \right] = 0$$

But

$$\left| \int_C \frac{\Delta s \, f(z)}{(z - s)^2(z - s - \Delta s)} \, dz \right| < |\Delta s| \int_C \frac{|f(z)|}{|z - s|^2|z - s - \Delta s|} \, |dz|$$

and since $f(z)$ is regular on C, we can say that $|f(z)|$ has an upper bound M on C. Furthermore, referring to Fig. 4-18, it is apparent that

$$|z - s| \geqq A$$

and that we can choose an increment Δs_1 such that

$$|z - s - \Delta s| > B \qquad \text{when } |\Delta s| < |\Delta s_1|$$

A and B are, respectively, the shortest lines from point s and from the circle of radius $|\Delta s_1|$ to points on C. Since s must be internal to the region enclosed by C, a circle of radius $|\Delta s_1|$ can always be found which will lie inside C. Thus, by virtue of Eq. (4-8), we have

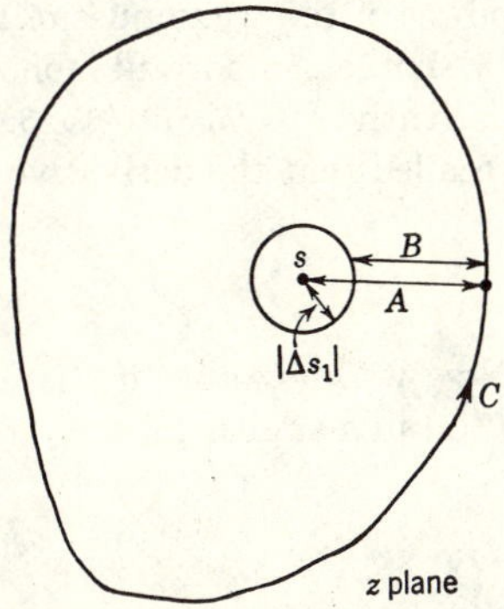

FIG. 4-18. Determination of lower bounds for $|z - s|$ and $|z - s - \Delta s|$.

$$|\Delta s| \int_C \frac{|f(z)|}{|z - s|^2|z - s - \Delta s|} \, |dz| < |\Delta s| \frac{ML}{A^2B}$$

when $|\Delta s| < |\Delta s_1|$. The right side has the limit zero as Δs approaches zero. Also, the term on the left has been shown to be greater than the term on the left of Eq. (4-29). Therefore, the limit given in Eq. (4-29) is correct. This completes the proof of Eq. (4-27a).

Equation (4-27b) is proved in a similar way, by treating

$$\lim_{\Delta s \to 0} \left[\frac{f'(s + \Delta s) - f'(s)}{\Delta s} - \frac{2}{2\pi j} \int_C \frac{f(z)}{(z - s)^3} \, dz \right] = 0$$

as the limit to be investigated. Equation (4-27a) is used for $f'(s + \Delta s)$ and $f'(s)$, and then the proof proceeds in exactly the same way as before.

The formula of Eqs. (4-26) and (4-27) are generally called the *Cauchy integral formulas.*

4-11. Implications of the Cauchy Integral Formulas. The Cauchy integral formulas lead to several general properties of analytic functions of far-reaching importance. One important conclusion is that, if $f(s)$ is regular at a point s_0, then each of its derivatives exists and is regular at s_0. To prove this statement, observe that if s_0 is regular it will always have a neighborhood in which contour C can be placed. Then Eqs. (4-27) establish the existence derivatives of all orders inside C, and thus in a neighborhood of s_0. Therefore, $f'(s)$, $f''(s)$, etc., are all regular at s_0. This being true for any regular point of $f(s)$, it follows that each derivative is regular throughout the region of regularity of $f(s)$. Thus, in no case can differentiation of an analytic function introduce a singularity which does not appear in the original function. This blanket appraisal of the properties of an analytic function and its derivatives is an example of the power of the techniques of analysis we are developing.

Another important conclusion deals with properties of the partial derivatives $\partial u/\partial \sigma$, $\partial u/\partial \omega$, $\partial v/\partial \sigma$, and $\partial v/\partial \omega$. Referring to Eq. (2-42a), it is recalled that the derivative can be written

$$f'(s) = \frac{\partial u}{\partial \sigma} + j \frac{\partial v}{\partial \sigma}$$

Now we know that if $f'(s)$ exists so also does $f''(s)$ exist. We can write $f''(s)$ in a similar fashion, giving

$$f''(s) = \frac{\partial^2 u}{\partial \sigma^2} + j \frac{\partial^2 v}{\partial \sigma^2}$$

Since this exists, we conclude that $\partial^2 u/\partial \sigma^2$ and $\partial^2 v/\partial \sigma^2$ exist and therefore that $\partial u/\partial \sigma$ and $\partial v/\partial \sigma$ are continuous. A similar conclusion follows for partial derivatives with respect to ω, if we start with

$$f'(s) = \frac{\partial v}{\partial \omega} - j \frac{\partial u}{\partial \omega}$$

In Sec. 2-8 it was pointed out that we could show that the real and imaginary parts of an analytic function satisfy the two-dimensional Laplace equation at all regular points. But in a footnote it was pointed out that we had no assurance that these second derivatives exist. Now we have the proof in a very general way. Given any analytic function, its real and imaginary parts are solutions of the two-dimensional Laplace equation at each regular point.

4-12. Morera's Theorem. The Cauchy integral theorem states that, if $f(s)$ is regular in a simply connected region R, the integral of the function over any closed curve in R will be zero. We shall now show that the converse is true, that if $f(s)$ is continuous in a region R, and if

$$\int_C f(s) \, ds = 0 \tag{4-30}$$

for every closed path C in R, then $f(s)$ is regular in R. This is a statement of Morera's theorem.*

To prove this theorem, we begin by observing that Theorem 4-2 is applicable, and so it is known that

$$F(s) = \int_{s_0}^{s} f(z) \, dz$$

is a single-valued function, regular in R, and having the derivative

$$\frac{dF(s)}{ds} = f(s)$$

Since $F(s)$ is regular in R, it is known from Sec. 4-11 that its derivative $f(s)$ is also regular in R. Thus, the statement of Morera's theorem is proved.

4-13. Use of Primitive Function to Evaluate a Contour Integral. Let $F(s)$ be regular in a simply connected region R, having the derivative

$$\frac{dF(s)}{ds} = f(s) \tag{4-31}$$

From Sec. 4-11 it is known that $f(s)$ is regular in R, and then from the Cauchy integral theorem and Theorem 4-2 we know that

$$\frac{d}{ds} \int_{s_0}^{s} f(z) \, dz = f(s) \tag{4-32}$$

where s_0 is some point in R. Comparison of Eqs. (4-31) and (4-32) establishes that

$$\int_{s_0}^{s} f(z) \, dz = F(s) + K \tag{4-33}$$

where K is any constant. Now suppose that we are required to find

$$\int_{s_1}^{s_2} f(z) \, dz$$

where s_1 and s_2 are both in R. Let s_0 be the number appearing in Eq. (4-33). Then we can write

$$\int_{s_1}^{s_2} f(s) \, ds = \int_{s_0}^{s_2} f(s) \, ds - \int_{s_0}^{s_1} f(s) \, ds$$

---

* To be an exact converse of the Cauchy integral theorem, R would be required to be simply connected; and Morera's theorem is often stated with this condition included. However, simple connectivity is not required in the proof of Morera's theorem.

and from Eq. (4-33) it follows that

$$\int_{s_1}^{s_2} f(s)\, ds = F(s_2) - F(s_1) \tag{4-34}$$

We state the result formally as a theorem, as follows:

Theorem 4-3.   If $F(s)$ is regular in a simply connected region R, and if s_1 and s_2 are two points in R, then

$$\int_{s_1}^{s_2} \frac{dF(s)}{ds}\, ds = F(s_2) - F(s_1)$$

This theorem is the complex-variable counterpart of the fundamental theorem of integral calculus of real variables. This theorem applies when integration is performed by finding the "antiderivative" and substituting limits. Whereas the corresponding real-variable theorem is the fundamental device used in real-variable integration, it is not true that Theorem 4-3 occupies an equally important position in contour integration. Most of the contour integrals of practical importance are amenable to evaluation either by the Cauchy integral formulas or through principles later to be developed from these formulas. There is no counterpart of the Cauchy integral formulas in the theory of real variables. This is one reason why complex-variable theory is so important. In Chap. 8 you will see examples of how the powerful methods available for evaluating contour integrals can be applied to real integrals by first regarding the real integral as a special case of a contour integral.

PROBLEMS

4-1. Refer to Fig. 4-2b. Let R be the open region enclosed by the nonsimple closed curve shown, and let R' be the corresponding closed region. Are R and R' connected regions? Explain.

4-2. Suppose that

$$f(s) = |s|^2$$

(a) Integrate $f(s)$ over the path shown in Fig. 4-9a.

(b) Integrate $f(s)$ over the path shown in Fig. 4-9b.

4-3. By converting to real integrals, integrate $f(s) = \sin s$ along each of the paths shown in Fig. P 4-3.

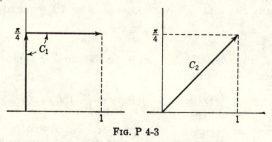

FIG. P 4-3

4-4. By converting to real integrals, integrate $f(s) = e^s$ along each of the contours given in Fig. P 4-3.

4-5. Using the paths C_1 and C_2 in Fig. P 4-5, evaluate the following integrals:

(a) $\int_{C_1} \operatorname{Re} s \, ds$

(b) $\int_{C_1} \operatorname{Im} s \, ds$

(c) $\int_{C_2} \operatorname{Re} s \, ds$

(d) $\int_{C_2} \operatorname{Im} s \, ds$

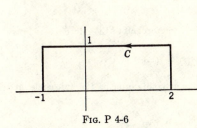 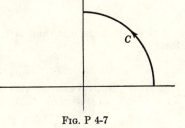

Fig. P 4-5

4-6. Using the contour shown in Fig. P 4-6, evaluate the following integrals:

(a) $\int_C e^{\operatorname{Re} s} \, ds$

(b) $\int_C e^{\operatorname{Im} s} \, ds$

(c) $\int_C e^{j \operatorname{Re} s} \, ds$

(d) $\int_C e^{j \operatorname{Im} s} \, ds$

(e) $\int_C \operatorname{Re} e^{js} \, ds$

(f) $\int_C \operatorname{Im} e^s \, ds$

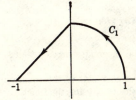 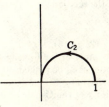

Fig. P 4-6   Fig. P 4-7

4-7. Perform each of the following integrations:

(a) $\int_C \dfrac{s}{|s|} \, ds$

(b) $\int_C \dfrac{\operatorname{Re} s}{|s|} \, ds$

(c) $\int_C \dfrac{\operatorname{Im} s}{|s|} \, ds$

for the path of radius R shown in Fig. P 4-7.

4-8. Prove that the following upper bounds are correct, using the contour C shown in Fig. P 4-8:

(a) $\left| \int_C |s^2| \, ds \right| < 7\pi/2$

(b) $\left| \int_C |\sin s| \, ds \right| < \pi \cosh 1$

(c) $\left| \int_C \operatorname{Im} e^{-s^2} \, ds \right| < 2e^{-1}$

(d) $\left| \int_C e^{-|s|^2} \, ds \right| < \pi e^{-1}$

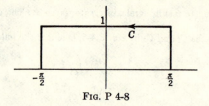

Fig. P 4-8

4-9. Prove the relation

$$\int_C \overline{f(s)} \, \overline{ds} = \overline{\int_C f(s) \, ds}$$

4-10. Prove Eq. (4-13).

4-11. In the proof of the Cauchy integral theorem, where do we use the condition that the function shall be regular on C?

4-12. By performing the integration over each side of the square in Fig. P 4-12, check the validity of the Cauchy integral theorem, using the function $f(s) = s^2$.

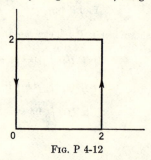

Fig. P 4-12

4-13. Let C designate the unit circle and C_1 the top half of the unit circle. Prove that

$$\int_C f(s) \, ds = \int_{C_1} [f(s) - f(-s)] \, ds$$

Furthermore, if $f(s)$ is analytic, with no singular points in or on the unit circle, and if $f(s) \equiv -f(-s)$, prove that

$$\int_{C_1} f(s) \, ds = 0$$

4-14. Use appropriate principles established in the text to evaluate

$$\int_C \left(\frac{1}{s-1} + \frac{1}{s+1} \right) ds$$

where C is a circle of radius 2 centered at the origin and directed counterclockwise.

4-15. Let C designate a square with sides parallel to the coordinate axes, intersecting these axes at $\sigma = \pm 4$ and $\omega = \pm 4$. The direction of integration is counterclockwise. Evaluate each of the following:

(a) $\displaystyle\int_C \frac{e^s}{s + j\pi} \, ds$ (b) $\displaystyle\int_C \frac{\cos s}{s} \, ds$

(c) $\displaystyle\int_C \frac{\sinh 2s}{s^4} \, ds$ (d) $\displaystyle\int_C \frac{e^{-s}}{s^2} \, ds$

4-16. Evaluate each of the following integrals in a counterclockwise sense around a simple closed path C consisting of a circle of radius 1, centered at $s = j$:

(a) $\int_C \frac{s^2 - 1}{s^2 + 1} ds$

(b) $\int_C \frac{s^2 - 1}{(s + 1)^2} ds$

(c) $\int_C \frac{s^2 + 1}{s^2 - 1} ds$

(d) $\int_C \frac{s^2 + j}{(s + j)^2} ds$

4-17. The nth-order Legendre polynomial can be given by the formula

$$P_n(s) = \frac{1}{2^n n!} \frac{d^n}{ds^n} (s^2 - 1)^n$$

By using an appropriate contour of integration, prove that $P_n(1) = 1$ for all n.

4-18. The nth-order Laguerre polynomial can be given by the formula

$$L_n(s) = e^s \frac{d^n}{ds^n} (s^n e^{-s})$$

Use appropriate contour integration to prove that $L_n(0) = n!$.

4-19. Take the function $f(s) = s^3$, and check the Cauchy integral formulas for $f(s)$, $f'(s)$, $f''(s)$, and $f'''(s)$ by actually performing the indicated integrations, using a circular path of integration.

4-20. Let C be a counterclockwise circular path passing through the origin, with center on the positive real axis and diameter π. Use appropriate Cauchy integral formulas to evaluate the integrals:

(a) $\int_C \frac{\cos s}{(s - \pi/2)^2} ds$

(b) $\int_C \frac{\cos s}{s - \pi/2} ds$

(c) $\int_C \frac{\cos s}{s^2 - \pi^2/4} ds$

4-21. The paths in the following integrals are described in Fig. P 4-21. Find the values of the integrals:

(a) $\int_{C_1} \frac{s}{(s - 1)(s - 3)} ds$

(b) $\int_{C_2} \frac{s}{(s - 1)(s - 3)} ds$

(c) $\int_{C_3} \frac{s}{\sin s} ds$

(d) $\int_\pi^j \sin s \, ds$

(e) $\int_{C_1} \frac{\sinh s}{s^2 + \pi^2/4} ds$

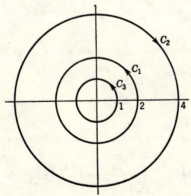

FIG. P 4-21

4-22. Using a unit counterclockwise circle as path C_1 or a path C_2 consisting of a square of side 2 with the origin of the coordinate axes at the center and sides parallel to the axes, evaluate each of the following integrals, in the simplest possible manner:

(a) $\displaystyle\int_{C_1}\left(\sin s^2 + \frac{1}{s}\right) ds$

(b) $\displaystyle\int_{C_1}\frac{e^{\sin s}}{s^2}\, ds$

(c) $\displaystyle\int_{C_2}|e^s|\, ds$

(d) $\displaystyle\int_{C_2}\frac{\sin^2 s}{s(2s+1)(2s-1)}\, ds$

4-23. Using the defining formula

$$\log s = \int_1^s \frac{dz}{z}$$

prove, by appropriate manipulations of the integral, that

$$\log \frac{1}{s} = -\log s$$

4-24. Use the integral definition of the logarithm to prove that

$$\log s^n = n \log s$$

4-25. Use the integral definition of the logarithm to prove that

$$\log s_1 s_2 = \log s_1 + \log s_2$$

4-26. Let C be a counterclockwise circular contour of radius $A < 1$ centered at the point $s = 1$. Prove that

$$\int_C \frac{\log |s|}{s-1}\, ds = \int_0^{2\pi} \sin^{-1} \frac{A \sin \theta}{\sqrt{1 + 2A \cos \theta + A^2}}\, d\theta$$

4-27. Let a function $G(s,z)$ be given such that, if s is fixed in a region R, the function is regular in z along a path C (not necessarily closed) and, if z is fixed at any point on C, the function is regular in s, for s in R. Define $g(s)$ by the formula

$$g(s) = \int_C F(s,z)\, dz \qquad s \text{ in } R$$

Use the appropriate Cauchy integral formula (assuming that an interchange of order of integration is permitted) to prove that

$$\frac{dg(s)}{ds} = \int_C \frac{\partial G(s,z)}{\partial s}\, dz$$

4-28. Let C be a circle centered at the origin and of radius R. Show that

$$\int_C \frac{s+a}{s(s+b)}\, ds$$

is independent of a and b, if $|b| < R$. Also, obtain this integral as a function of a and b, if $|b| > R$.

4-29. Use Theorem 4-3 to do Prob. 4-3.

4-30. Use the bilinear transformation to establish the following equality:

$$\int_{-j\sqrt{\frac{1-a}{1+a}}}^{j\sqrt{\frac{1-a}{1+a}}} \frac{ds}{1-s^2} = \int_{a-j\sqrt{1-a^2}}^{a+j\sqrt{1-a^2}} s\, ds$$

where $0 < a < 1$.

4-31. Let $f(s)$ be regular and bounded $|f(s)| < M$ at all points in the finite plane. Prove that $f(s)$ is a constant. (This is Liouville's theorem.) [HINT: Use the Cauchy integral formula for $f(z)$ at two points, say $s = 0$ and $s = s_1$, using for the contour of integration a circular path of radius R which is allowed to increase without limit. This will provide the information that $|f(s_1) - f(0)|$ is zero.]

4-32. Let a function $f(s)$ be undefined at a point s_0 and bounded and regular in a deleted neighborhood of this point. Prove that

$$\lim_{s \to s_0} f(s)$$

exists. (HINT: Use two concentric circles centered at s_0, and apply the Cauchy integral formula to the simply connected region obtained by joining these circles by a barrier. Then let the inner circle shrink to the point.) This result is known as Riemann's theorem.

4-33. Given the function

$$f(s) = \frac{s - a}{s + a}$$

use the appropriate Cauchy integral formula to prove that the nth derivative at zero is

$$f^{(n)}(0) = -2n! \left(-\frac{1}{a}\right)^n$$

HINT: In the integral, replace the variable of integration by its reciprocal.

CHAPTER 5

INFINITE SERIES

5-1. Introduction. In our quest of tools for determining the properties of functions we need a variety of methods for representing a function. We have the usual formulas (like sin s, e^s, $s^2 + 1$, etc.), but by employing other representations certain properties of the functions are put into evidence, and also other representations can lead to new functions. The Cauchy integral formula is one such representation, although its power is not yet fully brought out. The infinite-series representation is the next step.

As you undertake the study of this chapter, it is important to recognize that infinite series are used as *exact* representations for functions. The series form is important because certain properties of a function are placed in evidence in the series representation. Series are not used here in the sense of approximation, as when they are used for computation. Recognition of this fact is important.

5-2. Series of Constants. Consider an infinite sum of complex constants, which can be written in either of the two forms given on the two sides of the identity

$$\sum_{k=0}^{\infty} a_k = a_0 + a_1 + a_2 + \cdots \tag{5-1}$$

In order to study the series, we define a "function"

$$A_n = \sum_{k=0}^{n} a_k = a_0 + a_1 + \cdots + a_n \tag{5-2}$$

This is a function of n, having values only when n is an integer. A_n is called a *partial sum*. As n increases, A_n may do one of the two things illustrated in Fig. 5-1. If the sequence of values does not approach a limit, as at (a), the series is said to *diverge*. On the other hand, if it approaches a limit, as at (b), it is said to *converge*.

We are faced with a situation similar to the limit of a function discussed in Chap. 2. The only difference is that the function presently being considered takes on discrete values as a function of the integers represented by n. The definition of the limit of this function is similar to the previous one, modified to take care of the fact that now we are dealing with a limit as the independent variable n becomes infinite. Let ϵ be a small arbitrary positive number. The sequence A_n has a limit A if we can find a number N, which depends on ϵ, such that

$$|A_n - A| < \epsilon$$

when $n > N$ (5-3)

A convergent sequence is portrayed in Fig. 5-1b. A similarity with Fig. 2-5b will be noted, although in the present case the function yields only discrete points. For the ϵ shown, N would be 12. The importance of ϵ being arbitrary is to be emphasized. No matter how small ϵ is chosen to be, it must always be possible to find a number N. In general, N will increase as ϵ gets smaller.

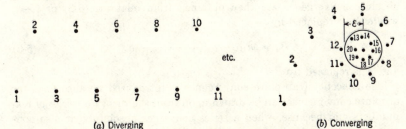

(a) Diverging   (b) Converging

FIG. 5-1. Examples of sequences of partial sums for diverging and converging series (numbers are subscripts on A_n). Points are plotted in the complex plane.

We now come to one of the theorems which will be left unproved. The theorem, which states the *Cauchy principle of convergence*, is as follows: A series like Eq. (5-1) converges if and only if, given an arbitrary small number $\epsilon > 0$, it is possible to find a number N such that

$$|A_n - A_m| < \epsilon$$

when $n > N$ and $m > N$ (5-4)

You should study these conditions in the light of Fig. 5-1b, so as to appreciate their reasonableness, in the absence of a proof.*

The Cauchy principle of convergence states necessary and sufficient conditions for convergence, and so either the Cauchy principle or the definition given earlier can be used interchangeably. The definition, conditions (5-3), is convenient because it puts the limit A into evidence. However, it is inconvenient in some cases because $A_n - A$ consists of an *infinite* number of terms. On the other hand, the conditions of the Cauchy principle are convenient when A is not needed

---

* E. T. Copson, "An Introduction to the Theory of Functions of a Complex Variable," Oxford University Press, New York, 1935.

explicitly and when it is convenient to use the fact that $A_n - A_m$ contains only a *finite* number of terms.

It is often convenient to consider the auxiliary series

$$\sum_{k=0}^{\infty} |a_k| = |a_0| + |a_1| + \cdots \qquad (5\text{-}5)$$

which is simpler than series (5-1) because all terms are positive and real. If series (5-5) converges, we can write

$$B = \sum_{k=0}^{\infty} |a_k|$$

and the original series is said to *converge absolutely*. Conditions for absolute convergence are then obtained from relations (5-3) or (5-4), where A_n is replaced by

$$B_n = |a_1| + |a_2| + \cdots + |a_n| \qquad (5\text{-}6)$$

and A is replaced by B.

The concept of absolute convergence is introduced because tests for absolute convergence can be drawn upon from the theory of series of real numbers and because, when a series converges absolutely, it also converges. This we shall prove as a theorem:

Theorem 5-1.   A series converges if it converges absolutely.

PROOF.   We are given the series

$$\sum_{k=0}^{\infty} a_k$$

and told that the absolute-value series converges, thus:

$$B = \sum_{k=0}^{\infty} |a_k| \qquad (5\text{-}7)$$

This means that, given a small positive number $\epsilon > 0$, we can find a number N such that

$$|B_n - B_m| < \epsilon$$
when $\qquad\qquad n,\, m > N \qquad (5\text{-}8)$

Let n be the greater of the two numbers m and n, and note that

$$|B_n - B_m| = |a_{m+1}| + |a_{m+2}| + \cdots + |a_n| \qquad (5\text{-}9)$$

Also, by the law of inequalities for sums we have

$$|a_{m+1} + a_{m+2} + \cdots + a_n| \leqq |a_{m+1}| + |a_{m+2}| + \cdots + |a_n| \quad (5\text{-}10)$$

But the term on the left above is $|A_n - A_m|$. Thus, by inequality (5-10) it follows that

$$|A_n - A_m| < \epsilon$$

when
$$n, m > N$$

Thus, it is proved that if a series converges absolutely it also converges. If the absolute-value series diverges, it is not necessary that the series will diverge. That is, the converse of Theorem 5-1 is not true.

We conclude this section by stating two tests for absolute convergence. These are taken from the theory of series of real numbers and their proofs can be found in the standard texts on advanced calculus.

1. *Cauchy Root Test.* A series

$$\sum_{k=0}^{\infty} a_k$$

converges absolutely (and therefore also converges) if

$$\varlimsup_{k \to \infty} \sqrt[k]{|a_k|} < 1 \qquad (5\text{-}11)*$$

and diverges if

$$\varlimsup_{k \to \infty} \sqrt[k]{|a_k|} > 1$$

2. *D'Alembert's Ratio Test.* The series converges absolutely (and therefore also converges) if

$$\lim_{k \to \infty} \left| \frac{a_{k+1}}{a_k} \right| < 1 \qquad (5\text{-}12)$$

and diverges if

$$\lim_{k \to \infty} \left| \frac{a_{k+1}}{a_k} \right| > 1$$

In either case, if the limit is 1, the test fails and gives no information. The root test is usually the more useful in analysis.

---

* The symbol $\varlimsup$ denotes the limit *superior*, which, loosely defined, is the limit approached by the *least upper bound* of the set of numbers $\sqrt[k]{|a_k|}$, where $k > n$ and n becomes infinite. The limit superior would enter in the case of a series like

$$1 + \frac{1}{2} + \frac{1}{4^2} + \frac{1}{2^3} + \frac{1}{4^4} + \frac{1}{2^5} + \cdots$$

In this case $\lim \sqrt[k]{a_k}$ does not exist, but $\varlimsup \sqrt[k]{a_k} = \frac{1}{2}$, and the series converges. When the limit exists, it is the same as the limit superior.

5-3. Series of Functions. In Sec. 5-2 we were dealing with series of constants. We can also consider series of functions of s such as

$$g_0(s) + g_1(s) + g_2(s) + \cdots = \sum_{k=0}^{\infty} g_k(s) \qquad (5\text{-}13)$$

The first observation is that for a fixed s such a series reverts back to the previous case; we get a series of constants. Thus, at a fixed value of s, convergence of a series of functions of s is defined and tested in the same way as for a series of constants. Thus, at $s = s_1$,

$$f(s_1) = \sum_{k=0}^{\infty} g_k(s_1) \qquad (5\text{-}14)$$

converges to a number $f(s_1)$ if, given an arbitrarily small $\epsilon > 0$, there exists a positive number $N(s_1)$ such that

$$|G_n(s_1) - f(s_1)| < \epsilon$$
when
$$n > N(s_1) \qquad (5\text{-}15)$$
and where $\qquad G_n(s_1) = g_0(s_1) + g_1(s_1) + \cdots + g_n(s_1) \qquad (5\text{-}16)$

The notation $N(s_1)$ is used to imply that this number might change if s is changed to another value. Now suppose that we try other values in a region R containing s_1 and find that the series converges at each value of s in R. The series is a function of s, and we can write

$$f(s) = \sum_{k=0}^{\infty} g_k(s) \qquad s \text{ in } R \qquad (5\text{-}17)$$

If the series diverges for some value of s, no function is defined by the series at that s. Convergence in R implies that, given an arbitrarily small $\epsilon > 0$, we can find $N(s)$ such that

$$|G_n(s) - f(s)| < \epsilon$$
when
$$n > N(s) \qquad (5\text{-}18)$$
and where $\qquad G_n(s) = g_0(s) + g_1(s) + g_2(s) + \cdots + g_n(s) \qquad (5\text{-}19)$

$N(s)$ is of course also a function of ϵ.

Relations (5-15) and (5-18) are quite similar, but constants are involved in the former and variables in the latter. Note that as s varies over the region R there may be a corresponding change in N even though ϵ remains fixed. Some series have the property that there can be a region R for which one constant N serves in the above capacity *for all values of s in R*. That is, N would depend on ϵ but not on s. Such a series is said to *converge uniformly* in R.

The idea of absolute convergence applies also to series of functions, in similarity with series of constants. Also, the Cauchy principle of convergence applies; i.e., given the usual arbitrarily small $\epsilon > 0$, the series converges in R if and only if

$$|G_n(s) - G_m(s)| < \epsilon \qquad (5\text{-}20)$$

when
$$n,\ m > N(s)$$

and converges uniformly in R if N is independent of s.

In order to illustrate these relatively abstract ideas, consider the following example: From algebra we have a formula for the sum of a geometric progression. Also, since complex numbers obey the ordinary rules of algebra, the formula can be applied to a geometric progression of complex numbers. Consider the special case of a geometric series where the ratio between terms is s. From the formula for the sum of n terms we have

$$\frac{1 - s^{n+1}}{1 - s} = 1 + s + s^2 + \cdots + s^n \qquad (5\text{-}21)$$

Now compare this with the infinite series

$$1 + s + s^2 + \cdots = \sum_{k=0}^{\infty} s^k$$

By either the root test or the ratio test this series converges absolutely (and therefore also converges) for $|s| < 1$. Therefore, for this range of s we can use the series to define a function, namely,

$$f(s) = \sum_{k=0}^{\infty} s^k \qquad |s| < 1 \qquad (5\text{-}22)$$

Equation (5-21) is a partial sum of the series in question, and

$$\lim_{n \to \infty} \frac{1 - s^{n+1}}{1 - s} = \frac{1}{1 - s} \qquad |s| < 1$$

Thus, we conclude that

$$f(s) = \frac{1}{1 - s} \qquad |s| < 1$$

Of course, $1/(1 - s)$ is also a function for $|s| \geq 1$, but $f(s)$ is *defined by the series*, and so $f(s) = 1/(1 - s)$ only for $|s| < 1$.

We already know the series converges, but let us try the definition of convergence as an exercise and to provide an opportunity to look at uniform convergence. We are to consider

$$\left| \frac{1}{1 - s} - \frac{1 - s^{n+1}}{1 - s} \right| = \left| \frac{\rho^{n+1}}{1 - s} \right| \leq \frac{\rho^{n+1}}{1 - \rho} \qquad (5\text{-}23)$$

where $s = \rho e^{j\phi}$. The inequality on the right follows because, when $|s| < 1$,

$$1 - \rho = 1 - |s| \leqq |1 - s|$$

If $n > N$, we have

$$\frac{\rho^{n+1}}{1 - \rho} < \frac{\rho^{N+1}}{1 - \rho}$$

and so, recognizing that ϵ is given, we shall choose N so that

$$\frac{\rho^{N+1}}{1 - \rho} < \epsilon$$

or

$$(N + 1) \log \rho < \log \epsilon(1 - \rho)$$

Note that $\log \rho$ is negative, because $\rho < 1$, and so dividing by $\log \rho$ reverses the inequality, giving

$$N + 1 > \frac{\log \epsilon(1 - \rho)}{\log \rho}$$

or

$$N(\rho) > \frac{\log \epsilon(1 - \rho)}{\log \rho} - 1 \tag{5-24}$$

Thus, from the given ϵ, a value of N can be found such that

$$\left| \frac{1}{1 - s} - \sum_{k=0}^{n} s^k \right| < \epsilon \tag{5-25}$$

when

$$n > N(\rho)$$

Now we have established that in the circular region

$$|s| < 1$$

the series converges. However, it does not converge uniformly in this region, because $N(\rho)$ approaches infinity as ρ approaches 1. No value N can be found which will do for all values of s in the region. The region of uniform convergence is

$$|s| \leqq \rho' < 1 \tag{5-26}$$

where ρ' is a constant. To prove this, note from Eqs. (5-24) that $N(\rho)$ increases with increasing ρ. Therefore, if

$$\rho \leqq \rho'$$

then

$$N(\rho) \leqq N(\rho')$$

The value $N = N(\rho')$, which is a constant, can be used in place of the variable $N(\rho)$ in relations (5-25). Therefore, since this value of N is satisfactory for all s in region (5-26), we can say that this is the region of uniform convergence. Note that the region of uniform convergence is closed.

This section is concluded with another theorem:

Theorem 5-2. If each of the functions $g_0(s)$, $g_1(s)$, etc., is continuous in a region R where the series

$$f(s) = \sum_{k=0}^{\infty} g_k(s)$$

converges uniformly, then $f(s)$ is continuous in R.

PROOF. To begin the proof, note that it is quite trivial to prove that, if each $g_k(s)$ is continuous at some point s_0 in R, then

$$G_n(s) = g_0(s) + g_1(s) + \cdots + g_n(s)$$

is also continuous at s_0. The sum of a *finite* number of continuous functions is also continuous. (If there is any doubt, you should prove this as an exercise.) We consider two points, s_0 and a general point s near s_0. Then pick an $\epsilon > 0$, and determine a fixed number n such that we can write the three sets of conditions

$$|f(s) - G_n(s)| < \frac{\epsilon}{3} \qquad (5\text{-}27a)$$

$$|f(s_0) - G_n(s_0)| < \frac{\epsilon}{3} \qquad (5\text{-}27b)$$

$$|G_n(s) - G_n(s_0)| < \frac{\epsilon}{3} \qquad |s - s_0| < \delta \qquad (5\text{-}27c)$$

We know that the number n can be found because s_0 and s lie in the region of uniform convergence. Also, we know a δ can be found for condition (5-27c) because $G_n(s)$ is continuous. In general δ would be a function of n, but n is fixed in this analysis. We are to investigate $|f(s) - f(s_0)|$, which we write as

$$|f(s) - f(s_0)| = |[f(s) - G_n(s)] - [f(s_0) - G_n(s_0)] + [G_n(s) - G_n(s_0)]|$$

By the usual rule for inequalities, the right-hand side is less than the sum of the absolute values of the three terms in brackets. But each of these is represented by one of the parts of relations (5-27) and is less than $\epsilon/3$. Therefore, we have shown that

$$|f(s) - f(s_0)| < \epsilon$$
$$\text{when} \qquad |s - s_0| < \delta$$

which constitutes a proof that $f(s)$ is continuous in R.

5-4. Integration of Series. The idea of uniform convergence was needed mainly to permit consideration of term-by-term integration of infinite series. If the series of continuous functions

$$f(s) = \sum_{k=0}^{\infty} g_k(s)$$

converges, we have shown that it is a continuous function in the region R of uniform convergence. We also know that continuity of a function is sufficient to ensure integrability. Thus, the integral

$$\int_C f(s) \, ds$$

exists for all paths in R. (Nothing needs to be said about regularity to ensure integrability.) We shall now prove the following theorem:

Theorem 5-3. If the given series converges uniformly to $f(s)$ on a path C of finite length, then

$$\int_C f(s) \, ds = \sum_{k=0}^{\infty} \int_C g_k(s) \, ds \qquad (5\text{-}28)$$

PROOF. The proof begins by observing that for a finite number of terms it is permissible to interchange integration and summation, as follows:

$$\sum_{k=0}^{n} \int_C g_k(s) \, ds = \int_C \left[\sum_{k=0}^{n} g_k(s) \right] ds \qquad (5\text{-}29)$$

Therefore we can write

$$\int_C f(s) \, ds - \sum_{k=0}^{n} \int_C g_k(s) \, ds = \int_C \left[f(s) - \sum_{k=0}^{n} g_k(s) \right] ds \qquad (5\text{-}30)$$

The series is uniformly convergent, and therefore, given an arbitrarily small $\epsilon > 0$, there can be found a number N such that

$$\left| f(s) - \sum_{k=0}^{n} g_k(s) \right| < \epsilon$$

when $n > N$ for all s on C. Also, the above quantity is the absolute value of the integrand on the right of Eq. (5-30). We can therefore use the upper bound of an integral, established in Sec. 4-4, to give

$$\left| \int_C \left[f(s) - \sum_{k=0}^{n} g_k(s) \right] ds \right| < \epsilon L$$

where L is the length of integration path. In view of Eq. (5-30), then also

$$\left| \int_C f(s) \, ds - \sum_{k=0}^{n} \int_C g_k(s) \, ds \right| < \epsilon L$$

when $n > N$. Since ϵ is arbitrarily small, ϵL is also an arbitrarily small number and so conditions are established whereby the right side of Eq. (5-28) converges to the left side.

It is almost trivial to point out that, if a series is uniformly convergent in a region R, it is also uniformly convergent on any curve C lying in R. Thus, a series may be integrated term by term in any region of uniform convergence.

Applications of integration of series will appear in the following section.

5-5. Convergence of Power Series. Series of the form

$$f(s) = \sum_{k=0}^{\infty} a_k(s - s_0)^k \tag{5-31}$$

are of great importance in the theory of functions of a complex variable. From the root and ratio tests we find that the series converges (in fact, converges absolutely) if, respectively,

$$|s - s_0| \, \overline{\lim_{k \to \infty}} \sqrt[k]{|a_k|} < 1 \qquad \text{or} \qquad |s - s_0| \lim_{k \to \infty} \left| \frac{a_{k+1}}{a_k} \right| < 1$$

Thus, if we define a number R_0 by either of the following:

$$\overline{\lim_{k \to \infty}} \sqrt[k]{|a_k|} = \frac{1}{R_0}$$

or

$$\lim_{k \to \infty} \left| \frac{a_{k+1}}{a_k} \right| = \frac{1}{R_0} \tag{5-32}$$

then the series converges when

$$|s - s_0| < R_0$$

This is a circular region, as illustrated in Fig. 5-2. Since the above analysis is based on absolute convergence, we are interested in knowing whether there is some point outside the circle where the series con-

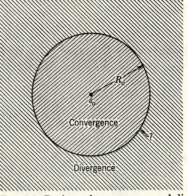

Fig. 5-2. Regions of convergence and divergence of a power series.

verges in the ordinary sense. Assume that s_1 is any point such that

$$\sum_{k=0}^{\infty} a_k(s_1 - s_0)^k$$

converges.   From the Cauchy formulation

$$\left| \sum_{k=m+1}^{n} a_k(s_1 - s_0)^k \right| < \epsilon \qquad n, m > N$$

where N is a number depending on the small arbitrary positive ϵ.   This relation is true for the particular case $n = m + 1 = k$, giving

$$|a_k(s_1 - s_0)^k| < \epsilon \qquad k > N$$

Each number of the set $|a_k(s_1 - s_0)^k|$, $k = 1, 2, \ldots, N$, is finite, and therefore in view of the previous relation there is a number K such that for all k

$$|a_k(s_1 - s_0)^k| < K$$

Now choose a general point s, where $|s - s_0| < |s_1 - s_0|$, and consider

$$\sum_{k=0}^{\infty} \left| a_k(s - s_0)^k \right| = \sum_{k=0}^{\infty} \left| a_k(s_1 - s_0)^k \right| \left| \frac{s - s_0}{s_1 - s_0} \right|^k < K \sum_{k=0}^{\infty} \left| \frac{s - s_0}{s_1 - s_0} \right|^k$$

The last series on the right converges, since $|s - s_0|/|s_1 - s_0| < 1$, and so the series converges absolutely at point s *if it converges at* s_1, where

$$|s - s_0| < |s_1 - s_0|$$

Now suppose that s_1 is outside the previously defined circle of radius R_0, and assume that the series converges at this point.   A point s can be found such that

$$R_0 < |s - s_0| < |s_1 - s_0|$$

leading to the contradiction that the series would have to converge absolutely outside the circle.   Thus, ordinary convergence is not possible at any point outside the circle.   However, ordinary convergence at some points on the circle is a possibility, since no contradiction ensues if s_1 is on the circle. R_0 is called the *radius of convergence* of the series.   The series converges absolutely inside the circle and diverges outside the circle, and may or may not converge on the circle.   These conclusions are summarized in Fig. 5-2.

Next we consider uniform convergence of power series and for that purpose shall need the following theorem:

Theorem 5-4.   If in a region R there is a sequence of positive constants M_k such that

$$|g_k(s)| \leqq M_k \tag{5-33}$$

and if

$$\sum_{k=0}^{\infty} M_k \tag{5-34}$$

converges, then

$$\sum_{k=0}^{\infty} g_k(s) \tag{5-35}$$

converges uniformly in R.

PROOF. Since series (5-34) converges, from the Cauchy principle of convergence we know that corresponding to $\epsilon > 0$ there exists a number N such that

$$\left| \sum_{k=0}^{n} M_k - \sum_{k=0}^{m} M_k \right| = \sum_{k=m}^{n} M_k < \epsilon$$

when $n \geqq m > N$. However, in view of the inequality rule for absolute values of sums and relation (5-33), it follows that

$$\left| \sum_{k=m}^{n} g_k(s) \right| \leqq \sum_{k=m}^{n} |g_k(s)| \leqq \sum_{k=m}^{n} M_k < \epsilon$$

when $n, m > N$, for all s in R. However, this is the same as

$$\left| \sum_{k=m}^{n} g_k(s) \right| = \left| \sum_{k=0}^{n} g_k(s) - \sum_{k=0}^{m} g_k(s) \right| < \epsilon$$

which is the Cauchy principle of convergence. Convergence is uniform because N is independent of s when s is in R. Theorem 5-4 provides the *Weierstrass M test* for uniform convergence.

Now return to the series of Eq. (5-31), having a known radius of convergence R_0. We write

$$|s - s_0| \leqq R_0' < R_0$$

and define

$$M_k = |a_k|(R_0')^k$$

giving

$$\sum_{k=0}^{\infty} M_k = \sum_{k=0}^{\infty} |a_k|(R_0')^k$$

Note that the series on the right is known to converge because $R_0' < R_0$. Also, if

$$|s - s_0| \leqq R_0'$$
$$|a_k| \, |s - s_0|^k \leqq M_k$$

and so, by Theorem 5-4,

$$\sum_{k=0}^{\infty} a_k(s - s_0)^k$$

converges uniformly in the region

$$|s - s_0| \leqq R_0' < R_0 \tag{5-36}$$

5-6. Properties of Power Series.   As an immediate consequence of the uniform convergence in region (5-36) we can say from Theorem 5-3 that

$$\int_C f(s)\ ds = \sum_{k=0}^{\infty} \int_C a_k(s - s_0)^k\ ds \tag{5-37}$$

where C is any curve of finite length inside region (5-36).   Furthermore, if C is a closed curve, each integral

$$\int_C a_k(s - s_0)^k\ ds = 0 \tag{5-38}$$

and so

$$\int_C f(s)\ ds = 0$$

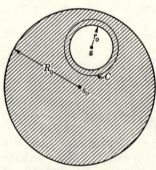

FIG. 5-3. Region of uniform convergence of series (5-40).

and from Morera's theorem it is concluded that $f(s)$ is regular in region (5-36).

Now let s be a point in region (5-36) and C a circle also lying in the region, as shown in Fig. 5-3.   From the Cauchy integral formula for the derivative,

$$\frac{df(s)}{ds} = \frac{1}{2\pi j} \int_C \frac{f(z)}{(z - s)^2}\ dz \tag{5-39}$$

and the integrand can be written

$$\frac{f(z)}{(z - s)^2} = \sum_{k=0}^{\infty} \frac{a_k(z - s_0)^k}{(z - s)^2} \tag{5-40}$$

The series in Eq. (5-40) converges uniformly in the doubly connected closed region shown shaded in Fig. 5-3.   This is true because

$$\left| \frac{a_k(z - s_0)^k}{(z - s)^2} \right| \leqq \frac{|a_k|(R_0')^k}{r_0^2}$$

and

$$\frac{1}{r_0^2} \sum_{k=0}^{\infty} |a_k|(R_0')^k$$

converges.   Therefore, we can use Theorem 5-3 (note that the theorem does not require C to be in a simply connected region) to obtain

$$\frac{1}{2\pi j} \int_C \frac{f(z)\ dz}{(z - s)^2} = \sum_{k=0}^{\infty} \frac{1}{2\pi j} \int_C \frac{a_k(z - s_0)^k}{(z - s)^2}\ dz$$

Each term under the summation is the derivative of $a_k(s - s_0)^k$, by the

Cauchy integral formulas, and so

$$\frac{df(s)}{ds} = \sum_{k=0}^{\infty} k a_k (s - s_0)^{k-1} \tag{5-41}$$

where s is in the open region

$$|s - s_0| < R_0'$$

Note that this region cannot be closed with an equality sign, because if s were to be on the boundary circle

$$|s - s_0| = R_0'$$

it would be impossible for C to be in the shaded region. Of course, R_0' can be arbitrarily close to R_0, the radius of convergence, and so we can state the following theorem:

Theorem 5-5. A function represented by a power series expanded about a point s_0, with radius of convergence R_0, is regular for $|s - s_0| < R_0$ and therefore possesses all derivatives for $|s - s_0| < R_0$. Furthermore, the nth derivative of $f(s)$ is given by the series obtained by term-by-term differentiation of the original series n times; and the radius of convergence of the series for the derivative is also R_0.

5-7. Taylor Series. In the previous section it was established that, within its circle of convergence, a power series defines a function which is regular at each point where it is defined. We also emphasized that the series is specified first and that the series defines the function. Now we approach the converse situation, where $f(s)$ is specified in some form other than the series

$$\sum_{k=0}^{\infty} a_k (s - s_0)^k$$

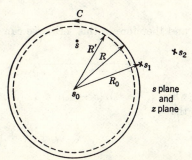

Fig. 5-4. Contour of integration used to develop the Taylor series, and region of convergence of that series.

We are to determine whether this form can be an expression for the function over any part of the complex plane. The way to proceed is to see whether the coefficients a_k can be determined from the given $f(s)$ and whether the series converges to $f(s)$.

Assume that $f(s)$ is analytic, select any point where the function is regular, and designate this point as s_0. In general, $f(s)$ will have some singular points s_1, s_2, etc., but, in view of the definition of regularity, there must be a neighborhood of s_0 in which there are no singular points. Let R_0 be the distance from s_0 to its nearest singular point. Referring to Fig. 5-4, let C be a circle centered at s_0, with radius $R < R_0$. Then, in the region

$$|s - s_0| < R$$

the Cauchy integral formula

$$f(s) = \frac{1}{2\pi j} \int_C \frac{f(z)}{z - s}\, dz \qquad (5\text{-}42)$$

is a valid representation for $f(s)$. The next step is to write

$$\frac{1}{z - s} = \frac{1}{z - s_0 + s_0 - s} = \frac{1}{z - s_0} \frac{1}{1 - (s - s_0)/(z - s_0)}$$

Although we have stipulated $|s - s_0| < R$ as the range of s, now choose a number $R' < R$ and restrict s to the region

$$|s - s_0| \leqq R' < R$$

In the integral of Eq. (5-42) z is confined to the circle $|z - s_0| = R$, and so

$$\left| \frac{s - s_0}{z - s_0} \right| \leqq \frac{R'}{R} < 1$$

In Sec. 5-3 it is shown that

$$\frac{1}{1 - s} = 1 + s + s^2 + \cdots \qquad |s| < 1$$

and therefore in this series we can replace s by $(s - s_0)/(z - s_0)$ to give

$$\frac{1}{1 - (s - s_0)/(z - s_0)} = 1 + \frac{s - s_0}{z - s_0} + \left(\frac{s - s_0}{z - s_0}\right)^2 + \cdots$$

when $|(s - s_0)/(z - s_0)| < 1$. Furthermore, since $R'/R < 1$, we know that

$$1 + \frac{R'}{R} + \left(\frac{R'}{R}\right)^2 + \cdots$$

converges and thus $(R'/R)^k$ can serve as M_k in the Weierstrass M test, proving that the above series converges uniformly for

$$\left| \frac{s - s_0}{z - s_0} \right| \leqq \frac{R'}{R}$$

and therefore for

$$|s - s_0| \leqq R'$$

when $|z - s_0| = R$. This latter condition is satisfied for the integral in Eq. (5-42), and so we can replace the integrand of Eq. (5-42) by the series

$$\frac{f(z)}{z - s} = f(z) \sum_{k=0}^{\infty} \frac{(s - s_0)^k}{(z - s_0)^{k+1}}$$

and then perform a term-by-term integration, with the result

$$f(s) = \frac{1}{2\pi j} \sum_{k=0}^{\infty} (s - s_0)^k \int_C \frac{f(z)}{(z - s_0)^{k+1}} \, dz \qquad |s - s_0| \leqq R'$$

This is a series in powers of $(s - s_0)^k$, where the integral factors are constants. R' can be as close as we like to R_0, and so the radius of convergence of this series is R_0. Finally, this result is conveniently written

$$f(s) = \sum_{k=0}^{\infty} a_k (s - s_0)^k \qquad |s - s_0| < R_0 \qquad (5\text{-}43)$$

where
$$a_k = \frac{1}{2\pi j} \int_C \frac{f(z)}{(z - s_0)^{k+1}} \, dz \qquad (5\text{-}44)$$

Certainly these coefficients exist, since the integrand is regular on the path of integration.

It is to be observed that Eq. (5-44) is very similar to the Cauchy integral formula for the kth derivative, differing only in the absence of the factor $k!$. Accordingly, we can write

$$a_k = \frac{f^{(k)}(s_0)}{k!} \qquad (5\text{-}45)$$

which is identical in form to the usual formula for the coefficients of a Taylor series in real variables. Accordingly, the series expansion given in Eq. (5-43) is called the *Taylor-series* expansion of $f(s)$ about point s_0.

In the derivation leading to Eq. (5-44) we designated C as a circle of radius R. However, Eq. (5-44) is invariant if C is distorted into any simple closed curve inside the circle of radius R but still enclosing s_0. Thus, in Eq. (5-44) we arrive at the final interpretation of C as any simple closed curve enclosing point s_0 but not large enough to enclose any points where $f(s)$ is singular.

By virtue of this proof we have shown that the series in Eq. (5-43) converges in the region $|s - s_0| < R_0$ and, furthermore, that it converges to the original function $f(s)$. The series now becomes a new representation for $f(s)$, valid in the circle of convergence. This development is important because it shows that for any analytic function a power-series expansion is possible about any point where the function is regular.

The information provided about the radius of convergence is especially important. In the process of arriving at Eq. (5-43) it was established that R_0, the radius of convergence of the series, is the distance from s_0 to the singular point closest to s_0. Thus, if the locations of singular points of $f(s)$ are known, it is immediately known from simple geometry in the complex plane what will be the radius of convergence for a Taylor expansion about any point; there is no need to carry out a convergence test.

As you develop an understanding of the implications of Eqs. (5-43) and (5-44), it is especially important to understand that Eq. (5-43) is a new representation of $f(s)$, differing from the original representation which is used for $f(z)$ in Eq. (5-44). The original representation can be a "closed-form" representation [like sin s, $s/(s+1)$, etc.], or it can be a series in powers of $s - s_0'$, where s_0' is some point *other than* s_0. However, in the latter case s_0 must lie in the region of convergence of the given series. Since the region of validity of Eq. (5-43) is generally different from the region of validity of the original representation, it is very important that the region of validity shall always be stipulated as part of the formula, as shown in Eq. (5-43).

In the above statement of possible original representations, the possibility that $f(s)$ may originally be represented by a series in powers of $s - s_0$ was omitted, in anticipation of a special consideration of this case. Suppose that $f(s)$ is defined by a convergent series

$$f(s) = \sum_{k=0}^{\infty} a_k'(s - s_0)^\kappa$$

having a finite radius of convergence. This function is a candidate for representation by Eq. (5-43), and accordingly we seek the a_k coefficients,

$$a_k = \frac{1}{2\pi j} \int_C \frac{f(z)}{(z - s_0)^{k+1}}\, dz = \frac{1}{2\pi j} \sum_{n=0}^{\infty} a_k' \int_C \frac{(z - s_0)^n}{(z - s_0)^{k+1}}\, dz \quad (5\text{-}46)$$

where C is a small circle centered at s_0 within the region of convergence. The interchange of integration and summation operations is justified by Theorem 5-3. Furthermore, from Eq. (4-24) it is known that

$$\int_C \frac{dz}{(z - s_0)^{k-n+1}} = \begin{cases} 0 & n \neq k \\ 2\pi j & n = k \end{cases}$$

Therefore, Eq. (5-46) yields

$$a_k = a_k'$$

This result may seem obvious and trivial, but it expresses the important principle that there is only *one* power-series expansion about a given point which converges to a given function, and this is the Taylor series. This fact is important because Eqs. (5-44) and (5-45) do not, in most practical cases, offer the simplest procedure for finding the coefficients. If some other procedure can be found to give a series which converges to the required function, this series must be identical with Eq. (5-43). In this statement there is no deprecation of Eqs. (5-44) and (5-45). On many occasions they are indispensable because of their generality, particularly in the subsequent proofs of general theorems.

As an illustration of the convenience of using alternative methods of obtaining a Taylor series, consider the function

$$f(s) = \frac{1}{(s-1)(s-2)} = \frac{1}{s^2 - 3s + 2}$$

expanded about the point $s_0 = 0$. We can perform a division algorithm as follows:

$$
\begin{array}{r}
\frac{1}{2} + \frac{3}{4}s + \frac{7}{8}s^2 \\
2 - 3s + s^2 \,\overline{\big)\, 1 \qquad\qquad\qquad} \\
\underline{1 - \frac{3}{2}s + \frac{1}{2}s^2} \\
\frac{3}{2}s - \frac{1}{2}s^2 \\
\underline{\frac{3}{2}s - \frac{9}{4}s^2 + \frac{3}{4}s^3} \\
\frac{7}{4}s^2 - \frac{3}{4}s^3 \\
\underline{\frac{7}{4}s^2 - \frac{21}{8}s^3 + \frac{7}{8}s^4} \\
\frac{15}{8}s^3 - \frac{7}{8}s^4
\end{array}
$$

For the finite number of steps shown,

$$f(s) = \tfrac{1}{2} + \tfrac{3}{4}s + \tfrac{7}{8}s^2 + \left(\frac{\frac{15}{8}s^3 - \frac{7}{8}s^4}{2 - 3s - s^2} \right)$$

where the quantity in parentheses is the remainder term. Without carrying out the details, it is apparent that for small $|s|$ the remainder approaches zero as the algorithm is continued; the series

$$\tfrac{1}{2} + \tfrac{3}{4}s + \tfrac{7}{8}s^2 + {}^{15}\!/_{16}s^3 + \cdots$$

converges to $f(s)$ for $|s| < 1$, and therefore it is the Taylor expansion about the origin. Since $f(s)$ is singular at $s = 1$ it is known that the radius of convergence is 1. From an inspection of the above series, it is apparent that, in the notation of Eq. (5-43),

$$a_k = \frac{2^{k+1} - 1}{2^{k+1}}$$

The same formula for a_k would be obtained from Eqs. (5-44) and (5-45), but with considerably greater difficulty.

5-8. Laurent Series. We begin this section in much the same way as the last. It is assumed that an analytic function $f(s)$ is given. However,

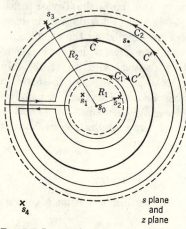

in this case the function is assumed to be regular in a doubly connected region bounded by two concentric circles, such as the dashed circles in Fig. 5-5. There are singular points outside this region, designated by $s_1, \ldots, s_4$. In this example R_1 and R_2 are respectively the distances from s_0 to s_2 and s_3. The point at the center, s_0, may or may not be singular. The region is made simply connected by the barrier shown at the left in the figure. The Cauchy integral formula can now be applied to represent $f(s)$ at point s. The closed path of integration can be a closed curve such as C' enclosing s, in the simply

FIG. 5-5. Integration contours used in developing the Laurent series, and region of convergence of that series.

connected region. Thus, $f(s)$ can be represented by

$$f(s) = \frac{1}{2\pi j} \int_{C'} \frac{f(z)}{z - s} \, dz \qquad (5\text{-}47)$$

The two portions of this path parallel to the barrier can be brought arbitrarily close together, and they ultimately cancel out. Therefore, the above can be rewritten

$$f(s) = \frac{1}{2\pi j} \int_{C_2} \frac{f(z)}{z - s} \, dz - \frac{1}{2\pi j} \int_{C_1} \frac{f(z)}{z - s} \, dz \qquad (5\text{-}48)$$

The development given for the integral of Eq. (5-42) applies to the above integral around C_2. Accordingly, we can immediately write

$$\frac{1}{2\pi j} \int_{C_2} \frac{f(z)}{z - s} \, dz = \frac{1}{2\pi j} \sum_{k=0}^{\infty} (s - s_0)^k \int_{C_2} \frac{f(z)}{(z - s_0)^{k+1}} \, dz \qquad |s - s_0| < R_2$$

$$(5\text{-}49)$$

A similar treatment can be applied to the integral around C_1. In this case we write

$$-\frac{1}{z - s} = \frac{1}{s - s_0 - z + s_0} = \frac{1}{s - s_0} \left[\frac{1}{1 - (z - s_0)/(s - s_0)} \right]$$

and, by arguments similar to those of the previous section, the quantity in brackets can be expressed by the series

$$\frac{1}{1 - (z - s_0)/(s - s_0)} = 1 + \frac{z - s_0}{s - s_0} + \left(\frac{z - s_0}{s - s_0}\right)^2 + \cdots$$

which converges uniformly for*

$$|s - s_0| \geqq R' > R > R_1$$

when $|z - s_0| = R$, the radius of C_1. Therefore, term-by-term integration is permitted, giving

$$-\frac{1}{2\pi j} \int_{C_1} \frac{f(z)}{z - s}\, dz = \frac{1}{2\pi j} \sum_{k=1}^{\infty} \frac{1}{(s - s_0)^k} \int_{C_1} (z - s_0)^{k-1} f(z)\, dz \quad (5\text{-}50)$$

Equations (5-49) and (5-50) define two functions

$$f_a(s) = \sum_{k=0}^{\infty} a_k (s - s_0)^k \qquad |s - s_0| < R_2$$

$$f_b(s) = \sum_{k=1}^{\infty} b_k \left(\frac{1}{s - s_0}\right)^k \qquad |s - s_0| > R_1$$

$$(5\text{-}51)$$

where

$$a_k = \frac{1}{2\pi j} \int_{C_2} \frac{f(z)}{(z - s_0)^{k+1}}\, dz$$

$$b_k = \frac{1}{2\pi j} \int_{C_1} (z - s_0)^{k-1} f(z)\, dz$$

$$(5\text{-}52)$$

and from Eq. (5-48) we see that

$$f(s) = f_a(s) + f_b(s) \qquad R_1 < |s - s_0| < R_2 \quad (5\text{-}53)$$

Note that $f_a(s)$ has no singularities inside circle R_2 and $f_b(s)$ has no singularities outside circle R_1. Equation (5-53) can also be written

$$f(s) = \sum_{k=-1}^{-\infty} b_{-k} (s - s_0)^k + \sum_{k=0}^{\infty} a_k (s - s_0)^k$$

when

$$R_1 < |s - s_0| < R_2$$

This form is of interest because it leads to further simplification, as we shall now see. Let C be any simple closed path lying between circles C_1 and C_2, as shown in Fig. 5-5. Each of these circles can be distorted into C

---

* $(R/R')^k$ serves as M_k in the Weierstrass M test, in similarity with the development following Eq. (5-42).

without passing over any singularities of $f(z)$. Therefore, from Eqs. (5-52),

$$a_k = \frac{1}{2\pi j} \int_{C_2} \frac{f(z)}{(z - s_0)^{k+1}} \, dz = \frac{1}{2\pi j} \int_C \frac{f(z)}{(z - s_0)^{k+1}} \, dz$$

$$b_{-k} = \frac{1}{2\pi j} \int_{C_1} \frac{f(z)}{(z - s_0)^{k+1}} \, dz = \frac{1}{2\pi j} \int_C \frac{f(z)}{(z - s_0)^{k+1}} \, dz$$

Thus, it is seen that a_k and b_{-k} are given by the same formula, and so $f(s)$ can be written

$$f(s) = \sum_{k=-\infty}^{\infty} a_k(s - s_0)^k \qquad R_1 < |s - s_0| < R_2 \qquad (5\text{-}54)$$

where

$$a_k = \frac{1}{2\pi j} \int_C \frac{f(z)}{(z - s_0)^{k+1}} \, dz \qquad (5\text{-}55)$$

The series expansion in Eq. (5-54) is called a *Laurent expansion* of the function. In a sense it is a generalization of the Taylor series; if $f(s)$ has no singularities inside circle R_2, then Eq. (5-55) gives zero for each a_k when $k < 0$, showing that in such a case the Laurent series reduces to the Taylor series.

Equations (5-44) and (5-55) are similar in appearance, but in the latter case no formula like Eq. (5-45) giving a_k as a derivative of $f(s)$ can be given. This is because, when $f(z)$ has singular points inside C, Eq. (5-55) is not a Cauchy integral formula.

By a proof exactly similar to the one given for the Taylor series, it can be shown that any series like Eq. (5-54) which converges to $f(s)$ must be the Laurent series, with coefficients given by Eq. (5-55). *For a given annular region*, the Laurent series is therefore the only expansion in positive and negative powers of $s - s_0$. Most frequently the coefficients of a Laurent series are obtained for a specific function by some process other than Eq. (5-55), by an algorithm division, for example. Other techniques of obtaining the coefficients are illustrated in Sec. 5-9.

5-9. Comparison of Taylor and Laurent Series. In general, an analytic function has a variety of representations in various regions in the s plane. Three specific representations have now been developed, the Cauchy integral formula, the Taylor series, and the Laurent series. From the viewpoint of developments in this chapter, the Cauchy integral formula representation is important because it leads directly to the Taylor and Laurent series. Therefore, continuing with emphasis on series, it is helpful to make a comparison between the two types.

We have seen that an analytic function has many series expansions. Particular circumstances determine which series is obtained in a given situation. These circumstances are specifically:

1. The choice of s_0

2. The required region of convergence of the series, in relation to locations of singular points

If s_0 is not a singular point, we can have both a Laurent and a Taylor expansion, but they are valid in different regions.

As an example consider

$$f(s) = \frac{1}{1 - s}$$

From previous treatment of this function we know that

$$\frac{1}{1 - s} = 1 + s + s^2 + \cdots \qquad |s| < 1 \qquad (5\text{-}56a)$$

$$\frac{1}{1 - s} = -\left(\frac{1}{s} + \frac{1}{s^2} + \cdots\right) \qquad |s| > 1 \qquad (5\text{-}56b)$$

Equation (5-56a) is a Taylor expansion with $s_0 = 0$. Its region of convergence is shown in Fig. 5-6a, showing a radius of 1 to the closest singularity. Equation (5-56b) is the Laurent series for the same function about the same point ($s_0 = 0$). It converges outside a circle passing through the singularity at $s = 1$, as shown in Fig. 5-6b.

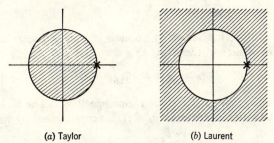

(a) Taylor   (b) Laurent

FIG. 5-6. Regions of convergence of Taylor and Laurent expansions of $1/(1 - s)$.

Now consider the general case of a function having more than one singular point in the finite plane. For each s_0 there can be at most one Taylor expansion; but more than one Laurent expansion is possible. The number of Laurent expansions depends on the number of singular points.

To illustrate this, consider

$$f(s) = \frac{1}{(s - 1)(s - 2)}$$

having the singular points shown by crosses in Fig. 5-7. Expansions are to be about the point $s_0 = 0$. There is one Taylor expansion, convergent in region 1, and there are two Laurent expansions, convergent, respectively, in regions 2 and 3. This is the function for which the Taylor expansion is obtained in Sec. 5-7 by the method of long division. Here we shall use a different method, starting with the recognition that the given function can be written

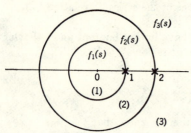

FIG. 5-7. Regions of convergence of expansions of $1/(s - 1)(s - 2)$ about the nonsingular point $s_0 = 0$.

$$f(s) = \frac{1}{1 - s} - \frac{\frac{1}{2}}{1 - s/2} \quad (5\text{-}57)$$

each of which can be expanded independently. In the manner of Eqs. (5-56), we have two series for each part, as follows:

$$\frac{1}{1 - s} = \begin{cases} 1 + s + s^2 + \cdots & |s| < 1 \\ -\left(\dfrac{1}{s} + \dfrac{1}{s^2} + \dfrac{1}{s^3} + \cdots\right) & |s| > 1 \end{cases} \quad (5\text{-}58)$$

and
$$\frac{-\frac{1}{2}}{1 - s/2} = \begin{cases} -\left(\dfrac{1}{2} + \dfrac{s}{4} + \dfrac{s^2}{8} + \dfrac{s^3}{16} + \cdots\right) & |s| < 2 \\ \dfrac{1}{s} + \dfrac{2}{s^2} + \dfrac{4}{s^3} + \dfrac{8}{s^4} + \cdots & |s| > 2 \end{cases} \quad (5\text{-}59)$$

For the Taylor series (region 1) we need $|s| < 1$ and therefore we add the first series of each of these sets to get

$$f_1(s) = \tfrac{1}{2} + \tfrac{3}{4}s + \tfrac{7}{8}s^2 + \tfrac{15}{16}s^3 + \cdots \quad |s| < 1 \quad (5\text{-}60)$$

which is identical with $f(s)$ for the region specified. The fact that this is actually the series representation of $f(s)$ is summarized by writing

$$f(s) = f_1(s) \qquad |s| < 1$$

In Sec. 5-7 a special subscript was not used to designate the series representation. In fact, the special designation is not ordinarily used; but in the present instance it is simpler to use this notation in order to stress the concept of having many representations for a single function.

For region 2 we have $1 < |s| < 2$, and so the appropriate series are picked from Eqs. (5-58) and (5-59) to obtain

$$f_2(s) = -\left(\cdots \frac{1}{s^3} + \frac{1}{s^2} + \frac{1}{s} + \frac{1}{2} + \frac{s}{4} + \frac{s^2}{8} + \cdots\right)$$
$$1 < |s| < 2 \quad (5\text{-}61)$$

This function is defined only for the region specified, but in that region it is identical with $f(s)$. Thus,

$$f(s) = f_2(s) \qquad 1 < |s| < 2$$

Finally, for region 3 the condition $|s| > 2$ is met by picking the second series in Eqs. (5-58) and (5-59) to get

$$f_3(s) = \frac{1}{s^2} + \frac{3}{s^3} + \frac{7}{s^4} + \frac{15}{s^5} + \cdots \qquad |s| > 2 \qquad (5\text{-}62)$$

and therefore

$$f(s) = f_3(s) \qquad |s| > 2$$

In summary, we have obtained a set of three series representations which, taken together, define $f(s)$ throughout the s plane, except on the two circles separating the regions. We can now write $f(s)$ in two alternative forms:

$$f(s) = \frac{1}{(s-1)(s-2)} \qquad (5\text{-}63)$$

or

$$f(s) = \begin{cases} f_1(s) & |s| < 1 \\ f_2(s) & 1 < |s| < 2 \\ f_3(s) & |s| > 2 \end{cases} \qquad (5\text{-}64)$$

Later on we shall see how the second representation also establishes values of $f(s)$ at nonsingular points on the boundary circles.

5-10. Laurent Expansions about a Singular Point. In recalling the steps that went into deriving the Laurent expansion you will note that s_0 could coincide with one of the singular points without invalidating the derivation. If s_0 should be a singular point, then a Laurent series is the only possible expansion; there is no Taylor series.

As an illustration consider the function of Eq. (5-63) again, but this time taking $s_0 = 1$, a singular point. In this case expansion will be in positive and negative powers of $s - 1$. Accordingly, it is convenient to write

$$\frac{1}{(s-1)(s-2)} = -\frac{1}{s-1}\left[\frac{1}{1-(s-1)}\right]$$

and then to expand the bracketed term in the series which we know, from Eq. (5-56a), as follows:

$$\frac{1}{(s-1)(s-2)} = -\frac{1}{s-1}[1 + (s-1) + (s-1)^2 + \cdots]$$
$$0 < |s-1| < 1$$

Thus, if we define

$$f_4(s) = -\left[\frac{1}{s-1} + 1 + (s-1) + (s-1)^2 + \cdots\right]$$
$$0 < |s-1| < 1 \qquad (5\text{-}65)$$

it is concluded that

$$f(s) = f_4(s) \qquad 0 < |s - 1| < 1 \qquad (5\text{-}66)$$

In a similar way we get a series for $|s - 1| > 1$ by writing

$$\frac{1}{(s-1)(s-2)} = \frac{1}{s-1}\left[\frac{1}{s-1} + \frac{1}{(s-1)^2} + \frac{1}{(s-1)^3} + \cdots\right]$$
$$|s - 1| > 1$$

or

$$f_5(s) = \frac{1}{(s-1)^2} + \frac{1}{(s-1)^3} + \frac{1}{(s-1)^4} + \cdots \qquad |s - 1| > 1 \qquad (5\text{-}67)$$

and

$$f(s) = f_5(s) \qquad |s - 1| > 1 \qquad (5\text{-}68)$$

Thus, in addition to Eqs. (5-63) and (5-64), we have a third way to specify $f(s)$, namely,

$$f(s) = \begin{cases} f_4(s) & 0 < |s - 1| < 1 \\ f_5(s) & |s - 1| > 1 \end{cases} \qquad (5\text{-}69)$$

where $f_4(s)$ and $f_5(s)$ are specified by the above series. Compare the regions of convergence of $f_4(s)$ and $f_5(s)$ as shown in Fig. 5-8 with the regions of Fig. 5-7. From these two sets of expansions of the same function you should get some feeling for the variations possible with different choices of s_0. Another set corresponding to f_4 and f_5 could be obtained using $s_0 = 2$. You can work this case out for yourself.

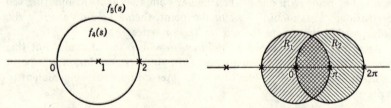

FIG. 5-8. Regions of convergence of expansions of $1/(s - 1)(s - 2)$ about the singular point $s_0 = 1$.

FIG. 5-9. Two regions of convergence for Laurent expansion of $1/\sin s$.

These examples have been relatively simple rational functions having a finite number of singularities. The singularities have been points where the functions become infinite. We shall now consider two examples of transcendental functions.

First look at

$$f(s) = \frac{1}{\sin s} \qquad (5\text{-}70)$$

which is singular at $\pm$ integral multiples of π, as shown in Fig. 5-9. There is an infinite number of Laurent expansions of this function, two of which will be obtained, for regions R_1 and R_2. First consider region R_1.

The integral formulas for the coefficients are difficult to evaluate, and so we shall take the reciprocal of the Taylor series for $\sin s$. Thus,

$$f_1(s) = \cfrac{1}{s - \cfrac{1}{3!}s^3 + \cfrac{1}{5!}s^5 - \cfrac{1}{7!}s^7 + \cdots}$$

$$= \frac{1}{s} + \frac{1}{3!}s + \left(\frac{1}{3!3!} - \frac{1}{5!}\right)s^3 + \left(\frac{1}{3!3!3!} - \frac{2}{3!5!} + \frac{1}{7!}\right)s^5 + \cdots$$

$$(5\text{-}71)$$

converges if $0 < |s| < \pi$. For region R_2 we use the fact that, except for change in sign, the Taylor series about $s = \pi$ has the same coefficients as the Taylor series about $s = 0$. Therefore, we can immediately write

$$f_2(s) = \frac{-1}{s - \pi} - \frac{1}{3!}(s - \pi) - \left(\frac{1}{3!3!} - \frac{1}{5!}\right)(s - \pi)^3$$

$$- \left(\frac{1}{3!3!3!} - \frac{2}{3!5!} + \frac{1}{7!}\right)(s - \pi)^5 \quad (5\text{-}72)$$

which converges for $0 < |s - \pi| < \pi$. This can be done for an infinite number of regions. For the two regions considered

$$\frac{1}{\sin s} = \begin{cases} f_1(s) & 0 < |s| < \pi \\ f_2(s) & 0 < |s - \pi| < \pi \end{cases} \quad (5\text{-}73)$$

As another example take

$$f(s) = \sin\frac{1}{s} \quad (5\text{-}74)$$

which is singular only at $s = 0$. The Laurent series is easily obtained by writing the Taylor series in $1/s$, giving

$$\sin\frac{1}{s} = \frac{1}{s} - \frac{1}{3!}\frac{1}{s^3} + \frac{1}{5!}\frac{1}{s^5} + \cdots \qquad |s| > 0 \quad (5\text{-}75)$$

In each of the cases presented so far, there was only one singular point encircled by the ring of convergence. The singular points in these examples are isolated. Now consider the function

$$f(s) = \frac{1}{\sin(1/s)}$$

which is singular at each point for which

$$s = \frac{1}{n\pi} \qquad n \text{ integer}$$

The point $s = 0$ is not an isolated singularity, as evidenced by the fact that any neighborhood of $s = 0$ must contain an infinite number of singular points. A Laurent expansion of this function about $s = 0$ is possible in any region described by

$$\frac{1}{(n + 1)\pi} < |s| < \frac{1}{n\pi}$$

but there is no region extending down to the singular point $s = 0$ in which a Laurent expansion exists.

5-11. Poles and Essential Singularities; Residues. In the previous section you were shown various examples of Laurent expansions about singular points. In general, if there is more than one singular point encircled by the ring of convergence, there are always an infinite number of negative-power terms. All cases are the same in this respect. However, if we confine our attention to *isolated singularities* and consider the specific Laurent expansion for which the singularity in question is the only one encircled by the ring of convergence, we find a significant difference between examples like Eqs. (5-65), (5-71), and (5-72), on the one hand, and Eq. (5-75). Examples of the first type have a finite number of negative-power terms (one, in these cases), whereas in the last example there are an infinite number of negative-power terms.

In order to discuss this more precisely, let us write a general case,

$$f(s) = f_a(s) + f_b(s) \qquad 0 < |s - s_0| < R_1 \qquad (5\text{-}76)$$

where

$$f_a(s) = \sum_{k=0}^{\infty} a_k(s - s_0)^k \qquad |s - s_0| < R_1$$

$$f_b(s) = \sum_{k=1}^{\infty} \frac{a_{-k}}{(s - s_0)^k} \qquad 0 < |s - s_0| \qquad (5\text{-}77)$$

Keep in mind that this is the expansion for which s_0 is the only singular point encircled by the ring of convergence. The difference between the various examples described in the previous paragraph is in terms of the function $f_b(s)$. Because of its importance in characterizing a function, $f_b(s)$ is called the *principal part* of $f(s)$ at the singularity s_0. A function will have a different principal part at each of its isolated singular points.

We can now define two classes of isolated singular points, as follows:

1. If the principal part has a finite number of terms, of highest negative power n, then s_0 is said to be a *pole of order n*.

2. If the principal part has an infinite number of terms, then s_0 is said to be an *essential singularity* of the *first kind*.

Careful attention should be given to the fact that the "principal-part" designation is used only for the particular Laurent expansion which is about a singular point and for which there is only one singular point encircled by the convergence ring. Other Laurent expansions have parts which look like $f_b(s)$, but they are not principal parts, and they do not yield information about the nature of the singular points.

If a singularity is not isolated, the above comments do not apply. Such a point is called an *essential singularity* of the *second kind*. It cannot be recognized by looking at the character of a principal part of a Laurent expansion, because such an expansion does not exist. An essential singularity of the second kind is recognized by showing that any neighborhood of the point must include other singular points.

The point at infinity can be a singular point of a function. Furthermore, such a singularity can be classified in the same way. In order to tell whether or not a function is singular at infinity, and to identify the nature of such a singularity, we consider the properties of $f(1/s)$ at $s = 0$. $f(s)$ is regular or singular at infinity in accordance with whether $f(1/s)$ is regular or singular at zero. If $f(1/s)$ has a pole of order n (or essential singularity) at $s = 0$, then $f(s)$ has, respectively, a pole of order n (or essential singularity) at infinity.

These remarks all apply only to functions which are single-valued in the neighborhood of the singular point in question. This fact was implied in Chap. 2 by the statement that the word *function* would imply a single-valued function. However, it is well to stress the importance of single-valuedness in considering classifications of singularities. The above classification exhausts the possibilities for single-valued functions; but when the concept of a function is extended to include the multivalued functions, we find a new class of singular points.

If $f(s)$ has a pole of order n, in its closest ring of convergence

$$f(s) = \frac{a_{-n}}{(s - s_0)^n} + \frac{a_{-n+1}}{(s - s_0)^{n-1}} + \cdots$$
$$+ \frac{a_{-1}}{s - s_0} + a_0 + a_1(s - s_0) + \cdots$$

and $$(s - s_0)^n f(s) = a_{-n} + a_{-n+1}(s - s_0) + \cdots$$
$$+ a_{-1}(s - s_0)^{n-1} + a_0(s - s_0)^n + \cdots \quad (5\text{-}78)$$

Thus, $(s - s_0)^n f(s)$ is described by a Taylor series and has no singularity at s_0. An essential singularity cannot be dealt with in this way, a fact which lends meaning to the word *essential*. Multiplying $f(s)$ by $(s - s_0)^n$ is a useful way to check whether a singular point is a pole or an essential singularity. The lowest integer n, such that $(s - s_0)^n f(s)$ has a limit as s approaches s_0, is the order of the pole; and if no such value of n exists, the point s_0 is an essential singularity.

Turning to the examples of the preceding section, $1/(s - 1)(s - 2)$ is seen to have a pole of order 1 at $s = 1$ because there is only one term to the principal part of Eq. (5-65). Also, if the function is multiplied by $s - 1$, the pole is removed. When a function has a rational algebraic factor in the denominator, we see that the resulting singularity is a pole; the factor can be canceled by putting a like factor in the numerator. However, transcendental functions can have poles too, as in the case of $1/\sin s$. Its Laurent series about the origin is given by Eq. (5-71) and shows a principal part consisting of one term. The singularity is a pole of order 1. We can confirm this by recalling that $s/\sin s$ approaches 1 as s approaches zero. Equation (5-75) provides an example of an essential singularity; the principal part has an infinite number of terms.

Now consider the formulas for the coefficients given by Eq. (5-55),

$$a_k = \frac{1}{2\pi j} \int_C \frac{f(s)}{(s - s_0)^{k+1}} \, ds$$

where C encircles s_0 (and no other singularity) in a counterclockwise direction. If $k = -1$, we get

$$a_{-1} = \frac{1}{2\pi j} \int_C f(s) \, ds \tag{5-79}$$

which places particular emphasis on a_{-1}, showing that, if we integrate around one singular point, the result is $2\pi j a_{-1}$. The important idea here is that no other coefficient in the series contributes to this integral. For this reason a_{-1} is called the *residue* at singularity s_0.

Others of the negative-order coefficients can be described in the guise of residues. For example:

a_{-2} is the residue of $(s - s_0)f(s)$
a_{-3} is the residue of $(s - s_0)^2 f(s)$
. . . .

as may be seen by looking at the expansion or at the integral formula for a_k.

Usually we wish to find the integral like Eq. (5-79) by knowing the residue; this is the essential idea of the "calculus of residues." Thus, we are interested in means *independent of the integral formula* to find a_{-1}. We have seen some methods in the last section for finding the Laurent series without using the integral formulas for the coefficients. Equation (5-78) provides a method which will always work for finding the residue at a *pole* of order n. As a Taylor series, its coefficients are given by the customary derivative formulas. Thus, a_{-1} is the coefficient of the $n - 1$ degree term in the expansion of $(s - s_0)^n f(s)$, and therefore, at an nth-order pole,

$$a_{-1} = \frac{1}{(n-1)!} \frac{d^{n-1}}{ds^{n-1}} (s - s_0)^n f(s) \bigg|_{s = s_0} \qquad (5\text{-}80)*$$

In a like manner, for the general coefficient a_k, we get

$$a_k = \frac{1}{(n+k)!} \frac{d^{n+k}}{ds^{n+k}} (s - s_0)^n f(s) \bigg|_{s = s_0} \qquad (5\text{-}81)$$

Formulas (5-80) and (5-81) are valid only at a pole; they do not apply if s_0 is an essential singularity.

5-12. Residue Theorem. A fundamental problem in the analysis of linear systems involves the evaluation of a contour integral over a closed path. We have already found some ways to do this for particular cases. Equation (5-79) provides a general procedure.

Let $f(s)$ be an analytic function having isolated singular points. Also, let C be a simple closed curve inside of which there are n singular points, as shown in Fig. 5-10. According to Sec. 4-7, the integral of

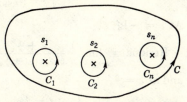

FIG. 5-10. Contour around a finite number of singular points replaced by contours around the individual singularities.

$f(s)$ around C can be written as the sum of n integrals, as follows:

$$\int_C f(s)\ ds = \int_{C_1} f(s)\ ds + \int_{C_2} f(s)\ ds + \cdots + \int_{C_n} f(s)\ ds \qquad (5\text{-}82)$$

Equation (5-79) applies at each singularity. In order to simplify notation, let the symbol d_k denote the residue at the kth singularity. Then, from Eq. (5-79)

$$\int_{C_k} f(s)\ ds = 2\pi j d_k$$

and Eq. (5-82) becomes

$$\int_C f(s)\ ds = 2\pi j \sum_{k=1}^{n} d_k \qquad (5\text{-}83)$$

---

* In the case of a simple pole, if $f(s)$ can be written as a ratio $p(s)/q(s)$, this equation can be written in an alternative form. We note that $q(s)$ has a simple zero and therefore can be written as the series

$$q(s) = (s - s_0) \left[q'(s_0) + \frac{q''(s_0)}{2} (s - s_0) + \cdots \right]$$

When this is substituted into Eq. (5-80), we get

$$a_{-1} = \frac{p(s_0)}{q'(s_0)}$$

This result is known as the *residue theorem*. It is an important theorem because of the ease with which residues can be found at poles. Residues at essential singularities are not so easily found, but often the Laurent series can be obtained by the method used to get Eq. (5-75).

As an application of the residue theorem, we shall find the integral

$$\int_C \frac{1/s}{\cos \pi(s - \frac{1}{4})} \, ds$$

where C is a counterclockwise simple closed curve enclosing the singular points $-\frac{1}{4}$, 0, $\frac{3}{4}$. The residue at the singularity at 0 is obtained from Eq. (5-78) as

$$\text{Residue} = \sqrt{\tilde{2}}$$

At $s = -\frac{1}{4}$

$$\text{Residue} = \lim_{s \to -\frac{1}{4}} \frac{1}{s} \frac{s + \frac{1}{4}}{\cos \pi(s - \frac{1}{4})} = \frac{1/s}{-\pi \sin \pi(s - \frac{1}{4})} \Big|_{s = -\frac{1}{4}} = -\frac{4}{\pi}$$

and at $s = \frac{3}{4}$

$$\text{Residue} = \lim_{s \to \frac{3}{4}} \frac{1}{s} \frac{s - \frac{3}{4}}{\cos \pi(s - \frac{1}{4})} = \frac{1/s}{-\pi \sin \pi(s - \frac{1}{4})} \Big|_{s = \frac{3}{4}} = -\frac{4}{3\pi}$$

For the last two cases we have used the Lhopital rule to evaluate an indeterminate form (see Prob. 5-48 for justification). The residue theorem gives the result

$$\int_C \frac{1/s}{\cos \pi(s - \frac{1}{4})} \, ds = j(\pi \sqrt{8} - 8 - \frac{8}{3})$$

If a function has a finite number of singular points and is singular at infinity, it is sometimes convenient to define a "residue at infinity." This residue is defined in such a way that the residue theorem can be generalized to apply to the Riemann sphere. On the Riemann sphere (see Fig. 3-17) a closed curve C has no outside; it has two "insides," one of which includes the point at infinity. If integration around C is in a positive sense with respect to one or more finite-plane poles which would be enclosed by C in the plane, then this integration encircles the point at infinity, on the Riemann sphere, in the negative sense. If the point at infinity is the only singular point so encircled, an extension of the residue theorem to this case would therefore yield

$$\int_C f(s) \, ds = -2\pi j[\text{residue at infinity}]$$

However, the above integral is $2\pi j$ times the sum of the finite-plane residues, leading to the *definition*

$$\text{Residue at infinity} = - \text{[sum of finite-plane residues]}$$

We have seen that the integral around a circular path enclosing a pole is independent of the radius of the path. If the path is a nonclosed arc of a circle, the integral along such an arc is in general a function of the radius. However, if a pole is simple, as the radius approaches zero, the integral approaches the product of the angle subtended by the arc multiplied by the residue (see Prob. 5-33).

5-13. Analytic Continuation. In the discussion of power series you were shown that a series like

$$\sum_{k=0}^{\infty} a_{1k}(s - s_1)^k$$

may have a region of convergence; and when it does, this region is a circle centered at s_1. Let the coefficients be such that there is a region of convergence of finite radius R_1. (The symbol R_1 will also be used to denote the region.) The series represents a function within its region of convergence; thus, we can define

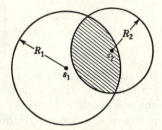

$$f_1(s) = \sum_{k=0}^{\infty} a_{1k}(s - s_1)^k \qquad |s - s_1| < R_1$$

This specific function is not defined on the circle $|s - s_1| = R_1$ or in the region external to R_1.

FIG. 5-11. Overlapping regions of convergence of two Taylor expansions.

Now look at $f_1(s)$ at some point such as s_2, inside but near the periphery of the circle of convergence. From Sec. 5-6 we know that $f_1(s)$ possesses all derivatives at s_2, and so we can write the new series

$$\sum_{k=0}^{\infty} \frac{1}{k!} \left. \frac{d^k f_1(s)}{ds^k} \right|_{s=s_2} (s - s_2)^k = \sum_{k=0}^{\infty} a_{2k}(s - s_2)^k$$

which will converge at least in a circle lying entirely within R_1. However, at the moment we have no reason to believe that the second series may not converge in a larger circle R_2 as shown in Fig. 5-11. The radius R_2 is controlled by coefficients which can be found, and so we see that R_2 is determined by the given function $f_1(s)$. We proceed, assuming that R_2 is known and large enough to carry region R_2 outside of region R_1.

In the second circle we have a second function

$$f_2(s) = \sum_{k=0}^{\infty} a_{2k}(s - s_2)^k \qquad |s - s_2| < R_2$$

Since the above is the Taylor expansion of $f_1(s)$, it must converge to $f_1(s)$ wherever $f_1(s)$ is defined and so in the shaded region

$$f_1(s) = f_2(s)$$

Thus, for the case assumed, $f_1(s)$ completely determines a function which is identical to it in the original region but which extends into a new region. Now suppose that this is done again, to obtain a succession of circles of convergence of a sequence of functions: $f_1(s)$, $f_2(s)$, $f_3(s)$, etc. In all but very exceptional cases an infinite sequence of such functions is possible, so that ultimately the finite plane is covered by overlapping circles.

Supposing this to be the case, we now define a function $f(s)$, where s covers the finite plane, by the sequence of formulas

$$f(s) = \begin{cases} f_1(s) & |s - s_1| < R_1 \\ f_2(s) & |s - s_2| < R_2 \\ f_3(s) & |s - s_3| < R_3 \\ \cdots\cdots\cdots\cdots \end{cases} \tag{5-84}$$

Each of the functions $f_1(s)$, $f_2(s)$, etc., is called an *element* of the analytic function $f(s)$. The sequence of elements which defines a function by analytic continuation is not unique. For example, $f_2(s)$ depends on the choice of s_2, which is arbitrary within the circle of convergence of $f_1(s)$, and so on.

We now envision the possibility that, starting with the same function $f_1(s)$, we might proceed in the same manner as above, but choosing a new set of points s_2', s_3', etc., for the centers of circular regions of convergence of a sequence of elements $g_2(s)$, $g_3(s)$, etc. Then we would obtain an analytic function $g(s)$ defined by

$$g(s) = \begin{cases} f_1(s) & |s - s_1| < R_1 \\ g_2(s) & |s - s_2'| < R_2' \\ g_3(s) & |s - s_3'| < R_3' \\ \cdots\cdots\cdots\cdots \end{cases} \tag{5-85}$$

Now suppose that there is a region G which is common to the two regions throughout which $f(s)$ and $g(s)$ are defined by the two processes of continuation just described. Certainly R_1 is part of G, and furthermore, since $f(s) = f_1(s)$ and $g(s) = f_1(s)$ for s in R_1, it follows that

$$f(s) = g(s) \qquad s \text{ in } R_1 \tag{5-86}$$

We proceed to show that $f(s)$ and $g(s)$ must be identical throughout G.

Since $f(s)$ and $g(s)$ are regular in G, each can be expanded in a Taylor series about some point s_a in R_1. Explicit expressions for these series can be written as follows:

$$f_a(s) = \sum_{k=0}^{\infty} a_{fk}(s - s_a)^k$$

$$g_a(s) = \sum_{k=0}^{\infty} a_{gk}(s - s_a)^k$$

(5-87)

Observe that these two series are elements, respectively, of $f(s)$ and $g(s)$, and not continuations of $f_1(s)$. This fact enables us to say that each series in Eq. (5-87) has a circle of convergence R_a extending to the

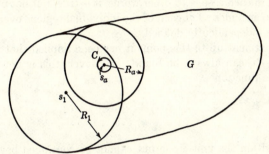

Fig. 5-12. Construction for proof of uniqueness of analytic continuation.

boundary of G, as shown in Fig. 5-12. Equation (5-45) provides the formulas for the coefficients in these series, giving

$$a_{fk} = \frac{1}{2\pi j} \int_C \frac{f(s)}{(s - s_a)^{k+1}} ds \qquad a_{gk} = \frac{1}{2\pi j} \int_C \frac{g(s)}{(s - s_a)^{k+1}} ds$$

where C is a small circle centered at s_a and lying wholly in R_a, as shown in Fig. 5-12. Path C can always be designated, because s_a is not permitted to lie on the boundary of R_a. By choosing C to lie in R_1 we can now use Eq. (5-86) to establish that the above two integrals are equal and hence that $a_{fk} = a_{gk}$. Therefore,

$$f_a(s) = g_a(s) \qquad s \text{ in } R_a$$

Also, each of the series $f_a(s)$ and $g_a(s)$ must be identical, within its circle of convergence, with the function from which it is derived. Therefore, we conclude that

$$f(s) = g(s) \qquad s \text{ in } R_a$$

But R_a extends outside of R_1, and thus it is seen that $f(s)$ and $g(s)$ are equal at least in part of G. This process can be repeated, each time getting a new circle which includes more of G than the previous one. All of G can be covered in this way, and thus we conclude that

$$f(s) = g(s) \qquad s \text{ in } G$$

This is an important result, for it shows that if a function is defined and *regular* at a point it is then uniquely determined in any region into which it can be analytically continued.

This brief treatment leaves many questions unanswered. In particular, in this proof we have shown that $f(s) = g(s)$ in G only if such a region G exists; and we did not prove the stronger property that continuation of $f_1(s)$ by either of the routes leading to $f(s)$ or $g(s)$ must yield the same region in each case. In other words, it is true, but not proved here, that region G includes all of each of the individual regions over which $f(s)$ and $g(s)$ are independently defined.

In the treatment up to this point it has been implied that a region G, larger than R_1, can always be found. However, this is not true. For example, the function

$$f_1(s) = \sum_{k=0}^{\infty} s^{k!} \tag{5-88}$$

converges within the unit circle but cannot be continued beyond. The unit circle is said to be a *natural boundary* of this function. Such examples as this are a curiosity so far as we are concerned.

When the radius of convergence of $f_1(s)$ is finite, there will always be at least one point on its circle of convergence which cannot be covered by any circle. Such an exceptional point will be on the boundaries of other circles, but never inside a circle. In Fig. 5-13 s_a is such a point. In attempting to approach s_a, the boundary of each region of convergence will pass through s_a. For the function

FIG. 5-13. Boundaries of convergence circles of elements of $f(s)$ passing through a singular point s_a.

$f(s)$ to be defined at a point, that point must lie inside the region of convergence of *at least one of the elements*. Since s_a lies in no such region, the function is not defined there and so s_a is a singular point.

You may have wondered, in the process of developing the sequence of elements, what would determine each radius of convergence. We now have the answer: each circle extends to a singular point of $f(s)$. Note that this is a singularity of $f(s)$, not of the element. This must be said because an element is not defined anywhere on the boundary of its region of convergence, and therefore no particular point on the boundary can be singled out as a singularity of the element.

To summarize these ideas, we now view an analytic function in a slightly different light from the original one, and, in fact, we extend the definition to include a new concept. Originally, in Chap. 2, an analytic function is defined as a function having at least one regular point. Now we see that regularity at one point is sufficient to give a Taylor expansion at that point and from that an analytic continuation to the finite plane, or to a natural boundary. We see that an analytic function can be defined as the collection of all its elements. *Only in some cases can it be written as a single formula.*

For a specific example let us consider the analytic continuation of

$$f_1(s) = 1 + s + s^2 + \cdots \qquad |s| < 1$$

From previous discussions we suspect that the continuation can be written

$$f(s) = \frac{1}{1 - s}$$

which is singular at $s = 1$. However, from the viewpoint of analytic continuation the location of this singularity would not be known. We use knowledge of it only to save the very considerable trouble of finding a sequence of elements and their respective radii of convergence. In a true analytic continuation process we would start only with the fact that the radius of convergence of $f_1(s)$ is 1. The start of one set of elements is shown in Fig. 5-14. There is at least an intuitive feeling that this system of circles can be continued to cover the finite s plane and that the point $s = 1$ is the only finite point excluded.

We can then write $f(s)$ in either of the forms

$$f(s) = \begin{cases} f_1(s) & |s| < R_1 = 1 \\ f_2(s) & |s - s_2| < R_2 \\ \cdots\cdots\cdots\cdots\cdots \end{cases} \tag{5-89a}$$

or

$$f(s) = \frac{1}{1 - s} \tag{5-89b}$$

Both representations express identical functional relationships. A formula like Eq. (5-89b) is called the *global definition*.

In this single example of analytic continuation we did not actually find a sequence of elements, intimating that to do so would be too much of a chore. The implication is that it would always be too difficult to accomplish, but this does not detract from the usefulness of the concept. Later on, particularly in Chap. 10, this will be an important conceptual tool.

Throughout the discussion of analytic continuation, we dealt only with the finite plane. In the Riemann-sphere interpretation this excludes only the point at infinity. The idea of analytic continuation would be somewhat more complete if the point at infinity could be included. This can easily be done by looking at $f(1/s)$ at the origin. If $f(1/s)$ can be continued to include the origin, then we say that $f(s)$ can be analytically continued to infinity.

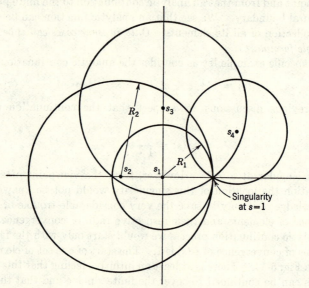

FIG. 5-14. Regions of convergence of a possible sequence of elements of the function $f(s) = 1/(1 - s)$.

5-14. Classification of Single-valued Functions. We are gradually developing the idea that singularities are important characteristics of a function. This can be carried a bit further by introducing some terminology whereby functions are classified according to kinds, numbers, and locations of singularities, as follows:

1. *Analytic.* An analytic function can now be defined more completely than in Chap. 2, as an element plus its analytic continuation. Now we see that a function like $f(s) = s$ for $|s| < 1$ and 0 for $1 < |s|$ is not analytic although it is regular everywhere except on the unit circle. Analytic, continuations of this function are possible across the unit circle, but they do not yield the function specified.

2. *Meromorphic.* A meromorphic function is an analytic function which has no essential singularities in the finite plane. It can have an essential singularity at infinity. The number of finite-plane poles can be infinite. An example is tan s.

3. *Rational.* A rational function belongs to a subclass of the meromorphic functions in which there are a finite number of finite-plane poles and at most a pole at infinity. A rational function can be expressed as a ratio of polynomials.

4. *Entire.* A function having no singularities in the finite plane is called an entire function. Unless it is a constant, it will have at least a pole at infinity, or it may have an essential singularity at infinity. Examples are: for a pole at infinity, a polynomial; for an essential singularity at infinity, sin s.

This classification can be tabulated as follows:

	Analytic (1)	Meromorphic (2)	Rational (3)	Entire (4)
Finite-plane singularities..	Any number, poles or essential	Any number, poles only	Finite number, poles only	None
Singularity at infinity.....	Pole or essential	Pole or essential	Pole only	Pole or essential

5-15. Partial-fraction Expansion. Suppose that a function $f(s)$ has n singular points at $s_1, s_2, \ldots, s_n$. The point at infinity may also be singular. We use the same starting point as in Sec. 5-8, namely, the Cauchy integral formula in the form

$$f(s) = \frac{1}{2\pi j} \left[\int_C \frac{f(z)}{z - s}\, dz - \sum_{v=1}^{n} \int_{C_v'} \frac{f(z)}{z - s}\, dz \right] \tag{5-90}$$

where s is in the region shown in Fig. 5-15, which is made simply connected by the barriers shown. Circle C is centered at the origin and has a radius large enough to enclose all singular points.

The integral

$$\frac{1}{2\pi j} \int_C \frac{f(z)}{z - s}\, dz$$

is treated by the same process used in Sec. 5-8 to arrive at Eq. (5-49). In the present case the former s_0 and C_2 are replaced, respectively, by 0 and C, giving

$$\frac{1}{2\pi j} \int_C \frac{f(z)}{z - s}\, dz = \frac{1}{2\pi j} \sum_{k=0}^{\infty} (s)^k \int_C \frac{f(z)}{z^{k+1}}\, dz$$

$$= \sum_{k=0}^{\infty} A_k s^k$$

$$= r(s) \tag{5-91}$$

Circle C can have an arbitrarily large radius, since all singularities except infinity are inside it, and so the power series for $r(s)$ has an infinite radius of convergence. Thus, $r(s)$ is an entire function.

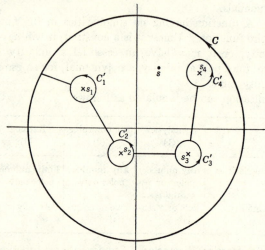

FIG. 5-15. Integration contours used in derivation of partial-fraction expansion.

Now consider the sum of integrals in the last term on the right of Eq. (5-90). The integral around C_1' is a typical case. Comparing with Eqs. (5-50) and (5-77), and recognizing that s_1 is singular, shows that

$$-\frac{1}{2\pi j}\int_{C_{1'}}\frac{f(z)}{z-s}\,dz = \sum_{k=1}^{N}\frac{a_{-k}}{(s-s_1)^k} \tag{5-92}$$

is the principal part of the Laurent expansion. N is the order of the pole; and if s_1 is an essential singularity, N becomes infinite. This term is defined for all values of s for which

$$0 < |s - s_1|$$

because it is a series in negative powers of $s - s_1$. The principal part just described will be called $g_1(s)$. A similar result is obtained at each of the singular points, giving

$$-\frac{1}{2\pi j}\int_{C_{1'}}\frac{f(z)}{z-s}\,dz = g_1(s)$$

$$-\frac{1}{2\pi j}\int_{C_{2'}}\frac{f(z)}{z-s}\,dz = g_2(s)$$

$$\cdots\cdots\cdots\cdots\cdots\cdots\cdots \tag{5-93}$$

$$-\frac{1}{2\pi j}\int_{C_{n'}}\frac{f(z)}{z-s}\,dz = g_n(s)$$

It is important to understand that each principal part arises from an expansion about a *different point*. The results are summarized by substituting Eqs. (5-91) and (5-93) in Eq. (5-90) to give

$$f(s) = g_1(s) + g_2(s) + \cdots + g_n(s) + r(s) \qquad (5\text{-}94)$$

This is another representation for $f(s)$, called the *partial-fraction* expansion. It is valid throughout the plane, and if any of the terms are series, the right-hand side of Eq. (5-94) converges everywhere except at the singular points.

If $f(s)$ is a rational function (ratio of two polynomials), all singularities are poles. Then each principal part consists of a finite number of terms. Also, a rational function has at most a pole at infinity, and the last term on the right of Eq. (5-94) will reduce to a polynomial, or to a constant if there is no pole at infinity.

The partial-fraction representation is particularly important in Laplace transform theory. It is important to know the actual coefficients in the expansion, and so methods of finding them are important. Equation (5-81) can be used to find the coefficients of all the principal parts at poles. For the function $r(s)$, Eq. (5-91) would be the general form. However, it is better to recognize $r(s)$ as a Taylor series for $f(s)$ − [sum of principal parts]. Thus, the principal parts should be found first, so that the entire function $r(s)$ can be obtained by applying the usual derivative formulas for coefficients of a Taylor series. In the case of a rational function, $r(s)$ can be found by using a division algorithm until the remainder is a rational function with numerator lower in degree than the denominator. This process yields $r(s)$ directly.

These two cases are illustrated by the following examples:

Example 1. Expand

$$f(s) = \frac{s^4 + 4s^3 + 4s^2}{s^3 + s^2 - s - 1}$$
$$= \frac{s^4 + 4s^3 + 4s^2}{(s-1)(s+1)^2}$$

in partial fractions.

Performing two steps of division gives

$$f(s) = s + 3 + \frac{2s^2 + 4s + 3}{(s-1)(s+1)^2}$$

The resulting rational function has no entire function in its expansion, because the function approaches zero as s becomes infinite. We need the principal parts of its Laurent expansions at -1 and $+1$. The required coefficients are easily found from Eq. (5-81). At $+1$ there is only one coefficient:

$$\frac{2s^2 + 4s + 3}{(s+1)^2}\bigg|_{s=1} = \frac{9}{4}$$

and so at this pole the principal part is

$$\frac{9/4}{s-1}$$

At -1 there are two terms in the principal part,

$$\frac{a_{-2}}{(s+1)^2} + \frac{a_{-1}}{s+1}$$

and the coefficients are

$$a_{-2} = \left.\frac{2s^2 + 4s + 3}{s-1}\right|_{s-1} = -\frac{1}{2}$$

$$a_{-1} = \left.\frac{d}{ds}\left(\frac{2s^2 + 4s + 3}{s-1}\right)\right|_{s=-1}$$

$$= \left.\frac{(s-1)(4s+4) - (2s^2 + 4s + 3)}{(s-1)^2}\right|_{s=-1} = -\frac{1}{4}$$

The partial-fraction expansion of the function is therefore

$$\frac{s^4 + 4s^3 + 4s^2}{s^3 + s^2 - s - 1} = \frac{9/4}{s-1} - \frac{1/2}{(s+1)^2} - \frac{1/4}{s+1} + 3 + s$$

Example 2.   Expand

$$f(s) = \frac{e^s}{s^3 + s^2 - s - 1} = \frac{e^s}{(s-1)(s+1)^2}$$

in partial fractions.

The principal parts will be obtained first.   At pole $s = 1$ the principal part is

$$\left.\frac{e^s}{(s+1)^2}\right|_{s-1} = \frac{e}{4}$$

At pole $s = -1$ there are two terms having coefficients

$$a_{-2} = \left.\frac{e^s}{s-1}\right|_{s=-1} = -\frac{e^{-1}}{2}$$

$$a_{-1} = \left.\frac{d}{ds}\frac{e^s}{s-1}\right|_{s=-1} = \left.\frac{(s-2)e^s}{(s-1)^2}\right|_{s=-1} = -\frac{3}{4}e^{-1}$$

The sum of the two principal parts is

$$\frac{e/4}{s-1} - \frac{3/4e}{s+1} - \frac{1/2e}{(s+1)^2} = \frac{(e^2-3)s^2 + 2(e^2-1)s + (e^2+5)}{4e(s-1)(s+1)^2}$$

Subtracting from $f(s)$ gives the entire function

$$r(s) = \frac{4e^{s+1} - (e^2-3)s^2 - 2(e^2-1)s - (e^2+5)}{4e(s-1)(s+1)^2}$$

The numerator of the above expression has zeros of appropriate orders at $s = 1$ and -1, and so $r(s)$ has no poles, in the finite plane. The Taylor series for $r(s)$ is not required. Thus

$$\frac{e^s}{(s-1)(s+1)^2} = \frac{1}{4e}\left[\frac{e^2}{s-1} - \frac{3}{s+1} - \frac{2}{(s+1)^2}\right.$$
$$\left. + \frac{4e^{s+1} - (e^2-3)s^2 - 2(e^2-1)s - (e^2+5)}{(s-1)(s+1)^2}\right]$$

is the required partial-fraction expansion. Of course, the last term could be expanded in a Taylor series about $s = 0$.

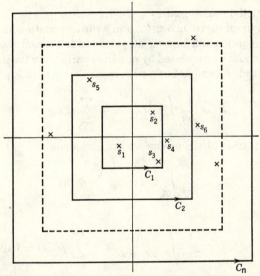

FIG. 5-16. Sequence of contours for derivation of partial-fraction expansion of meromorphic functions.

5-16. Partial-fraction Expansion of Meromorphic Functions (Mittag-Leffler Theorem). The development of Sec. 5-15 does not apply to the general meromorphic functions, because meromorphic functions have an infinite number of poles. However, these poles are isolated, and this fact makes it possible in some cases to get a partial-fraction type of expansion which takes the form of an infinite series.

As a result of being infinite in number and isolated, the sequence of singularities extends to infinity. (This conclusion is based on the Boltzano-Weierstrass theorem, not proved in this text.[*]) As shown in Fig. 5-16, let there be a sequence of square closed contours $C_1, C_2, \ldots,$

[*] E. T. Copson, "An Introduction to the Theory of Functions of a Complex Variable," p. 17, Oxford University Press, New York, 1935.

each one enclosing more singular points than the previous one, and with no singular points on the contours. Also, let $G_n(s)$ denote the sum of the principal parts for all poles between C_{n-1} and C_n. Thus, the sum of principal parts for all poles inside C_n is

$$\sum_{v=1}^{n} G_v(s)$$

It is also assumed that $f(s)$ is regular at $s = 0$ and that there is a constant M such that

$$|f(s)| < M$$

for s on all of the square paths C_1, C_2, ...

Choose a typical curve such as C_n, and write an equation like Eq. (5-90). Here the summation will be over all poles inside C_n, and the integrals in the summation may be replaced by principal parts, as in the previous case. Thus, for s inside C_n and not at a pole, in similarity with Eq. (5-90),

$$f(s) = \frac{1}{2\pi j} \int_{C_n} \frac{f(z)}{z - s} \, dz + \sum_{v=1}^{n} G_v(s) \qquad (5\text{-}95)$$

Since $f(s)$ is regular at the origin, we can also write

$$f(0) = \frac{1}{2\pi j} \int_{C_n} \frac{f(z)}{z} \, dz + \sum_{v=1}^{n} G_v(0) \qquad (5\text{-}96)$$

and therefore

$$f(s) - f(0) = \frac{1}{2\pi j} \int_{C_n} \frac{s}{z} \frac{f(z)}{z - s} \, dz + \sum_{v=1}^{n} [G_v(s) - G_v(0)] \qquad (5\text{-}97)$$

Now allow n to increase. The summation becomes an infinite series, and we are interested to know whether or not it converges. Convergence is investigated by looking at

$$\left| \sum_{v=1}^{n} [G_v(s) - G_v(0)] - [f(s) - f(0)] \right| = \frac{1}{2\pi} \left| \int_{C_n} \frac{s}{z} \frac{f(z)}{z - s} \, dz \right| \qquad (5\text{-}98)$$

Let s be inside the curve designated by $C_{N'}$, and let n be greater than some number $N > N'$. Also, let A_k be half the side of square C_k. From the geometry of Fig. 5-16 it is found that for $n > N$

$$\left| \frac{s}{z} \right| < \sqrt{2} \frac{A_{N'}}{A_n}$$

$$\frac{1}{|z - s|} < \frac{1}{A_N - A_{N'}}$$

and we also know that on C_n

$$|f(s)| < M$$

Thus, if $n > N$

$$\frac{1}{2\pi} \left| \int_{C_n} \frac{s}{z} \frac{f(z)}{z-s} \, dz \right| \leqq \frac{1}{2\pi} \int_{C_n} \left| \frac{s}{z} \right| \frac{|f(z)|}{|z-s|} \, dz < \frac{\sqrt{2} \, M}{2\pi} \frac{A_{N'}(8A_n)}{A_n(A_N - A_{N'})}$$

$$= \frac{8MA_{N'}}{\sqrt{2} \, \pi(A_N - A_{N'})} \tag{5-99}$$

Now note that, if an arbitrary small number ϵ is given, then

$$\frac{8MA_{N'}}{\sqrt{2} \, \pi(A_N - A_{N'})} < \epsilon$$

if

$$A_N > A_{N'} + \frac{8MA_{N'}}{\sqrt{2} \, \pi\epsilon} \tag{5-100}$$

By way of the numbering on the sequence of curves $C_1, C_2, \ldots$, this will establish the number N such that

$$\frac{1}{2\pi} \left| \int_{C_n} \frac{s}{z} \frac{f(z)}{z-s} \, dz \right| < \epsilon$$

when

$$n > N$$

Thus, we have convergence and

$$f(s) = f(0) + \sum_{v=1}^{\infty} [G_v(s) - G_v(0)] \tag{5-101}$$

Furthermore, by relation (5-100), N is a function of N' and ϵ. If s remains within square $C_{N'}$, then one value of N will serve. We conclude that convergence is uniform for s within any square (and consequently within a circle also).

In this development each G function consists of one or more principal parts. Therefore, in the infinite sum there will be no change if we go back to the notation $g_v(s)$ for a single principal part. This is desirable because the groupings to form G_v were arbitrary and necessary only to account for the possibility that we might not always be able to increase the size of the square by an amount to include only one additional pole. It is more appropriate in the final result to revert to the previous notation, giving the *Mittag-Leffler theorem*, which states that $f(s)$ can be represented by

$$f(s) = f(0) + \sum_{v=1}^{\infty} [g_v(s) - g_v(0)] \tag{5-102}$$

if $f(s)$ is uniformly bounded on a set of curves such as shown in Fig. **5-16.**

If the series

$$\sum_{v=1}^{\infty} g_v(s) \qquad \text{and} \qquad \sum_{v=1}^{\infty} g_v(0)$$

converge, then

$$\sum_{v=1}^{\infty} [g_v(s) - g_v(0)] = \sum_{v=1}^{\infty} g_v(s) - \sum_{v=1}^{\infty} g_v(0)$$

and we get the simpler form

$$f(s) = \sum_{v=1}^{\infty} g_v(s) \qquad\qquad (5\text{-}103)$$

by equating constant terms and terms which vary with s.

Equation (5-102) is not valid if $f(s)$ has a pole at $s = 0$. Now suppose that $f(s)$ has such a pole, with principal part $g_0(s)$. Then

$$f_1(s) = f(s) - g_0(s) \qquad\qquad (5\text{-}104)$$

has no pole at the origin and can be expanded in accordance with Eq. (5-102) to give

$$f_1(s) = f_1(0) + \sum_{v=1}^{\infty} [g_v(s) - g_v(0)] \qquad\qquad (5\text{-}105)$$

and

$$f(s) = f_1(0) + g_0(s) + \sum_{v=1}^{\infty} [g_v(s) - g_v(0)] \qquad\qquad (5\text{-}106)$$

We note here that $g_v(s)$ is the same for $f_1(s)$ as for $f(s)$ because subtracting a principal part at $s = 0$ has no effect on principal parts at other poles. Now, if

$$\sum_{v=1}^{\infty} g_v(s) \qquad \text{and} \qquad \sum_{v=1}^{\infty} g_v(0)$$

converge, Eq. (5-105) can be replaced by

$$f_1(s) = \sum_{v=1}^{\infty} g_v(s)$$

and then, referring to Eq. (5-104),

$$f(s) = \sum_{v=0}^{\infty} g_v(s) \qquad\qquad (5\text{-}107)$$

Thus, for the case where the series of principal parts converges, the restriction that there shall be no pole at the origin can be dropped.

Example 3.   Find the partial-fraction expansion of

$$f(s) = \frac{1}{\sin s}$$

This function has first-order poles at $s = \pm n\pi$ for integral values of n. The square contours may then be as shown in Fig. 5-17, and we need to

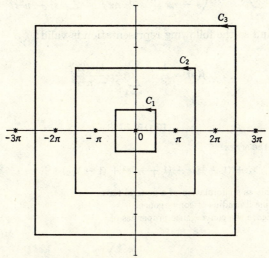

FIG. 5-17. Contours used in obtaining the partial-fraction expansion of $1/\sin s$.

check whether or not $f(s)$ is uniformly bounded on these contours.   This is checked by looking at

$$|\sin s|^2 = |\sin \sigma \cosh \omega + j \cos \sigma \sinh \omega|^2 = \sin^2 \sigma + \sinh^2 \omega$$

On a vertical side $\sigma = \pm(2n + 1)\pi/2$, and on a horizontal side $\omega = \pm(2n + 1)\pi/2$.   Therefore, respectively,

$$|\sin s| \geq \begin{cases} \left| \sin \dfrac{(2n + 1)\pi}{2} \right| = 1 & \text{vertical side} \\[2ex] \left| \sinh \dfrac{(2n + 1)\pi}{2} \right| > \sinh \dfrac{\pi}{2} > 1 & \text{horizontal side} \end{cases}$$

Thus, on each contour shown in Fig. 5-17,

$$|f(s)| \leq 1$$

which establishes that the conditions of the Mittag-Leffler theorem are satisfied.

At each pole the residue is

$$\lim_{s \to n\pi} \frac{s - n\pi}{\sin s} = \frac{1}{\cos n\pi} = (-1)^n$$

It can be shown that

$$\sum_{n=1}^{\infty} (-1)^n \left(\frac{1}{s - n\pi} + \frac{1}{s + n\pi} \right) = \sum_{n=1}^{\infty} \frac{(-1)^n (2s)}{s^2 - n^2\pi^2}$$

converges, and so the following representation is valid:

$$f(s) = \frac{1}{s} + \sum_{n=1}^{\infty} \frac{(-1)^n (2s)}{s^2 - n^2\pi^2} \tag{5-108}$$

PROBLEMS

5-1. Given the series

$$\tfrac{3}{2} + (1 - \tfrac{1}{4})s + (1 + \tfrac{1}{8})s^2 + (1 - \tfrac{1}{16})s^3 + \cdots$$

(a) Write this as a summation of a general term.
(b) Determine its radius of convergence.

5-2. Investigate the convergence properties of:

(a) $\displaystyle\sum_{n=1}^{\infty} ns^n$ (b) $\displaystyle\sum_{n=0}^{\infty} 2^n s^n$

(c) $\displaystyle\sum_{n=0}^{\infty} n! s^n$ (d) $\displaystyle\sum_{n=0}^{\infty} n^{\sqrt{n}} s^n$

5-3. Investigate the convergence properties of:

(a) $\displaystyle\sum_{n=2}^{\infty} \frac{s^n}{n^2 - 1}$ (b) $\displaystyle\sum_{n=1}^{\infty} \frac{n^2}{2^n} (s - 3)^n$

(c) $\displaystyle\sum_{n=2}^{\infty} \frac{s^n}{(\log n)^n}$ (d) $\displaystyle\sum_{n=0}^{\infty} \frac{n!}{2^n} s^n$

(e) $\displaystyle\sum_{n=1}^{\infty} \frac{n!}{n^n} s^n$

5-4. Determine the regions of convergence and uniform convergence of the series:

(a) $\displaystyle\sum_{n=0}^{\infty} e^{ns}$ (b) $\displaystyle\sum_{n=0}^{\infty} \left(\frac{s - 1}{s + 1} \right)^n$

5-5. If each of the series of constants

$$\sum_{k=0}^{\infty} a_k \qquad \sum_{k=0}^{\infty} b_k \qquad \sum_{k=0}^{\infty} c_k$$

converges, and if $c_k = a_k + b_k$, prove that

$$\sum_{k=0}^{\infty} c_k = \sum_{k=0}^{\infty} a_k + \sum_{k=0}^{\infty} b_k$$

5-6. Referring to the notation in Prob. 5-5, assume that

$$\sum_{k=0}^{\infty} c_k$$

converges. Show, by a counterexample, that it is not necessarily true that

$$\sum_{k=0}^{\infty} a_k \qquad \text{and} \qquad \sum_{k=0}^{\infty} b_k$$

converge.

5-7. Show that the series in Eq. (5-88) converges for $|s| < 1$.

5-8. Prove that, if two series

$$\sum_{k=0}^{\infty} g_k(s) \qquad \text{and} \qquad \sum_{k=0}^{\infty} h_k(s)$$

converge uniformly in respective regions R_1 and R_2, then the series

$$\sum_{k=0}^{\infty} g_k(s) + h_k(s)$$

converges uniformly in the region common to R_1 and R_2 (the intersection of R_1 and R_2).

5-9. Given the real series

$$x - \frac{x^2}{2} + \frac{x^3}{3} - \frac{x^4}{4} + \cdots$$

prove that this series converges for $-1 < x \leq 1$ and converges uniformly for $-1 < x' \leq x \leq x'' < 1$.

5-10. Starting with the power series for $1/(1 + s)$, obtain the binomial expansion

$$(1 + s)^{-n} = \sum_{k=0}^{\infty} \frac{(n + k - 1)!}{(n - 1)!k!} (-s)^k \qquad \text{integer } n > 0$$

(HINT: Consider term-by-term differentiation.)

5-11. By following the pattern

$$\int_0^s (1 + z)\, dz = \frac{(1 + s)^2}{2} - \frac{1}{2} \qquad \int_0^s \left[\frac{(1 + z)^2}{2} - \frac{1}{2} \right] dz = \frac{(1 + s)^3}{3!} - \frac{s}{2} - \frac{1}{3!}$$

etc., derive the binomial formula (for integer $n > 0$)

$$(1 + s)^n = \sum_{k=0}^n \frac{n!}{k!(n - k)!}\, s^k$$

5-12. Starting with the power series for $1/(1 + s)$, obtain a series for $\log (1 + s)$. Justify your steps, and state the region of convergence of the new series, explaining how this region of convergence was determined.

5-13. Given the series

$$\cos 2s = 1 - \frac{2^2 s^2}{2!} + \frac{2^4 s^4}{4!} + \cdots$$

obtain series expansions for

(a) $\sin^2 s$ (b) $\cos^2 s$ (c) $\sin s \cos s$

(HINT: Consider term-by-term differentiation.)

5-14. Given the power series

$$\sin s = s - \frac{s^3}{3!} + \frac{s^5}{5!} \cdots$$

$$\cos s = 1 - \frac{s^2}{2!} + \frac{s^4}{4!} \cdots$$

(a) Determine the radius of convergence of each series.

(b) Starting with these series, and without squaring the cosine series or without using the derivative formulas for the Taylor coefficients, obtain three terms of the power series in s for

$$\frac{1}{\cos^2 s}$$

and find its radius of convergence by any justifiable method.

5-15. Do Prob. 5-14 by evaluating the first three coefficients of the Taylor series.

5-16. By any method obtain series expansions, in powers of $s + 1$, for the following:

(a) $\dfrac{1}{1 - e^s}$ (b) $\dfrac{1}{1 + s^2}$

(c) $\dfrac{1}{s^2}$ (d) $\dfrac{1}{4s - s^2}$

In each case specify the radius of convergence.

5-17. Use the Cauchy integral formula for the nth derivative of a function $f(s)$ to prove that for a Taylor series

$$f(s) = \sum_{n=0}^{\infty} a_n (s - s_1)^n$$

the coefficients obey the inequality

$$|a_n| \leq \frac{M_r}{r^n}$$

where $M_r = \max |f(s)|$ on $|s - s_1| = r < $ [radius of convergence].

5-18. Given a polynomial

$$f(s) = \sum_0^N a_n s^n$$

Use appropriate ideas relating to series to show that it can be written

$$f(s) = \sum_{k=0}^N A_k (s - 1)^k$$

where

$$A_k = \sum_{n=k}^N \frac{n! a_n}{k!(n - k)!}$$

5-19. Given a Laurent series

$$\sum_{k=-\infty}^{\infty} a_k s^k$$

which converges for $R_1 < |s| < R_2$. Determine its region of uniform convergence.

5-20. For the function

$$\frac{s^2 + s + 3}{s^3 + 2s^2 + s + 2}$$

obtain the following expansions, and in each case establish the region of convergence:

(a) Taylor expansion about $s = 0$
(b) Taylor expansion about $s = -1$
(c) Laurent expansions (two of them) about $s = 0$
(d) Laurent expansion about each singular point

5-21. Carry out the tasks specified in Prob. 5-20, but for the reciprocal of the function specified in that problem.

5-22. You are given the function

$$f(s) = \frac{1}{(2s + 1)(s - 1)^2}$$

For each of the following cases specify the region (or regions, if more than one series is possible) of Taylor and/or Laurent expansions. The cases are for expansions about the following points:

(a) $s = 0$ (b) $s = 1$
(c) $s = -\frac{1}{2}$ (d) $s = -2$

5-23. The function

$$\frac{1}{\sin (1/s)}$$

is singular at the origin. Show that the Laurent expansion about the origin for this function has zero radius of convergence (i.e., such an expansion does not exist).

5-24. Obtain the appropriate series expansion for

$$f(s) = \sin \frac{\pi}{s - 1}$$

(a) Expanded about $s = 0$ (b) Expanded about $s = 1$
(c) Expanded about $s = 2$

5-25. For each of the following functions, locate the singular points, and identify whether they are poles (and of what order) or essential singularities (and of what kind):

(a) $\dfrac{e^s}{s}$

(b) $e^{1/s}$

(c) $e^{-1/s}$

(d) $\dfrac{s^2}{(s^2 + 1)^2}$

(e) $\sin s$

(f) $\dfrac{1}{\sinh s}$

5-26. For each of the following specify whether the function is regular or singular at infinity, and if it is singular, specify whether it is a pole or an essential singularity and, if the latter, what kind:

(a) $\dfrac{e^s}{1 - \sin s}$

(b) $\dfrac{s^3 - 2}{s^2 + 2}$

(c) $e^{s^2} - e^s$

(d) $\tan s$

(e) $\dfrac{s^3 - 2s^2 + s}{s^3 + 3s^2}$

(f) $\dfrac{\sin s}{s}$

5-27. Find the residues at the indicated singular points for the following:

(a) $\dfrac{\sin s}{s^3}$ at $s = 0$

(b) $\dfrac{1}{s^3 - s^2}$ at $s = 1$

(c) $\dfrac{1 - e^{-2s}}{s^4}$ at $s = 0$

(d) $\dfrac{\cos s}{\sin^2 s}$ at $s = \pi$

(e) $\dfrac{e^{2s}}{(s - 1)^2}$ at $s = 1$

(f) $\dfrac{\tan s}{(1 - e^s)^2}$ at $s = 0$

5-28. Given the function

$$f(s) = \frac{1}{s \sin s}$$

(a) Locate and classify its singularities.

(b) Evaluate the residues at these singular points.

(c) Evaluate the integral of $f(s)$ in a counterclockwise direction around a circle of radius 5, centered at the origin.

5-29. (a) Use the method of residues to evaluate the integral

$$\int_C \frac{s \, ds}{1 - e^s}$$

where C is the rectangular path shown in Fig. P 5-29.

(b) Let I_0 designate the answer to part a. In terms of I_0, what is the above integral if the contour is changed (1) to C_1 and (2) to C_2?

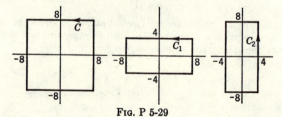

FIG. P 5-29

5-30. Evaluate the integral

$$\int_C \frac{s^3 + 1}{(s - 1)(s^2 + 4)} \, ds$$

where C is a counterclockwise circle centered at 2 and of (a) radius 2; (b) radius 4.

5-31. Let C be the unit circle, with counterclockwise sense of integration. Evaluate each of the following:

(a) $\int_C \frac{e^{-s}}{s^2} \, ds$ (b) $\int_C \frac{ds}{s \sin s}$ (c) $\int_C s e^{1/s} \, ds$

5-32. Evaluate the integral

$$\int_C \frac{ds}{\cosh s}$$

over each of the following counterclockwise paths:

(a) C_a, a unit circle centered at $s = 0$
(b) C_b, a unit circle centered at $s = j$
(c) C_c, a circle of radius 2 centered at $s = 0$

5-33. Let C designate a semicircular arc of radius R, centered at a simple pole s_0 and subtending an angle θ_0. Let A_0 be the residue at the pole. Prove that

$$\lim_{R \to 0} \int_C f(s) \, ds = j\theta_0 A_0$$

5-34. You are given the series

$$\sum_{k=0}^{\infty} \frac{k! - 1}{k!} s^k$$

Determine its radius of convergence, and obtain a "global" representation. Also, obtain a series for the element of this function expended about point $s = -\frac{1}{2}$.

5-35. Obtain a global formula for the function defined by the series

$$\sum_{n=0}^{\infty} (-1)^n e^{ns}$$

5-36. Show that the two series

$$f_1(s) = \sum_{k=0}^{\infty} \frac{2^{k+1} - 3}{(-2)^k} (s - 1)^k$$

and

$$f_2(s) = \sum_{k=0}^{\infty} \left[\left(-\frac{1}{2} \right)^k - 2 \left(-\frac{1}{3} \right)^k \right] (s - 2)^k$$

are elements of the same analytic function. (For a hint, see Prob. 5-10.)

5-37. Obtain the partial-fraction expansion of the function given in Prob. 5-22.

5-38. Obtain the partial-fraction expansion of the function given in Prob. 5-20.

5-39. Obtain the partial-fraction expansion of the reciprocal of the function given in Prob. 5-20.

5-40. Obtain two terms of the Taylor expansion for $r(s)$, in powers of s, for Example 2 in Sec. 5-15.

5-41. Obtain the partial-fraction expansion of

$$\frac{\sin s}{s(s - 1)(s - 2)^2(s - 3)^3}$$

including two terms in the Taylor expansion of $r(s)$, in powers of s.

5-42. Obtain the partial-fraction expansion of

$$f(s) = \frac{\cos s}{(s - \pi/2)(s - \pi)(s - 2\pi)}$$

including two terms of the Taylor expansion of $r(s)$, in powers of s.

5-43. Derive the formula

$$\tan s = \sum_{\substack{n=1 \\ (\text{odd})}}^{\infty} \frac{1}{n^2\pi^2} \frac{8s}{1 - (2s/n\pi)^2}$$

5-44. Derive the formula

$$\cot s = \frac{1}{s} - \sum_{n=1}^{\infty} \frac{1}{n^2\pi^2} \frac{2s}{1 - (s/n\pi)^2}$$

5-45. Obtain the Mittag-Leffler expansion for

$$f(s) = \frac{\sin s}{\cos 2s}$$

5-46. From the formula given in Prob. 5-44 derive the infinite-product representation

$$\sin s = s \prod_{n=1}^{\infty} \left(1 - \frac{s^2}{n^2\pi^2} \right)$$

[HINT: Observe that $\cot s = d(\log \sin s)/ds$.]

5-47. From the formula given in Prob. 5-43 derive the infinite-product representation

$$\cos s = \prod_{\substack{n=1 \\ (\text{odd})}}^{\infty} \left(1 - \frac{4 s^2}{n^2\pi^2} \right)$$

[HINT: Observe that $\tan s = -d(\log \cos s)/ds$.]

5-48. Suppose an analytic function $f(s) = p(s)/q(s)$ has a removable singularity at a point s_0 due to $p(s)$ and $q(s)$, each having a zero of order n at s_0. Prove that $f(s)$ approaches the limit

$$\lim_{s \to s_0} f(s) = \frac{p^{(n)}(s)}{q^{(n)}(s)} \bigg|_{s = s_0}$$

Note that this has the appearance of Lhopital's rule applied to a function of a complex variable.

CHAPTER 6

MULTIVALUED FUNCTIONS

6-1. Introduction. Having established the concept of a single-valued function, $w = f(s)$, we now naturally ask whether such a function can always have an inverse whereby s can be specified as a function of w. In those cases where several values of s yield identical values of w we are in trouble, for then the inverse cannot be single-valued, and in the true sense of the word an inverse function does not exist. The main task of this chapter is to develop a method of analysis which will permit "multivalued functions" to be treated at least partially like single-valued functions.

We can draw some examples from the realm of real variables. The function

$$y = \sin x$$

is single-valued, but the inverse

$$x = \sin^{-1} y$$

is multivalued, as illustrated in Fig. 6-1a. The same comments can be made about

$$y = x^2$$

and its inverse

$$x = y^{\frac{1}{2}}$$

which is shown in Fig. 6-1b. Probably your experiences with the square-root function, and the problem of choosing signs in the case of real variables, has pointed up the need for isolating these cases.

In each of the above two examples a given function is single-valued, and its inverse is multivalued. There are other cases which are multivalued "both ways." An example is

$$y^2 = x^2 - 1$$

which is shown graphically in Fig. 6-1c.

When dealing with complex variables we sometimes find multivaluedness which does not appear in the real-variable counterpart. For example, in Chap. 4 we met the multivalued function log s. However,

169

log x (where $x > 0$ and real) is not multivalued.   Thus it is apparent that graphical illustrations like those of Fig. 6-1 are inadequate for the general case of a function of a complex variable.   It is hoped that ultimately you will conclude that multivalued functions are simpler to understand when the variable is complex than when it is real.   This simplification comes about through the concept of a Riemann surface.   You met this briefly in the discussion of log s in Chap. 4 and will see much more of it in this chapter.

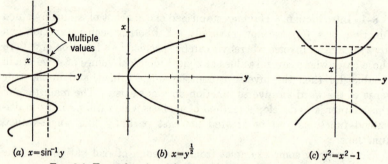

(a) $x = \sin^{-1} y$ (b) $x = y^{\frac{1}{2}}$ (c) $y^2 = x^2 - 1$

FIG. 6-1. Examples of multivalued functions of a real variable.

6-2. Examples of Inverse Functions Which Are Multivalued.   Perhaps the simplest multivalued function is the inverse of

$$w = s^2 \tag{6-1}$$

which will be written

$$s = w^{\frac{1}{2}}$$

The exponent $\frac{1}{2}$ in the above equation is *defined* as a notation which implies the inverse of Eq. (6-1).

The necessary ideas for studying the inverse of Eq. (6-1) were anticipated in Figs. 3-3 and 3-4.   If the two w planes of those figures are regarded as being identical, areas A' and B' are identical and the above functional relationship would carry this area into area A or area B of the s plane.   From the formula alone there would be no way to differentiate between areas A and B.

We could continue to regard w as the independent variable when analyzing the inverse of those functions which have previously been considered.   There would be some advantage in doing this, particularly in considering mapping properties, because then labels on the planes would remain unchanged.   However, there are advantages in always using s as the independent variable; and since we shall be considering functions other than the inverses of previously treated single-valued functions, we shall continue to use s as the independent variable.

Accordingly, the inverse of Eq. (6-1) is now written with s and w interchanged, as follows:

$$w = s^{1/2} \tag{6-2}*$$

This function is described by Figs. 3-3 and 3-4 *if the w- and s-plane labels are interchanged,* so that now there will be two s planes which map onto a single w plane.

We shall now exploit the idea of having *two* s planes. If somehow a distinction can be made between these two s planes, we could then regard overlying points in the two planes as being *different,* and the function $w = s^{1/2}$ would appear to be single-valued. To do this necessitates overcoming an obstacle introduced by the wedge-shaped cuts along the negative real axis. The difficulty is surmounted by the ingenious device of imagining the two planes to be attached along the cut edges. Referring to Fig. 6-2, for each pair of edges consisting of one edge from each plane, one solid line and one dashed line fit together. Then curves such as C and C' do not cross a cut but pass continuously from one plane to the other. When the two s planes are joined in this way, they form a *Riemann surface.* Each of the s planes is called a *sheet* of the Riemann surface; and the cut in each sheet is called a *branch cut.* A point like s_b in Fig. 6-2 is called a *branch point.* That portion of the function described when s is in one sheet is called a *branch* of the function.

Suppose that there are two points s_1 and s_1' similarly located in the two sheets of Fig. 6-2. The Riemann-surface interpretation allows them to be regarded as different points. In this way $w_1 = f(s_1)$ and $w_1' = f(s_1')$ are clearly distinct because ang $s_1' = 2\pi +$ ang s_1. With this interpretation $f(s)$ becomes single-valued. Many theorems originally proved for single-valued functions now become applicable in the multivalued case.

Since it is important to be able to identify a point with a particular sheet, it is necessary to have a method of keeping track of this. We do so by considering the angular position of a line drawn from the branch point to the point in question.† In the case of Fig. 6-2 this is merely the angle of the variable s. In sheet 1 this angle (ϕ) lies in the range $-\pi < \phi \leq \pi$, and in sheet 2 the range is $\pi < \phi \leq 3\pi$.

In order further to explain these concepts, consider neighborhoods of points s and s', where the unprimed value is always in sheet 1 and the

---

* The notation $w = \sqrt{s}$ is purposely avoided. In this chapter the $\sqrt{}$ symbol will be reserved for use with positive real numbers, and when the symbol is used, a positive sign will be understood. In Chap. 10 the symbol $\sqrt{s}$ is used, but with a specific meaning defined there.

† Polar coordinates centered at the origin can be used to identify which sheet a point is in only when there is a branch point at the origin. When a branch point is at some other point, as in some of the later examples, an auxiliary polar-coordinate system is centered at the branch point in order to accomplish this task.

primed one is in sheet 2. Each of these neighborhoods will be transformed into neighborhoods of corresponding points in the w plane. A few particular cases are considered, beginning with points s_1 and s_1'. There is no possibility of the neighborhood of s_1' becoming confused with the neighborhood of s_1. This permits us to use the definition of continuity without being bothered by multivaluedness.

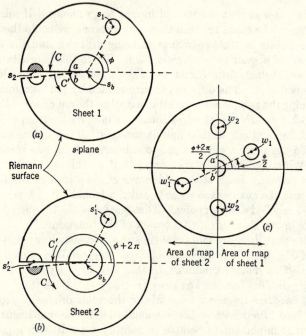

FIG. 6-2. Riemann-surface interpretation of the function $w = s^{1/2}$.

A point like s_2 on a "solid-line" edge of a branch cut cannot have a neighborhood wholly in one sheet. Its neighborhood must be in two sheets, as indicated by the two shaded areas in Fig. 6-2. This neighborhood goes into a neighborhood of w_2 in the w plane. The corresponding point s_2' has a neighborhood consisting of the two nonshaded circular segments, which transforms into a neighborhood of w_2'. Although the neighborhoods of s_2 and s_2' are each in two sheets, the function is single-valued in each neighborhood.*

We now come to the unique feature of a branch point. If we try to put a small circle around s_b in sheet 1, we find that points a and b cannot

---

* Later on it is shown that choice of the branch cut is arbitrary. For a different choice, say along the positive real axis, s_1 and s_1' would each have a neighborhood in a single sheet.

be connected; from point a we must proceed into sheet 2. If points a and b are allowed to approach each other, the corresponding points in the w plane approach a' and b', which are at the ends of a semicircle, as shown in Fig. 6-2c. A small circle which encircles a branch point only once cannot transform into a closed figure in the function plane. Two or more circuits (two in this example) around a branch point are required to give a closed figure in the function plane. Branch points are designated by an *order* number. The order is *one less* than the number of circuits around it required to give a closed figure in the function plane.

The above description brings to light other distinctive features of a branch point. Unlike points such as s_1 and s_2, a branch point cannot be assigned to any one sheet of the Riemann surface, and therefore it cannot have a neighborhood lying in only one sheet. That is, it is impossible to define a neighborhood of a branch point in which the function is single-valued.

The fact that encircling a branch point only once does not close the figure traced in the function plane can be used to test whether or not a given point is a branch point. As an example, we shall test whether $s = 0$ and $s = 1$ are branch points of the function

$$w = s^{1/2}$$

At $s = 0$ we write

$$s = \rho^{j\phi} \qquad w = re^{j\theta}$$

giving

$$r^2 e^{j2\theta} = \rho e^{j\phi}$$

and

$$r = \sqrt{\rho} \qquad \theta = \frac{\phi}{2}$$

If ϕ is increased by 2π, so that point $s = 0$ is encircled once, θ will increase by π, which will carry w only halfway around the origin. Thus, $s = 0$ is a branch point. Now look at the pair of points $s = 1$ and $w = 1$. In their neighborhoods we write

$$s = 1 + \rho e^{j\phi} \qquad w = 1 + re^{j\theta}$$

and

$$1 + 2re^{j\theta} + r^2 e^{j2\theta} = 1 + \rho e^{j\phi}$$

As r is made very small, the r^2 term approaches zero faster than r and so the above approaches

$$2re^{j\theta} \doteq \rho e^{j\phi}$$

showing that point $w = 1$ is encircled only once when $s = 1$ is encircled once by a small circle. Thus, $s = 1$ is not a branch point.

We have seen that the function described by Eq. (6-2) has a branch point at $s = 0$. If the Riemann-sphere interpretation is introduced, we can also identify a branch point at the point infinity. A small circular path enclosing the point at infinity on the Riemann sphere becomes a large circle in the flat plane. Thus, to test whether the point at infinity is a branch point, we look at the figure traced in the function plane as we follow one circuit around a large circle (approaching infinite radius) in the s plane. If the function-plane figure does not close, the point at infinity is a branch point. The point at infinity can also be investigated by examining $f(1/s)$ at the origin.

Thus it is concluded that the function $w = s^{1/2}$ has branch points of order 1, at $s = 0$ and at infinity. They are located at ends of the branch cut. Every multivalued function has a branch point at each end of a branch cut. As we shall see later, some functions have branch cuts extending between pairs of finite branch points.

We can learn a bit more about inverse functions by considering the inverse of

$$s = w^3 \tag{6-3}$$

which is conventionally written

$$w = s^{1/3} \tag{6-4}$$

where the exponent $\frac{1}{3}$ is defined to mean the inverse of Eq. (6-3). In this case the Riemann surface has three sheets, each of which maps onto one-third of the w plane. With a little thought you will see that it is necessary to encircle the point $s = 0$ three times in order to get a closed figure in the w plane. Also, it is evident that infinity is a branch point and that both branch points are of order 2. The interconnection of edges of branch cuts is illustrated in Fig. 6-3 by curves C, C', and C'' and by the sequence of numbers. Points 2 and 3, 4 and 5, and 6 and 1 are, respectively, connected together.

As a final example in this section consider a multivalued function having an inverse which is also multivalued. The case in point is

$$w^2 = s^2 - 1 \tag{6-5}$$

which can be written $w^2 = (s - 1)(s + 1)$ in order to show that, if either point $s = 1$ or $s = -1$ is encircled twice, then point $w = 0$ is encircled once. Thus, points -1 and $+1$ are first-order branch points in the s plane. These are the branch points of w as a function of s. To get the branch points in the w plane, for s as a function of w, we write

$$s^2 = (w + j)(w - j)$$

which shows that in the w plane there are branch points at j and $-j$.

The complete representation of this function requires Riemann surfaces of two sheets for each variable, as shown in Fig. 6-4. Branch cuts are indicated by the double lines. This situation is too complicated to admit a complete graphical picture. We shall consider only the transformation from s to w. Branch points at $s = +1$ and -1 are enclosed by four circles, which go into the four semicircles with similar labels in the w plane. A few rectangular-coordinate lines in the s plane and their traces in the w plane are also shown.

The distortion of the s plane in going to the w plane can be visualized by thinking of the two sides of the branch cut being pulled apart in the direction of the arrows, while the branch points move together. Further interpretation of the mapping of this function is obtained by considering

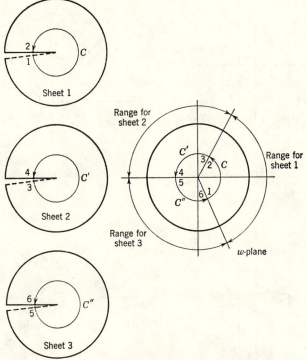

FIG. 6-3. Riemann-surface interpretation of the function $w = s^{1/3}$.

what happens in going from one sheet to the other at the branch cut. For example, when going from A to A' in the s plane the corresponding w point goes along the line with similar labels in the w plane. At A' the branch cut is encountered in the s plane, and a continuation of this line must be $B'B$ in the other sheet. The corresponding line is also shown in the w plane. Thus, in going along a vertical line in the s plane which "crosses" a branch cut (not actually crossing, but transferring to the other plane) we find that we go to the w-plane branch point at $w = j$ and transfer there to the other surface in the w plane. This is consistent with the interpretation of $w = +j$ and $-j$ as branch points in the w plane. Lines CC' and $D'D$ show what happens in a more general case.

6-3. The Logarithmic Function. The function

$$w = \log s \tag{6-6}$$

was introduced in Sec. 4-9. An introductory discussion of the reason for defining the Riemann surface, and a perspective portrayal of the Riemann surface for the logarithmic function, is given in Fig. 4-15.

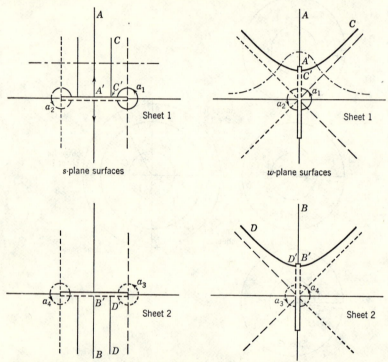

Fig. 6-4. Riemann surfaces in the w and s planes for the function $w^2 = s^2 - 1$.

Also, the function $w = e^s$ is treated graphically in Chap. 3; and if we write $s = e^w$ instead, we have $w = \log s$ as its inverse. Thus, Fig. 3-13 describes the logarithmic function if the s and w labels are interchanged. The graphical interpretation of Eq. (6-6) is shown in Fig. 6-5, which may profitably be compared with Figs. 3-13 and 3-14.

It is to be emphasized that there are an infinity of sheets in this Riemann surface. Each sheet maps onto an infinite strip of the w plane. If the branch cut is along the negative real axis, the w-plane strips have edges at odd multiples of π, as shown.

A path encircling the origin in the Riemann surface of the s plane produces a vertical line in the w plane. Thus, for a finite number of encirclements the path in the w plane does not close. But by admitting the point at infinity in the w plane this vertical line may be said to close at infinity, after an infinite number of encirclements. Thus, it is reasonable to designate the s-plane points at zero and infinity as branch points of infinite order.

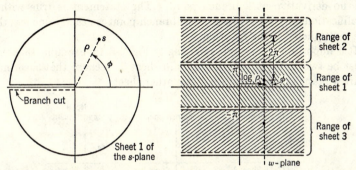

FIG. 6-5. Riemann-surface interpretation of the function $w = \log s$.

6-4. Differentiability of Multivalued Functions.

We recall that if a function is to have a derivative it is necessary that the limit

$$\lim_{s \to s_0} \frac{f(s) - f(s_0)}{s - s_0}$$

shall exist. In general, this limit cannot exist if $f(s)$ is not a single-valued function of s, because the choice of $f(s)$ would not be unique. However, let s_0 be defined to be in one sheet of a Riemann surface, like s_1 in Fig. 6-2, and let s be in its neighborhood in that sheet. Under these conditions, $f(s) - f(s_0)$ is a unique function of s, and the limit of the differential quotient can exist. When it does exist, it is called the derivative of the function. Now suppose that s_0 coincides with a branch point. As previously shown, at a branch point it is impossible to define a neighborhood in which the function is single-valued. Therefore, the differential quotient has no unique value, and no limit can be taken. On the basis of the above arguments it is concluded that no derivative can exist at a branch point. Thus, in addition to poles and essential singularities we have a third class of singularities, the *branch points*.

In considering the question of differentiability at a branch point it is necessary to mention that a certain point may be a branch point with respect to several sheets of the Riemann surface but may be a regular point in other sheets. You will find an example of this phenomenon in Sec. 6-8.

As we discuss the general case, let an n-valued function have a branch point of order m at s_0. From the known properties of a branch point we can say that s_0 must lie in $m + 1$ sheets. However, it is possible, as in Fig. 6-6, for n to be greater than $m + 1$, in which case there will be $n - m - 1$ other sheets in which the corresponding point is not a branch point. The function may have a derivative at these other points. Thus, we need to include the above qualification when stating that a function has no derivative at a branch point. The statement is true without qualification only if the order of the branch point is exactly one less than the multiplicity of the function.

It is important to observe, referring to Fig. 6-6, for example, that there must be two or more branch points in sheet 3, because that sheet must share a branch cut with at least one other sheet.

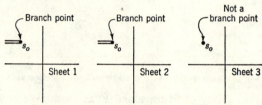

Fig. 6-6. Example of a first-order branch point for a three-valued function. No derivative can exist at s_0 in sheets 1 and 2, but it can exist at s_0 in sheet 3.

Two important conclusions are derived from the above discussion, one being that at points other than branch points the derivative of a multivalued function might possibly be multivalued, and the other that there can be no derivative at a branch point.

Although we keep in mind the definition of the derivative as a limit process in the various sheets, we can use already existing knowledge of the derivative to obtain the derivative in the various sheets. All cases of interest can be written in the implicit form

$$g(s) = h(w) \tag{6-7}$$

where $g(s)$ and $h(w)$ are analytic functions of s and w, respectively. At corresponding points s and w satisfying Eq. (6-7), and where the derivatives $g'(s)$ and $h'(w)$ exist, it follows from the theory of derivatives that

$$g'(s) = h'(w) \frac{dw}{ds}$$

giving

$$\frac{dw}{ds} = \frac{g'(s)}{h'(w)} \tag{6-8}$$

The function

$$s = w^2$$

is an example of Eq. (6-7), and from it we get the derivative of $w = s^{\frac{1}{2}}$ by using Eq. (6-8), as follows:

$$\frac{dw}{ds} = \frac{1}{2w} = \frac{1}{2s^{\frac{1}{2}}}$$

or

$$\frac{d(s^{\frac{1}{2}})}{ds} = \frac{1}{2s^{\frac{1}{2}}}$$

This result is in agreement with the usual rules for differentiation. Note that the derivative is multivalued. Let ϕ be in the range $-\pi < \phi \leqq \pi$ so that in sheets 1 and 2 s is designated, respectively, by $\rho e^{j\phi}$ and $\rho e^{j(\phi+2\pi)}$. Then the derivative is, respectively,

$$\frac{1}{2\sqrt{\rho}\, e^{j\phi/2}}$$

and

$$\frac{1}{2\sqrt{\rho}\, e^{j(\phi/2+\pi)}} = -\frac{1}{2\sqrt{\rho}\, e^{j\phi/2}}$$

in these two sheets. In these formulas $\sqrt{\rho}$ is always interpreted as positive.

Another example is

$$s = e^{w}$$

from which we get

$$\frac{d(\log s)}{ds} = \frac{1}{e^{w}} = \frac{1}{s}$$

The general point s is given by

$$s = \rho e^{j(\phi+2n\pi)}$$

where n is an integer depending on the sheet in question. In a philosophical sense the derivative

$$\frac{1}{s} = \frac{1}{\rho}\, e^{-j(\phi+2n\pi)} = \frac{1}{\rho}\, e^{-j\phi}$$

is different in each sheet. However, the above formula shows that the same numerical value is obtained. All values "differ" only by an integral multiple of 2π in the angle.

Finally, let us consider the derivative dw/ds for the multivalued function defined implicitly by Eq. (6-5). We get

$$\frac{dw}{ds} = \frac{s}{w} = \frac{s}{(s-1)^{\frac{1}{2}}(s+1)^{\frac{1}{2}}}$$

which is double-valued. In each of these cases we observe that the formula indicates nonexistence of the derivative at a branch point.

6-5. Integration around a Branch Point. As a further illustration of the usefulness of a Riemann surface, consider the integrals

$$\int_C s^{1/2}\, ds \qquad \text{and} \qquad \int_{C'} s^{1/2}\, ds$$

where C is a counterclockwise closed curve encircling the origin once

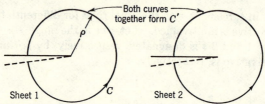

Fig. 6-7. Separation of paths of integration in the sheets of a Riemann surface.

and C' encircles the origin twice. For simplicity, take each curve as a circle of radius ρ, as shown in Fig. 6-7. For the first case, taking $s = \rho e^{j\phi}$,

$$\int_C s^{1/2}\, ds = \int_{-\pi}^{\pi} \sqrt{\rho}\, e^{j\phi/2}\, (j\rho e^{j\phi})\, d\phi$$

$$= \tfrac{2}{3} \sqrt{\rho^3}\, e^{j3\phi/2} \Big|_{-\pi}^{\pi} = -j\tfrac{4}{3} \sqrt{\rho^3}$$

For the second case,

$$\int_{C'} s^{1/2}\, ds = \tfrac{2}{3} \sqrt{\rho^3}\, e^{j3\phi/2} \Big|_{-\pi}^{3\pi} = 0$$

Two tentative conclusions are reached from this example. First, the integral once around a branch point is dependent on the radius of the path. This was not the case for integration around poles or essential singularities, where path dimensions are unimportant. The second point of interest is that in going around a second time the integral is the *negative* of the first, giving a total of zero. With other types of singularities, the same value is obtained for each of any number of encirclements. Of course, the difference arises because the two sheets of the Riemann surface are not equivalent; and this is precisely the point being stressed.

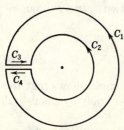

Fig. 6-8. Effect of radius of integration path.

As a final comparison of a branch point with a pole (or essential singularity) let us investigate how it is that one integral is independent of the path radius and the other is not. Refer to Fig. 6-8, which shows two closed curves C_1 and C_2 (closed except for the small gap) enclosing an isolated singular point in a region where the function is otherwise regular.   There possibly is a branch cut between C_3 and C_4.   The Cauchy integral theorem gives

$$\int_{C_1} f(s)\ ds - \int_{C_2} f(s)\ ds + \int_{C_3} f(s)\ ds + \int_{C_4} f(s)\ ds = 0 \qquad (6\text{-}9)$$

Furthermore, if $f(s)$ is single-valued, we can let C_3 and C_4 approach each other and, in the limit,

$$\int_{C_3} f(s)\ ds = -\int_{C_4} f(s)\ ds \qquad (6\text{-}10)$$

Then Eq. (6-9) becomes

$$\int_{C_1} f(s)\ ds = \int_{C_2} f(s)\ ds \qquad (6\text{-}11)$$

showing the integral to be independent of radius of the path.   However if the paths enclose a branch point, there must be a branch cut, which we shall put between the paths C_3 and C_4.

Now C_3 and C_4 are on opposite sides of the branch cut, and therefore they cannot be brought together.   Thus, although Eq. (6-9) is still true for the multivalued function, Eq. (6-10) is not

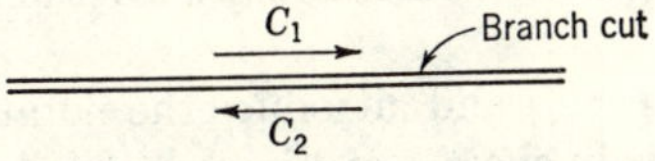

FIG. 6-9. Integration paths on two sides of a branch cut.

necessarily true, and consequently Eq. (6-11) is also not true, in general. It is suggested that you perform the integrations along all these paths for $w = s^{1/2}$ to confirm these statements.

One of the essential ideas to be gained from this discussion is the fact that integrations in opposite directions along opposite sides of a branch cut do not cancel.   Thus, if there are two arcs such as C_1 and C_2 in Fig. 6-9, in the general case we must expect that

$$\int_{C_1} f(s)\ ds + \int_{C_2} f(s)\ ds \neq 0$$

in contrast to the situation if there were no branch cut between the two paths.   Here we are assuming that the integration paths are actually on the edges of the branch cut, and, in fact, one of them must be in a different sheet of the Riemann surface, since each sheet has only one edge at the branch cut.   That is, the two arcs form the same line when the two sheets are superimposed.

As another example of integration around a branch point, consider the integral

$$\int_C \frac{ds}{(s^2 - 1)^{1/2}}$$

where C is the path shown in Fig. 6-10.   The sheet shown in this figure will be regarded as sheet 1 of Fig. 6-4.   Pertinent parts of Fig. 6-4 are

redrawn in Fig. 6-10. The integrand is multivalued, being the reciprocal
of w as given by Eq. (6-5). Branch points and the branch cut are shown
in Fig. 6-10. No singularities will be passed over if the path is distorted
to C', and so the integral around C' will be the same as around C. C' consists of two segments of opposite edges of the branch cut and circles of
radius ρ around each branch point. Let C_1 be the circle around point

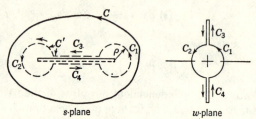

FIG. 6-10. Integration around branch points of the function $w = (s^2 - 1)^{1/2}$.

$s = 1$, and introduce the notation $s - 1 = \rho e^{j\phi}$, $ds = j\rho e^{j\phi}\,d\phi$. Since C_1
is in sheet 1 of Fig. 6-4, it follows that $-\pi < \phi \leqq \pi$, and

$$(s^2 - 1)^{1/2} = \sqrt{\rho}\, e^{j\phi/2} \sqrt[4]{(2 + \rho \cos \phi)^2 + \rho^2 \sin^2 \phi}\; e^{j\frac{1}{2} \tan^{-1}\frac{\rho \sin \phi}{2+\rho \cos \phi}}$$

The condition $\rho < 1$ ensures that $(s^2 - 1)^{1/2}$ will have a positive real part.
The absolute value of the above expression is greater than $\sqrt{\rho(2 - \rho)}$,
and so if the small gaps in the circles are negligible, the following upper
bound is obtained:

$$\left| \int_{C_1} \frac{ds}{(s^2 - 1)^{1/2}} \right| < \frac{2\pi \sqrt{\rho}}{\sqrt{2 - \rho}}$$

In an exactly similar way it can be shown that

$$\left| \int_{C_1} \frac{ds}{(s^2 - 1)^{1/2}} \right| < \frac{2\pi \sqrt{\rho}}{\sqrt{2 - \rho}}$$

These estimates will suffice for the integrals around C_1 and C_2.

Now consider paths C_3 and C_4, on which we shall write $s = r$, where r
varies from $1 - \rho$ to $-1 + \rho$. On C_3 we then have

$$(s^2 - 1)^{1/2} = j \sqrt{1 - r^2}$$

since C_3 is in sheet 1. On C_4, which moves into sheet 2 upon being
brought into coincidence with C_3, we have

$$(s^2 - 1)^{1/2} = -j \sqrt{1 - r^2}$$

It follows that*

$$\int_{C_1+C_4} \frac{ds}{(s^2-1)^{1/2}} = -j\int_{1-\rho}^{-1+\rho} \frac{dr}{\sqrt{1-r^2}} + j\int_{-1+\rho}^{1-\rho} \frac{dr}{\sqrt{1-r^2}}$$

$$= j2\int_{-1+\rho}^{1-\rho} \frac{dr}{\sqrt{1-r^2}}$$

$$= j2[\sin^{-1}(1-\rho) - \sin^{-1}(-1+\rho)]$$

Finally, we conclude from the equivalence of paths C and C' ($= C_1 + C_2 + C_3 + C_4$) that

$$\left| \int_C \frac{ds}{(s^2-1)^{1/2}} - j2[\sin^{-1}(1-\rho) - \sin^{-1}(-1+\rho)] \right|$$

$$= \left| \int_{C_1+C_2} \frac{ds}{(s^2-1)^{1/2}} \right| < \frac{4\pi\sqrt{\rho}}{\sqrt{2-\rho}}$$

Now let ρ approach zero. The right-hand side approaches zero, and therefore

$$\int_C \frac{ds}{(s^2-1)^{1/2}} = j2\pi$$

If C had been in sheet 2, the sign would have been negative. Note that, although the integrand becomes infinite at the branch points, these are not poles. We see that the integral around these branch points approaches zero as the radius approaches zero. This would not be true for integration around a pole.

As another example of integration around a branch point, consider

$$\int_C \log s\, ds$$

where C is a small counterclockwise circular path around the origin. We again designate a point on this circle by

$$s = \rho e^{j\phi} \qquad \log s = \log \rho + j\phi$$

giving $\qquad \displaystyle\int_C \log s\, ds = j\rho \log \rho \int_0^{2\pi} e^{j\phi}\, d\phi - \rho \int_0^{2\pi} \phi e^{j\phi}\, d\phi$

For finite ρ the first integral on the right is zero, and the second one is not zero, showing that again in this case an integral around a circular path centered at a branch point depends on the radius and that the integral approaches zero as the radius approaches zero.

Sometimes integration is wanted over a portion of a small circle around a branch point, particularly a semicircle. Consider each of the integrals

$$\int_C s^n\, ds \qquad \int_C s^n \log s\, ds$$

* In evaluating this integral by substituting limits we must use the same branch of $\sin^{-1} r$ for each limit.

where C is a circular arc, as shown in Fig. 6-11a, and where n is a real number (not necessarily an integer) in the range $-1 < n$. With the same notation as before, it is evident that

$$\int_C s^n \, ds = j\rho^{n+1} \int_0^{\phi_1} e^{j(n+1)\phi} \, d\phi \tag{6-12a}$$

$$\int_C s^n \log s \, ds = j\rho^{n+1} \log \rho \int_0^{\phi_1} e^{j(n+1)\phi} \, d\phi - \rho^{n+1} \int_0^{\phi_1} \phi e^{j(n+1)\phi} \, d\phi \tag{6-12b}$$

Since $-1 < n$, the factor ρ^{n+1} in Eq. (6-12a) approaches zero as ρ approaches zero. The first integral on the right of Eq. (6-12b) is not necessarily zero. However, the factor $\rho^k \log \rho$ approaches zero as ρ approaches zero when $k > 0$. This can be established by applying the Lhopital rule. Thus, in each case the following is true,

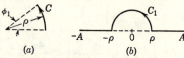

FIG. 6-11. Integration contours consisting of circular arcs.

$$\left. \begin{array}{l} \displaystyle\lim_{\rho \to 0} \int_C s^n \, ds = 0 \\[2ex] \displaystyle\lim_{\rho \to 0} \int_C s^n \log s \, ds = 0 \end{array} \right\} \quad -1 < n \qquad \begin{array}{l}(6\text{-}13a)\\[2ex](6\text{-}13b)\end{array}$$

for any circular arc of radius ρ centered at the branch point at $s = 0$. For Eq. (6-13a) the origin is a branch point only if n is not an integer. We note that these branch points do not need to be at the origin. The same conclusion is reached if s is replaced by $s - s_0$, where s_0 is a branch point.

Now suppose that $F(s)$ is an analytic function which is regular at s_0, and let n be restricted to the range $-1 < n < 0$. Each of the functions

$$f(s) = (s - s_0)^n F(s) \quad \text{and} \quad f(s) = \log (s - s_0) F(s) \tag{6-14}$$

will have a branch point at s_0 and an integral, over any circular arc centered at s_0, which approaches zero as the radius approaches zero. This can be seen by expanding $F(s)$ in a Taylor series about s_0, integrating term by term, and then applying the above result to each term.

By allowing $F(s)$ in Eqs. (6-14) to be singular at s_0, we note that a branch point can also be a pole or an essential singularity. The functions

$$\frac{1}{s^{1/2}} \left(\frac{1}{s} \right) \qquad (\log s) e^{1/s}$$

are two examples. If $F(s)$ should be singular at s_0, it would expand in a Laurent series with one or more negative-power terms. Then the condition $-1 < n$ stipulated in Eqs. (6-13) will not necessarily be satisfied for every term, and we cannot conclude that the integral over the path stipulated will approach zero.

If a function can be written in either of the forms in Eq. (6-14), and if s_0 is a regular point of $F(s)$, then the branch point s_0 is also called an *algebraic singularity* or a *logarithmic singularity* in the two cases, respectively.

A particularly interesting result is obtained if s_0 is an algebraic or logarithmic singularity and the arc is part of a more extensive contour. Since the integral approaches zero as the radius of the arc approaches zero, this means that we can integrate "through" such a singular point without need to consider the circular arc at all. For example, referring to Fig. 6-11b, suppose that s_0 is at the origin, and let C_1 be the complete contour. In view of the above results it follows that

$$\lim_{\rho \to 0} \int_{C_1} f(s) \, ds = -\lim_{\rho \to 0} \left[\int_{-A}^{-\rho} f(\sigma) \, d\sigma + \int_{\rho}^{A} f(\sigma) \, d\sigma \right]$$

In this chapter we are not in a position to give practical examples leading to integrations of this type. However, such integrals do arise in the solution of partial differential equations by the Laplace transform method and also in some of the theorems presented in Chap. 7.

6-6. Position of Branch Cut. In the examples so far given the branch cuts were taken as straight lines between branch points. However, a branch cut can be any simple arc connecting two branch points; branch points are unique, but branch cuts are not. It is always best to choose the simplest branch cuts possible, as we have done in the examples.

6-7. The Function $w = s + (s^2 - 1)^{1/2}$. A particularly interesting multivalued function arises from the inverse of

$$s = \frac{1}{2} \left(w + \frac{1}{w} \right)$$

which reduces to

$$w^2 - 2ws + 1 = 0$$

or
$$w = s + (s^2 - 1)^{1/2} \tag{6-15}$$

This expression indicates that there are branch points at $s = +1$ and -1, since there the function is single-valued. There is no branch point at infinity because w goes either to zero or to infinity as s approaches infinity.

It is suggested that you now look at Figs. 3-11 and 3-12. Taking into account the interchange of symbols w and s, it is seen that the branch points and branch cut joining them (between -1 and $+1$ on the real axis) were anticipated in those figures. The important features are redrawn in the present notation in Fig. 6-12a, where the branch cut goes from $s = -1$ to $s = +1$ by the most direct route. Another viewpoint is shown in Fig. 6-12b, where the branch cut goes via the point at infinity. Although this particular branch cut goes to infinity, it does not terminate there; it comes back on the other side, confirming that the point at infinity is not a branch point.

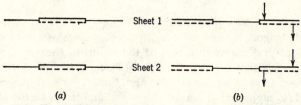

FIG. 6-12. Two positions of branch cuts for the function of Eq. (6-15).

6-8. Locating Branch Points. By now presumably you appreciate that branch points are important in the appraisal and analysis of multi-valued functions. So far we have considered relatively simple functions, for which the branch points are easily located.

In Sec. 6-7 we used the characteristic that the function has fewer values at a branch point than at other points. In that case, the function was single-valued at the branch point. However, the function

$$w = [s + (s^2 - 1)^{1/2}]^{1/2}$$

is four-valued, in general, and double-valued at $s^2 = 1$. Thus, the points $s = \pm 1$ are branch points at which this particular function is not single-valued.

In many cases an explicit expression is not available from which branch points can be recognized in the above manner. For example, it is often known only that a multivalued function

$$w = g(s)$$

is the inverse of a single-valued function

$$s = f(w)$$

When this is the case, it is possible to find branch points of $g(s)$ from the properties of $f(w)$.

Let w_0 be a regular point of $f(w)$ at which the first n derivatives are zero. Such a point in the w plane is called a *saddle point of order n*.* Since w_0 is a regular point, a Taylor-series expansion is possible,

$$s = f(w) = a_0 + a_{n+1}(w - w_0)^{n+1} + a_{n+2}(w - w_0)^{n+2} + \cdots \quad (6\text{-}16)$$

---

* For a discussion of saddle points see E. A. Guillemin, "The Mathematics of Circuit Analysis," pp. 298–302, John Wiley & Sons, Inc., New York, 1949.

When w is near w_0, this can be reduced to the approximate equality

$$s - a_0 \doteq a_{n+1}(w - w_0)^{n+1}$$

or $$w - w_0 \doteq (a_{n+1})^{-1/(n+1)}(s - a_0)^{1/(n+1)}$$

This relationship indicates the presence of a branch point of the inverse function at $s = a_0$. The branch point is seen to be of order n, because point $s = a_0$ must be encircled $n + 1$ times for point w_0 to be encircled once, this being accurately true only as w is a vanishing distance from w_0.

If at least the first derivative of $f(w)$ is zero at a point w_0, we conclude that this is sufficient for point $a_0 = f(w_0)$ in the s plane to be a branch point of the inverse function $g(s)$. Since this condition is merely sufficient, it will not necessarily find all the branch points of $g(s)$.

It has been observed that the order (n) of a branch point may be smaller than one less than the number of sheets in the Riemann surface. In that case, point $s = a_0$ will be a regular point in some sheets. As a pertinent case, take

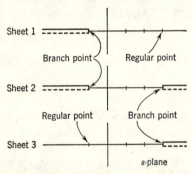

Fig. 6-13. Possible arrangement of branch cuts for the inverse of $s = w^3 - 3w + 1$.

$$s = w^3 - 3w + 1$$

$$\frac{ds}{dw} = 3w^2 - 3 = 0 \qquad \text{when } w = \pm 1$$

and $$\frac{d^2s}{dw^2} = \pm 6 \qquad \text{when } w = \pm 1$$

Thus, since the second derivative is not zero, it is seen that $s = -1$ and $s = 3$ are first-order branch points, each at the end of a branch cut between two sheets only. However, since this is a cubic equation, the inverse is triple-valued and there must be three sheets in the s plane. These three sheets, and one appropriate set of branch cuts, are shown in Fig. 6-13. Points $s = 3$ and $s = -1$ are regular points, respectively, in sheets 1 and 3. This is a specific numerical example of the principle originally illustrated in Fig. 6-6.

To find another sufficient condition for a branch point, suppose that $f(w)$ has a pole of order $n + 1$ at w_0. The Laurent series about w_0 is

$$s = f(w) = b_{n+1}(w - w_0)^{-(n+1)} + b_n(w - w_0)^{-n} + \cdots \qquad (6\text{-}17)$$

When w is near w_0, this can be reduced to the approximate equality

$$s \doteq b_{n+1}(w - w_0)^{-(n+1)}$$

or $$w - w_0 \doteq (b_{n+1})^{1/(n+1)}(s)^{-1/(n+1)} \qquad (6\text{-}18)$$

Equation (6-18) implies a branch point of order n at the point infinity in the s plane. No branch point occurs if the pole is simple. This gives another sufficient condition for a function $g(s)$ to have a branch point, namely, if $f(w)$ has a pole of order greater than 1. The order of the branch point is one less than the order of the pole.

This condition can be extended to include poles of $f(w)$ at infinity, but Eq. (6-18) does not handle that case. When $f(w)$ has a pole of order $n + 1$ at infinity, for large $|w|$ the approximation

$$s = f(w) \doteq Aw^{n+1}$$

can be used. A is a constant. Solving for w gives

$$w \doteq (A)^{-1/(n+1)}(s)^{1/(n+1)} \cdot \qquad (6\text{-}19)$$

which implies a branch point of order n at the point infinity in the s plane. When $n = 0$ [a first-order pole of $f(w)$ at infinity], there is no branch point of its inverse at infinity.

Finally, when $f(w)$ has an essential singularity at some point w_0, we should expect its inverse to have an infinite-order branch point, because then the Laurent series about a finite point has an infinite number of negative-power terms like Eq. (6-17). If there is an essential singularity at infinity, the Taylor series has an infinite number of positive-power terms in w and this also indicates the presence of an infinite-order branch point. Although it is true that the inverse function will have branch points related to essential singularities of $f(w)$, the function $f(w)$ is of no aid in finding the branch points, because $f(w)$ approaches no unique point on the Riemann sphere at an essential singularity. (A function is not infinite at an essential singularity, except when approached from a certain direction.) The function $s = e^w$ is a pertinent case. Its inverse, log s, has infinite-order branch points at zero and infinity, but these values are not derivable from the essential singularity of e^w, which is at the point infinity in the w plane.

According to these briefly outlined ideas, we can find branch points of a function $g(s)$ in many cases by looking at the function $f(w)$ of which $g(s)$ is the inverse. However, this will not find all the branch points, if $f(w)$ has an essential singularity; but it does suffice in many cases. The procedure is particularly useful for the important case where $f(w)$ is a rational function.

6-9. Expansion of Multivalued Functions in Series. An expansion of $g(s)$ in a Taylor series about a point s_0 is possible if s_0 is in one sheet of the Riemann surface. Of course, $g(s)$ must be regular at s_0 and the expansion, which is single-valued, will represent the function only in the sheet in which s_0 is located. Branch points are singular, and so radii of convergence will be limited by branch points in the same way as by other singularities.

As an example, we shall expand

$$g(s) = s^{1/2} \tag{6-20}$$

about the point $s = 1$ (for sheet 1) and $s = e^{j2\pi}$ (for sheet 2). The derivatives are

$$g'(s) = \tfrac{1}{2}s^{-1/2}$$
$$g''(s) = -\tfrac{1}{4}s^{-3/2} \tag{6-21}$$
$$g'''(s) = \tfrac{3}{8}s^{-5/2}$$
$$. \; . \; . \; . \; . \; . \; . \; . \; .$$

giving the following values for the two sheets:

	Sheet 1	Sheet 2
s_0	1	$e^{j2\pi}$
$g(s_0)$	1	-1
$g'(s_0)$	$\tfrac{1}{2}$	$-\tfrac{1}{2}$
$g''(s_0)$	$-\tfrac{1}{4}$	$\tfrac{1}{4}$
$g'''(s_0)$	$\tfrac{3}{8}$	$-\tfrac{3}{8}$

The following two series are obtained:

$$g(s) = 1 + \tfrac{1}{2}(s - 1) - \tfrac{1}{8}(s - 1)^2 + \tfrac{1}{16}(s - 1)^3$$
$$+ \cdots \quad \text{about } s_0 = 1 \tag{6-22a}$$
$$g(s) = -1 - \tfrac{1}{2}(s - 1) + \tfrac{1}{8}(s - 1)^2 - \tfrac{1}{16}(s - 1)^3$$
$$+ \cdots \quad \text{about } s_0' = e^{j2\pi} \tag{6-22b}$$

The radius of convergence is 1, because of the branch point at $s = 0$.

The function

$$g(s) = \log s \tag{6-23}$$

provides another example. Points on all the infinite number of sheets corresponding to point $s = 1$ on sheet 1 can be written $e^{j2\pi n}$, where n is a positive or negative integer. The derivatives are

$$g'(s) = s^{-1}$$
$$g''(s) = -s^{-2}$$
$$g'''(s) = 2s^{-3}$$
$$. \; . \; . \; . \; . \; . \; . \; . \; .$$

and at $s_0 = e^{j2\pi n}$

$$g(s_0) = j2\pi n$$
$$g'(s_0) = e^{-j2\pi n} = 1$$
$$g''(s_0) = -e^{-j4\pi n} = -1$$
$$g'''(s_0) = 2e^{-j2\pi n} = 2$$

and the series for the various sheets are represented by the one formula

$$g(s) = j2\pi n + (s - 1) - \tfrac{1}{2}(s - 1)^2 + \tfrac{1}{3}(s - 1)^3 + \cdots \quad (6\text{-}24)$$

As we should expect, on the various sheets the function differs only in the imaginary part. The radius of convergence is 1, as determined by the branch point at $s = 0$, and as can also be seen by looking at the coefficients.

These two examples show how the Riemann-surface concept serves to clarify the writing of series for multivalued functions. Without the Riemann surface a Taylor series would be impossible.

6-10. Application to Root Locus. The theory of multivalued inverse functions is of practical value in predicting stability as a function of gain in a single-loop feedback system. In the system portrayed in Fig. 6-14 the output/input function is

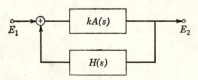

FIG. 6-14. A single-loop feedback system.

$$F(s) = \frac{E_2}{E_1} = \frac{kA(s)}{1 - kA(s)H(s)} \quad (6\text{-}25)$$

where $kA(s)$ is the transmission function of the amplifier and $H(s)$ is the transmission function of the feedback path. The real number k is introduced to represent a gain adjustment.

The roots of the equation

$$1 - kA(s)H(s) = 0 \quad (6\text{-}26)$$

are the characteristic values which determine the form of the natural response. Roots having positive real parts indicate instability. We are interested in determining the locus of roots of Eq. (6-26) as k varies. This is one type of problem in the category of *root-locus* problems wherein loci of roots are plotted as functions of some variable parameter.

Now define the functional relationship

$$w = A(s)H(s) \quad (6\text{-}27)$$

and consider the multivalued inverse function

$$s = f(w) \quad (6\text{-}28)$$

In general $A(s)H(s)$ is a ratio of polynomials. Roots of Eq. (6-26) are all those values of s determined from Eq. (6-28) when $w = 1/k$. As k varies, w ranges over the real axis in the w plane. Therefore, we shall learn much about the roots by finding the lines in the s plane which transform into the real axis of the w plane. These lines are the root loci. Knowledge of properties of multivalued functions, and particularly an understanding of behavior near branch points, helps materially in finding the root loci.

Branch points of $f(w)$ on the real axis and at infinity are significant. From previous discussions we know that the portion of the real axis in the neighborhood of a branch point w_0 on the real axis transforms into a system of radial "spokes," as shown in Fig. 6-15, for the case where the corresponding point s_0 is finite. The number of spokes is $2(n + 1)$, where n is the order of the branch point. Alternate spokes correspond to portions of the real axis to the right and left of w_0 if w_0 is finite, and to portions of the real axis extending to $+\infty$ and $-\infty$ if w_0 is at infinity. Thus, at any finite point s_0 corresponding to a branch point w_0 the root-loci curves will intersect in the manner of Fig. 6-15.

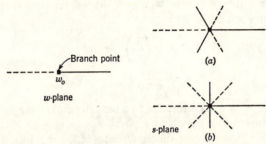

FIG. 6-15. Mapping of the real axis in the neighborhood of a real branch point in the s plane. (a) Second-order branch point; (b) third-order branch points.

Now suppose that $A(s)H(s)$ has a pole of order n at infinite s. Then infinity in the w plane is a branch point of order $n - 1$. Consider a Laurent expansion of $A(s)H(s)$ about the origin, in the region outside all singularities,

$$w = \cdots + \frac{a_{-2}}{s^2} + \frac{a_{-1}}{s} + a_0 + a_1 s + \cdots + a_{n-1}s^{n-1} + a_n s^n \quad (6\text{-}29)$$

Behavior at infinity can be roughly estimated by keeping only the last term,

$$w \doteq a_n s^n \quad (6\text{-}30)$$

and taking $w = re^{jv\pi}$, where v is any integer, which ensures that w will be real. Corresponding values of s are

$$s \doteq \left(\frac{1}{a_n}\right)^{1/n} r^{1/n} e^{jv\pi/n} \quad (6\text{-}31)$$

This shows that loci approach infinity along the equally spaced radial lines specified by Eq. (6-31) for successive values of v. Alternate lines correspond to positive and negative parts of the real axis, corresponding, respectively, to even and odd values of v. This analysis does not locate the center of this star of radiating lines. To get this, look at the two-term approximation

$$w \doteq a_{n-1}s^{n-1} + a_n s^n \qquad (6\text{-}32)$$

and try
$$s = \rho e^{j\phi} + B$$

which, for constant ϕ and variable ρ, represents a line radiating from point $s = B$. Substituting in Eq. (6-32) gives

$$re^{jv\pi} = a_n(\rho^n e^{jn\phi} + nB\rho^{n-1}e^{j(n-1)\phi} + \cdots) + a_{n-1}(\rho^{n-1}e^{j(n-1)\phi} + \cdots)$$
$$= a_n\rho^n e^{jn\phi} + (na_nB + a_{n-1})\rho^{n-1}e^{j(n-1)\phi} + \cdots \qquad (6\text{-}33)$$

With increasing ρ we see that the angle of the right-hand side approaches an integral multiple of π most rapidly if the second term is zero. Thus, the asymptotes are given by

$$s = B + \left(\frac{1}{a_n}\right)^{1/n} r^{1/n}e^{jv\pi/n}$$

where
$$B = -\frac{a_{n-1}}{na_n} \qquad (6\text{-}34)$$

The point $s = \infty$ can correspond to a branch point in the *finite* w plane, say w_0. In such a case the above analysis can be applied to the function $1/(w - w_0)$ in order to locate the asymptotes of the root loci. The above treatment applies only if w_0 is real.

These ideas may be reinforced by considering some examples. First consider

$$w = A(s)H(s) = \frac{s(s + 2)}{(s + 1)^3}$$

By inspection it is seen that there are a third-order pole at $s = -1$ and therefore a second-order branch point at $w = \infty$.

Other branch points are obtained by setting the derivative equal to zero,

$$\frac{dw}{ds} = \frac{(s + 1)^2[(s + 1)(2s + 2) - 3(s^2 + 2s)]}{(s + 1)^6} = 0$$

giving
$$s^2 + 2s - 2 = 0$$
$$s = -1 + \sqrt{3}, \; -1 - \sqrt{3}$$

Corresponding values of w, the branch points, are, respectively,

$$w = 0.385, \; -0.385$$

A test will show that these are first-order branch points.

The Riemann surface in the w plane has three sheets, and therefore we expect to find three lines in the s plane, corresponding to the real axis in each of the three sheets. It is perhaps preferable to think about the positive and negative parts of the real axis separately. We start at the origin in the w plane and move along the positive real axis, assuming simultaneous motion in all three sheets. Corresponding points will move from zeros of $A(s)H(s)$ in the s plane (the points $s = \infty$, -2, and 0). Arrows in Fig. 6-16 indicate this progress of s-plane points as they trace out the locus, while w moves toward infinity along the positive real axis. Note the behavior at the branch point, showing a right-angle turn in each of the s-plane loci. Without precisely defining the correspondence of s-plane points with points in the three surfaces, it is impossible to specify which locus turns upward at $s = 0.732$. The one coming from $s = \infty$ was chosen arbitrarily. A similar analysis applies for the negative real axis, yielding the loci shown dashed.

The system would be unstable for all values of $w = 1/k$ corresponding to loci points in the right-half s plane. Suppose that k starts at $+\infty$ and moves toward zero. Then w moves in the direction described above,

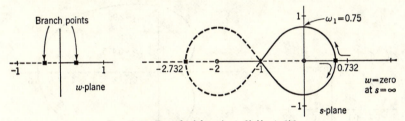

FIG. 6-16. Root loci for $s(s + 2)/(s + 1)^3$.

branches at point $s = 0.732$, and at some value of $w = w_1$ (where $s = j\omega_1$, $-j\omega_1$) crosses over into the left half plane. The corresponding value $k_1 = 1/w_1$ is the upper limit of gain for stable operation.

In this case k_1 can be found rather easily, although a similar computation will often involve the solution of a high-degree equation. At point $s = j\omega_1$ we know that w is a real number $1/k$. From the original equation k and ω are obtained as solutions of

$$\frac{1}{k} = \frac{j\omega(2 + j\omega)}{(1 + j\omega)^3}$$

which gives the two equations

$$k\omega^2 = -(1 - 3\omega^2)$$
$$2k\omega = (3\omega - \omega^3)$$

when real and imaginary parts are equated. Eliminating k yields the equation

$$\omega^4 + 3\omega^2 - 2 = 0$$

having the positive solution

$$\omega^2 = 0.561$$

The negative solution is of no value because ω must be real. Thus, we have found $\omega_1 = \sqrt{0.561} = 0.749$. The corresponding value of k is

$$k = \frac{3 - \omega^2}{2} = \frac{2.44}{2} = 1.22$$

With the aid of the root-locus diagram we see that the system is stable if

$$k < 1.22$$

As you appraise this development, you may wonder why this solution for limiting value of k could not proceed without help of the root-locus diagram. It is certainly true that the critical value of k is found by a purely algebraic process. However, this algebraic process would merely give

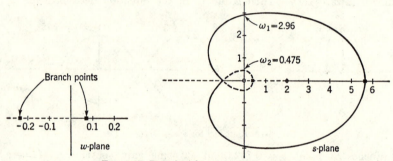

Fig. 6-17. Root loci for $s(s - 2)/(s + 1)^3$.

the value of k when s is purely imaginary. It would not tell in which direction s is moving as k varies. In other words, the root locus tells us which way to place the above inequality sign. It also provides assurance that there will be no other critical value of k. The root-locus plot is a kind of picture of the function which provides certain key information at a glance. Cases which are complicated bring advantages of the root locus into sharper focus.

Before going to a significantly more complicated case, consider what changes ensue when the previous case is replaced by the nonminimum phase function

$$w = A(s)H(s) = \frac{s(s - 2)}{(s + 1)^3}$$

An analysis exactly paralleling the previous one shows that the inverse function has branch points at $w = -0.236$ and $+0.070$, for which the corresponding critical points in the s plane are, respectively, 0.355 and 5.65. The locus is shown in Fig. 6-17. In this case there are two values of k for which there is a transition across the $j\omega$ axis. To find them, we set

$$\frac{1}{k} = \frac{j\omega(-2 + j\omega)}{(1 + j\omega)^3}$$

and proceed as before to solve for k and ω. With the stipulation that each shall be real, we get

$$\omega^4 - 9\omega^2 + 2 = 0$$

which has roots $\omega^2 = 8.77$ and 0.225. Both of these are positive, and so we get

$$\omega_1 = 2.96 \qquad \omega_2 = 0.475$$

with corresponding values

$$k_1 = 2.89 \qquad k_2 = -1.39$$

Negative k of course represents a reversal of connections somewhere in the system. Inspection of the direction of motion on the locus shows that we move in the direction of stability with decreasing k at both k_1 and k_2 (decreasing in an algebraic sense). Therefore, the system would be stable in the range

$$-1.39 < k < 2.89$$

As a final example, consider

$$w = A(s)H(s) = \frac{s(s + 1)}{(s + 2)^5}$$

There is a fourth-order branch point at infinite w, due to the multiple pole at $s = -2$. There is also a second-order branch point at $w = 0$, due to the multiple zero at infinite s. The other branch points are found as before, by setting derivatives of $A(s)H(s)$ equal to zero. By routine algebra it is found that the first derivative is zero at $s = \pm 0.82$, giving corresponding branch points at $w = 0.00837$ and -0.0646. The second derivative is not zero at either of the above values of s, and so it follows that these are first-order branch points.

The branch point at $w = 0$ is of second order and causes the loci to radiate toward infinity asymptotically along 60° radial lines, as shown in Fig. 6-18. The point of intersection of these asymptotes is found from Eq. (6-34) after suitable manipulation of the given function to cast it in a suitable form for recognizing the coefficients a_n and a_{n-1}. Equation (6-29) is written for a function having a pole at infinity, but in the present case there is a zero at infinity. Accordingly, we look at the reciprocal, in expanded form,

$$\frac{1}{A(s)H(s)} = \frac{s^5 + 10s^4 + 40s^3 + 80s^2 + 80s + 32}{s^2 + s}.$$

By division, the two highest-order terms are

$$s^3 + 9s^2 + \cdots$$

and so the intercept of the asymptotes is at

$$B = -\tfrac{9}{3} = -3$$

This information is enough to establish the general form for the loci shown in the figure. Two critical values of k are found by finding real

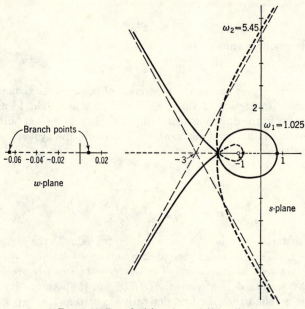

Fig. 6-18. Root loci for $s(s + 1)/(s + 2)^5$.

values of k and ω which are solutions of

$$\frac{1}{k} = \frac{j\omega(1 + j\omega)}{(2 + j\omega)^3}$$

Routine algebra yields the following pair of equations,

$$\omega^4 - 40\omega^2 + (80 - k) = 0$$
$$\omega^4 - \frac{80 - k}{10}\,\omega^2 + \frac{32}{10} = 0$$

which combine to give

$$\omega^6 - 30\omega^4 + 32 = 0$$

having approximate roots $\omega^2 = 1.05, 29.7, -1.02$. Positive values give intercepts on the ω axis as follows:

$$\omega_1 = \sqrt{1.05} = 1.025 \qquad \omega_2 = \sqrt{29.7} = 5.45$$

The corresponding values of k are, respectively,

$$k_1 = 29 \qquad k_2 = -225$$

and the system is found to be stable in the range

$$-225 < k < 39$$

In each of these examples, along with the limiting values of gain, we also obtained corresponding values of ω. These values of ω are the imaginary parts of the characteristic values of the system and, in the light of information given in Chap. 1, give the frequency at which the system would oscillate as it goes into the region of instability. Thus, if k moves farther than -225 in the negative direction, the system will begin to oscillate at an angular frequency of 1.025. Moving in the other direction at $k = 39$ would give oscillations at an angular frequency of 5.45.

Root locus is one of those areas in which complex-variable theory makes a great contribution. Central in this contribution are the properties of branch points and the concept of conformal mapping. From the theory of mapping we know that the locus curves can cross only at zeros, poles, or branch points. In fact, since zero and infinity in the w plane are on the real axis, the locus curves are conveniently regarded as curves going from zeros to poles (two curves from each zero to each pole, corresponding to positive and negative parts of the real w axis) and passing through and branching at s-plane traces of the *real* branch points. Branching takes place at equal angles, in the manner already indicated. The knowledge that loci can cross only at branch points, zeros, or poles is very helpful in estimating the curves qualitatively from a very small amount of data. In these examples we have deduced the essential characteristics of the curves from the computation of singular points and the critical points where they cross the $j\omega$ axis. These are relatively simple calculations, and so by using this viewpoint a relatively large amount of qualitative information is obtained economically.

PROBLEMS

6-1. Referring to Fig. 6-4, how do rectangular coordinates in the w plane transform to the s plane?

6-2. For the function

$$w^2 = s^2 - 1$$

show traces in the s plane of small circles around the branch points at $\pm j$ in the w plane.

6-3. Consider the function

$$s = w^2 + w$$

and write w as an explicit function of s. Then sketch w-plane loci corresponding to the circles $|s| = \frac{1}{8}, \frac{1}{4}, \frac{1}{2}$.

6-4. Prove by using the test for encirclements that the function given in Eq. (6-15) has no branch point at infinity.

6-5. The Riemann surface for

$$w = \sin^{-1} s$$

has an infinity of sheets. Show three sheets, and indicate how they are joined at appropriately chosen branch cuts. Sketch some curves in the w plane corresponding to rectangular coordinates in two of the s-plane sheets.

6-6. Investigate the mapping properties of

$$w = e^{(s)^{\frac{1}{2}}}$$

6-7. Obtain three terms of the Taylor expansion of the inverse of $s = w^3 - 3w + 1$, expanded about $s = 3$ at the point in the sheet in which $s = 3$ is not a branch point.

6-8. Considering $w = f(s)$ to be the inverse of $s = w^3 - 3w + 1$, obtain three terms of each of three Taylor series for $f(s)$, expanded about point $s = 0$. In each case specify the radius of convergence. Roots of this equation are $w = -1.8794$, $+0.3473$, $+1.5321$.

6-9. Obtain the integral

$$\int_C s^{\frac{1}{2}} \, ds$$

by converting to a real integral, where C is a counterclockwise circle of radius unity, centered at the origin. Do this for the following cases:

(a) C begins and ends at point $s = -1$, starting in the sheet for which $(-1)^{\frac{1}{2}} = j$.

(b) C begins and ends at point $s = 1$, starting in the sheet for which $(1)^{\frac{1}{2}} = 1$.

(c) Repeat part b, but using the sheet for which $(1)^{\frac{1}{2}} = -1$.

6-10. Consider the function

$$f(s) = \frac{e^s}{[1 + (s + 1)^{\frac{1}{2}}]^2}$$

and let C_1 and C_2 be circles, of radius 0.5 centered at the origin, in the sheets which correspond, respectively, to $(1)^{\frac{1}{2}} = 1$ and $(1)^{\frac{1}{2}} = -1$. Let C_1' and C_2' be similar circles of radius 2, in the same respective sheets as C_1 and C_2. (C_1' and C_2' cannot be completely closed, because each encounters a branch cut.) Assuming counterclockwise integration, find:

(a) $\int_{C_1} f(s) \, ds$ (b) $\int_{C_2} f(s) \, ds$

(c) $\int_{C_1'} f(s) \, ds$ (d) $\int_{C_1'} f(s) \, ds$

6-11. Find the integral

$$\int_C g(s) \, ds$$

where $g(s)$ is the inverse of $s = w^3 - 3w + 1$, and where C is a counterclockwise circle centered at the branch point at $s = 3$ and of radius 2. Integration starts and ends at point $s = 1$. (HINT: Convert to a real integral, and, by looking at the graphical interpretation of an integral as an area, convert to an integral which can be evaluated.)

6-12. The function

$$s = f(w) = 3w^4 + 4w^3 - 6w^2 - 12w + 9$$

has zeros at $w = 0.6767$, 1.0961, $-1.5530 \pm j1.2775$. The inverse

$$w = g(s)$$

is multivalued.

(a) Locate the branch points of $g(s)$.

(b) In the w plane sketch the loci corresponding to the positive and negative real and imaginary axes of the s plane. Along with this, clearly indicate the saddle points (traces of the branch points) and the zeros of $f(w)$. Also, show the pertinent asymptotes.

(c) Consider three circles of radii 0.5, 1.0, and 1.2 centered at the origin in the w plane. Sketch their traces in the s plane.

6-13. For the multivalued function

$$f(s) = \frac{1}{1 - s^{1/2}}$$

(a) Obtain the Laurent expansion about point $s = 1$, in the sheet for which $(1)^{1/2} = 1$.

(b) Obtain the Taylor expansion about point $s = 1$, in the sheet for which $(1)^{1/2} = -1$.

6-14. For the multivalued function

$$f(s) = \frac{1}{\log s - 1}$$

(a) Obtain the Laurent expansion about point $s = e$, in the sheet for which $\log e = 1$.

(b) Obtain the Taylor expansion about point $s = e$, in the sheet for which $\log e = 1 + j2\pi$.

6-15. For the two functions

(a) $f_a(s) = \sin s^{1/2}$ (b) $f_b(s) = \cos s^{1/2}$

obtain Taylor-series expansions about $s = \pi$, and specify the radius of convergence of each. Also, consider the possibility of expanding about $s = 0$ in each case. Are these functions both of the same kind?

6-16. One formula for the nth-order Tchebysheff polynomial is

$$T_n(s) = \frac{(-2)^n n!}{(2n)!} (1 - s^2) \frac{d^n}{ds^n} (1 - s^2)^{n - 1/2}$$

By using the known series

$$(1 + s)^{1/2} = 1 + \tfrac{1}{2}s - \tfrac{1}{8}s^2 + \tfrac{1}{16}s^3 + \cdots$$

obtained from Eqs. (6-22), employ the Cauchy integral formulas to prove that

$$\begin{array}{ll} T_n(0) = 1 & n \text{ even} \\ T_n(0) = 0 & n \text{ odd} \end{array}$$

6-17. Sketch the root locus for the following functions:

$$(a)\ \frac{s(s+1)}{(s+2)^3} \qquad (b)\ \frac{s(s-1)}{(s+2)^3} \qquad (c)\ \frac{s(s+1)}{(s+2)^4}$$

and determine the range of gain for stable operation as well as the natural frequencies at the points of transition to instability.

6-18. Let the loop-gain function of a feedback system be

$$\frac{k}{(s^2+4s+5)^2-4}$$

Sketch the root locus, and determine the range of k for stable operation.

6-19. The following nonminimum phase system function

$$A(s)H(s) = \frac{s-1.5}{(s+0.5)(s+1.5)}$$

is used in a feedback loop. Determine the critical points from which the root locus can be estimated, and sketch the root locus. Determine the range of k for stability and the natural frequency at the point of transition from stability to instability.

6-20. An open-loop unstable system is described by the function

$$A(s)H(s) = \frac{s+1.5}{(s+0.5)(s-1.5)}$$

Determine the critical points from which the root locus can be estimated, and sketch the locus. Determine the range of k for closed-loop stability and the natural frequency at the point of transition from stability to instability.

CHAPTER 7

SOME USEFUL THEOREMS

7-1. Introduction. Much of the theory of linear system performance deals with the properties of a certain limited class of functions of a complex variable. In this chapter we shall state and prove a few of the theorems which apply to the more important of these classes of functions. These theorems therefore are grouped with that part of the book given over to the basic theory. Of course, they are only a small part of the many theorems which can be proved; but they are theorems of particular interest because of their wide applicability.

7-2. Properties of Real Functions. A *real function* of a complex variable s is defined as a function which is real when s is real. As illustrated in Chap. 1, polynomials with real coefficients are important in the study of linear systems. Such polynomials are real functions. Other examples of real functions are e^s and $\cos s$.

Real functions have certain simple but important properties, which will now be discussed. Our attention will be confined to functions which are real, single-valued, and analytic, henceforth referred to as *real analytic functions*. We begin with the following theorem:

Theorem 7-1. If $f(s)$ is a real analytic function, then all its derivatives are real analytic functions.

PROOF. Let σ_0 be a real point at which the derivative exists. The differential quotient can be written in two ways, as follows:

$$\lim_{\Delta s \to 0} \frac{f(\sigma_0 + \Delta s) - f(\sigma_0)}{\Delta s} = \lim_{\Delta \sigma \to 0} \frac{f(\sigma_0 + \Delta \sigma) - f(\sigma_0)}{\Delta \sigma}$$

The limit on the left is $df(s)/ds$, and the limit on the right is $df(\sigma)/d\sigma$. Thus,

$$\frac{df(s)}{ds}\bigg|_{s=\sigma_0} = \frac{df(\sigma)}{d\sigma}\bigg|_{\sigma=\sigma_0} \tag{7-1}$$

But $f(\sigma)$ is real, and so its derivative with respect to the real variable σ is also real. The process may be continued for all derivatives.

The second theorem is as follows:

Theorem 7-2.   If $f(s)$ is single-valued, regular, and real over a segment of the real axis, then at all regular points

$$f(\bar{s}) = \overline{f(s)} \qquad (7\text{-}2)$$

and $f(s)$ is a real analytic function.

PROOF.   Since the function has points of regularity, by definition it is analytic.   If σ_0 is a real regular point interior to the defined interval of

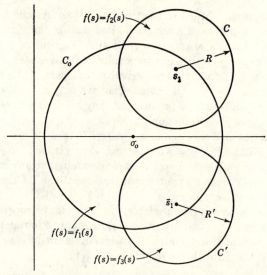

FIG. 7-1. Analytic continuation of a real function into symmetrical (mirror-image) regions.

regularity, we can obtain an element of $f(s)$ by expanding as a Taylor series about σ_0,

$$f_1(s) = f(\sigma_0) + f'(\sigma_0)(s - \sigma_0) + \frac{f''(\sigma_0)}{2}(s - \sigma_0)^2 + \cdot \cdot \cdot$$

within circle C_0 in Fig. 7-1.   The subscript on $f_1(s)$ implies an element of $f(s)$.   By a proof similar to that of Theorem 7-1, each coefficient is real. Therefore, for the general term

$$\frac{f^{(n)}(\sigma_0)}{n!}(\bar{s} - \sigma_0)^n = \frac{f^{(n)}(\sigma_0)}{n!}\overline{(s - \sigma_0)^n}$$

It follows that

$$f_1(\bar{s}) = \overline{f_1(s)} \qquad (7\text{-}3)$$

which may also be written

$$f(\bar{s}) = \overline{f(s)} \qquad s \text{ in } C_0$$

Now choose a complex point s_1 within this circle of convergence, and expand in a new Taylor series about s_1, to give a second element,

$$f_2(s) = f_1(s_1) + f_1'(s_1)(s - s_1) + \frac{f_1''(s_1)}{2}(s - s_1)^2 + \cdots \qquad (7\text{-}4)$$

Also expand about $\bar{s}_1$ to give a third element,

$$f_3(s) = f_1(\bar{s}_1) + f_1'(\bar{s}_1)(s - \bar{s}_1) + \frac{f_1''(\bar{s}_1)}{2}(s - \bar{s}_1)^2 + \cdots \qquad (7\text{-}5)$$

These will converge, respectively, inside circles C and C'. At the moment we do not know whether $R = R'$. However, all derivatives of $f(s)$ can themselves be expanded in series which converge inside C_0. The coefficients in these series are real, and so Eq. (7-3) applies to each derivative, giving

$$f_1'(\bar{s}_1) = \overline{f_1'(s_1)}$$
$$f_1''(\bar{s}_1) = \overline{f_1''(s_1)}$$
$$\cdots \cdots \cdots$$

and so
$$f_3(\bar{s}) = \overline{f_2(s)} \qquad s \text{ in } C \qquad (7\text{-}6)$$

Therefore, whenever s is in the circle of convergence of $f_2(s)$, $\bar{s}$ will be in the circle of convergence of $f_3(s)$, showing that $R = R'$. The two circles of convergence are mirror images of one another, reflected in the real axis.

This process of analytic continuation is repeated indefinitely, to cover the whole s plane. Always a pair of symmetrically located circles will be obtained, showing that for the general case

$$f(\bar{s}) = \overline{f(s)} \qquad (7\text{-}7)$$

A point $s = \sigma_1$, outside the original region where $f(s)$ is specified as real, is a special case of the general point in Eq. (7-7). But $\bar{\sigma}_1 = \sigma_1$, and so $f(\sigma_1) = \overline{f(\sigma_1)}$, and this is possible only if $f(\sigma_1)$ is real.

There are three important corollaries to this last theorem:

Corollary 1. At all regular points of a real analytic function,

$$f(\bar{s}) = \overline{f(s)}$$

Corollary 2. Complex zeros and singular points of a real analytic function occur in conjugate pairs.

Corollary 3. Corresponding coefficients in two Laurent expansions about a pair of conjugate points of a real analytic function are complex conjugates.

The part of Corollary 2 relating to zeros follows directly from Eq. (7-2); if s_a is a zero of $f(s)$, then $\bar{s}_a$ must also be a zero of $f(s)$. To prove the part relating to singular points, observe that

$$f'(\bar{s}) = \overline{f'(s)} \qquad (7\text{-}8)$$

and suppose that s_b is singular and that $\bar{s}_b$ is not singular. From Eq. (7-8) we would get

$$f'(s_b) = \overline{f'(\bar{s}_b)}$$

The right-hand side would exist, while the left-hand side would not, showing a contradiction, which proves that $\bar{s}_b$ must be singular if s_b is singular.

In order to prove Corollary 3, think of the formulas for the coefficients of the Laurent expansions about the pair of conjugate points s_c and $\bar{s}_c$, using the cir-

Fig. 7-2. Integration contours for evaluation of coefficients of Laurent expansions at a pair of conjugate points.

cular integration contours shown in Fig. 7-2. The coefficients of the nth terms are, respectively,

$$\frac{1}{2\pi j} \int_C \frac{f(z)}{(z - s_c)^{n+1}}\, dz \qquad \text{for point } s_c$$

$$-\frac{1}{2\pi j} \int_{C'} \frac{f(z')}{(z' - \bar{s}_c)^{n+1}}\, dz' \qquad \text{for point } \bar{s}_c$$

Note the minus sign in the second case, which arises because C' is oriented in the negative sense. In these two integrals, respectively, let

$$z = s_c + re^{j\theta} \qquad dz = jre^{j\theta}\, d\theta$$
$$z' = \bar{s}_c + re^{-j\theta} \qquad dz' = -jre^{j\theta}\, d\theta$$

and then the above two integrals become, respectively,

$$\frac{1}{2\pi} \int_0^{2\pi} \frac{f(s_c + re^{j\theta})}{r^n e^{jn\theta}}\, d\theta \qquad \text{and} \qquad \frac{1}{2\pi} \int_0^{2\pi} \frac{f(\bar{s}_c + re^{-j\theta})}{r^n e^{-jn\theta}}\, d\theta$$

The integrand of the second integral is the conjugate of the integrand of the first, and both integrals are with respect to the real variable θ. Therefore, the integrals are themselves conjugates, and we write the equation

$$-\frac{1}{2\pi j} \int_{C'} \frac{f(z')}{(z' - \bar{s}_c)^{n+1}}\, dz' = \frac{1}{2\pi j} \overline{\int_C \frac{f(z)}{(z - s_c)^{n+1}}\, dz}$$

which shows that the coefficients bear a conjugate relationship.

7-3. Gauss Mean-value Theorem (and Related Theorems).

In this section we shall prove the following theorem:

Theorem 7-3.   Let $f(s)$ be regular inside and on a circle C with center at $s_0 = \sigma_0 + j\omega_0$. Then, in the notation $f(s) = u(\sigma,\omega) + jv(\sigma,\omega)$, if u_a and v_a are, respectively, the averages (with respect to θ) of u and v on C, it is true that

$$u(\sigma_0,\omega_0) = u_a$$
$$v(\sigma_0,\omega_0) = v_a$$

PROOF.   The conditions stated in the theorem permit us to write the Cauchy integral formula for $f(s_0)$,

$$f(s_0) = \frac{1}{2\pi j} \int_C \frac{f(s)}{s - s_0} \, ds \tag{7-9}$$

But C is a circle of radius r, and so we have $s - s_0 = re^{j\theta}$ and $ds = jre^{j\theta} \, d\theta$, giving

$$f(s_0) = \frac{1}{2\pi} \int_0^{2\pi} f(s_0 + re^{j\theta}) \, d\theta \tag{7-10}$$

This is equivalent to the pair of expressions

$$u(\sigma_0,\omega_0) = \frac{1}{2\pi} \int_0^{2\pi} u(\sigma_0 + r \cos \theta, \, \omega_0 + r \sin \theta) \, d\theta$$
$$v(\sigma_0,\omega_0) = \frac{1}{2\pi} \int_0^{2\pi} v(\sigma_0 + r \cos \theta, \, \omega_0 + r \sin \theta) \, d\theta \tag{7-11}$$

The above integrals are, respectively, the averages u_a and v_a, taken with respect to θ.

Another useful theorem can be derived, as follows:

Theorem 7-4.   Let $f(s)$ be regular in and on a circle C with center at $s_0 = \sigma_0 + j\omega_0$. Also, let M_u and m_u be, respectively, the maximum and minimum of $u(\sigma,\omega)$ on C (and similarly let M_v and m_v be the respective maximum and minimum of v on C).   Then

$$m_u \leqq u(\sigma_0,\omega_0) \leqq M_u$$
$$m_v \leqq v(\sigma_0,\omega_0) \leqq M_v \tag{7-12}$$

with the equality signs holding simultaneously in both equations (if at all) if and only if $f(s)$ is identically constant.

PROOF.   The functions $u(\sigma,\omega)$ and $v(\sigma,\omega)$ on C are functions of the single variable θ.   The average of such a function must lie between its minimum and maximum; thus the inequality signs are established.   Furthermore, the average is equal to the maximum (or minimum) if and only if the function is constant, and therefore the equality signs apply simultaneously on each side of each equation. We have yet to prove that the equality signs apply simultaneously in *both* equations and that this is true if and only if $f(s)$ is identically constant. Now assume that $u(\sigma,\omega)$ is constant on C, and let $f(s)$ be written in a Taylor series, for s on C, as follows:

$$f(s) = f(s_0) + f'(s_0)re^{j\theta} + \frac{f''(s_0)}{2} r^2 e^{j2\theta} + \cdots \qquad (7\text{-}13)$$

from which the real component on C is seen to be

$$u(\sigma,\omega) = u(\sigma_0,\omega_0) + |f'(s_0)|r \cos(\theta + \alpha_1) + \frac{|f''(s_0)|}{2} r^2 \cos(2\theta + \alpha_2) + \cdots$$

where α_1, α_2, etc., are, respectively, the angles of $f'(s_0)$, $f''(s_0)$, etc. The above expression is constant if and only if each coefficient of the θ-dependent terms is zero. Thus, all derivatives must be zero at $f(s_0)$, and so $f(s)$ is identically constant. A similar proof would apply for the imaginary component. Thus, the theorem is proved.

A corresponding theorem can be proved for the function $|f(s)|$, as follows:

Theorem 7-5. If $f(s)$ is regular in and on a circle C with center at s_0, and if M and m are, respectively, the maximum and minimum of $|f(s)|$ on C, then

$$m \leqq |f(s_0)| \leqq M \qquad (7\text{-}14)$$

if $f(s)$ has no zero inside C, and

$$|f(s_0)| \leqq M \qquad (7\text{-}15)$$

if $f(s)$ has a zero inside C, the equality signs holding simultaneously (if at all) if and only if $f(s)$ is identically constant.

PROOF. If M is the maximum of $f(s)$ on C, from Eq. (7-10) we get

$$|f(s_0)| \leqq \frac{1}{2\pi} \int_0^{2\pi} |f(s_0 + re^{j\theta})|\, d\theta \leqq M$$

Furthermore, if $f(s_0)$ has no zero inside C, a similar relation for the function $1/f(s)$ gives

$$\left|\frac{1}{f(s_0)}\right| \leqq \frac{1}{2\pi} \int_0^{2\pi} \frac{1}{|f(s_0 + re^{j\theta})|}\, d\theta \leqq \frac{1}{m}$$

Thus, if $f(s)$ has no zero inside C, we have shown that

$$m \leqq f(s_0) \leqq M \qquad (7\text{-}16)$$

If $f(s)$ is identically constant, $m = M$ and the equality signs apply on both sides of Eq. (7-16). Now suppose $f(s)$ is not identically constant. Since there is no zero of $f(s)$ inside C, we know that $f(s_0)$ cannot be zero, and one other term in Eq. (7-13) must be different from zero. This equation then shows that $|f(s)|$ cannot be constant on C, and therefore that it must have an average on C lying between m and M. Thus, we have

$$|f(s_0)| \leqq \frac{1}{2\pi} \int_0^{2\pi} |f(s_0 + re^{j\theta})| \, d\theta < M$$

A similar development establishes that

$$m < |f(s_0)|$$

if $f(s)$ is not identically constant. We have proved that neither equality sign applies in Eq. (7-14) if $f(s)$ is not identically constant, and that both equality signs hold if $f(s)$ is identically constant.

If $f(s)$ has a zero inside C, the Cauchy integral formula does not apply to $1/f(s)$, and so we get Eq. (7-15) instead of Eq. (7-14). If $f(s)$ is identically constant (and therefore zero), the equality sign obviously holds. Now assume that $f(s)$ is not identically constant. If $f(s_0) \neq 0$, the previous argument yields the information that $|f(s_0)| < M$. However, it is now possible to have $f(s_0) = 0$, in which case Eq. (7-13) shows that $|f(s)|$ will be constant on C if all derivatives except the first are zero at s_0. Thus, for this particular case the earlier argument that $|f(s)|$ cannot be constant on C does not apply. However, since $f(s_0) = 0$, we must have $|f(s_0)| < M$ unless $M = 0$, in which case $f(s)$ is identically constant, contrary to the original assumption. Thus, the equality sign in Eq. (7-15) applies if and only if $f(s)$ is identically constant and equal to zero.

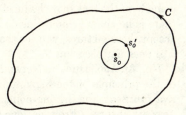

FIG. 7-3. A simple closed curve in the s plane lying in a region where $f(s)$ is regular.

7-4. Principle of the Maximum and Minimum.

Theorems 7-4 and 7-5 give estimates of a function at the center of a circle, in terms of values on the circle. We now extend these two theorems to regions having boundaries which are noncircular simple closed curves, like Fig. 7-3.

Theorem 7-6. Let R be the region consisting of C and its interior, and let $f(s)$ be regular and not identically constant in R. Then the maximum values of $|f(s)|$, $u(\sigma,\omega)$, and $v(\sigma,\omega)$ in R occur on boundary C; and the minimum values of $u(\sigma,\omega)$ and $v(\sigma,\omega)$ in R occur on boundary C. If $f(s)$ has no zero in R, $|f(s)|$ also attains its minimum in R on boundary C.

PROOF. The proof is exactly the same for all three functions. Suppose that we assume that there is an internal point s_0 at which $|f(s_0)|$ attains the maximum of $|f(s)|$ for s in R. By Theorem 7-5 there is a point s_0' on a circle centered at s_0 having the property

$$|f(s_0)| < |f(s_0')|$$

But this contradicts the assumption that $|f(s_0)|$ is the maximum in R, and so no internal point can be a maximum of $|f(s)|$. If $f(s)$ has no zero in R, we assume that $|f(s_0)|$ is a minimum and are again led to a contradiction. Similar results are obtained for the functions $u(\sigma,\omega)$ and $v(\sigma,\omega)$ by using Theorem 7-4.

The principles of the maximum and minimum have very simple geometric interpretations. Think of the contour C in the s plane and its trace in the $f(s)$ plane. Two cases are illustrated in Fig. 7-4, at (b), where

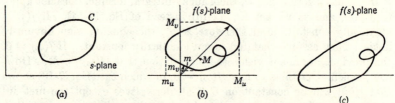

FIG. 7-4. Geometrical illustration of principle of the maximum and minimum.

$f(s)$ has no zero inside C, and at (c), where there is such a zero. We are using here the fact that the inside of C maps into the inside of its trace, if there is no singular point inside C, a property yet to be proved. If you accept these pictures, you see that for case c the minimum of $|f(s)|$ is zero d does not occur on C.

7-5. An Application to Network Theory. An example of the usefulness of the principle of the maximum can be taken from the class of *positive real functions*, the driving-point impedance and admittance functions of network theory. From various considerations of the physical properties of networks it can be shown that such a function $Z(s)$ is analytic and has the following properties:

$$\begin{aligned} \text{Re}\,[Z(s)] &\geqq 0 && \text{when Re } s \geqq 0 \\ Z(s) \text{ is real} && \text{when } s \text{ is real} \end{aligned} \tag{7-17}$$

These conditions define a positive real function. We shall now demonstrate the interesting fact that, if $Z(s)$ is positive real, then

$$|\text{ang } Z(s)| \leqq |\text{ang } s|$$

when

$$-\frac{\pi}{2} \leqq \text{ang } s \leqq \frac{\pi}{2} \tag{7-18}$$

The proof depends on the principle of the maximum and on the transformation properties of the function

$$w = \frac{z - 1}{z + 1} \tag{7-19}$$

which are presented in detail in Chap. 3. We need the following properties of this function:

1. The j axis of the z plane becomes the unit circle of the w plane.

2. The right half of the z plane goes into the inside of the unit circle in the w plane.

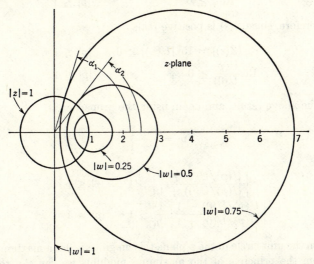

FIG. 7-5. Pertinent properties in the z plane of the transformation $w = (z - 1)/(z + 1)$.

3. A circle $|w| = b$ in the w plane goes into a circle in the z plane having intercepts $(1 + b)/(1 - b)$ and $(1 - b)/(1 + b)$ on the positive real axis

4. The circle $|z| = 1$ is orthogonal to all circles of the family described in (3).

Figure 7-5 illustrates these properties. For proofs you should consult Chap. 3. A further property of the transformation is evident from the figure. This property is:

5. If we let $z_1 = e^{j\alpha_1}$, $z_2 = e^{j\alpha_2}$ be two values of z on the unit circle, and if

$$|w_2| < |w_1|$$

then it is evident from Fig. 7-5 that

$$|\alpha_2| < |\alpha_1|$$

This property results from the fact that circle $|w_2| = $ constant lies inside circle $|w_1|$ = constant and that each of these circles is normal to the circle $|z| = 1$.

Now use the transformation

$$p = \frac{s/R - 1}{s/R + 1} \tag{7-20}$$

where R is a positive real number, and adopt the notation $U(p) = Z(s)$, from which we have $U(0) = Z(R) \geqq 0$. Equation (7-20) is of the form $w = (z - 1)/(z + 1)$, and so

$$\text{Re } s \geqq 0 \qquad\qquad |p| \leqq 1$$

and therefore, since $Z(s)$ is positive real,

$$\text{Re } [Z(s)] = \text{Re } [U(p)] \geqq 0 \qquad |p| \leqq 1$$

and
$$\text{Re } \frac{U(p)}{U(0)} \geqq 0 \qquad\qquad |p| \leqq 1$$

From the above result, and again using the properties of

$$w = \frac{z - 1}{z + 1}$$

we get

$$\left| \frac{U(p)/U(0) - 1}{U(p)/U(0) + 1} \right| \leqq 1 \qquad |p| \leqq 1$$

and
$$\left| \frac{U(p)/U(0) - 1}{U(p)/U(0) + 1} \right| \frac{1}{|p|} \leqq 1 \qquad |p| = 1$$

Thus, on the unit circle in the p plane this magnitude has a maximum of 1, and from the principle of the maximum modulus we can say that the last relation is true for $|p| \leqq 1$. Of course, this theorem applies only if the function in question is regular for $|p| \leqq 1$. That this is the case can be seen by observing (1) that $\text{Re } [U(p)] \geqq 0$, preventing the factor $U(p)/U(0) + 1$ from becoming zero, (2) that any pole of $U(p)$ will not be a pole of the bilinear function, and (3) that the pole of $1/p$ is canceled by the numerator.* Thus, we can also write

$$\left| \frac{U(p)/U(0) - 1}{U(p)/U(0) + 1} \right| \leqq |p| \qquad\qquad |p| \leqq 1$$

Now write this result in terms of s variables, as follows:

$$\left| \frac{Z(s)/Z(R) - 1}{Z(s)/Z(R) + 1} \right| \leqq \left| \frac{s/R - 1}{s/R + 1} \right| \qquad \text{Re } s \geqq 0 \tag{7-21}$$

* It can be shown that $U(p)$ can have poles only on the circle $|p| = 1$.

In the earlier notation we can let

$$z_1 = \frac{s}{R} \qquad z_2 = \frac{Z(s)}{Z(R)}$$

and from property 5 of the function

$$\frac{z-1}{z+1}$$

$$|\text{ang } z_2| \leqq |\text{ang } z_1| \tag{7-22}$$

but since R and $Z(R)$ are real, this is the same as

$$|\text{ang } Z(s)| \leqq |\text{ang } s| \tag{7-23}$$

which is the required result.

7-6. The Index Principle. Let $f(s)$ be an analytic function having a zero of order n_p at $s = s_p$. Then we can write

$$f(s) = (s - s_p)^{n_p} F(s) \tag{7-24}$$

where $F(s)$ is nonzero and regular at $s = s_p$. Observing that the derivative is

$$f'(s) = n_p(s - s_p)^{n_p-1} F(s) + (s - s_p)^{n_p} F'(s)$$

we get the new function

$$\frac{f'(s)}{f(s)} = \frac{n_p}{s - s_p} + \frac{F'(s)}{F(s)} \tag{7-25}$$

Recalling that $F(s_p)$ is not zero, it is evident from the above that $f'(s)/f(s)$ has a simple pole with residue n_p at $s = s_p$.

Now suppose that $f(s)$ has a pole of order m_q at $s = s_q$. Following the same pattern as above, we can write

$$f(s) = \frac{1}{(s - s_q)^{m_q}} G(s) \tag{7-26}$$

where $G(s)$ is nonzero and regular at $s = s_q$. In this case the derivative is

$$f'(s) = \frac{-m_q}{(s - s_q)^{m_q+1}} G(s) + \frac{1}{(s - s_q)^{m_q}} G'(s)$$

and $\qquad \dfrac{f'(s)}{f(s)} = \dfrac{-m_q}{s - s_q} + \dfrac{G'(s)}{G(s)} \tag{7-27}$

Thus, in this case the function $f'(s)/f(s)$ has a simple pole with residue $-m_q$ at $s = s_q$.

With Eqs. (7-25) and (7-27) established, we now consider a positively oriented contour C in the s plane enclosing N zeros of orders

$n_1, n_2, \ldots, n_N$ and M poles of orders $m_1, m_2, \ldots, m_M$. By using the calculus of residues, it follows that

$$\int_C \frac{f'(s)}{f(s)}\,ds = j2\pi \left(\sum_{p=1}^{N} n_p - \sum_{q=1}^{M} m_q \right) \tag{7-28}$$

Since

$$\frac{d[\log f(s)]}{ds} = \frac{f'(s)}{f(s)}$$

the above integral can also be evaluated in terms of this antiderivative if curve C can be interpreted as lying in a simply connected region of regularity of $f'(s)/f(s)$. This we do by identifying a starting point s_1 and an ending point s_1' which are infinitesimally close, but on opposite sides of a barrier, as shown in Fig. 7-6. Thus we can say

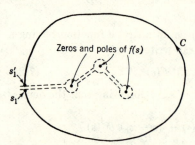

$$\int_C \frac{f'(s)}{f(s)}\,ds = \log f(s_1') - \log f(s_1)$$
$$= \log |f(s_1')| - \log |f(s_1)|$$
$$+ j[\text{ang } f(s_1') - \text{ang } f(s_1)] \tag{7-29}$$

Fig. 7-6. Use of a barrier to allow C to lie in a simply connected region of regularity of $f'(s)/f(s)$.

But $\log |f(s_1')| = \log |f(s_1)|$, and the imaginary component is the increase in angle of $f(s)$ as s moves around C from s_1 to s_1'. Therefore, combining Eqs. (7-28) and (7-29) gives

Increase in angle of $f(s)$ in going around a curve C

$$= 2\pi \left(\sum_{p=1}^{N} n_p - \sum_{q=1}^{M} m_q \right) \tag{7-30}$$

If both sides of the above equation are divided by 2π, the result is the *number of times* $f(s)$ passes around the origin. This result can be stated as follows:

Theorem 7-7. If s travels a simple closed curve C in the s plane, and if there are no singular points other than poles inside C, and no singular or zero points on C, then the number of times $f(s)$ passes around the origin is equal to the weighted sum of the number of zeros inside C minus the weighted sum of the number of poles inside C, where in each case the weighting number is the order of the zero or pole. In this statement positive encirclement is counterclockwise.

This theorem admits of a simple geometrical interpretation. Referring to Fig. 7-7, we can see that, if a point a_k is inside C, as C is traced in a positive sense the angle of a quantity $(s - a_k)^{\alpha_k}$ will increase by $2\pi\alpha_k$. But if a_k is a point outside C, there is no net angle change as C is traced. Similarly, if b_k is another point inside C, the quantity $(s - b_k)^{-\beta_k}$ will decrease in angle by an amount $2\pi\beta_k$ but will have no net angle change if b_k is outside C. Now, if $f(s)$ is a rational function,

$$f(s) = \frac{(s - a_1)^{\alpha_1}(s - a_2)^{\alpha_2} \cdots (s - a_n)^{\alpha_n}}{(s - b_1)^{\beta_1}(s - b_2)^{\beta_2} \cdots (s - b_m)^{\beta_m}}$$

we can apply the above reasoning to each factor of the numerator and denominator. If a_k or b_k is inside C, the corresponding term contributes $+2\pi\alpha_k$ or, respectively, $-2\pi\beta_k$ to the total angle change of $f(s)$. From this observation we are led to the conclusion stated in Theorem 7-7, but the above analysis applies only if $f(s)$ is a rational function.

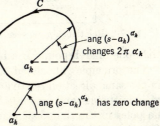

FIG. 7-7. Effect, on angle of $f(s)$, of a zero or pole being inside or outside a simple closed curve C.

7-7. Applications of the Index Principle, Nyquist Criterion. Let $f(s)$ be an analytic function, and in the s plane choose a point s_0 inside a simple closed curve C. Furthermore, assume that there are no poles or other singular points inside or on C. We form the new function

$$w(s) = f(s) - f(s_0)$$

which has a zero at s_0. There may be some other values $s_1, s_2, \ldots, s_n$ inside C such that

$$f(s_1) = f(s_2) = \cdots = f(s_n) = f(s_0)$$

Then $w(s)$ has $n + 1$ zeros inside C, and by Theorem 7-7 we know that the w point will encircle the origin $n + 1$ times as s travels over curve C. The corresponding $f(s)$-plane curve is the same as the $w(s)$-plane curve, except that the origin is shifted an amount $f(s_0)$. An illustration is shown in Fig. 7-8.

Point s_0 is a typical point inside C, and from the above we conclude that the inside of C maps into the inside of the trace of C in the $f(s)$ plane. Note, however, that this is true only if $f(s)$ is regular inside and on C.

Another interesting observation shows that, when C is traced with its area on the left, the area is also on the left in the $f(s)$ plane.

If point $f(s_0)$ is encircled more than once, as in the case of Fig. 7-8, the inverse function which transforms from the $f(s)$ plane to the s plane must be multivalued and that part of the area which is enclosed twice must lie in two sheets of a Riemann surface. Consequently, a branch cut must be crossed twice in the course of a complete traversal of the curve. This is possible only if there is a branch point inside the double part of the curve. Higher numbers of encirclements would correspondingly indicate more

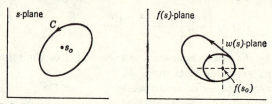

FIG. 7-8. Relationship between enclosed areas in the s plane and the $f(s)$ plane.

than one branch point, or a higher-order branch point, inside the curve.

The principle enunciated above, that in both planes curves are traced in the same direction relative to their enclosed areas, is useful in plotting of immittance- and response-function loci. These are functions having no poles in the right half plane. In practical cases these functions are finite when s is infinite, so that the $j\omega$ axis of the s plane goes into a bounded closed curve in the function plane. We can consider the right half plane to be "enclosed" by the $j\omega$ axis plus a large semicircle to the right,

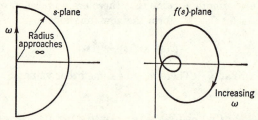

FIG. 7-9. Locus of the $j\omega$ axis for a positive real function.

as implied by Fig. 7-9. Then the right-half s plane maps into the interior of the function plane; and, furthermore, the s-plane area is encompassed in a clockwise direction, and therefore the same will be true in the function plane. Thus, we have shown that for real frequencies (imaginary s) the loci of immittance functions and response functions of stable systems are traced out in such a way that the curve moves in a clockwise direction as ω increases. This idea comes in handy in plotting experimental data; the constraint on direction of encirclement can sometimes help in determining where the locus goes when the data points are too few to determine the curve from the points themselves.

The *Nyquist criterion* for stability of a feedback system is another extension of the index principle. In Chap. 6 it is pointed out that the response function for a system like Fig. 6-14 is*

$$F(s) = \frac{kA(s)}{1 + kA(s)H(s)}$$

Assume that $A(s)$ and $H(s)$ are the response functions of stable systems. Then, the single-loop system is stable if there are no zeros of $1 + kA(s)H(s)$ in the right half plane. As in the previous example, if $A(s)H(s)$ is finite at infinite s, the $j\omega$ axis may be regarded as enclosing the right-half s plane; and we know that this right half plane will map onto the inside of a locus in the plane of the function

$$w(s) = 1 + kA(s)H(s)$$

If this locus does not encircle the origin of the w plane, there are no zeros in the right half of the s plane and the system is stable. The process of plotting this locus and checking for encirclements of the origin is the Nyquist test for stability.

Note that the origin of the w plane goes into the point -1 in the plane of the function $kA(s)H(s)$. Therefore, we can just as well plot the latter function, taking $s = j\omega$, and check whether or not the resulting locus encircles the point -1. This is the usual form in which the Nyquist criterion is employed.

This theory, presented here in only its simplest form, can be extended to the case where either $A(s)$ or $H(s)$ is open-loop unstable (see Prob. 7-18).

7-8. Poisson's Integrals. Recall the Cauchy integral formula

$$f(s) = \frac{1}{2\pi j} \int_C \frac{f(z)}{z - s}\, dz \tag{7-31}$$

where s is a point inside a closed curve C and $f(s)$ is regular in the closed region consisting of C and its interior. This formula is interesting because it shows that specification of $f(s)$ on C also determines $f(s)$ inside C. Further interpretation of this formula can be obtained by specifying C to be a circle centered at the origin, with no loss in generality.

In Eq. (7-31) let

$$\begin{aligned} z &= Re^{j\theta} \\ dz &= jRe^{j\theta}\, d\theta = jz\, d\theta \\ s &= \rho e^{j\phi} \end{aligned} \tag{7-32}$$

* A $+$ sign appears in the denominator in order to conform to the usual treatments of the Nyquist criterion. This change means that a $-$ sign would appear in the circle which symbolizes combination of the two signals in Fig. 6-14.

as shown in Fig. 7-10. Now write the factor $1/(z - s)$ appearing in the integrand of Eq. (7-31) as follows,

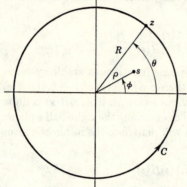

$$\frac{1}{z - s} = \frac{1}{z}\frac{z}{z - s}$$

and develop it by the following sequence of steps:

$$\frac{1}{z - s} = \frac{z(\bar{z} - \bar{s})}{z(z - s)(\bar{z} - \bar{s})}$$

$$= \frac{z\bar{z} - z\bar{s} + s\bar{s} - s\bar{s}}{z(z - s)(\bar{z} - \bar{s})}$$

$$= \frac{z\bar{z} - s\bar{s} + \bar{s}(s - z)}{z(z - s)(\bar{z} - \bar{s})}$$

$$= \frac{R^2 - \rho^2}{z|z - s|^2} + \frac{1}{z - z\bar{z}/\bar{s}}$$

FIG. 7-10. Definition of s and z for derivation of Poisson's integrals.

If this is substituted in Eq. (7-31), we get

$$f(s) = \frac{1}{2\pi j}\int_C \frac{(R^2 - \rho^2)f(z)}{|z - s|^2}\frac{dz}{z} + \frac{1}{2\pi j}\int_C \frac{f(z)}{z - z\bar{z}/\bar{s}}\,dz$$

From Eq. (7-32) it is seen that $dz/z = j\,d\theta$, and so the first integral reduces to a single real integral. The second integral is zero because the singular point at $z\bar{z}/\bar{s} = R^2/\bar{s}$ lies outside the circle. Thus, we have the result

$$f(s) = \frac{1}{2\pi}\int_0^{2\pi} \frac{(R^2 - \rho^2)f(Re^{j\theta})}{|Re^{j\theta} - s|^2}\,d\theta \tag{7-33}$$

This may appear to be a more complicated formula than Eq. (7-31), but it is amenable to a particular interpretation because, as will now be shown, it can be written as the sum of two real integrals. In Eq. (7-33), z is on circle C, and accordingly it is appropriate to adopt the notation

$$f(Re^{j\theta}) = \text{Re}\,[f(Re^{j\theta})] + j\,\text{Im}\,[f(Re^{j\theta})]$$
$$f(s) = u(\sigma,\omega) + jv(\sigma,\omega)$$

giving the pair of equations

$$u(\sigma,\omega) = \frac{1}{2\pi}\int_0^{2\pi} \frac{(R^2 - \rho^2)\,\text{Re}\,[f(Re^{j\theta})]}{|Re^{j\theta} - s|^2}\,d\theta \tag{7-34a}$$

$$v(\sigma,\omega) = \frac{1}{2\pi}\int_0^{2\pi} \frac{(R^2 - \rho^2)\,\text{Im}\,[f(Re^{j\theta})]}{|Re^{j\theta} - s|^2}\,d\theta \tag{7-34b}$$

Thus it is seen that the real (respectively imaginary) component of $f(s)$ can be determined at points inside C in terms of a real integral of the real (respectively imaginary) component of the function on C.

A second pair of relations is obtained by writing

$$\frac{1}{z - s} = \frac{1}{z}\left(1 + \frac{s}{z - s}\right)$$

which can be modified as follows:

$$
\begin{aligned}
\frac{1}{z - s} &= \frac{1}{z} + \frac{s(\bar{z} - \bar{s})}{z(z - s)(\bar{z} - \bar{s})} \\
&= \frac{1}{z} + \frac{s\bar{z} - s\bar{s} + z\bar{s} - z\bar{s}}{z(z - s)(\bar{z} - \bar{s})} \\
&= \frac{1}{z} - \frac{z\bar{s} - \bar{z}s - \bar{s}(z - s)}{z(z - s)(\bar{z} - \bar{s})} \\
&= \frac{1}{z} - j\frac{2 \operatorname{Im} (z\bar{s})}{z|z - s|^2} - \frac{1}{z - z\bar{z}/\bar{s}}
\end{aligned}
$$

When this is substituted in Eq. (7-31), the last term yields zero because the pole is not enclosed by the contour. Also,

$$\frac{1}{2\pi j}\int_C \frac{f(z)}{z}\,dz = f(0)$$

and
$$\operatorname{Im} (z\bar{s}) = R\rho \sin (\theta - \phi)$$

Equation (7-31) then gives

$$f(s) = f(0) + \frac{1}{\pi j}\int_0^{2\pi} \frac{R\rho \sin (\theta - \phi)f(Re^{j\theta})}{|Re^{j\theta} - s|^2}\,d\theta \qquad (7\text{-}35)$$

Again the factor multiplying $f(Re^{j\theta})$ in the integrand is real, and so real and imaginary components can be separated, as was done in the previous case. The following pair of formulas is obtained:

$$u(\sigma,\omega) = u(0,0) + \frac{1}{\pi}\int_0^{2\pi} \frac{R\rho \sin (\theta - \phi) \operatorname{Im} [f(Re^{j\theta})]}{|Re^{j\theta} - s|^2}\,d\theta \quad (7\text{-}36a)$$

$$v(\sigma,\omega) = v(0,0) - \frac{1}{\pi}\int_0^{2\pi} \frac{R\rho \sin (\theta - \phi) \operatorname{Re} [f(Re^{j\theta})]}{|Re^{j\theta} - s|^2}\,d\theta \quad (7\text{-}36b)$$

Equations (7-36) have an interpretation similar to Eqs. (7-34), but an interchange of real and imaginary components has occurred in the integrand. The integrals in Eqs. (7-34) and (7-36) are called *Poisson's integrals.*

Further interpretation is possible. Note that Eq. (7-34a) and Eq. (7-36b) together give the real and imaginary components of $f(s)$ in terms of only the real part of $f(Re^{j\theta})$. Similarly, a combination of Eqs. (7-34b) and (7-36a) gives $f(s)$ in terms of the imaginary component of $f(Re^{j\theta})$. Two more formulas are therefore possible, as follows:

$$f(s) = jv(0,0) + \frac{1}{2\pi} \int_0^{2\pi} \frac{[(R^2 - \rho^2) - j2R\rho \sin (\theta - \phi)] \operatorname{Re} [f(Re^{j\theta})]}{|Re^{j\theta} - s|^2} d\theta$$

$$(7\text{-}37a)$$

$$f(s) = u(0,0) + \frac{1}{2\pi} \int_0^{2\pi} \frac{[2R\rho \sin (\theta - \phi) + j(R^2 - \rho^2)] \operatorname{Im} [f(Re^{j\theta})]}{|Re^{j\theta} - s|^2} d\theta$$

$$(7\text{-}37b)$$

These formulas are important because they show that $f(s)$ is completely determined inside C by having either the real or imaginary component of $f(Re^{j\theta})$ specified on C.

The derivation of these formulas is based on the condition that $f(s)$ should be regular on the circle $|s| = R$. Now suppose that a function $h(\theta)$ is given, without specifying it to be the real or imaginary component of an analytic function, and let this function $h(\theta)$ be used in either one of Eqs. (7-37) in place of $\operatorname{Re} [f(Re^{j\theta})]$ or $\operatorname{Im} [f(Re^{j\theta})]$. If $h(\theta)$ has isolated singular points (discontinuities, for example) but is integrable, then either of Eqs. (7-37) will yield a function which can be shown to be regular for $|s| < R$. However, on the circle $|s| = R$ this function will be singular at the points $s = Re^{j\theta}$, where $h(\theta)$ is singular. Thus, the integral formulas in Eqs. (7-37) define functions for $|s| < R$ which were not included in the original derivation.

Further insight into the above question is acquired by considering the general function $h(\theta)$ and looking at its Fourier series

$$h(\theta) = \sum_{n=-\infty}^{\infty} c_n e^{jn\theta} = c_0 + \sum_{n=1}^{\infty} c_n e^{jn\theta} + \sum_{n=1}^{\infty} c_{-n} e^{-jn\theta}$$

which is assumed to exist. Since $h(\theta)$ is real, it is known that c_0 is real and also that

$$c_{-n} = \bar{c}_n$$

Therefore, $$h(\theta) = \left(\frac{c_0}{2} + \sum_{n=1}^{\infty} c_n e^{jn\theta} \right) + \left(\frac{c_0}{2} + \sum_{n=1}^{\infty} \bar{c}_n e^{-jn\theta} \right) \qquad (7\text{-}38)$$

is a possible representation, which is useful because the second term is the conjugate of the first.

If the series

$$\frac{c_0}{2} + \sum_{n=1}^{\infty} c_n e^{jn\theta}$$

converges, so also will

$$\frac{c_0}{2} + \sum_{n=1}^{\infty} \frac{c_n}{R^n} s^n$$

converge in and on the circle $s = Re^{j\theta}$, since if s is on the circle the two series are identical and we know that a power series converges inside any circle on which it converges. Now we can use this series to define the analytic function

$$f(s) = 2\left(\frac{c_0}{2} + \sum_{n=1}^{\infty} \frac{c_n}{R^n} s^n\right) \tag{7-39}$$

which will have the property, put into evidence by Eq. (7-38), that

$$h(\theta) = \frac{f(Re^{j\theta}) + \overline{f(Re^{j\theta})}}{2} = \mathrm{Re}\,[f(Re^{j\theta})]$$

Note that Eq. (7-39) is an alternative form for Eq. (7-37a). It also gives $f(s)$ in terms of its real component on the unit circle, although in this case the real component enters by way of the coefficients of the Fourier series.

The series presentation is useful because it provides a condition on a given $h(\theta)$ function (namely, that the positive-power series shall converge) which is sufficient to make it serve as the real component, on a circle, of some analytic function which will be regular inside and on the circle. If the positive-power series does not converge, the $f(s)$ function will be singular at some point on the circle $|s| = R$.

In place of Eq. (7-39) we could have written the different function

$$f(s) = j2\left(\frac{c_0}{2} + \sum_{n=1}^{\infty} \frac{c_n}{R^n} s^n\right) \tag{7-40}$$

In this case, Eq. (7-38) gives

$$h(\theta) = \frac{f(Re^{j\theta})}{2j} + \left[\overline{\frac{f(Re^{j\theta})}{2j}}\right]$$

$$= \frac{f(Re^{j\theta}) - \overline{f(Re^{j\theta})}}{2j} = \mathrm{Im}\,[f(Re^{j\theta})]$$

and Eq. (7-40) becomes an alternative to Eq. (7-37b) for finding $f(s)$ from the imaginary component on a circle. Again convergence of the positive-power series of Eq. (7-38) is seen to be sufficient for $h(\theta)$ to serve as the imaginary component of $f(Re^{j\theta})$, such that $f(s)$ will be regular on the unit circle.

A digression into the subject of Fourier series would be out of place here, and so we shall close this discussion by considering the square-wave function

$$h(\theta) = \begin{cases} 1 & 0 < \theta < \pi \\ 0 & \pi < \theta < 2\pi \end{cases}$$

This is not continuous and possesses no derivative at the points of discontinuity at $\theta = 0$, π.   The series representation is

$$
h(\theta) = \frac{4}{\pi}\left(\sin\theta + \tfrac{1}{3}\sin 3\theta + \tfrac{1}{5}\sin 5\theta + \cdots \right)
$$
$$
= \frac{2}{j\pi}\left(e^{j\theta} + \tfrac{1}{3}e^{j3\theta} + \tfrac{1}{5}e^{j5\theta} + \cdots - e^{-j\theta} - \tfrac{1}{3}e^{-j3\theta} - \tfrac{1}{5}e^{-j5\theta} - \cdots \right)
$$

and it converges.   However, the series

$$
\frac{4}{\pi}\left(e^{j\theta} + \tfrac{1}{3}e^{j3\theta} + \tfrac{1}{5}e^{j5\theta} + \cdots \right)
$$

does not converge for $\theta = 0$ or $\theta = \pi$.   Likewise, the complex series

$$
s + \tfrac{1}{3}s^3 + \tfrac{1}{5}s^5 + \cdots
$$

diverges for $s = \pm 1$ but converges for $|s| < 1$.

If the above $h(\theta)$ is used for Re $[f(Re^{j\theta})]$ or Im $[f(Re^{j\theta})]$ in Eqs. (7-37), the integrals will still exist and will yield functions which are regular at

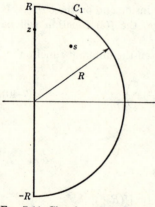

Fig. 7-11. Closed contour C consisting of arc C_1 and j axis from $-R$ to R.

points inside the circle.   However, this analytic function cannot be analytically continued to all points on the boundary; it will be singular at points where $h(\theta)$ is singular.

7-9. Poisson's Integrals Transformed to the Imaginary Axis.* The imaginary axis is a degenerate case of a circle, and so we might reasonably expect to be able to write formulas similar to the Poisson's integrals by employing an integration along the imaginary (or possibly the real) axis.   However, we cannot expect the formulas to look the same, because R, which appears in the formulas, must be infinite.   Also, since integration will then be carried to infinity, it will be necessary to restrict $f(s)$ to being regular at infinity and at all points on or to the right of the j axis.

The easiest way to proceed is to start again, by applying the Cauchy integral formula to the path of Fig. 7-11.   The integrand can be written

$$
\frac{1}{z - s} = \frac{1}{z - s}\frac{z + \bar{s}}{z + \bar{s}} = \frac{z + \bar{s} + s - s}{(z - s)(z + \bar{s})}
$$

---

* Reference to Chap. 8 may be helpful in dealing with the improper integrals occurring in this and subsequent sections.

and by writing $s + \bar{s} = 2\sigma$ and $s - \bar{s} = j2\omega$ the above becomes

$$\frac{1}{z - s} = \frac{2\sigma}{z^2 - \sigma^2 - \omega^2 - j2\omega z} + \frac{1}{z + \bar{s}} \tag{7-41}$$

In Fig. 7-11 let C_1 be the arc of radius R, and let C be the closed curve traced in a clockwise direction. The point $z = -\bar{s}$ is not enclosed by C, and so

$$\int_C \frac{f(z)\,dz}{z + \bar{s}} = 0 \tag{7-42}$$

Therefore, writing $z = jy$ on the portion of the integration on the imaginary axis, and noting that the direction of integration is in the negative sense, we get

$$f(s) = -\frac{1}{2\pi j} \int_C \frac{f(z)}{z - s}\,dz = \frac{1}{\pi} \int_{-R}^R \frac{\sigma f(jy)\,dy}{\sigma^2 + \omega^2 + y^2 - 2\omega y}$$
$$- \frac{1}{\pi j} \int_{C_1} \frac{\sigma f(z)\,dz}{z^2 - \sigma^2 - \omega^2 - j2\omega z} \tag{7-43}$$

Since $f(z)$ is regular at infinity, it must approach a constant there. Therefore, as R approaches infinity, the integral over C_1 behaves as follows,

$$\lim_{R \to \infty} \int_{C_1} \frac{\sigma f(z)\,dz}{z^2 - \sigma^2 - \omega^2 - j2\omega z} = \lim_{R \to \infty} j \int_{-\pi/2}^{\pi/2} \frac{\sigma f(Re^{j\theta}) Re^{j\theta}\,d\theta}{R^2 e^{j2\theta}} = 0$$

where in the second integral we have used $z = Re^{j\theta}$, $dz = jRe^{j\theta}\,d\theta$ and all terms except z^2 in the denominator are omitted because they become negligible as R approaches infinity. Therefore, in the limit,

$$f(s) = \frac{1}{\pi} PV \int_{-\infty}^{\infty} \frac{\sigma f(jy)}{\sigma^2 + (\omega - y)^2}\,dy$$

The principal value occurs here because the integration is always from $-R$ to R.* However, since $f(jy)$ approaches a constant at infinity and the denominator goes to infinity as y^2, the integral will converge without taking the principal value. Thus, we have the final form

$$f(s) = \frac{1}{\pi} \int_{-\infty}^{\infty} \frac{\sigma f(jy)}{\sigma^2 + (\omega - y)^2}\,dy \tag{7-44}$$

By now we have become accustomed to pairs of formulas, and so a second formulation for $f(s)$ is to be expected. To get this, we write

$$\frac{1}{z - s} = \frac{1}{z - s}\frac{z + \bar{s}}{z + \bar{s}} = \frac{z + \bar{s} + z - z + s - s}{(z - s)(z + \bar{s})}$$

* See Sec. 8-4.

and again use $s + \bar{s} = 2\sigma$ and $s - \bar{s} = j2\omega$ to give

$$\frac{1}{z - s} = \frac{2z - j2\omega}{z^2 - \sigma^2 - \omega^2 - j2\omega z} - \frac{1}{z + \bar{s}} \tag{7-45}$$

Equation (7-42) applies, showing that the integral of the last term is zero for contour C of Fig. 7-11. Therefore, $f(s)$ is given by

$$f(s) = -\frac{1}{2\pi j} \int_C \frac{f(z)}{z - s}\, dz = \frac{1}{\pi j} \int_{-R}^{R} \frac{(w - y)f(jy)\, dy}{\sigma^2 + \omega^2 + y^2 - 2\omega y}$$
$$- \frac{1}{\pi j} \int_{C_1} \frac{(z - j\omega)f(z)\, dz}{z^2 - \sigma^2 - \omega^2 - j2\omega z} \tag{7-46}$$

As the radius R approaches infinity, the integral over C_1 approaches

$$\lim_{R \to \infty} \frac{1}{\pi j} \int_{\pi/2}^{-\pi/2} \frac{Re^{j\theta}f(Re^{j\theta})jRe^{j\theta}}{R^2 e^{j2\theta}}\, d\theta = \lim_{R \to \infty} \frac{1}{\pi} \int_{\pi/2}^{-\pi/2} f(Re^{j\theta})\, d\theta = -f(\infty)$$

Therefore, for this case we get

$$f(s) = f(\infty) + \frac{1}{\pi j} PV \int_{-\infty}^{\infty} \frac{(w - y)f(jy)\, dy}{\sigma^2 + (\omega - y)^2} \tag{7-47}$$

Here the principal value must be retained, because the integrand approaches zero only at the rate $1/y$ and $f(\infty)$ is not necessarily zero.

Equations (7-44) and (7-47) break down into two pairs, using real and imaginary components, as follows:

$$u(\sigma,\omega) = \frac{1}{\pi} \int_{-\infty}^{\infty} \frac{\sigma\, \mathrm{Re}\, [f(jy)]}{\sigma^2 + (\omega - y)^2}\, dy \tag{7-48a}$$

$$v(\sigma,\omega) = \frac{1}{\pi} \int_{-\infty}^{\infty} \frac{\sigma\, \mathrm{Im}\, [f(jy)]}{\sigma^2 + (\omega - y)^2}\, dy \tag{7-48b}$$

and

$$u(\sigma,\omega) = u(\infty,0) + \frac{1}{\pi} \int_{-\infty}^{\infty} \frac{(\omega - y)\, \mathrm{Im}\, [f(jy)]}{\sigma^2 + (\omega - y)^2}\, dy \tag{7-49a}$$

$$v(\sigma,\omega) = v(\infty,0) - \frac{1}{\pi} \int_{-\infty}^{\infty} \frac{(\omega - y)\, \mathrm{Re}\, [f(jy)]}{\sigma^2 + (\omega - y)^2}\, dy \tag{7-49b}$$

Finally, combining a real and imaginary component from each pair, we get two more formulas:

$$f(s) = jv(\infty,0) + \frac{1}{\pi} \int_{-\infty}^{\infty} \frac{[\sigma - j(\omega - y)]\, \mathrm{Re}\, [f(jy)]}{\sigma^2 + (\omega - y)^2}\, dy \tag{7-50}$$

and

$$f(s) = u(\infty,0) + \frac{1}{\pi} \int_{-\infty}^{\infty} \frac{(\omega - y + j\sigma)\, \mathrm{Im}\, [f(jy)]}{\sigma^2 + (\omega - y)^2}\, dy \tag{7-51}$$

In appraising these results it is necessary to understand that $f(s)$ must be regular on and to the right of the j axis and that in these formulas s is in the right half plane.

With these results the utility of this work begins to become apparent. If we are dealing with a frequency-domain response function, ω is the real frequency variable (imaginary component of the generalized frequency s). The last two formulas show that, if *either* the real or the imaginary component of a response function is specified along the j axis (i.e., for real frequencies), then the response function is completely determined. Of course, these formulas give $f(s)$ only in the right half plane. But by analytic continuation we know that the function would be completely determined.

7-10. Relationships between Real and Imaginary Parts, for Real Frequencies. In the previous section $u(\sigma,\omega)$ and $v(\sigma,\omega)$ are given in terms of $\text{Re}\,[f(jy)] = u(0,y)$ and $\text{Im}\,[f(jy)] = v(0,y)$. The integrands contain functions of the real frequency variable y, but the functions obtained from the integrals are functions of the two variables σ and ω. For practical cases, in which we are interested in functions of real frequency, it is important to know whether we can set $\sigma = 0$, in order to relate $u(0,\omega)$ to $v(0,\omega)$, or vice versa.

From Eqs. (7-49),

$$u(0,\omega) - u(0,\infty) = \lim_{\sigma \to 0} \frac{1}{\pi} PV \int_{-\infty}^{\infty} \frac{(\omega - y)v(0,y)}{\sigma^2 + (\omega - y)^2} \, dy \qquad (7\text{-}52a)$$

$$v(0,\omega) - v(0,\infty) = -\lim_{\sigma \to 0} \frac{1}{\pi} PV \int_{-\infty}^{\infty} \frac{(\omega - y)u(0,y)}{\sigma^2 + (\omega - y)^2} \, dy \qquad (7\text{-}52b)$$

where we have written $u(0,\infty)$ in place of $u(\infty,0)$ and similarly for v. This is permissible because $f(s)$ has a finite value at infinite s and therefore must be the same for $s = \infty$ and $s = j\infty$.

There is no doubt about these limits existing under the conditions for which the formulas apply; but the limits cannot be found from these formulas by setting $\sigma = 0$ in the integrands, because then we get

$$\frac{1}{\pi} PV \int_{-\infty}^{\infty} \frac{v(0,y)}{\omega - y} \, dy \qquad \text{and} \qquad -\frac{1}{\pi} PV \int_{-\infty}^{\infty} \frac{u(0,y)}{\omega - y} \, dy$$

respectively, for the two cases. These integrals do not exist, because the integrands become infinite at $y = \omega$. In order to find a satisfactory formulation, note that

$$PV \int_{-\infty}^{\infty} \frac{\omega - y}{\sigma^2 + (\omega - y)^2} \, dy = 0$$

as can be seen by carrying out the indicated integration. Therefore, from the respective integrals in Eqs. (7-52) we can subtract

$$\frac{1}{\pi} PV \int_{-\infty}^{\infty} \frac{(\omega - y)v(0,\omega)}{\sigma^2 + (\omega - y)^2} \, dy \qquad \text{and} \qquad \frac{1}{\pi} PV \int_{-\infty}^{\infty} \frac{(\omega - y)u(0,\omega)}{\sigma^2 + (\omega - y)^2} \, dy$$

without changing the results, and then

$$\lim_{\sigma \to 0} PV \int_{-\infty}^{\infty} \frac{(\omega - y)v(0,y)}{\sigma^2 + (\omega - y)^2} \, dy = \lim_{\sigma \to 0} PV \int_{-\infty}^{\infty} \frac{(\omega - y)[v(0,y) - v(0,\omega)]}{\sigma^2 + (\omega - y)^2} \, dy$$

$$\lim_{\sigma \to 0} PV \int_{-\infty}^{\infty} \frac{(\omega - y)u(0,y)}{\sigma^2 + (\omega - y)^2} \, dy = \lim_{\sigma \to 0} PV \int_{-\infty}^{\infty} \frac{(\omega - y)[u(0,y) - u(0,\omega)]}{\sigma^2 + (\omega - y)^2} \, dy$$

Now we are in a better position because, when $\sigma = 0$, these integrands have no pole at $y = \omega$. The numerator is zero at that point. Thus, if we are permitted to take the limit inside the integral sign, an integrable function will be obtained. As a result of $f(\infty)$ being a constant and the fact that

$$\frac{1}{\sigma^2 + (\omega - y)^2} \leq \frac{1}{(\omega - y)^2}$$

it can be shown (see Chap. 8) that the above two integrals are uniformly convergent for $\sigma \geq 0$. This being the case, the $\sigma \to 0$ limit can be taken inside the integral, and the results are

$$u(0,\omega) - u(0,\infty) = \frac{1}{\pi} PV \int_{-\infty}^{\infty} \frac{v(0,y) - v(0,\omega)}{\omega - y} \, dy \qquad (7\text{-}53a)$$

$$v(0,\omega) - v(0,\infty) = -\frac{1}{\pi} PV \int_{-\infty}^{\infty} \frac{u(0,y) - u(0,\omega)}{\omega - y} \, dy \qquad (7\text{-}53b)$$

We shall now develop an equivalent way to write these formulas. Since they are similar, only the first one will be treated. The integral in Eq. (7-53a) can be written

$$PV \int_{-\infty}^{\infty} \frac{v(0,y) - v(0,\omega)}{\omega - y} \, dy = \lim_{\substack{\epsilon \to 0 \\ A \to \infty}} \int_{-A}^{\omega - \epsilon} + \int_{\omega - \epsilon}^{\omega + \epsilon} + \int_{\omega + \epsilon}^{A}$$

The integrand has a removable singularity at $y = \omega$, and so the integral from $\omega - \epsilon$ to $\omega + \epsilon$ approaches zero as ϵ approaches zero. For the other two integrals we have

$$\int_{-A}^{\omega - \epsilon} \frac{v(0,y) - v(0,\omega)}{\omega - y} \, dy = \int_{-A}^{\omega - \epsilon} \frac{v(0,y)}{\omega - y} \, dy + v(0,\omega) \log \frac{\epsilon}{A + \omega}$$

$$\int_{\omega + \epsilon}^{A} \frac{v(0,y) - v(0,\omega)}{\omega - y} \, dy = \int_{\omega + \epsilon}^{A} \frac{v(0,y)}{\omega - y} \, dy + v(0,\omega) \log \frac{A - \omega}{\epsilon}$$

and therefore

$$\int_{-A}^{\omega - \epsilon} + \int_{\omega + \epsilon}^{A} = \int_{-A}^{\omega - \epsilon} \frac{v(0,y)}{\omega - y} \, dy + \int_{\omega + \epsilon}^{A} \frac{v(0,y)}{\omega - y} \, dy + v(0,\omega) \log \frac{A - \omega}{A + \omega}$$

As A approaches infinity, the last term approaches zero and so

$$PV \int_{-\infty}^{\infty} \frac{v(0,y) - v(0,\omega)}{\omega - y} \, dy = \lim_{\substack{\epsilon \to 0 \\ A \to \infty}} \int_{-A}^{\omega-\epsilon} \frac{v(0,y)}{\omega - y} \, dy + \int_{\omega+\epsilon}^{A} \frac{v(0,y)}{\omega - y} \, dy$$

$$= (PV)_2 \int_{-\infty}^{\infty} \frac{v(0,y)}{\omega - y} \, dy$$

The symbol $(PV)_2$ signifies the principal value in two senses, with respect to the infinite limits and also with respect to the pole at $y = \omega$. Thus, Eqs. (7-53) can be changed to the forms

$$u(0,\omega) - u(0,\infty) = \frac{1}{\pi} (PV)_2 \int_{-\infty}^{\infty} \frac{v(0,y)}{\omega - y} \, dy \qquad (7\text{-}54a)$$

$$v(0,\omega) - v(0,\infty) = -\frac{1}{\pi} (PV)_2 \int_{-\infty}^{\infty} \frac{u(0,y)}{\omega - y} \, dy \qquad (7\text{-}54b)$$

The functions defined by the above integrals are called *Hilbert transforms*. Interestingly enough, they look almost like what would be obtained by putting $\sigma = 0$ in Eqs. (7-52). The only difference is the double principal-value designation, and it was to arrive at this feature that the extra work was required.

If u and v are the real and imaginary components of a real function

$$f(s) = u(\sigma,\omega) + jv(\sigma,\omega)$$

then, since $f(\bar{s}) = \overline{f(s)}$, it follows that

$$u(\sigma,\omega) = u(\sigma,-\omega)$$
$$v(\sigma,\omega) = -v(\sigma,-\omega)$$

That is, u is an *even* function of ω, and v is an *odd* function of ω. These properties can be used to modify the integrals of Eqs. (7-53), as follows:

$$PV \int_{-\infty}^{\infty} \frac{v(0,y) - v(0,\omega)}{\omega - y} \, dy$$

$$= \lim_{A \to \infty} \left[\int_{-A}^{0} \frac{v(0,y) - v(0,\omega)}{\omega - y} \, dy + \int_{0}^{A} \frac{v(0,y) - v(0,\omega)}{\omega - y} \, dy \right]$$

$$= \lim_{A \to \infty} \int_{0}^{A} \left[\frac{v(0,-y) - v(0,\omega)}{\omega + y} + \frac{v(0,y) - v(0,\omega)}{\omega - y} \right] dy$$

$$= 2 \int_{0}^{\infty} \frac{y[v(0,y)] - \omega[v(0,\omega)]}{\omega^2 - y^2} \, dy$$

In a similar fashion it can be shown that

$$PV \int_{-\infty}^{\infty} \frac{u(0,y) - u(0,\omega)}{\omega - y} \, dy = 2\omega \int_{0}^{\infty} \frac{u(0,y) - u(0,\omega)}{\omega^2 - y^2} \, dy$$

All these formulas are subject to the condition that $f(\infty)$ shall exist. Since we have shown that v is an odd function of ω if $f(s)$ is real, the only way for v to be finite at $\omega = \infty$ is for it to be zero there. Furthermore, since u is an even function, it is not necessarily zero at $\omega = \infty$. Accordingly, we can now write

$$u(0,\omega) - u(0,\infty) = \frac{2}{\pi} \int_0^\infty \frac{y[v(0,y)] - \omega[v(0,\omega)]}{\omega^2 - y^2} \, dy \qquad (7\text{-}55a)$$

$$v(0,\omega) = -\frac{2\omega}{\pi} \int_0^\infty \frac{u(0,y) - u(0,\omega)}{\omega^2 - y^2} \, dy \qquad (7\text{-}55b)$$

A similar procedure applied to the integrals of Eqs. (7-54) yields

$$u(0,\omega) - u(0,\infty) = \frac{2}{\pi} PV \int_0^\infty \frac{yv(0,y)}{\omega^2 - y^2} \, dy \qquad (7\text{-}56a)$$

$$v(0,\omega) = -\frac{2\omega}{\pi} PV \int_0^\infty \frac{u(0,y)}{\omega^2 - y^2} \, dy \qquad (7\text{-}56b)$$

It is emphasized that the last two sets of formulas apply only if $f(s)$ is a real analytic function. In Eqs. (7-56) the PV designation refers to integration through the pole at $y = \omega$.

Equation (7-55b) can be changed to still another form by first writing

$$v(0,\omega) = \frac{2}{\pi} \int_0^\infty \frac{u(0,y) - u(0,\omega)}{y/\omega - \omega/y} \frac{dy}{y}$$

and then making the change of variable $\alpha = \log(y/\omega)$, giving

$$v(0,\omega) = \frac{2}{\pi} \int_{-\infty}^\infty \frac{u(0,\omega e^\alpha) - u(0,\omega)}{e^\alpha - e^{-\alpha}} \, d\alpha = \frac{1}{\pi} \int_{-\infty}^\infty \frac{u(0,\omega e^\alpha) - u(0,\omega)}{\sinh \alpha} \, d\alpha$$

If the last integral is written as the sum of two integrals, one from $-\infty$ to 0 and the other from 0 to ∞, and if a sign change is made in the variable of integration of the first integral, the resulting form is

$$v(0,\omega) = \frac{1}{\pi} \int_0^\infty \frac{u(0,\omega e^\alpha) - u(0,\omega e^{-\alpha})}{\sinh \alpha} \, d\alpha$$

This can be integrated by parts, using the relation

$$\frac{d[\log \coth (\alpha/2)]}{d\alpha} = -\frac{1}{\sinh \alpha}$$

with the following result:

$$v(0,\omega) = \lim_{\substack{\epsilon \to 0 \\ A \to \infty}} \left\{ [u(0,\omega e^{-\alpha}) - u(0,\omega e^\alpha)] \log \coth \frac{\alpha}{2} \right\}_\epsilon^A$$

$$+ \frac{1}{\pi} \int_0^\infty \frac{du(0,\omega e^\alpha)}{d\alpha} \log \coth \frac{\alpha}{2} \, d\alpha$$

$$- \frac{1}{\pi} \int_0^\infty \frac{du(0,\omega e^{-\alpha})}{d\alpha} \log \coth \frac{\alpha}{2} \, d\alpha \qquad (7\text{-}57)$$

Each of the three terms on the right of Eq. (7-57) will be considered separately. The first term approaches zero at the upper limit, as A approaches infinity, because $u(0,\infty)$ and $u(0,0)$ are both finite and $\log \coth \infty = \log 1 = 0$. At the lower limit it is convenient to replace $\log \coth (\alpha/2)$ by $\log \cosh (\alpha/2) - \log \sinh (\alpha/2)$, giving

$$[u(0,\omega e^{-\alpha}) - u(0,\omega e^{\alpha})] \log \cosh \frac{\alpha}{2} + [u(0,\omega e^{-\alpha}) - u(0,\omega e^{\alpha})] \log \sinh \frac{\alpha}{2}$$

The function $u(0,\omega e^{\alpha})$ is a continuous function of ωe^{α}, because it is the real part of an analytic function which is regular on the $jy = j\omega e^{\alpha}$ axis. Since e^{α} is a continuous function of α, it follows that $u(0,\omega e^{\alpha})$ is a continuous function of α. Therefore, each of the above bracketed expressions approaches zero. This means that the term containing $\log \cosh (\alpha/2)$ is zero when α goes to zero. The other term is indeterminate, because $\log \sinh (\alpha/2)$ approaches infinity. The limit is found by first multiplying and dividing by $\sinh (\alpha/2)$, to give

$$\left[\frac{u(0,\omega e^{-\alpha}) - u(0,\omega e^{\alpha})}{\sinh (\alpha/2)} \right] \left[\sinh \frac{\alpha}{2} \log \sinh \frac{\alpha}{2} \right]$$

After replacing α by the lower limit ϵ, the left-hand bracketed expression can be evaluated by the Lhopital rule as follows,

$$\lim_{\epsilon \to 0} \frac{\dfrac{du(0,\omega e^{-\epsilon})}{d\epsilon} - \dfrac{du(0,\omega e^{\epsilon})}{d\epsilon}}{\cosh (\epsilon/2)} = \lim_{\epsilon \to 0} \frac{-2 \dfrac{du(0,\omega e^{\epsilon})}{d\epsilon}}{\cosh (\epsilon/2)} = -2 \frac{du(0,\omega e^{\epsilon})}{d\epsilon} \bigg|_{\epsilon=0}$$

which exists. Furthermore, the second factor approaches zero as ϵ approaches zero, and so the entire expression approaches zero. Thus, under the conditions stipulated on u, the first term in Eq. (7-57) is zero. Now look at the last integral in Eq. (7-57), and make appropriate variable changes to get

$$-\frac{1}{\pi} \int_0^{\infty} \frac{du(0,\omega e^{-\alpha})}{d\alpha} \log \coth \frac{\alpha}{2} \, d\alpha = -\frac{1}{\pi} \int_0^{\infty} \frac{du(0,\omega e^{-\alpha})}{d\alpha} \log \coth \left| \frac{-\alpha}{2} \right| \, d\alpha$$

$$= \frac{1}{\pi} \int_{-\infty}^0 \frac{du(0,\omega e^{\alpha})}{d\alpha} \log \coth \left| \frac{\alpha}{2} \right| \, d\alpha$$

Combining this with the other integral in Eq. (7-57) gives the final result

$$v(0,\omega) = \frac{1}{\pi} \int_{-\infty}^{\infty} \frac{du(0,\omega e^{\alpha})}{d\alpha} \log \coth \left| \frac{\alpha}{2} \right| \, d\alpha \qquad (7\text{-}58)$$

or the equivalent form

$$v(0,\omega) = \frac{1}{\pi} \int_{-\infty}^{\infty} \frac{du(0,y)}{d(\log y)} \log \coth \left| \frac{\alpha}{2} \right| \, d\alpha$$

Although this formula may appear to be more complicated than the others, it is useful because it yields readily to interpretation. Let y be the usual frequency variable, and think of u plotted as a function of $\log y$, as in Fig. 7-12a. The horizontal axis can also be used for the variable

$$\alpha = \log y - \log \omega$$

by allowing its origin to change with ω. The integration called for in Eq. (7-58) is performed for some given value of ω, which would fix the origin of the α axis in Fig. 7-12a. The integrand consists of two factors, the slope of this curve and the factor

$$\log \coth \left| \frac{\alpha}{2} \right|$$

which is sketched in Fig. 7-12b, with the α origin shown coincident with the origin in Fig. 7-12a. The nature of Fig. 7-12b is such that the integral is predominantly influenced by the range of integration near $\alpha = 0$. As ω is allowed to change, the point $\alpha = 0$ shifts, carrying the curve of Fig. 7-12b with it. Thus, we see that the value of $v(0,\omega)$ is roughly proportional to the slope of the u versus $\log y$ curve near the point $y = \omega$. You may find it instructive to sketch curves for a simple case, to confirm this conclusion that v is large in regions where u is rapidly changing.

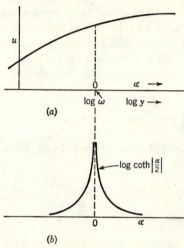

FIG. 7-12. Graphical interpretation of Eq. (7-58).

A result similar to Eq. (7-58) can be obtained in which the derivative is with respect to the frequency variable, rather than with respect to the logarithm. In Eq. (7-58) change the variable back to y, by making the substitutions

$$\frac{du(0,\omega e^{\alpha})}{d\alpha} = \frac{du(0,y)}{dy} \frac{dy}{d\alpha} = y \frac{du(0,y)}{dy}$$

$$d\alpha = \frac{dy}{y}$$

$$\coth \left| \frac{\alpha}{2} \right| = \left| \frac{e^{\alpha/2} + e^{-\alpha/2}}{e^{\alpha/2} - e^{-\alpha/2}} \right| = \left| \frac{e^{\alpha} + 1}{e^{\alpha} - 1} \right| = \left| \frac{y + \omega}{y - \omega} \right|$$

The result is

$$v(0,\omega) = \frac{1}{\pi} \int_0^\infty \frac{du(0,y)}{dy} \log \left| \frac{y + \omega}{y - \omega} \right| dy \qquad (7\text{-}59)$$

7-11. Gain and Angle Functions. The formulas developed in the previous sections are valid for relating real and imaginary components of analytic functions which are regular on and to the right of the j axis. If $g(s)$ is a network response function, it is often convenient to deal with the derived function

$$f(s) = \log g(s) \qquad (7\text{-}60)$$

whose real and imaginary components are the *gain* and *angle* functions, respectively. This function raises an interesting question because its singularities are relatively mild, being logarithmic singularities (branch points of the logarithm) at zeros and poles of $g(s)$. As a first observation, we immediately see that the previous formulas do not apply if $g(s)$ has zeros in the right half plane. This is the genesis of the often-quoted statement that gain and angle functions are related only for minimum-phase networks.

Suppose that $g(s)$ has zeros on the $j\omega$ axis or at infinity. This is an important practical possibility, and therefore the formulas would have serious limitations if such zeros are not admissible. But since the logarithm at a pole behaves the same as at a zero, we can also simultaneously treat poles on the j axis or at infinity. Zeros and poles occurring on the imaginary axis can

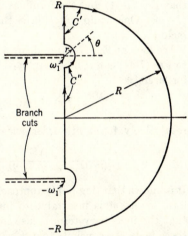

Fig. 7-13. Contour for deriving formulas relating real and imaginary parts of functions having logarithmic singularities on the j axis.

conveniently be considered by going back to the original formulation.

Now suppose that points $j\omega_1$ and $-j\omega_1$ are nth-order zeros or nth-order poles of $g(s)$. Each of these points is a branch point of $\log g(s)$, and for each we can extend the branch cut to the left so that a closed curve C' can be formed with indentations into the right half plane, as shown in Fig. 7-13. C' lies in one sheet of the Riemann surface and is the same as C of Fig. 7-11 except for these indentations. The manipulations leading to Eqs. (7-43) and (7-46) are still valid if in each case the real integral from $-R$ to R is replaced by a contour integral along C'', from $-jR$ to jR.

This contour includes the indentations; and that is why the integral cannot remain real.

In Chap. 6 we showed that an integral over a small circle around a branch point of the logarithmic function approaches zero as the radius of the circle approaches zero. It will also be zero, in the limit, for the semicircle of Fig. 7-13. Thus, we can let the radius of these semicircles approach zero, and the result will be identical with Eqs. (7-43) and (7-46). The integral from $-R$ to R can be taken through the branch point.*

Since Eqs. (7-43) and (7-46) have now been shown to be valid even when there are logarithmic singularities due to zeros or poles of $g(s)$ on the j axis, we conclude that all the formulas developed from them are also valid. Of course, throughout this discussion proper behavior at infinity was assumed. This leads us to the question of whether or not we can allow $g(s)$ to have a logarithmic singularity at infinity. We shall find that Eq. (7-44) and the formulas derived from it are valid, but not Eq. (7-47). One could not expect the latter to be applicable, because it includes $f(\infty)$.

We look at Eq. (7-43) and consider whether or not the process leading to Eq. (7-44) can still be employed. The important consideration is that for large $|z|$

$$f(z) \approx \pm n \log |z| \pm jn \tan^{-1} \frac{y}{x}$$

respectively, for an nth-order pole or zero at infinity. Thus, for large $|y|$,

$$\frac{f(jy)}{\sigma^2 + \omega^2 + y^2 - 2\omega y} \approx \pm \frac{n \log |y|}{y^2} \pm j \frac{n}{y^2} \frac{\pi}{2}$$

It is known that $(\log |y|)/y$ approaches zero as y becomes infinite. Therefore, $\log |y|/y^2$ is integrable to infinity. Also, the factor $1/y^2$ ensures that the imaginary term is also integrable to infinity. On the basis of these facts, we conclude that R can be allowed to approach infinity and the first integral of Eq. (7-43) will converge. For the second integral, around C_1, we let $z = Re^{j\theta}$, and as R becomes infinite, this integral approaches

$$\int_{\pi/2}^{-\pi/2} \frac{n \log R + jn\theta}{R^2 e^{j2\theta}} (jRe^{j\theta}) \, d\theta$$

which approaches zero, because $(\log R)/R$ approaches zero as R becomes infinite. Thus, the limit processes leading to Eq. (7-44) are valid. It follows that Eqs. (7-48) are valid, but not Eqs. (7-49). Therefore, the formulas developed in Sec. 7-10 do not apply to the logarithm of a response function that has a zero or pole at infinity.

---

* Refer to Sec. 6-5 for a detailed account of integration through a branch point.

PROBLEMS

7-1. Check the relation

$$f(\bar{s}) = \overline{f(s)}$$

for the following:

    (a) $f(s) = s^n$ *(b)* $f(s) = \sin s$ *(c)* $f(s) = \cosh s$

7-2. If $f(s)$ is a real analytic function, show that $\operatorname{Re}[f(j\omega)]$ and $\operatorname{Im}[f(j\omega)]$ are, respectively, even and odd functions of ω.

7-3. Assuming that $f(s)$ is a rational function with real coefficients, prove that $f(\bar{s}) = \overline{f(s)}$, without relying on Theorem 7-2.

7-4. Find the residues at each pair of poles, for the function

$$f(s) = \frac{2s^3 + 8s^2 + 30s + 20}{(s^2 + 2s + 5)^2}$$

and observe that the residues are themselves conjugates.

7-5. Find the residues of

$$f(s) = \frac{\cos s}{\sinh s}$$

at the conjugate pair of poles nearest the origin, and observe that the residues are themselves conjugates.

7-6. If $f(s)$ is a real analytic function, prove that

$$F(s) = f(s)f(-s)$$

is a real analytic function of a complex variable ω (where $s = j\omega$), and furthermore show that

$$F(j\omega) = |f(j\omega)|^2$$

where ω is real.

7-7. Prove Schwarz's lemma, a statement of which follows: Let $f(s)$ be regular inside and on the circle $|s| = R$, and let it have the property

$$|f(Re^{j\theta})| < M$$

and $f(0) = 0$. Then it is true that

$$|f(s)| < \left| \frac{s}{R} \right| M \qquad |s| < R$$

7-8. Prove that, if a rational function is positive real, the coefficients of the numerator and denominator polynomial must be real and positive.

7-9. Prove that, if $f(s)$ is a positive real function, any pole on the j axis must be simple and must have a positive real part.

7-10. If $f(s)$ is given by

$$f(s) = \frac{1}{s+1} + \frac{a+jb}{s+2+j} + \frac{a-jb}{s+2-j}$$

where a and b are real, determine the ranges of a and b such that $f(s)$ will be positive real.

7-11. Compute and plot loci of

$$f(s) = 4 + 6s + 4s^2 + s^3$$

for s on each of the three circles $|s| = 1$, $|s| = 2$, $|s| = 3$, determining how many zeros are inside each circle.

7-12. Use the index principle to prove the fundamental theorem of algebra, namely, the number of zeros of a polynomial is equal to the degree of the polynomial.

7-13. Use a Nyquist plot to determine the range of k for stable operation for a feedback loop having the system function

$$A(s)H(s) = \frac{s(s+2)}{(s+1)^3}$$

and compare with the result obtained by the root-locus method in Sec. 6-10. (Note the change in sign of k.)

7-14. Do Prob. 7-13 for the system function

$$A(s)H(s) = \frac{s(s+1)}{(s+2)^5}$$

and compare the result with Sec. 6-10.

7-15. Do Prob. 6-17 by making a Nyquist plot.

7-16. Do Prob. 6-18 by making a Nyquist plot.

7-17. Do Prob. 6-19 by making a Nyquist plot.

7-18. Derive and state a modification of the Nyquist criterion where the system is open-loop unstable, meaning that $A(s)H(s)$ has right-half-plane poles. In this theorem you will use a number N, the sum of the multiplicities of the right-half-plane poles of $A(s)H(s)$.

7-19. Referring to Prob. 7-18, use the Nyquist plot to solve Prob. 6-20.

7-20. Let the function

$$h(\theta) = \begin{cases} 1 - \dfrac{2}{\pi}\theta & 0 \leqq \theta \leqq \pi \\ 1 + \dfrac{2}{\pi}\theta & -\pi \leqq \theta \leqq 0 \end{cases}$$

be given, representing the real part of $f(s)$ on the unit circle. Obtain a power-series expansion about the origin, for $f(s)$. Is $f(s)$ regular at all points on the unit circle?

Let the function

$$h(\theta) = \begin{cases} 1 & 0 < \theta < \pi \\ -1 & -\pi < \theta < 0 \end{cases}$$

be given, representing the imaginary part of $f(s)$ on the unit circle. Obtain a power-series expansion about the origin for $f(s)$. Is $f(s)$ regular at all points on the unit circle?

7-21. By considering the function $f(s) = e^s$, prove that, for $0 < \sigma < 1$,

$$\frac{1}{2\pi} \int_0^{2\pi} \frac{(1 - \sigma^2 - j2\sigma \sin \theta)e^{\cos \theta} \cos (\sin \theta)}{1 + \sigma^2 - 2\sigma \cos \theta} \, d\theta = e^\sigma$$

By considering the function $f(s) = s$, prove that, for $0 < \omega < 1$,

$$\frac{1}{2\pi} \int_0^{2\pi} \frac{[-2\omega \cos \theta + j(1 - \omega^2)] \sin \theta}{1 + \omega^2 - 2\omega \sin \theta} \, d\theta = j\omega$$

7-22. Check both of Eqs. (7-55) on the function

$$f(s) = \frac{1}{1+s}$$

7-23. Check both of Eqs. (7-55) on the function

$$f(s) = \frac{s}{1+s}$$

7-24. Check both of Eqs. (7-55) on the function

$$f(s) = \frac{1+s}{2+s}$$

7-25. Check both of Eqs. (7-55) on the function

$$f(s) = \frac{1}{(1+s)^2}$$

CHAPTER 8

THEOREMS ON REAL INTEGRALS

8-1. Introduction. The application of the theory of functions of a complex variable to the analysis of linear systems is the main subject of this text. One of the most fruitful ways to go from the system equations to formulations in terms of complex variables is by way of the Laplace and Fourier integrals. However, a reasonably complete understanding of this process requires a good understanding of the Fourier integral theorem. The Fourier integral theorem takes us into some rather subtle mathematical principles because it is stated in terms of *improper integrals*. Therefore, before going on to the study of the Laplace transform, we shall turn our attention away from complex variables long enough to present some preparatory material on improper real integrals.

A modern mathematical treatment of this subject should be given in terms of Lebesgue integration. However, for engineering applications a satisfying degree of rigor can be attained within the conceptual framework of Riemann integration. This will necessitate restricting the treatment to a limited class of functions and in some cases will result in proofs which are partially intuitive. In fact, the proofs (or partial proofs) given in this chapter are offered mainly to help illuminate the principles involved. An understanding of the concepts in the various theorems is more important than the proofs themselves. But this understanding cannot be acquired without some sort of proof, even though it may not be the most general one possible.

8-2. Piecewise Continuous Functions of a Real Variable. In his use of the calculus of real variables in engineering problems the applied scientist may easily fall into the habit of thinking only of the most "well-behaved" functions, functions which are continuous and have continuous derivatives. These are functions like e^x and x^n, upon which we can operate unquestionably with the various standard processes of the calculus, such as the rules for differentiation and integration by parts.

For certain good reasons it is customary to consider certain discontinuous functions as idealized excitations for linear systems. Two examples are the step function and the periodic square wave. These are conceptually simple (it is easy to draw their graphs and to visualize what they look like).   However, the derivative does not exist at points of discontinuity, and this fact gives us reason to be careful when using discontinuous functions in the usual processes of the calculus.

A function $g(x)$ is said to be *piecewise continuous in an interval* if in that interval all points of discontinuity are isolated and if the function approaches a finite limit from each side of the point of discontinuity. Isolation of the discontinuity means that there is an interval around each discontinuous point such that the function is continuous at all other points in this interval.   If a function is piecewise continuous over the interval $-\infty < x < \infty$, we say that it is a *piecewise continuous* function.

Let $f(x)$ have an isolated discontinuity at x_0.   The single-sided limits mentioned in the above paragraph are defined formally by the following formulas,

$$\lim_{\epsilon \to 0} f(x_0 + \epsilon) = f(x_0+)$$
$$\lim_{\epsilon \to 0} f(x_0 - \epsilon) = f(x_0-)$$

$$(8\text{-}1)$$

where ϵ is positive.   On the other hand, if $f(x_0)$ exists and both limits equal $f(x_0)$, the function is continuous at x_0.   The symbols written on the right of the above two equations will occasionally be used to imply the limit process shown.   Since the

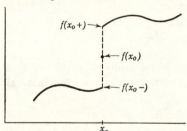

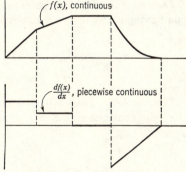

Fig. 8-1. Definition of a function at a discontinuity.

Fig. 8-2. Example of a continuous function having a piecewise continuous derivative.

limits exist on each side of the discontinuity, and the function is continuous otherwise, it follows that a piecewise continuous function is bounded in any finite interval.

In many cases it is satisfactory to leave the function undefined at a point of discontinuity.   However, when Fourier integral formulations are used, it is convenient to define a piecewise continuous function at a point of discontinuity as the average of the above two limits, as follows:

$$f(x_0) = \frac{f(x_0+) + f(x_0-)}{2}$$

$$(8\text{-}2)$$

These ideas are illustrated graphically in Fig. 8-1.

We shall also deal with functions which are continuous but which have a piecewise continuous first derivative. An example is shown in Fig. 8-2. A continuous function does not necessarily have a piecewise continuous derivative. This question is discussed later, and an example is shown in Fig. 8-3.

8-3. Theorems and Definitions for Real Integrals. It is important briefly to consider a few fundamental theorems as applied to piecewise continuous functions. For the most part these theorems will look familiar, but we emphasize that now we are considering a fairly general class of functions. As you study these theorems, you should recognize how it is possible for them to be true under the conditions stated.

Theorem 8-1.* If a function $f(x)$ is piecewise continuous in the interval $a \leqq x \leqq b$, then both the integrals

$$\int_a^b f(x)\, dx \quad \text{and} \quad \int_a^b |f(x)|\, dx$$

exist and

$$\left| \int_a^b f(x)\, dx \right| \leqq \int_a^b |f(x)|\, dx \tag{8-3}$$

Theorem 8-2. If, in the interval $a \leqq x \leqq b$, $h(x)$ is a continuous function having a piecewise continuous derivative

$$f(x) = \frac{dh(x)}{dx}$$

then

$$\int_a^b f(x)\, dx = h(b) - h(a) \tag{8-4}$$

Theorem 8-3. *Integral as a Function of the Upper Limit.* If $f(x)$ is piecewise continuous, $a \leqq x \leqq b$, and t is a number in this same interval, then

$$F(t) = \int_a^t f(x)\, dx \tag{8-5}$$

is a continuous function of t and

$$\frac{dF(t)}{dt} = f(t) \tag{8-6}$$

at all points where $f(t)$ is continuous in the interval a, b.

Theorem 8-4. *Integration by Parts.* If two functions $f(x)$ and $g(x)$ are each continuous in an interval $a \leqq x \leqq b$, with piecewise continuous derivatives, then

$$\int_a^b f(x) \frac{dg(x)}{dx}\, dx = f(b)g(b) - f(a)g(a) - \int_a^b g(x) \frac{df(x)}{dx}\, dx \tag{8-7}$$

* In this and following theorems, the interval $a \leqq x \leqq b$ can be replaced by $a < x < b$ if the pertinent functions approach limits at a and b.

Theorem 8-5. Integral over a Set of Measure 0. Over the interval $a \leq x \leq b$ let there be isolated points $x_1, x_2, \ldots$ at which $f(x)$ is nonzero. Let $f(x)$ be zero everywhere else in the interval. Then

$$\int_a^b f(x) = 0$$

This theorem states, in effect, that the integral of any function is independent of the definition of the function at one or more *isolated* points. For example, if we integrate over a discontinuity, the integral is independent of how we define the function at the point of discontinuity. The set of points defined in the theorem is an example of a *set of measure* 0.*

8-4. Improper Integrals. Improper integrals are of two kinds. In the first kind the integrand remains finite, but the interval of integration extends to infinity. In the second kind the interval of integration remains finite, but within that interval the integrand becomes infinite.

Improper Integral of the First Kind. Let $f(x)$ be piecewise continuous in the interval $a \leq x \leq A$, where A is as large as we like. If the limit

$$I_1 = \lim_{A \to \infty} \int_a^A f(x)\, dx \tag{8-8}$$

exists, we define this as the integral

$$I_1 = \int_a^\infty f(x)\, dx \tag{8-9}$$

Whenever you see an integral written like Eq. (8-9), it should be interpreted by the limit appearing in Eq. (8-8). This is an improper integral of the *first kind*. Recalling the definition of a limit, it is evident that an improper integral of the first kind converges if, and only if, corresponding to an arbitrary $\epsilon > 0$ there exists a number X such that

$$\left| \int_a^A f(x)\, dx - I_1 \right| < \epsilon$$

when $$A > X$$

This formulation serves as the precise definition of convergence. Note the similarity with convergence of a series.

In order to illustrate improper integrals of the first kind, consider the three cases $f(x) = 1/x$, $f(x) = e^{-x}$, and $f(x) = \sin x$, each integrated from 1 to ∞. In the first case

$$\int_1^A \frac{dx}{x} = \log A - \log 1 = \log A$$

---

* A set of measure 0 is more general than this, but this is the only case we shall encounter.

This does not approach a limit, as A becomes infinite, and so the integral

$$\int_1^\infty \frac{dx}{x}$$

does *not* exist.   In the second case

$$\int_1^A e^{-x}\, dx = -e^{-A} + e^{-1}$$

As A approaches infinity, the above approaches $1/e$ and so we have

$$\int_1^\infty e^{-x}\, dx = \frac{1}{e}$$

This improper integral converges.   Finally, for the third example,

$$\int_1^A \sin x\, dx = -\cos A + \cos 1$$

which does not approach a limit as A becomes infinite.   Note that the first example does not approach a limit because $\log A \to \infty$.   On the other hand, in the third case $\cos A$ remains finite, but it approaches no limit.

In the Fourier integral theorem we encounter improper integrals like

$$\int_{-\infty}^\infty f(x)\, dx$$

Such an integral is defined by using the above definition twice, once at each limit.   Thus, when it exists,

$$\int_{-\infty}^\infty f(x)\, dx = \lim_{A \to \infty} \int_0^A f(x)\, dx + \lim_{B \to \infty} \int_{-B}^0 f(x)\, dx \qquad (8\text{-}10)$$

In some cases this limit does not exist, except for the special case $A = B$. In other words, integration is then always performed over an interval centered at the origin.   This is called the *principal value* of the integral and is written

$$PV \int_{-\infty}^\infty f(x)\, dx = \lim_{A \to \infty} \int_{-A}^A f(x)\, dx \qquad (8\text{-}11)$$

If the limit in Eq. (8-10) exists, it is the same as the principal value.   As an example, consider $f(x) = \sin x$.   The infinite integral does not exist, but the principal value does exist (having the value zero), because $\sin x$ is an odd function, and the integral from $-A$ to A is always zero.

Improper Integral of the Second Kind.   Let $f(x)$ be piecewise continuous in the interval $a \leqq x \leqq c < b$, where c is arbitrarily close to b, and where $f(b)$ is infinite.   If

$$I_2 = \lim_{c \to b} \int_a^c f(x)\, dx \qquad (8\text{-}12)$$

exists, we define this as the integral

$$I_2 = \int_a^b f(x)\, dx \qquad\qquad (8\text{-}13)$$

and we say the integral converges. This is an improper integral of the *second kind*. Again referring to the definition of a limit, we find that such an integral converges if, and only if, corresponding to an arbitrary $\epsilon > 0$ there exists a positive number δ such that

$$\left| \int_a^c f(x)\, dx - I_2 \right| < \epsilon$$

when $$0 < b - c < \delta$$

A similar statement would be possible if the integrand were infinite at the lower limit.

Two examples will suffice to illustrate this kind of improper integral, namely, $f(x) = 1/x$ and $f(x) = 1/\sqrt{x}$. Integration is from 0 to 1. The integrand becomes infinite at 0 in each case. For the first example

$$\int_c^1 \frac{dx}{x} = \log 1 - \log c$$

which approaches infinity as c approaches zero, showing that the integral diverges. The second example gives a converging integral because

$$\int_c^1 \frac{dx}{\sqrt{x}} = 2(\sqrt{1} - \sqrt{c})$$

approaches 2 as c approaches zero.

Absolute Convergence of Improper Integrals. For either type of improper integral, if

$$\int_a^\infty |f(x)|\, dx \text{ (first kind)} \qquad \text{or} \qquad \int_a^b |f(x)|\, dx \text{ (second kind)}$$

exists, then the integral is said to *converge absolutely*. Absolute convergence is a more stringent property of a function than convergence. For example,

$$\int_0^\infty \frac{\sin x}{x}\, dx \text{ converges} \qquad \text{but} \qquad \int_0^\infty \left| \frac{\sin x}{x} \right| dx \text{ does not}$$

The concept of absolute convergence is sometimes useful, particularly as it appears in the statement of sufficient conditions for applicability of the Fourier integral theorem. The following theorem is useful:

Theorem 8-6. If a function $f(x)$ has an absolutely converging improper integral, then the corresponding integral of $f(x)$ also converges.

Cauchy Principle of Convergence. As in the case of infinite series, there is an alternative way to state necessary and sufficient conditions for convergence of an integral, in which the limit value is not needed.   Using the same terminology as for series, this is called the *Cauchy principle of convergence*.   We shall state it for each of the two types of improper integrals, in the notation of Eqs. (8-8) and (8-13).

The integrals in question converge if, and only if, given an arbitrary small $\epsilon > 0$ for type 1 there exists a number X' such that

$$\left| \int_{A'}^{A} f(x) \, dx \right| < \epsilon$$

when $A, A' > X'$

and for type 2 there exists a number δ' such that

$$\left| \int_{c'}^{c} f(x) \, dx \right| < \epsilon$$

when $0 < (b - c), (b - c') < \delta'$

Recall that in type 1 the upper limit is infinite, and so X' is a large number. In the second type the integrand is infinite at the upper limit b, and δ' is small.   (The lower limit a is assumed to be less than the upper limit b.)

8-5. Almost Piecewise Continuous Functions.   In defining piecewise continuous functions it was stipulated that each discontinuity should be finite.   Having now considered improper integrals of the second kind, we can admit certain cases where the function becomes infinite at isolated points.   First we observe that, if $n < 1$,

$$\int_{0}^{b} x^{-n} \, dx = \frac{x^{1-n}}{1 - n} \bigg|_{0}^{b} = \frac{b^{1-n}}{1 - n}$$

exists.

Suppose that x_k is a singular point of $f(x)$ but that in some neighborhood

$$|x - x_k| < \delta$$

the function has the property

$$|f(x)| < \frac{M}{|x - x_k|^n} \qquad n < 1 \qquad (8\text{-}14)$$

where M is some constant.   Then, since the right-hand side of the inequality is an integrable function, it follows that $|f(x)|$ and $f(x)$ are integrable over x_k.   As an example, consider

$$f(x) = \frac{\sin x}{\sqrt{x}}$$

where integration is from 0 to 1.   For this interval,

$$\left| \frac{\sin x}{\sqrt{x}} \right| < \frac{1}{\sqrt{x}}$$

and therefore the integral exists.

The objective is to define a class of functions slightly less restricted than the piecewise continuous functions, but sufficiently restricted so that Theorems 8-1 through 8-4 will still apply. If integration over a finite interval encounters a finite number of points like x_k, and if the function is otherwise piecewise continuous, it will be integrable over each singular point and therefore will be integrable over the interval.

We now define almost piecewise continuity as follows: A function $f(x)$ is *almost piecewise continuous in a finite interval* if there are a finite number of points $x_1, \ldots, x_k, \ldots, x_N$ such that at each of them behavior of the function is described by relation (8-14) and if the function is piecewise continuous in every interval not containing these points. If a function is almost piecewise continuous in every finite interval it will be called an *almost piecewise continuous function.*

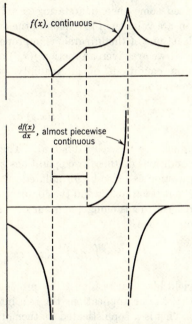

Reference to these two types of functions will be made repeatedly. Therefore, in the interest of brevity, abbreviations will often be used henceforth. We shall designate a piecewise continuous function by PC and an almost piecewise continuous function by APC.

You will note that an APC function is defined in such a way as to ensure integrability over any finite interval. Whether or not it is integrable over the infinite interval depends on convergence conditions as previously described for improper integrals of the first kind. By defining the class of APC functions we make it possible to eliminate concern about convergence of the second kind.

FIG. 8-3. A continuous function having an almost piecewise continuous derivative.

Further insight into the meaning of the APC property is afforded by Fig. 8-3, which is to be compared with Fig. 8-2. Figure 8-2 is an example of a continuous function which has a PC derivative. By inventing the APC class of functions we admit cases like Fig. 8-3, where $f(x)$ has one or more points x_k where the derivative becomes infinite while the function itself remains finite.

This section is concluded by pointing out that we have indeed defined an APC function in such a way that Theorems 8-1 through 8-4 can be read with the word *almost* preceding the words *piecewise continuous* in each instance. Proof of this statement is left to you as an exercise.

8-6. Iterated Integrals of Functions of Two Variables (Finite Limits). In the development of the theory of the Laplace transform there are many instances in which a function of two variables is integrated first with respect to one variable and then with respect to the other. Adequate treatment of the subject sometimes requires inversion of the order of integration, a process which is not always justified. Therefore, we shall need some theorems stating certain conditions which are sufficient to ensure validity of inversion of the order of integration. In most cases at least one of the integrals is improper.

If we are given a function $f(x,y)$ of two real variables which meets certain conditions of integrability, we can perform the following two operations,

$$\int_c^d dy \int_a^b f(x,y)\, dx \qquad \int_a^b dx \int_c^d f(x,y)\, dy$$

which are basically different. The question then arises: Under what conditions do these two operations give results which are identical? The question can also be formulated for the case where $f(x,z)$ is a function of the real variable x and the complex variable z and where integration with respect to z is along a contour C. Then we want to know when the two operations

$$\int_C dz \int_a^b f(x,z)\, dx \qquad \int_a^b dx \int_C f(x,z)\, dz$$

yield identical results. The problem is further complicated by the fact that in our applications the real integral can have infinite limits.

This is a sophisticated mathematical problem; and we shall make no attempt to give it a completely general treatment. We begin by recalling that in typical semielementary texts it is proved that the order of integration can be inverted if the limits are fixed and if $f(x,y)$ *is simultaneously continuous in both variables.* But our problems do not fit into this simple category: our functions can be discontinuous, and the integrals can be improper.

This is one of the questions anticipated in the introduction for which Riemann integration is really inadequate. Accordingly, we shall be satisfied with a "proof" which is largely intuitive. In general, we shall be satisfied to deal with the question of whether

$$\int_c^d dy \int_a^b f(x,y)\, dx \qquad \text{and} \qquad \int_a^b dx \int_c^d f(x,y)\, dy$$

are equal for those cases where, except at a finite number of points, $f(x,y)$ is almost piecewise continuous in each variable when the other is held constant. An example would be

$$f(x,y) = \frac{1}{\sqrt{|x + y - 2|} \, \sqrt{|x - y|}}$$

which is APC in x and y, with singular points on the loci $x = y$ and $x = 2 - y$. Except when $y = 1$ or $x = 1$, the function is APC in x, with y held constant, and in y, with x held constant. If we put $x = 1$, we get

$$f(1,y) = \frac{1}{|y - 1|}$$

and when $y = 1$,

$$f(x,1) = \frac{1}{|x - 1|}$$

neither of which is APC. The loci of singular points for this case are shown in Fig. 8-4a.

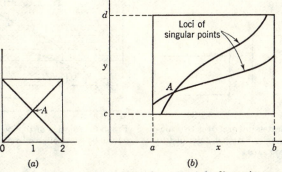

FIG. 8-4. Illustrations of how the points where $f(x,y)$ is discontinuous or becomes infinite can be described by a locus in the xy plane. (a) The example given in the text; (b) a more general case.

Thus, we intuitively expect that $f(x,y)$ will be characterized by loci of singular points as indicated for a more general case in Fig. 8-4b. These lines could be loci of points where $f(x,y)$ is discontinuous by finite jumps, but this case is trivial, and so we assume that they are loci in the same sense as in Fig. 8-4a. An infinite discontinuity of the type permitted by the APC condition is experienced as the locus is crossed in a direction parallel to either axis. However, we do not expect APC behavior as point A is crossed in either direction. Of course, much more complicated situations than this might be envisioned in which the loci of singular points would form a network of lines covering the rectangle and in which more than two lines might intersect at a point.

In preparation for the intuitive proof, we need the following lemma:

Lemma A. If we consider a rectangle R, with sides parallel to the coordinate axes, and if this rectangle is subdivided into a set of subrectangles R_n, then if the iterated integral of $f(x,y)$ exists over R, it equals the sum of the iterated integrals over R_n. That is,

$$\underbrace{\int dy \int dx}_{R} = \sum_{R_n} \int dy \int dx$$

and

$$\underbrace{\int dx \int dy}_{R} = \sum_{R_n} \int dx \int dy$$

PROOF. For a proof, we shall show this to be true for the single subdivision shown in Fig. 8-5a. It will be quite evident that a similar proof holds for Fig. 8-5b and that the proof can be applied repeatedly to take care of further subdivisions.

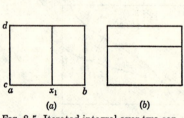

FIG. 8-5. Iterated integral over two contiguous rectangles with sides parallel to the coordinate axes.

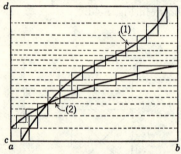

FIG. 8-6. Isolation of the loci of singular points by a set of small rectangles.

From the definition of an integral, we have

$$\int_c^d dy \int_a^b f\,dx = \int_c^d dy \int_a^{x_1} f\,dx + \int_c^d dy \int_{x_1}^b f\,dx$$

and

$$\int_a^b dx \int_c^d f\,dy = \int_a^{x_1} dx \int_c^d f\,dy + \int_{x_1}^b dx \int_c^d f\,dy$$

Each of the integrals on the right is an iterated integral over one of the subrectangles, and hence the lemma is proved.

Now let the loci of singular points be enclosed by an array of rectangles, as shown in Fig. 8-6. These rectangles can be made arbitrarily small. Corresponding to this, the rest of the area is divided into strips, as indicated in the figure. The strips can be either horizontal or vertical. Let R_n represent members of the set of rectangles in which there are no singular points, and let S_n represent members of the set of rectangles which do contain singular points.   According to Lemma A,

$$\int_c^d dy \int_a^b f(x,y)\, dx - \sum_{R_n} \int dy \int f(x,y)\, dx = \sum_{S_n} \int dy \int f(x,y)\, dx$$

and

$$\int_a^b dx \int_c^d f(x,y)\, dy - \sum_{R_n} \int dx \int f(x,y)\, dy = \sum_{S_n} \int dx \int f(x,y)\, dy$$

Each of the summations over the R_n set of rectangles exists, and they are equal, because $f(x,y)$ is simultaneously continuous in x and y over these rectangles.   Therefore, if the right-hand sides of the above equations can be shown to approach zero, as the rectangles are made smaller while becoming infinite in number, we shall have established that

$$\int_c^d dy \int_a^b f(x,y)\, dx \qquad \text{and} \qquad \int_a^b dx \int_c^d f(x,y)\, dy$$

both exist.   Furthermore, since it has been pointed out that

$$\sum_{R_n} \int dy \int f(x,y)\, dx = \sum_{R_n} \int dx \int f(x,y)\, dy$$

we also have

$$\left| \int_c^d dy \int_a^b f(x,y)\, dx - \int_a^b dx \int_c^d f(x,y)\, dy \right|$$
$$\leqq \left| \sum_{S_n} \int dy \int f(x,y)\, dx \right| + \left| \sum_{S_n} \int dx \int f(x,y)\, dy \right| \quad (8\text{-}15)$$

We shall outline what is involved in proving that each term on the right side of Eq. (8-15) approaches zero.   Two typical rectangles, labeled 1 and

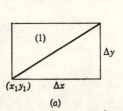
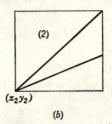

Fig. 8-7. Approximation of the locus of singular points, within a rectangle, by straight lines (a) where there is a single locus line, (b) where two lines intersect.

2, are magnified in Fig. 8-7.   They are also slightly modified to the extent that it is assumed that the rectangles are small enough to allow the loci to be approximated by straight lines.   In these two rectangles, owing to

APC properties, we can use the following estimates:

$$|f(x,y)| < \frac{M}{|x - x_1 - k_1(y - y_1)|^m} \qquad \text{over (1)}$$

$$|f(x,y)| < \frac{N}{|x - x_2 - k_2(y - y_2)|^n \, |x - x_2 - k_3(y - y_2)|^p} \qquad \text{over (2)}$$

in which M and N are constants and m, n, and p are each between 0 and 1. The numbers k_1, k_2, and k_3 are reciprocals of the slopes of the lines. The first of these can be integrated with no difficulty, and an upper bound can be found for the second (see Prob. 8-27). We find

$$\left| \int_{y_1}^{y_1 + \Delta y} dy \int_{x_1}^{x_1 + \Delta x} f(x,y)\, dx \right| < (\text{constant})(\Delta y)^{2-m}$$

$$\left| \int_{y_2}^{y_2 + \Delta y} dy \int_{x_2}^{x_2 + \Delta x} f(x,y)\, dx \right| < (\text{constant})(\Delta y)^{2-n-p}$$

Similar results are obtained for integration in the reverse order. As the rectangles are made smaller, their number increases in proportion to $1/\Delta y$. Thus, the contribution of the integrals of type 1 varies as Δy^{1-m}. But $1 - m > 0$, and so this contribution can approach zero as Δy approaches zero.* The number of rectangles of type 2 remains constant as Δy goes to zero. Thus, their contribution goes to zero, because $2 - n - p > 0$. If the function has finite discontinuities on these loci, there is no difficulty in showing that the contribution of the S_n rectangles approaches zero.

In this brief discussion we have omitted many details. In particular, we did not take into consideration the special case where the loci may be either horizontal or vertical. This causes no difficulty but does not fit into this analysis because it would give k a limiting value of 0 or ∞. However, to a degree we have established the following:

Theorem 8-7.   If $f(x,y)$ is almost piecewise continuous in x and y in the intervals $a \leqq x \leqq b$ and $c \leqq y \leqq d$, except at possibly a finite number of points, then

$$\int_c^d dy \int_a^b f(x,y)\, dx \qquad \text{and} \qquad \int_a^b dx \int_c^d f(x,y)\, dy$$

both exist, and they are equal.

An example may be helpful. Let

$$f(x,y) = \frac{x}{\sqrt{|x - y|}}$$

---

* This brief intuitive argument assumes that a single multiplying constant M in the above appraisal will suffice at all points. Otherwise, this argument is not valid. In view of the intuitive nature of this proof, it would be inconsistent to pursue this question in detail.

which is APC, with a singular point at $x = y$. The locus of the singular point is a line at 45°. Let us check whether or not

$$\int_0^1 dy \int_0^1 \frac{x\,dx}{\sqrt{|x-y|}} \quad \text{and} \quad \int_0^1 dx \int_0^1 \frac{x\,dy}{\sqrt{|x-y|}}$$

are identical. For the first one,

$$\int_0^1 \frac{x\,dx}{\sqrt{|x-y|}} = \int_0^y \frac{x\,dx}{\sqrt{y-x}} + \int_y^1 \frac{x\,dx}{\sqrt{x-y}} = \tfrac{4}{3} y^{3/2}$$
$$+ \tfrac{2}{3}(1+2y)\sqrt{1-y}$$

and, for the second one,

$$\int_0^1 \frac{x\,dy}{\sqrt{|x-y|}} = \int_0^x \frac{x\,dy}{\sqrt{x-y}} + \int_x^1 \frac{x\,dy}{\sqrt{y-x}} = 2x^{3/2} + 2x\sqrt{1-x}$$

by ordinary processes of integration. Integrate the first from 0 to 1, with respect to y, giving

$$\tfrac{4}{3}\int_0^1 y^{3/2}\,dy + \tfrac{2}{3}\int_0^1 (1+2y)\sqrt{1-y}\,dy$$
$$= \tfrac{4}{3}\int_0^1 y^{3/2}\,dy + \tfrac{2}{3}\int_0^1 (3-2w)\sqrt{w}\,dw$$
$$= \tfrac{4}{3}\int_0^1 y^{3/2}\,dy - \tfrac{4}{3}\int_0^1 w^{3/2}\,dw$$
$$+ 2\int_0^1 \sqrt{w}\,dw = 2\int_0^1 \sqrt{w}\,dw = \tfrac{4}{3}$$

where $w = 1 - y$ is used as a variable change. The second one, integrated with respect to x, gives

$$2\int_0^1 x^{3/2}\,dx + 2\int_0^1 x\sqrt{1-x}\,dx = 2\int_0^1 x^{3/2}\,dx + 2\int_0^1 (1-w)\sqrt{w}\,dw$$
$$= 2\int_0^1 x^{3/2}\,dx - 2\int_0^1 w^{3/2}\,dw$$
$$+ 2\int_0^1 \sqrt{w}\,dw = 2\int_0^1 \sqrt{w}\,dw = \tfrac{4}{3}$$

where $w = 1 - x$. Both results are the same, in agreement with Theorem 8-7.

8-7. Iterated Integrals of Functions of Two Variables (Infinite Limits). As an extension of the previous section, suppose that

$$I(y) = \int_0^\infty f(x,y)\,dx \tag{8-16}$$

converges for some range of y, say $c \leqq y \leqq d$. The question is whether or not

$$\int_c^d dy \int_0^\infty f(x,y)\,dx \quad \text{and} \quad \int_0^\infty dx \int_c^d f(x,y)\,dy$$

are equal if $f(x,y)$ meets the conditions given in Theorem 8-7, for every finite rectangle. You may recognize here a resemblance to the term-by-term integration of series, as discussed in Chap. 5.

Suppose that, given a small arbitrary positive number ϵ, it is possible to find a number X, which depends only on ϵ, such that

$$\left| \int_{A'}^{A} f(x,y)\ dx \right| < \epsilon$$

when
$$A, A' > X$$

for all y in the range $c \leqq y \leqq d$. If this is possible, the integral is said to be uniformly convergent with respect to y, in the range specified. (This is the Cauchy version of the definition of uniform convergence.)

Now assume that the integral with respect to x is uniformly convergent, and observe that the integral can be written as a series

$$\int_0^\infty f(x,y)\ dx = \sum_{n=0}^\infty B_n(y) \tag{8-17}$$

where
$$B_n(y) = \int_{\beta_n}^{\beta_{n+1}} f(x,y)\ dx \tag{8-18}$$

and the sequence $0 = \beta_0 < \beta_1 < \beta_2 < \beta_3 \cdots$ is any infinite ascending sequence of numbers starting from zero. Let such a sequence be chosen, and choose a number N such that

$$\beta_N > X$$

Then, because the sequence of β's is ascending, it is true that

$$\beta_{v+1}, \beta_u > X$$
when
$$u, v > N$$

We arbitrarily let v be the larger of the two numbers u and v, if they are not equal. Note also that

$$\int_{\beta_u}^{\beta_{v+1}} f(x,y)\ dx = \sum_{n=u}^{v} B_n(y)$$

However, owing to uniform convergence of the integral and the appropriate choice of u and v indicated above,

$$\left| \int_{\beta_u}^{\beta_{v+1}} f(x,y)\ dx \right| < \epsilon$$
when
$$u, v > N$$
and consequently

$$\left| \sum_{n=u}^{v} B_n(y) \right| < \epsilon$$
when
$$u, v > N$$

Thus, it is seen that uniform convergence of the original integral ensures uniform convergence of any series constructed from it in accordance with Eqs. (8-17) and (8-18). From Theorem 5-3, it is known that a uniformly convergent series can be integrated term by term, and so we have

$$\int_c^d dy \int_0^\infty f(x,y)\, dx = \int_c^d dy \sum_{n=0}^\infty B_n(y)$$

$$= \sum_{n=0}^\infty \int_c^d B_n(y)\, dy$$

$$= \sum_{n=0}^\infty \int_c^d dy \int_{\beta_n}^{\beta_{n+1}} f(x,y)\, dx$$

$$= \sum_{n=0}^\infty \int_{\beta_n}^{\beta_{n+1}} dx \int_c^d f(x,y)\, dy$$

We now recall that the sequence $\beta_0 < \beta_1 \cdots$ is arbitrary, and under this condition the summation of the last expression yields

$$\int_0^\infty dx \int_c^d f(x,y)\, dy$$

Thus, we have proved the following theorem:

Theorem 8-8.   If $f(x,y)$ meets the conditions given in Theorem 8-7, for every finite rectangle in the xy plane, and if the integral

$$\int_0^\infty f(x,y)\, dx$$

converges uniformly with respect to the upper limit, in the range $c \leqq y \leqq d$, then

$$\int_c^d dy \int_0^\infty f(x,y)\, dx = \int_0^\infty dx \int_c^d f(x,y)\, dy$$

A similar theorem can be proved regarding inversion of the integration order of

$$\int_C dz \int_0^\infty f(x,z)\, dx$$

where $f(x,z)$ is a function of a real variable x and complex variable z. However, there is a slight difference: almost piecewise continuity has not been defined for a function of a complex variable, and so this is avoided by requiring $f(x,z)$ to be continuous in z for all values of x for which the function is defined. The proof follows by recognizing that a contour integral can be written as the sum of four real integrals and then by applying Theorem 8-8 to each of the real integrals. The result is stated as follows:

Theorem 8-9.   If $f(x,z)$ is APC in x for fixed z and continuous in z for each x for which the function is defined, and if

$$\int_0^\infty f(x,z) \, dx$$

converges uniformly with respect to the upper limit, for z on a curve C, then

$$\int_C dz \int_0^\infty f(x,z) \, dx = \int_0^\infty dx \int_C f(x,z) \, dz$$

8-8. Limit under the Integral for Improper Integrals.   In dealing with the Laplace integral we shall have occasional need to know whether or not, given a function of two variables $f(x,z)$, we can write

$$\lim_{z \to z_0} \int_0^\infty f(x,z) \, dx = \int_0^\infty \lim_{z \to z_0} f(x,z) \, dx$$

In order to appreciate that this is a question of interest, consider the integral

$$\int_0^\infty \frac{\sin y}{y} \, dy$$

which is evaluated in Sec. 8-12.   Now let $z > 0$ be real, and let $y = zx$, giving

$$\int_0^\infty \frac{\sin zx}{x} \, dx = \frac{\pi}{2} \qquad z > 0$$

This is a function of z having the limit

$$\lim_{z \to 0} \int_0^\infty \frac{\sin zx}{x} \, dx = \frac{\pi}{2}$$

However, if we take the limit before integrating, we get

$$\int_0^\infty \lim_{z \to 0} \frac{\sin zx}{x} \, dx = 0$$

showing that different results are obtained depending on whether we take the limit first or integrate first.   Therefore, conditions are wanted which are sufficient to permit an interchange of the limit and the integral.

It is a simple matter to show that these limits can be different only at points where the integral is a discontinuous function of z.   In the above example, if z approaches zero from the negative side, the integral approaches $-\pi/2$, confirming the above statement for this example.

The following theorem is adequate:

Theorem 8-10.   Let a function $f(x,z)$ be of the form

$$f(x,z) = g(x,z)h(x)$$

where z is complex and $g(x,z)$ is continuous in each variable when the other variable is held constant and where $h(x)$ is APC. If the integral

$$\int_0^\infty f(x,z)\, dx$$

converges uniformly with respect to the infinite limit, in a region R, it is true that

$$\lim_{z \to z_0} \int_0^\infty f(x,z)\, dx = \int_0^\infty \lim_{z \to z_0} f(x,z)\, dx$$

if z_0 lies in R, and z lies in R as it approaches z_0.

We shall omit the proof of this theorem with the comment that two considerations are necessary. Since the integrand is APC in variable x, it is necessary to consider the validity of the result in the light of singular points of the integrand at finite values of x, where the integral is improper. By restricting behavior at these points to being of the APC type these singular points cause no difficulty. Then it is necessary to consider the effect of the infinite limit, and in doing this it is found that uniform convergence is sufficient. The proof of this is similar to the proof of continuity of an infinite series of continuous functions, Theorem 5-2.

The example given earlier does not meet the conditions of the theorem. The integral

$$\int_0^\infty \frac{\sin zx}{x}\, dx$$

converges uniformly for $z \geqq z' > 0$, but not for $z \geqq 0$.

8-9. M Test for Uniform Convergence of an Improper Integral of the First Kind. Since uniform convergence of an improper integral occasionally arises as a needed condition, it is important to be able to test a given integral for this property. The situation is much the same as for series, and so for an intuitive understanding you are referred to Theorem 5-4.

The case of an improper integral of the first kind is dealt with by the following theorem:

Theorem 8-11. Let $M(x)$ be a function which is positive for all x and such that

$$|f(x,z)| \leqq M(x)$$

for all z in a region R. Then, if

$$\int_0^\infty M(x)\, dx$$

exists, it can be concluded that

$$\int_0^\infty f(x,z)\, dx$$

converges uniformly with respect to the infinite limit, for z in R.

8-10. A Theorem for Trigonometric Integrals. The integrals

$$\int_a^b f(x) \sin yx\ dx$$

$$\int_a^b f(x) \cos yx\ dx$$

are functions of y. In the proof of the Fourier integral theorem we shall need to know the limit approached by these functions as y becomes infinite. An intuitive preview of the answer can be gleaned from the observation that, as y increases, the areas of $\sin yx$ and $\cos yx$, over any finite interval of x, approach zero because the period approaches zero. Ultimately this action becomes dominant, causing the integrals to approach zero.

In proving that these integrals approach zero we are faced with a slightly complicated situation because the oscillatory nature of the integrand is essential to the behavior of the integral, and so an appraisal must be used which does not depend upon the absolute value of the integrand. We shall outline the proof, assuming that $f(x)$ is piecewise continuous between a and b. It is possible to subdivide the interval of integration into N increments with end points x_k and to establish a set of constants A_k which will determine the staircase function

$$g(x) = \begin{cases} A_1 & a < x < x_1 \\ A_2 & x_1 < x < x_2 \\ \cdots\cdots\cdots\cdots\cdots \\ A_N & x_{N-1} < x < b \end{cases}$$

which will approximate $f(x)$ in the sense that, for any small arbitrary positive number ϵ, it will be true that

$$\int_a^b |f(x) - g(x)|\ dx < \frac{\epsilon}{2}$$

For all values of y it is true that

$$\left| \int_a^b f(x) \sin yx\ dx \right| - \left| \int_a^b g(x) \sin yx\ dx \right|$$

$$\leqq \left| \int_a^b [f(x) - g(x)] \sin yx\ dx \right| \leqq \int_a^b |f(x) - g(x)|\ dx < \frac{\epsilon}{2}$$

and from this we have

$$\left| \int_a^b f(x) \sin yx\ dx \right| < \frac{\epsilon}{2} + \left| \int_a^b g(x) \sin yx\ dx \right|$$

But the integral on the right is subject to the appraisal

$$\left| \int_a^b g(x) \sin yx \, dx \right| = \left| \sum_{k=1}^N A_k \int_{x_k}^{x_{k+1}} \sin yx \, dx \right|$$

$$= \left| \sum_{k=1}^N A_k \frac{\cos yx_k - \cos yx_{k+1}}{y} \right| < \frac{2NM}{|y|}$$

where M is an upper bound of the set of A_k numbers. The numbers N and M are fixed, and so if

$$|y| > \frac{4NM}{\epsilon}$$

we have

$$\frac{2NM}{|y|} < \frac{\epsilon}{2}$$

and therefore

$$\left| \int_a^b f(x) \sin yx \, dx \right| < \epsilon$$

Since ϵ is arbitrarily small, this shows that the integral approaches zero. Obviously the same result would have been obtained if $\cos yx$ had been used. This completes the proof for the case where $f(x)$ is piecewise continuous and the limits on the integral are finite. However, the proof can readily be extended to include improper integrals if $f(x)$ is APC and if it is absolutely integrable to infinity. Suppose that the APC function $f(x)$ becomes infinite at x_0 and that

$$\int_0^\infty |f(x)| \, dx$$

exists. Then, by virtue of the definition of convergence of an improper integral, we can choose numbers X_1, X_2, and X_3, where

$$X_1 < x_0 < X_2$$

and such that

$$\left| \int_{X_1}^{X_2} f(x) \sin yx \, dx \right| < \int_{X_1}^{X_2} |f(x)| \, dx < \frac{\epsilon}{4}$$

$$\left| \int_{X_2}^\infty f(x) \sin yx \, dx \right| < \int_{X_2}^\infty |f(x)| \, dx < \frac{\epsilon}{4}$$

Then we can use the previous result to show that

$$\left| \int_0^{X_1} f(x) \sin yx \, dx \right| < \frac{\epsilon}{4} \qquad \text{when } y > y_1$$

and

$$\left| \int_{X_2}^{X_3} f(x) \sin yx \, dx \right| < \frac{\epsilon}{4} \qquad \text{when } y > y_2$$

If y_0 is equal to the larger of the two numbers y_1 and y_2, then we have

$$\left| \int_0^\infty f(x) \sin yx \, dx \right| < \epsilon$$

for
$$y > y_0$$

and again we have shown that the integral approaches zero as y increases.

In the most general case, where $f(x)$ is not necessarily APC, this result is known as the *Riemann-Lebesgue theorem for trigonometric integrals*. We now state the result for the restricted conditions used here, as follows:

Theorem 8-12.   If $f(x)$ is piecewise continuous, in the interval from a to b, then

$$\lim_{|y| \to \infty} \int_a^b f(x) \sin yx \, dx = 0$$

$$\lim_{|y| \to \infty} \int_a^b f(x) \cos yx \, dx = 0$$

(8-19a)

Furthermore, if $f(x)$ is APC, and if

$$\int_0^\infty |f(x)| \, dx$$

exists, then

$$\lim_{|y| \to \infty} \int_0^\infty f(x) \sin yx \, dx = 0$$

$$\lim_{|y| \to \infty} \int_0^\infty f(x) \cos yx \, dx = 0$$

(8-19b)

8-11. Theorems on Integration over Large Semicircles.

In later work there are frequent occurrences of improper integrals falling into one or the other of the following two categories:

Case a $\qquad\qquad PV \int_{-\infty}^\infty H(jy) \, dy$

Case b $\qquad\qquad PV \int_{-\infty}^\infty H(jy) e^{jvb} \, dy$

where $H(z)$ is an analytic function of z. It is possible for $H(z)$ to be multivalued, but if this is the case, it is important to arrange branch cuts so that the jy axis will lie wholly in one sheet of the Riemann surface. Branch points and essential singularities, but not poles, can occur on the jy axis.

The general plan of procedure is to recognize these integrals as limits of complex-plane contour integrals, as follows:

Case a $\qquad PV \int_{-\infty}^\infty H(jy) \, dy = -j \lim_{R \to \infty} \int_C H(z) \, dz$ (8-20)

Case b $\qquad PV \int_{-\infty}^\infty H(jy) e^{jvb} \, dy = -j \lim_{R \to \infty} \int_C H(z) e^{zb} \, dz$ (8-21) where $b < 0$ and C is the contour shown in Fig. 8-8a, extending along the j axis from $-jR$ to $+jR$.* Then the path is closed by a contour C_1, as shown in Fig. 8-8b, or by $C_1 + C'$, as shown in Fig. 8-8c, if $H(z)$ has a branch cut extending to infinity. Then we can use the calculus of residues to write, for case a,

$$-j \int_C H(z)\, dz = j \int_{C_1} H(z)\, dz - 2\pi[\text{sum of residues inside closed curve}]$$
(8-22)

or $$- j \int_C H(z)\, dz = j \int_{C_1} H(z)\, dz + j \int_{C'} H(z)\, dz$$
$$- 2\pi[\text{sum of residues inside closed curve}] \quad (8\text{-}23)$$

A similar expression can be written for case b. In these equations the signs are predicated upon the usual convention in which a positive direction of integration is one in which the enclosed area is on the left.

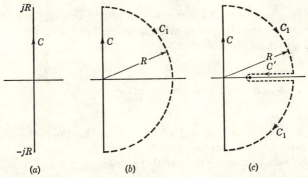

Fig. 8-8. Various contours used to evaluate an integral with two infinite limits.

As R is allowed to approach infinity, as required in Eqs. (8-20) and (8-21), the right-hand side of Eq. (8-22) reduces to an integral over a semicircle of radius approaching infinity, minus 2π times the sum of all residues in the right half plane. In making this statement we assume that the number of poles is finite, and so ultimately all right-half-plane poles will lie within the closed curve, as R increases.

Presently we shall stipulate properties on $H(z)$ which are sufficient to ensure that

$$\lim_{R \to \infty} \int_{C_1} H(z)\, dz = 0 \qquad \text{or} \qquad \lim_{R \to \infty} \int_{C_1} H(z)e^{zb}\, dz = 0$$

Then only the sum of residues remains, unless $H(z)$ is multivalued, necessitating additional arcs such as C' as part of the closed path. It was stated

* The more usual notation, as used in mathematics books, is obtained if z is replaced by jz, with a corresponding 90° rotation of the contours. The form used here is more natural for practical problems arising from the Fourier integral theorem.

that there should be a finite number of poles of $H(z)$ in the right half plane. However, if $H(z)$ is meromorphic and can be expanded in a Mittag-Leffler expansion (see Sec. 5-16), then it can be shown that the present theory is applicable by virtue of a term-by-term integration of the series.

The above discussion sets the stage to show why it is important to investigate integrals over large semicircular arcs. We continue to consider the two cases, showing what conditions on $H(z)$ will make the integrals over the semicircular arcs approach zero as the radius approaches infinity.

We now assume that

$$\lim_{R \to \infty} RH(Re^{j\theta}) = 0, \text{ uniformly } -\frac{\pi}{2} \leqq \theta \leqq \frac{\pi}{2} \qquad \text{for case } a \quad (8\text{-}24)$$

and $$\lim_{R \to \infty} H(Re^{j\theta}) = 0, \text{ uniformly } -\frac{\pi}{2} \leqq \theta \leqq \frac{\pi}{2} \qquad \text{for case } b \quad (8\text{-}25)$$

where in each case zero is approached *uniformly with respect to* θ. This means that corresponding to an arbitrarily small positive ϵ we can in each case find a number R_0 such that

$$R|H(Re^{j\theta})| < \epsilon \qquad \text{for case } a \qquad (8\text{-}26)$$

when $R > R_0$ and $-\pi/2 \leqq \theta \leqq \pi/2$ and

$$|H(Re^{j\theta})| < \epsilon \qquad \text{for case } b \qquad (8\text{-}27)$$

when $R > R_0$ and $-\pi/2 \leqq \theta \leqq \pi/2$. Beyond this point the proofs for the two cases are different.

Consider case a; making the variable change

$$z = Re^{j\theta} \qquad dz = jRe^{j\theta}\, d\theta$$

we get $$\int_{C_1} H(z)\, dz = -j \int_{-\pi/2}^{\pi/2} Re^{j\theta} H(Re^{j\theta})\, d\theta$$

and so, invoking relation (8-26),

$$\left| \int_{C_1} H(z)\, dz \right| \leqq \int_{-\pi/2}^{\pi/2} R|H(Re^{j\theta})|\, d\theta < \pi\epsilon \qquad (8\text{-}28)$$

if $R > R_0$, where R_0 is the value defined in relation (8-26). Since $\pi\epsilon$ is arbitrarily small, it is established that

$$\lim_{R \to \infty} \int_{C_1} H(z)\, dz = 0 \qquad (8\text{-}29)$$

Proceeding in a similar way for case b, we have

$$\int_{C_1} H(z)e^{bz}\, dz = -j \int_{-\pi/2}^{\pi/2} Re^{j\theta} H(Re^{j\theta}) e^{bR \cos \theta} e^{jbR \sin \theta}\, d\theta$$

and $$\left| \int_{C_1} H(z)e^{bz}\, dz \right| \leqq \int_{-\pi/2}^{\pi/2} R|H(Re^{j\theta})| e^{bR \cos \theta}\, d\theta \qquad (8\text{-}30)$$

Now let R be greater than R_0, so that $|H(Re^{j\theta})| < \epsilon$, by relation (8-27).

Also, note that the exponential is an even function of θ, so that for $R > R_0$ we can write

$$\left| \int_{C_1} H(z)e^{bz}\, dz \right| \leq 2R\epsilon \int_0^{\pi/2} e^{bR \cos \theta}\, d\theta \tag{8-31}$$

At this point we refer to Fig. 8-9 for an estimate of $\cos \theta$, namely,

$$\cos \theta \geq 1 - \frac{2}{\pi}\theta \geq 0 \qquad 0 \leq \theta \leq \frac{\pi}{2}$$

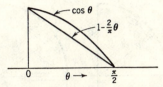

FIG. 8-9. Replacing $\cos \theta$ by a linear function which is always less than $\cos \theta$.

Therefore, assuming $b < 0$,

$$e^{bR \cos \theta} \leq e^{bR}e^{-2bR\theta/\pi} \qquad 0 \leq \theta \leq \frac{\pi}{2}$$

This estimate is used on the right side of relation (8-31) to give

$$2R\epsilon \int_0^{\pi/2} e^{bR \cos \theta}\, d\theta \leq 2R\epsilon e^{bR} \int_0^{\pi/2} e^{-2bR\theta/\pi}\, d\theta = \frac{\pi\epsilon e^{bR}}{-b}\left(e^{-bR} - 1\right)$$

$$= \frac{\pi\epsilon}{|b|}\left(1 - e^{bR}\right) < \frac{\pi\epsilon}{|b|}$$

Thus it has been shown that, if the arbitrary small positive number $\pi\epsilon/|b|$ is chosen, there is a number R_0 such that

$$\int_{C_1} H(z)e^{bz}\, dz < \frac{\pi\epsilon}{|b|} \qquad b < 0$$

when $R > R_0$, and thus it is proved that

$$\lim_{R \to \infty} \int_{C_1} H(z)e^{bz}\, dz = 0 \qquad b < 0 \tag{8-32}$$

In appraising these conclusions, we must keep in mind the restrictions on $H(z)$ and also the condition $b < 0$ in case b. Exactly similar procedures apply, under appropriate conditions, leading to the use of a semicircle in the left half plane. It is found that if

$$\lim_{R \to \infty} RH(Re^{j\theta}) = 0, \text{ uniformly } \frac{\pi}{2} \leq \theta \leq \frac{3\pi}{2} \qquad \text{for case } a \tag{8-33}$$

and $\qquad \displaystyle\lim_{R \to \infty} H(Re^{j\theta}) = 0, \text{ uniformly } \frac{\pi}{2} \leq \theta \leq \frac{3\pi}{2} \qquad \text{for case } b \tag{8-34}$

then, referring to Fig. 8-10 for the definition of C_2, we get

$$\lim_{R \to \infty} \int_{C_1} H(z)\,dz = 0 \qquad \text{for case } a \qquad (8\text{-}35)$$

$$\lim_{R \to \infty} \int_{C_1} H(z)e^{bz}\,dz = 0, \, b > 0 \qquad \text{for case } b \qquad (8\text{-}36)$$

Note that $b > 0$ when the semicircle is in the left half plane.

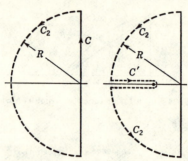

Fig. 8-10. Alternative contours to be used in certain cases for evaluating integrals with two infinite limits.

If there is a branch cut, C_1 or C_2 refers to the complete semicircular arc, excluding the gaps at the branch cuts and excluding whatever additional arcs (such as C' in Fig. 8-10) may be needed in order to keep the complete closed contour in one sheet of the Riemann surface.

These results are so important that we now state them formally as two theorems:

Theorem 8-13.   If $H(z)$ is an analytic function having the property

$$\lim_{R \to \infty} RH(Re^{j\theta}) = 0 \qquad \left\{ \begin{array}{l} -\dfrac{\pi}{2} \leqq \theta \leqq \dfrac{\pi}{2} \\[1mm] \text{or} \\[1mm] \dfrac{\pi}{2} \leqq \theta \leqq \dfrac{3\pi}{2} \end{array} \right. \qquad (8\text{-}37)$$

uniformly with respect to θ, then

$$\lim_{R \to \infty} \int_{C_1} H(z)\,dz = 0 \qquad (8\text{-}38)$$

or

$$\lim_{R \to \infty} \int_{C_2} H(z)\,dz = 0 \qquad (8\text{-}39)$$

where C_1 and C_2 are semicircles in the right and left half planes, respectively, centered at the origin and of radius R.

Theorem 8-14 (*Jordan's Lemma*).   If $H(z)$ is an analytic function having the property

$$\lim_{R \to \infty} H(Re^{j\theta}) = 0 \quad \left\{ \begin{array}{l} -\dfrac{\pi}{2} \leqq \theta \leqq \dfrac{\pi}{2} \\ \text{or} \\ \dfrac{\pi}{2} \leqq \theta \leqq \dfrac{3\pi}{2} \end{array} \right. \tag{8-40}$$

uniformly with respect to θ, then, if b is a nonzero real number,

$$\lim_{R \to \infty} \int_{C_1} H(z)e^{bz}\, dz = 0 \qquad \text{if } b < 0 \tag{8-41}$$

or

$$\lim_{R \to \infty} \int_{C_2} H(z)e^{bz}\, dz = 0 \qquad \text{if } b > 0 \tag{8-42}$$

where C_1 and C_2 are semicircles in the right and left half planes, respectively, centered at the origin and of radius R.

These theorems have been presented in terms of semicircles in the right and left half planes.   However, by means of a variable change to provide a rotation in the complex plane, it is possible to state these theorems for semicircles having any arbitrary orientation.   In fact, these theorems are usually stated for semicircles in the top and bottom half planes.   For the generalization of Jordan's lemma to any semicircle centered at the origin, see Prob. 8-24.   Also, it scarcely needs to be pointed out that for the conditions stated in Theorems 8-13 and 8-14 the integrals over arcs smaller than the semicircles also approach zero.   Very often, as in Example 3 in the following section, an integral over a quarter circle is encountered.

8-12. Evaluation of Improper Real Integrals by Contour Integration. The two theorems presented in the previous section find many applications in later chapters.   However, rather than refer ahead to these applications, we present here three arbitrarily chosen examples.

In each example we start with a real integral in the variable y.   Then, in order to fit the examples to the theorems, it is convenient to write the given integrals with jy as the variable.   Doing this yields integrations over semicircles in the right and left half planes, in similarity with the theorems as stated.

Example 1.   Evaluate the integral

$$\int_{-\infty}^{\infty} \frac{dy}{a^2 + y^2}$$

The integrand can be converted to a function of jy by writing $y^2 = -(jy)^2$, and therefore we formulate the problem as follows,

$$\int_{-\infty}^{\infty} \frac{dy}{a^2 - (jy)^2} = -j \lim_{R \to \infty} \int_C \frac{dz}{a^2 - z^2}$$

where path C is defined in Fig. 8-8. At infinity the integrand behaves in the manner described in Theorem 8-13, for all directions of approach to infinity. Thus, integrals over both C_1 and C_2 approach zero as R approaches infinity. C_1 is arbitrarily selected, which means that $C + C_1$ encloses a right-half-plane pole at $z = a$, with residue

$$-\frac{1}{a + z}\bigg|_{z=a} = -\frac{1}{2a}$$

Thus, if $H(z) = 1/(a^2 - z^2)$,

$$\lim_{R \to \infty} \int_C H(z)\, dz = -\lim_{R \to \infty} \int_{C_1} H(z)\, dz - 2\pi j\left(-\frac{1}{2a}\right)$$

and, using Theorem 8-13 to eliminate the integral over C_1, we have

$$\int_{-\infty}^{\infty} \frac{dy}{a^2 + y^2} = \frac{\pi}{a}$$

The same result would be obtained by integrating over path C_2. Then integration is in the positive sense, but the residue at the pole at $-a$ is $1/2a$. These two sign changes cancel, giving the same result. Note that this integral can be integrated directly, by substituting limits in the inverse tangent function.

Example 2. Evaluate the integral

$$\int_0^{\infty} \frac{\sin y}{y}\, dy$$

The function $(\sin y)/y$ is an even function, and so

$$\int_0^{\infty} \frac{\sin y}{y}\, dy = \frac{1}{2} \int_{-\infty}^{\infty} \frac{\sin y}{y}\, dy = \frac{1}{2} \int_{-\infty}^{\infty} \frac{e^{jy} - e^{-jy}}{j2y}\, dy$$

$$= \lim_{R \to \infty} \left(-j \int_C \frac{e^z - e^{-z}}{4z}\, dz\right) = \lim_{R \to \infty} -j \int_C \frac{\sinh z}{2z}\, dz$$

This would suggest a separate integration of each term in the integrand, but then each integrand would have a pole at the origin, on contour C, and so Theorem 8-14 will not apply. However, $(\sinh z)/z$ has no pole in the finite plane, and so the path C can be distorted to one side of the origin, as

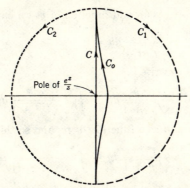

FIG. 8-11. Contours used in evaluating $\int_0^\infty \frac{\sin y}{y}\,dy$.

shown in Fig. 8-11, without changing the integral. Thus,

$$\int_C \frac{\sinh z}{2z}\,dz = \int_{C_0} \frac{\sinh z}{2z}\,dz = \underbrace{\int_{C_0} \frac{e^z}{4z}\,dz}_{A} - \underbrace{\int_{C_0} \frac{e^{-z}}{4z}\,dz}_{B}$$

The quantity b of Theorem 8-14 is $+1$ and -1 for integrals A and B, respectively. Therefore

$$\lim_{R \to \infty} \int_{C_2} \frac{e^z}{4z}\,dz = 0 \qquad \lim_{R \to \infty} \int_{C_1} \frac{e^{-z}}{4z}\,dz = 0$$

and so for integrals A and B the path C_0 is closed, respectively, to the left (C_2) and right (C_1). Path $C_0 + C_1$ encloses no pole, and so integral B approaches zero as R becomes infinite. Path $C_0 + C_2$ encloses the pole at $z = 0$ with residue $\frac{1}{4}$. Thus,

$$\lim_{R \to \infty} \int_C \frac{\sinh z}{2z}\,dz = \lim_{R \to \infty} \int_{C_0} \frac{e^z}{4z}\,dz = j\frac{\pi}{2}$$

and finally $\qquad \int_0^\infty \frac{\sin y}{y}\,dy = \lim_{R \to \infty} -j \int_C \frac{\sinh z}{2z}\,dz = \frac{\pi}{2}$

The same result would be obtained by using a path distorted to the left of the origin, rather than C_0. A double sign change would occur, because the direction of integration would be changed, and because of a change in sign of the residue.

Example 3. Evaluate the integral

$$\int_{-\infty}^{\infty} \frac{\cos y}{\sqrt{y^2 + 1}}\,dy$$

Since $(\sin y)/\sqrt{y^2 + 1}$ is an odd function of y, we have

$$\int_{-\infty}^{\infty} \frac{\sin y}{\sqrt{y^2 + 1}}\, dy = 0$$

This fact allows us to write

$$\int_{-\infty}^{\infty} \frac{\cos y}{\sqrt{y^2 + 1}}\, dy = \int_{-\infty}^{\infty} \frac{e^{-jy}}{\sqrt{y^2 + 1}}\, dy$$

We want the integrand as a function of jy, and so this is changed to

$$\int_{-\infty}^{\infty} \frac{e^{-jy}}{\sqrt{y^2 + 1}}\, dy = \int_{-\infty}^{\infty} \frac{e^{-jy}}{\sqrt{1 - (jy)^2}}\, dy$$

$$= \lim_{R \to \infty} \int_{C} \frac{e^{-z}}{(z^2 - 1)^{\frac{1}{2}}}\, dz$$

Now, referring to Fig. 8-12, we have

$$\int_{C} + \int_{C_1} + \int_{C'} = 0$$

But the integral over C_1 is zero, by Theorem 8-14, and so we look at

$$\int_{C'} \frac{e^{-z}}{(z^2 - 1)^{\frac{1}{2}}}\, dz = \lim_{a \to 0} 2 \int_{R}^{a+1} \frac{e^{-z}}{\sqrt{x^2 - 1}}\, dx$$

$$+ \lim_{a \to 0} \int_{0}^{2\pi} \frac{e^{-1} e^{-a(\cos\theta + j\sin\theta)} jae^{j\theta}}{\sqrt{a}\,(2e^{j\theta} + ae^{j\theta})^{\frac{1}{2}}}\, d\theta$$

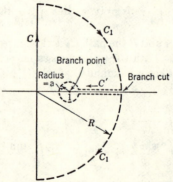

Fig. 8-12. Contours used in evaluating $\displaystyle\int_{-\infty}^{\infty} \frac{\cos y}{y^2 + 1}\, dy$.

The last integral on the right approaches zero because the integrand approaches zero as $\sqrt{a}$. Now we have

$$\int_{-\infty}^{\infty} \frac{\cos y}{\sqrt{y^2 + 1}}\, dy = -\lim_{R \to \infty} \int_{C'} \frac{e^{-z}}{(z^2 - 1)^{\frac{1}{2}}}\, dz$$

$$= 2 \int_{1}^{\infty} \frac{e^{-x}}{\sqrt{x^2 - 1}}\, dx$$

In this example we did not attain a numerical result directly. However, this integral is a special case of one of the integral representations of the Hankel function $H_0^{(1)}(jw)$, which is*

$$H_0^{(1)}(jw) = \frac{1}{\pi j} \int_1^\infty \frac{e^{-wx}}{\sqrt{x^2 - 1}} \, dx$$

Thus, we have the result†

$$\int_{-\infty}^\infty \frac{\cos y}{\sqrt{y^2 + 1}} \, dy = 2\pi j H_0^{(1)}(j)$$
$$= 2\pi (0.268) = 1.685$$

Suppose that we had not succeeded in reducing the new integral to a known function. Then a numerical approximation could have been used. It is instructive to observe that the new form is more amenable to calculation than the original. The new form is better because the oscillating cosine of the original form has been replaced by the rapidly decaying exponential e^{-x}. The algebraic singularity at $x = 1$ in the new form would offer no difficulty, because near $x = 0$ the integrand can be approximated by

$$\frac{e^{-x}}{\sqrt{x^2 - 1}} \doteq \frac{1}{\sqrt{x - 1}}$$

which can be integrated analytically over a small interval.

PROBLEMS

8-1. For each of the following, obtain an explicit formula for the function:

(a) $f_1(t) = \int_0^t a \sin (b - x) \, dx$ (b) $f_2(t) = \int_0^t a \sin (b - x) \, da$

(c) $f_3(t) = \int_0^t a \sin (b - x) \, db$ (d) $f_4(t) = \int_0^t a \sin (b - t) \, db$

(e) $f_5(t) = \int_0^{\sin t} a \sin t \, da$ (f) $f_6(t) = \int_{-x}^t a \sin (b - x) \, db$

8-2. Consider the two functions

$$f(t) = \begin{cases} 2 & 0 \leqq t \leqq 1 \\ 3 - t & 1 \leqq t \leqq 2 \end{cases} \qquad g(t) = \begin{cases} 2 + t & 0 \leqq t \leqq 1 \\ 4.5 - 1.5t & 1 \leqq t \leqq 2 \end{cases}$$

Evaluate

$$\int_0^2 f(t) \frac{dg(t)}{dt} \, dt$$

directly, and also by Theorem 8-4.

---

* See Courant and Hilbert, "Methods of Mathematical Physics," p. 479, Interscience Publishers, Inc., New York, 1953.

† Numerical values for $H_0^{(1)}(jw)$ are found in Jahnke and Emde, "Tables of Functions," Dover Publications, New York, 1945.

8-3. Determine which of the following improper integrals exist. It is not necessary to evaluate those which do exist, but you should give logical proofs of existence or nonexistence.

(a) $\int_0^\infty e^{-x^2}\, dx$

(b) $\int_0^\infty \frac{\sin^2 x}{x}\, dx$

(c) $\int_0^\infty \frac{\sin x}{x}\, dx$

(d) $\int_{-\infty}^0 e^x\, dx$

(e) $\int_{-\infty}^\infty e^{\sqrt{1+x^2}} \cos x\, dx$

(f) $\int_{-\infty}^\infty x \cos x\, dx$

8-4. Do Prob. 8-3 for the following integrals:

(a) $\int_0^1 \frac{dx}{\sqrt{x}}$

(b) $\int_0^1 \frac{dx}{x}$

(c) $\int_0^1 \log x\, dx$

(d) $\int_0^2 \frac{1}{\sqrt{|x-1|}}\, dx$

(e) $\int_0^1 \frac{\log x}{\sqrt{x}}\, dx$

(f) $\int_0^1 e^{1/x}\, dx$

8-5. Assuming that

$$\int_0^\infty f(x)\, dx \quad \text{and} \quad \int_0^\infty g(x)\, dx$$

both converge, prove that

$$\int_0^\infty [f(x) + g(x)]\, dx$$

converges and is equal to the sum of the first two integrals. Also, by means of a counterexample, show that convergence of the last integral does not imply convergence of the other two.

8-6. For the following functions, catalogue them according to whether they are continuous, piecewise continuous, almost piecewise continuous, or none of these. Answer for each of the intervals $0 \leq x \leq 10$ and $0 < x \leq 10$:

(a) $x \sin \frac{1}{x}$

(b) $\tan x$

(c) $\sqrt{|\cot x|}$

(d) $\log x$

(e) $\sin \frac{1}{x}$

(f) $\frac{1}{x}$

8-7. Use the fact that

$$\int_0^\infty \frac{\sin t}{t}\, dt = \frac{\pi}{2}$$

to show that

$$\frac{2}{\pi} \int_0^\infty \frac{\sin \omega x}{x}\, dx$$

is a function of ω which is -1 for $\omega < 0$ and $+1$ for $\omega > 0$.

8-8. Check whether or not the question marks may be replaced by equality signs in the following:

(a) $\int_0^1 dy \int_0^1 x \log |x - y|\, dx$? $\int_0^1 dx \int_0^1 x \log |x - y|\, dy$

(b) $\int_0^1 dy \int_0^1 \frac{dx}{\sqrt{|x - y|}}$? $\int_0^1 dx \int_0^1 \frac{dy}{\sqrt{|x - y|}}$

(c) $\int_0^2 dy \int_0^2 f(x - y)\, dx$? $\int_0^2 dx \int_0^2 f(x - y)\, dy$

where $f(x) = 1/\sqrt{|x|}$ when $0 < |x| < 1$ and $f(x) - 1/\sqrt{2 - x}$ when $1 < |x| < 2$.

(d) $\int_0^1 dy \int_0^2 \frac{dx}{\sqrt{|1 - x^2 y^2|}}$? $\int_0^2 dx \int_0^1 \frac{dy}{\sqrt{|1 - x^2 y^2|}}$

In each case determine whether the integrand is APC in each variable, as required in Theorem 8-7.

8-9. Determine whether the integral

$$\int_0^\infty y e^{-yx}\, dx$$

converges, and converges uniformly, in the intervals

(a) $0 \leq y \leq 1$ (b) $1 \leq y \leq 2$

8-10. Referring to Prob. 8-9, compare the following:

(a) $\lim_{y \to 0} \int_0^\infty y e^{-yx}\, dx$ and $\int_0^\infty \lim_{y \to 0} y e^{-yx}\, dx$

(b) $\lim_{y \to 1} \int_0^\infty y e^{-yx}\, dx$ and $\int_0^\infty \lim_{y \to 1} y e^{-yx}\, dx$

and relate your findings to the convergence properties obtained in Prob. 8-9.

8-11. Referring to Prob. 8-9, compare the following:

(a) $\int_0^1 dy \int_0^\infty y e^{-yx}\, dx$ and $\int_0^\infty dx \int_0^1 y e^{-yx}\, dy$

(b) $\int_1^2 dy \int_0^\infty y e^{-yx}\, dx$ and $\int_0^\infty dx \int_1^2 y e^{-yx}\, dy$

and relate your findings to the convergence properties obtained in Prob. 8-9.

8-12. For each of the following cases check whether or not the question mark can be replaced by an equality sign:

(a) $\int_0^1 dy \int_0^\infty e^{-x} \sin xy\, dx$? $\int_0^\infty dx \int_0^1 e^{-x} \sin xy\, dy$

(b) $\int_0^1 dy \int_0^\infty x e^{-yx}\, dx$? $\int_0^\infty dx \int_0^1 x e^{-yx}\, dy$

Relate your conclusions to convergence properties of the integrals.

8-13. If $f(x)$ has the property that

$$\int_0^\infty |f(x)|\, dx$$

$$e^{-j\pi/2} = +j$$

converges, prove that

$$\int_0^\infty f(x+y)\,dx$$

converges uniformly, $a \leq y \leq y_0$, where a and y_0 are arbitrary real numbers.

8-14. For each of the following integrals, use the M test to determine a range of y for which the integral

$$\int_0^\infty f(x,y)\,dx$$

converges uniformly:

(a) $f(x,y) = \left(\dfrac{\sin xy}{x}\right)^2$

(b) $f(x,y) = \cos(x^2 + y^2)e^{-(x+y)}$

(c) $f(x,y) = \dfrac{e^{xy}}{x^2 + y^2}$

8-15. Let $f(x)$ be APC, and assume that

$$\int_0^\infty |f(x)|^2\,dx$$

converges.   Use appropriate theorems in the text to prove that

$$\lim_{y\to 0}\int_0^\infty f(x+y)f(x)\,dx = \int_0^\infty |f(x)|^2\,dx$$

8-16. Use contour integration to evaluate the following integrals:

(a) $\displaystyle\int_{-\infty}^\infty \dfrac{\sin y}{1+y^2}\,dy$

(b) $\displaystyle\int_{-\infty}^\infty \dfrac{e^{jy}}{1+y^2}\,dy$

(c) $\displaystyle\int_{-\infty}^\infty \dfrac{dy}{(1+y^2)^2}$

8-17. Prove the identity

$$\int_0^\infty \frac{\sin y + y\cos y}{1+y^2}\,dy = \int_0^\infty \frac{e^{-x}}{1+x}\,dx$$

8-18. Evaluate the integral

$$\int_0^\infty \frac{dx}{\sqrt{x}\,(1+x^2)}$$

by the following two methods.

(a) Contour integration on the integral as it stands.   (HINT: Put a branch cut along the positive real axis.   After obtaining the answer in this way, decide whether a branch cut along the negative real axis could have been used.)

(b) Let $x = s^2$, and obtain a new integral, which is then evaluated by contour integration.

8-19. Evaluate the integral

$$\int_0^\infty \frac{dx}{\sqrt[3]{x}\,(1+x)}$$

where $\sqrt[3]{x}$ is positive.   (HINT: Use a branch cut along the positive real axis, and then decide whether or not you could get the same result by using a branch cut along the negative real axis.)

8-20. Use the principle of contour integration in the complex plane to evaluate the integral

$$\int_0^\infty \frac{\sin \pi y}{y(1-y^2)}\,dy$$

8-21. Verify both of Eqs. (7-54) for the function

$$f(s) = \frac{1}{1+s}$$

(HINT: Use the principles established in Sec. 8-12, with a semicircular indentation around the pole at $y = \omega$. The PV_2 integral does not include the integral over this semicircle, and so the portion due to it must be subtracted out.)

8-22. Verify both of Eqs. (7-54) for the function

$$f(s) = \frac{1}{(1+s)^2}$$

(See hint in Prob. 8-21.)

8-23. Let $Z(s)$ be the impedance of a capacitor in parallel with a series RL branch. Take each element value as unity. The real component of this impedance (for real frequencies) is

$$u(\omega) = \frac{1}{1 - \omega^2 + \omega^4}$$

and the imaginary component is

$$v(\omega) = -\frac{\omega^3}{1 - \omega^2 + \omega^4}$$

Show that both of Eqs. (7-55) are satisfied in this case.

8-24. Define a semicircle C' centered at the origin and of radius R, and give conditions on $H(z)$ and specify the orientation of the semicircle such that

$$\lim_{R \to \infty} \int_{C'} H(z)e^{cz}\, dz = 0$$

where c is the complex number $|c|e^{j\gamma}$.

8-25. As an alternative to the method given in the text for finding

$$\int_0^\infty \frac{\sin y}{y}\, dy$$

show that

$$\int_0^\infty \frac{\sin y}{y}\, dy = \lim_{\epsilon \to 0} \frac{1}{2j}\left(\int_{-\infty}^{-\epsilon} \frac{e^{jy}}{y}\, dy + \int_\epsilon^\infty \frac{e^{jy}}{y}\, dy \right)$$

and then use Jordan's lemma and the results of Prob. 5-33 to evaluate the quantity on the right.

8-26. Let a rational function $F(s)$ have at least a second-order zero at infinite s. This means that the degree of the denominator must be at least 2 greater than the degree of the numerator. Use an appropriate theorem to prove that the summation of the residues over all the poles is zero.

8-27. In the discussion leading to Theorem 8-7, the following properties of iterated integrals over the rectangles in Fig. 8-7 are stated:

$$\int_{y_1}^{y_1+\Delta y} dy \int_{x_1}^{x_1+\Delta x} \frac{dx}{|x - x_1 - k_1(y - y_1)|^m} < (\text{constant})(\Delta y)^{2-m}$$

$$\int_{y_2}^{y_2+\Delta y} dy \int_{x_2}^{x_2+\Delta x} \frac{dx}{|x - x_2 - k_2(y - y_2)|^n |x - x_2 - k_3(y - y_2)|^p}$$
$$< (\text{constant})(\Delta y)^{2-n-p}$$

Carry out these integrations, establishing the correctness of the above results.

CHAPTER 9

THE FOURIER INTEGRAL

9-1. Introduction. In this chapter we deal with perhaps one of the most important single topics in the theory of the Laplace transform. In Chap. 1 you were given a nonrigorous development which formally states that under certain conditions, given a function $f(t)$, we can form the function

$$\mathcal{F}(j\omega) = \int_{-\infty}^{\infty} f(t)e^{-j\omega t}\, dt \tag{9-1}$$

and then recover $f(t)$ by the inversion formula

$$f(t) = \frac{1}{2\pi} PV \int_{-\infty}^{\infty} \mathcal{F}(j\omega)e^{j\omega t}\, d\omega \tag{9-2}$$

This is a statement of the bare essentials of the Fourier integral theorem. In these formulas ω and t are real, and so these are real integrals. In linear system analysis, t is usually time, but of course this is incidental, and t is regarded as merely a real variable in this discussion.

With the background of the present chapter we are now prepared to give a proof which is mathematically more rigorous than the proof given in Chap. 1, and also we shall have a precise statement of sufficient conditions on $f(t)$ to ensure validity of the Fourier integral theorem.

9-2. Derivation of the Fourier Integral Theorem. Sufficient conditions will be developed as we progress; and then the results will be collected and stated as a theorem at the end. The first condition is that $f(t)$ shall be APC and the integral

$$\int_{-\infty}^{\infty} |f(t)|\, dt$$

shall exist. Then, by Theorem 8-11, the integral

$$\int_{-\infty}^{\infty} f(t)e^{-j\omega t}\, dt$$

will converge uniformly with respect to the upper and lower limits, for all ω.

The proof of the theorem consists in proving Eq. (9-2) to be true. As a first step we substitute $\mathcal{F}(j\omega)$, as given by Eq. (9-1), into the integral of

Eq. (9-2).   By omitting the factor $1/2\pi$, the right side of Eq. (9-2) becomes

$$PV \int_{-\infty}^{\infty} \mathfrak{F}(j\omega)e^{j\omega t} \, d\omega = \lim_{A \to \infty} \int_{-A}^{A} e^{j\omega t} \, d\omega \int_{-\infty}^{\infty} f(\tau)e^{-j\omega\tau} \, d\tau$$

The pair of iterated integrations can be inverted, by virtue of Theorem 8-8, giving

$$
\begin{aligned}
PV \int_{-\infty}^{\infty} \mathfrak{F}(j\omega)e^{j\omega t} \, d\omega &= \lim_{A \to \infty} \int_{-\infty}^{\infty} f(\tau) \, d\tau \int_{-A}^{A} e^{j\omega(t-\tau)} \, d\omega \\
&= \lim_{A \to \infty} 2 \int_{-\infty}^{\infty} f(\tau) \frac{\sin A(t-\tau)}{t-\tau} \, d\tau \\
&= \lim_{A \to \infty} 2 \int_{-\infty}^{\infty} f(u+t) \frac{\sin Au}{u} \, du \qquad (9\text{-}3)
\end{aligned}
$$

The last step is arrived at by changing the variable in accordance with $u = \tau - t$ and regarding t as constant.

Now assume that t is a value at which $f(t-)$ and $f(t+)$ exist.   We write this integral in three parts, as follows:

$$2 \int_{-\infty}^{\infty} f(u+t) \frac{\sin Au}{u} \, du = I_1(A,t) + I_2(A,t) + I_3(A,t) \qquad (9\text{-}4)$$

where

$$I_1(A,t) = 2 \int_{-\infty}^{-\delta} f(u+t) \frac{\sin Au}{u} \, du$$

$$I_2(A,t) = 2 \int_{-\delta}^{\delta} f(u+t) \frac{\sin Au}{u} \, du \qquad (9\text{-}5)$$

$$I_3(A,t) = 2 \int_{\delta}^{\infty} f(u+t) \frac{\sin Au}{u} \, du$$

The Riemann theorem for trigonometric integrals (Theorem 8-12) applies to $I_1(A,t)$ and $I_3(A,t)$ because $f(u+t)/u$ is absolutely integrable over intervals which avoid the origin.   Now we have

$$\lim_{A \to \infty} I_1(A,t) = 0$$

$$\lim_{A \to \infty} I_3(A,t) = 0$$

and so, from Eqs. (9-3) and (9-4), we are left with

$$PV \int_{-\infty}^{\infty} \mathfrak{F}(j\omega)e^{j\omega t} \, d\omega = \lim_{A \to \infty} 2 \int_{-\delta}^{\delta} f(u+t) \frac{\sin Au}{u} \, du \qquad (9\text{-}6)$$

If $f(t)$ is discontinuous at the value of t in question, then $f(u+t)$ is discontinuous at $u = 0$ and this would lead to difficulty in a later step if the integral were left in this form.   This trouble can be avoided by writing it as the sum of two integrals,

$$2 \int_{-\delta}^{\delta} f(u + t) \frac{\sin Au}{u} \, du = 2 \int_{-\delta}^{0} f(u + t) \frac{\sin Au}{u} \, du$$

$$+ 2 \int_{0}^{\delta} f(u + t) \frac{\sin Au}{u} \, du$$

$$= 2 \int_{0}^{\delta} [f(u + t) + f(-u + t)] \frac{\sin Au}{u} \, du$$

$$(9\text{-}7)$$

Now change the variable to $w = Au$, giving

$$2 \int_{-\delta}^{\delta} f(u + t) \frac{\sin Au}{u} \, du = 2 \int_{0}^{\delta A} \left[f\left(t + \frac{w}{A}\right) + f\left(t - \frac{w}{A}\right) \right] \frac{\sin w}{w} \, dw$$

Let A approach infinity, without justifying the step, to get a clue to the answer. Justification will come later. This yields the tentative guess that the right-hand side of the above equation is

$$2[f(t+) + f(t-)] \int_{0}^{\infty} \frac{\sin w}{w} \, dw = \pi[f(t+) + f(t-)] \qquad (9\text{-}8)$$

Now we shall prove that this is correct.

The sine integral is the starting point, and we proceed by modifying it as follows:

$$\int_{0}^{\infty} \frac{\sin w}{w} \, dw = \lim_{A \to \infty} \int_{0}^{A\delta} \frac{\sin w}{w} \, dw = \lim_{A \to \infty} \int_{0}^{\delta} \frac{\sin Au}{u} \, du$$

Therefore, the term on the left of Eq. (9-8) can also be written as follows:

$$\lim_{A \to \infty} 2 \int_{0}^{\delta} [f(t+) + f(t-)] \frac{\sin Au}{u} \, du$$

A check on the correctness of the tentative answer given by Eq. (9-8) is obtained by computing the difference between the above expression and the right side of Eq. (9-7). By omitting the factor 2, this difference is

$$\lim_{A \to \infty} \left[\int_{0}^{\delta} \frac{f(t + u) - f(t+)}{u} \sin Au \, du + \int_{0}^{\delta} \frac{f(t - u) - f(t-)}{u} \sin Au \, du \right]$$

As u approaches zero, each of the quantities $f(t + u) - f(t+)$ and $f(t - u) - f(t-)$ approaches zero, because $f(t)$ is piecewise continuous. Thus, there is the possibility of the above quantities, when divided by u, being integrable over the interval indicated. However, they will be integrable only if the numerators approach zero as fast as, or faster than, u.

As a sufficient condition assume that we can find numbers K and α such that

$$|f(t + u) - f(t+)| < K_1 u^{\alpha_1}$$
$$|f(t - u) - f(t-)| < K_2 u^{\alpha_2}$$

(9-9)

when $u < \delta$

and where α_1 and α_2 are each greater than zero. This will ensure integrability. For example, then,

$$\left| \int_0^\delta \frac{f(t + u) - f(t+)}{u}\, du \right| \leqq \int_0^\delta K_1 u^{\alpha_1 - 1}\, du = \frac{K_1 \delta^{\alpha_1}}{\alpha_1}$$

and similarly for the other integral. Relations (9-9) are called *Lipshitz conditions* of order α and are a property of $f(t)$ which we now require, in addition to the original condition of being APC.

Assuming that conditions (9-9) are satisfied, Theorem 8-12 can now be used to arrive at the results

$$\lim_{A \to \infty} \int_0^\delta \frac{f(t + u) - f(t+)}{u} \sin Au\, du = 0$$

$$\lim_{A \to \infty} \int_0^\delta \frac{f(t - u) - f(t-)}{u} \sin Au\, du = 0$$

This confirms the earlier guess that the right side of Eq. (9-8) is a correct evaluation of the right side of Eq. (9-6), and therefore also of the left side of Eq. (9-3). In other words, we have proved, for the conditions assumed, that

$$\frac{f(t+) + f(t-)}{2} = \frac{1}{2\pi} PV \int_{-\infty}^{\infty} \mathfrak{F}(j\omega) e^{j\omega t}\, d\omega$$

(9-10)

at any point where $f(t)$ satisfies Lipshitz conditions of order $\alpha > 0$. Of course, at any point where $f(t)$ is continuous, $[f(t+) + f(t-)]/2 = f(t)$, and so Eq. (9-2) is an adequate representation if

$$f(t) = \frac{f(t+) + f(t-)}{2}$$

is used for the definition of $f(t)$ at points of finite discontinuity.

A few words about the physical meaning of the Lipshitz condition may be enlightening. Existence of right- and left-hand derivatives at each value of t would be sufficient to carry out the concluding steps of the proof, without explicitly stating the Lipshitz condition. In fact, existence of the single-sided derivative is equivalent to the Lipshitz condition of order 1. Most practical functions, like the examples in Fig. 9-1, do have right- and left-hand derivatives. However, we would like to be a bit more general and include functions like Fig. 9-2. Here there are points where the function remains finite, may or may not be discontinuous, but where the single-sided derivative becomes infinite. These functions

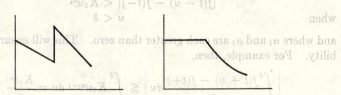

FIG. 9-1. Examples of functions having finite single-sided derivatives.

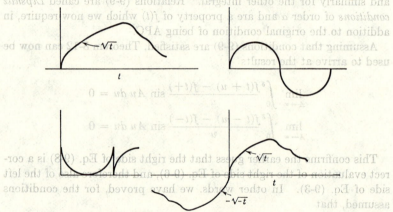

FIG. 9-2. Examples of bounded functions which have derivatives (or single-sided derivatives) which become infinite.

satisfy the Lipshitz condition, and so the Fourier integral theorem is valid for them.*

We now state the Fourier integral theorem as follows:

Theorem 9-1. Let $f(t)$ be a function which is almost piecewise continuous, which satisfies the Lipshitz condition of order $\alpha > 0$ at each point where the function is finite, and for which the integral

$$\int_{-\infty}^{\infty} |f(t)| \, dt$$

converges. Then,

$$\mathfrak{F}(j\omega) = \int_{-\infty}^{\infty} f(t) e^{-j\omega t} \, dt$$

defines the function $\mathfrak{F}(j\omega)$ by an integral which converges uniformly with

* The function $1/\sqrt{-\log |t|}$ does not satisfy a Lipshitz condition at the origin.

respect to the infinite limits, for all real ω. Furthermore, $f(t)$ is related to $\mathfrak{F}(j\omega)$ by

$$\frac{f(t+) + f(t-)}{2} = \frac{1}{2\pi} PV \int_{-\infty}^{\infty} \mathfrak{F}(j\omega)e^{j\omega t}\,d\omega$$

at all points where $f(t)$ is not infinite. The first integral is called the Fourier integral, and the second integral is called the *inversion integral*.

9-3. Some Properties of the Fourier Transform. For further appraisal of the Fourier integral theorem we consider two examples. As the first one, take the *even* function

$$f(t) = e^{-a|t|} \qquad a > 0 \tag{9-11}$$

for which

$$\begin{aligned}
\mathfrak{F}(j\omega) &= \int_{-\infty}^{0} e^{(a-j\omega)t}\,dt + \int_{0}^{\infty} e^{-(a+j\omega)t}\,dt \\
&= \frac{1}{a - j\omega} + \frac{1}{a + j\omega} \\
&= \frac{2a}{a^2 + \omega^2} \tag{9-12}
\end{aligned}$$

As the second example, consider the *odd* function

$$f(t) = \begin{cases} -e^{at} & t < 0 \\ e^{-at} & t > 0 \end{cases} \tag{9-13}$$

which gives

$$\begin{aligned}
\mathfrak{F}(j\omega) &= -\int_{-\infty}^{0} e^{(a-j\omega)t}\,dt + \int_{0}^{\infty} e^{-(a+j\omega)t}\,dt \\
&= \frac{1}{a - j\omega} + \frac{1}{a + j\omega} \\
&= -j\frac{2\omega}{a^2 + \omega^2} \tag{9-14}
\end{aligned}$$

In each case note that $\mathfrak{F}(j\omega)$ is an analytic function given by a comparatively simple formula which has meaning for all values of ω, real or complex. On the other hand, if $\omega = x + jy$, the Fourier integral becomes*

$$\int_{-\infty}^{\infty} f(t)e^{yt}e^{-jxt}\,dt$$

Introduction of the factor e^{yt} may prevent this integral from converging for certain values of y. Thus, although the integral may not converge for all complex values of ω, we seem to get a function from the integral which is defined for all values of ω, complex as well as real. At least this is true for the examples given.

The above facts point up the reason for making a distinction in terminology between the *Fourier integral* and the function $\mathfrak{F}(j\omega)$. The latter

---

* Here we are making an exception to the practice otherwise followed in this text, that ω shall be real.

is called the *Fourier transform* of $f(t)$. It is a function of ω in which no integral need appear. The transform is a function which can be represented by the integral for certain ranges of complex ω but which can exist for other values of ω.

We shall now briefly consider a few of the universally important properties of the Fourier transform function $\mathfrak{F}(j\omega)$, as they are related to properties of $f(t)$.

1. *Properties of Real and Imaginary Parts of* $\mathfrak{F}(j\omega)$. In the first example $f(t) \equiv f(-t)$, and we found that $\mathfrak{F}(j\omega)$ is a real function of ω having the property $\mathfrak{F}(j\omega) \equiv \mathfrak{F}(-j\omega)$. In the second example $f(t) \equiv -f(-t)$, and the corresponding $\mathfrak{F}(j\omega)$ is imaginary and has the property $\mathfrak{F}(j\omega) \equiv -\mathfrak{F}(-j\omega)$. These are specific examples of a general property which can be derived by writing $f(t)$ as the sum of even and odd parts:

$$f(t) = f_e(t) + f_o(t) \tag{9-15}$$

where
$$f_e(t) = \frac{f(t) + f(-t)}{2} \qquad f_o(t) = \frac{f(t) - f(-t)}{2} \tag{9-16}$$

Then the Fourier integral can be written

$$\int_{-\infty}^{\infty} f(t)e^{-j\omega t}\, dt = \int_0^{\infty} [f(-t)e^{j\omega t} + f(t)e^{-j\omega t}]\, dt$$
$$\mathfrak{F}(j\omega) = 2\int_0^{\infty} f_e(t) \cos \omega t\, dt - j2\int_0^{\infty} f_o(t) \sin \omega t\, dt \tag{9-17}$$

Thus, when $f(t)$ is real, the real part of $\mathfrak{F}(j\omega)$ is always an even function of ω, and the imaginary part is always an odd function of ω. Furthermore, if $f(t)$ is even, $\mathfrak{F}(j\omega)$ is real (and also even), and if $f(t)$ is odd, $\mathfrak{F}(j\omega)$ is imaginary (and also odd). A good way to summarize these properties is to write

$$\mathfrak{F}(j\omega) = \overline{\mathfrak{F}(-j\omega)} \qquad \omega \text{ real} \tag{9-18}$$

In the terminology of Chap. 7, $\mathfrak{F}(j\omega)$ is a real function of $j\omega$.

This general case, for a real function $f(t)$, can be summarized by defining two transform functions, as follows:

$$\mathfrak{F}(j\omega) = \mathfrak{F}_r(j\omega) + j\mathfrak{F}_i(j\omega) \tag{9-19}$$

where
$$\mathfrak{F}_r(j\omega) = 2\int_0^{\infty} f_e(t) \cos \omega t\, dt$$
$$\mathfrak{F}_i(j\omega) = -2\int_0^{\infty} f_0(t) \sin \omega t\, dt \tag{9-20}$$

2. *Differentiability of the Fourier Transform*. If $f(t)$ satisfies the conditions of the Fourier integral theorem, as stated by Theorem 9-1, we know that

$$\mathfrak{F}(j\omega) = \int_{-\infty}^{\infty} f(t)e^{-j\omega t}\, dt$$

converges uniformly for all ω. Uniform convergence is sufficient to allow us to write

$$\mathfrak{F}(j\omega) - \mathfrak{F}(j\omega_0) = \int_{-\infty}^{\infty} f(t)(e^{-j\omega t} - e^{-j\omega_0 t})\, dt$$

and to take the limit as ω approaches ω_0. This is by virtue of Theorem 8-10. The integrand approaches zero, and so it follows that the transform is a continuous function of the *real* variable ω.

Now apply the Fourier integral theorem to the function $tf(t)$, and adopt the notation

$$\mathfrak{F}_1(j\omega) = \int_{-\infty}^{\infty} tf(t)e^{-j\omega t}\, dt$$

This function can be integrated under the integral with respect to ω, by Theorem 8-8, and so we can write

$$\int_0^{\omega} \mathfrak{F}_1(j\Omega)\, d\Omega = j \int_{-\infty}^{\infty} f(t)(e^{-j\omega t} - 1)\, dt$$
$$= j[\mathfrak{F}(j\omega) - \mathfrak{F}(0)]$$

Now differentiate with respect to ω, to obtain

$$\mathfrak{F}_1(j\omega) = j\, \frac{d\mathfrak{F}(j\omega)}{d\omega} \tag{9-21}$$

Since $\mathfrak{F}_1(j\omega)$ exists and is continuous, we see that $\mathfrak{F}(j\omega)$ has a continuous derivative with respect to the real variable ω. A similar result can be obtained if $t^2 f(t)$ satisfies the conditions of the Fourier integral theorem, in which case the second derivative of $\mathfrak{F}(j\omega)$ is found to be continuous. We summarize by stating that if $t^n f(t)$ is absolutely integrable from $-\infty$ to ∞, then the nth derivative of $\mathfrak{F}(j\omega)$ with respect to ω is continuous and is given by

$$\frac{d^n \mathfrak{F}(j\omega)}{d\omega^n} = \frac{1}{j^n}\, \mathfrak{F}_n(j\omega) \tag{9-22}$$

In the above, ω must be real because uniform convergence is assured only for real ω. Thus, in Eq. (9-21) we can say that the derivative exists for real ω but not necessarily for complex ω. Consequently, this equation *does not* establish that $\mathfrak{F}(j\omega)$ is analytic.

To show that $\mathfrak{F}(j\omega)$ is not necessarily analytic, consider the function

$$f(t) = \frac{\sin bt}{t(t^2 - \pi^2/b^2)}$$

By routine integration, the Fourier transform is found to be

$$\mathfrak{F}(j\omega) = \begin{cases} \dfrac{b^2}{\pi}\left(1 + \cos \pi\, \dfrac{\omega}{b}\right) & 0 \leq |\omega| \leq b \\ 0 & |\omega| > b \end{cases}$$

For this example, $\mathfrak{F}(j\omega)$ has a continuous first derivative, but the second derivative is not continuous, and higher-order derivatives do not exist. If we check the absolute integrability of $t^n f(t)$, we find that absolute integrability is retained when $n = 1$ but not when $n = 2$. Hence, it is to be expected that the first derivative will be continuous but not necessarily the second.

The case where $f(t)$ is identically zero for $|t|$ greater than some fixed value is of particular interest. For such a function the Fourier integral has finite limits, and so $t^n f(t)$ is absolutely integrable for all n. We conclude that in this case its Fourier transform is indefinitely differentiable with respect to ω.

Many other properties of the Fourier transform function can be derived, but the same information can be obtained more easily from the Laplace transform, because the theory of functions of a complex variable is then more extensively applicable.

9-4. Remarks about Uniqueness and Symmetry. For a given $f(t)$, its transform $\mathfrak{F}(j\omega)$ is uniquely determined by the Fourier integral of $f(t)$. Therefore, each Fourier integrable function has one and only one transform function $\mathfrak{F}(j\omega)$ as its "mate." We now ask whether or not two different $f(t)$ functions could ever produce the same $\mathfrak{F}(j\omega)$ function. Suppose that $f_1(t)$ and $f_2(t)$ are two different functions but that the transform of each is the same function $\mathfrak{F}(j\omega)$. Since $\mathfrak{F}(j\omega)$ is the same for each, an identical inversion integral is obtained for $f_1(t)$ and $f_2(t)$. Therefore, from Eq. (9-10) it follows that $f_1(t)$ and $f_2(t)$ must satisfy the equation

$$\frac{f_1(t+) + f_1(t-)}{2} = \frac{f_2(t+) + f_2(t-)}{2} \qquad (9\text{-}23)$$

At any continuous point of a function $f(t)$,

$$\frac{f(t+) + f(t-)}{2} = f(t) \qquad (9\text{-}24)$$

and therefore Eq. (9-23) tells us that

$$f_1(t) \equiv f_2(t)$$

except possibly at isolated points where one or the other function might be defined arbitrarily. The inversion integral can give no information about whether or not $f_1(t)$ and $f_2(t)$ are equal at such isolated points. This is to be expected, because the set of such isolated points is a set of measure 0. We recall from Theorem 8-5 that an integral is unaffected if the integrand is defined arbitrarily over a set of measure zero. Thus, the Fourier integral can yield the same $\mathfrak{F}(j\omega)$ function for two $f(t)$ functions which differ over a set of measure 0, and the inversion integral will be insensitive to this difference. This possible difference between two functions having the same transform is primarily of academic interest.

In fact, if we continue to agree to define a function at a discontinuity as the mean of the limits approached from the two sides, then $f(t)$ is uniquely related to $\mathfrak{F}(j\omega)$.

This property of uniqueness is very important. It means that tables of paired functions can be constructed which are solutions of the pair of integral equations

$$\mathfrak{F}(j\omega) = \int_{-\infty}^{\infty} f(t)e^{-j\omega t}\,dt \tag{9-25a}$$

$$f(t) = \frac{1}{2\pi}PV \int_{-\infty}^{\infty} \mathfrak{F}(j\omega)e^{j\omega t}\,d\omega \tag{9-25b}$$

where Eq. (9-24) is understood to define $f(t)$ in the second equation. Because these relationships occur in pairs, and because we have established uniqueness, it follows that if one of these integrals is known the other one is automatically known. For example, in Sec. 8-9 we obtained the formula

$$\frac{2a}{a^2 + \omega^2} = \int_{-\infty}^{\infty} e^{-a|t|}e^{-j\omega t}\,dt \tag{9-26}$$

In view of the above property we immediately know the value of an additional integral, namely,

$$e^{-a|t|} = \frac{1}{2\pi}\int_{-\infty}^{\infty} \frac{2a}{a^2 + \omega^2}\,e^{j\omega t}\,d\omega \tag{9-27}$$

The usual PV designation is not needed here because the integral converges as it stands.

In this way we see that the Fourier integral theorem has the effect of doubling the size of a given table of integrals like Eq. (9-25a). If we have a table of integrals representing integrals like Eq. (9-25a), we also implicitly have a table of integrals like Eq. (9-25b). Except where the PV designation is needed, these two integrals are essentially the same. Let us consider this statement a bit further. Equation (9-27) can be made to look like Eq. (9-25a) by replacing ω by t and t by $-\omega$ to give

$$\frac{\pi}{a}e^{-a|\omega|} = \int_{-\infty}^{\infty} \frac{1}{a^2 + t^2}\,e^{-j\omega t}\,dt \tag{9-28}$$

If Eqs. (9-27) and (9-28) are typical of all cases, it would appear that to obtain one pair of functions satisfying Eq. (9-25a) is to obtain a second pair. In this case the two pairs are:

$f(t)$	$\mathfrak{F}(j\omega)$
$e^{-a\lvert t\rvert}$	$\dfrac{2a}{a^2 + \omega^2}$
$\dfrac{1}{a^2 + t^2}$	$\dfrac{\pi}{a}e^{-a\lvert\omega\rvert}$

One naturally asks whether this illustrates a generally valid conclusion. As was mentioned above, the integral of Eq. (9-25b) cannot be made to look like the integral of Eq. (9-25b) unless the *PV* designation can be omitted from the latter. Also, this example is a special case in the sense that $\mathfrak{F}(j\omega)$ is real. Consider the general case of Eq. (9-25b), in which the integral converges without taking the principal value. Then, if t is replaced by $-t$, we can write

$$f(-t) = \frac{1}{2\pi} \int_{-\infty}^{\infty} \mathfrak{F}(j\omega)e^{-j\omega t}\,d\omega$$

and if t and ω are now interchanged we get

$$2\pi f(-\omega) = \int_{-\infty}^{\infty} \mathfrak{F}(jt)e^{-j\omega t}\,dt \tag{9-29}$$

If $\mathfrak{F}(jt)$ is real, this last equation looks like Eq. (9-25a), except for incidental differences in notation. If $\mathfrak{F}(jt)$ is complex, $f(-\omega)$ is neither even nor odd and, in the notation of Eqs. (9-19) and (9-20), the above equation reduces to the pair

$$\pi[f(-\omega) + f(\omega)] = \int_{-\infty}^{\infty} \mathfrak{F}_r(jt)e^{-j\omega t}\,dt \tag{9-30a}$$

$$\pi[f(-\omega) - f(\omega)] = j \int_{-\infty}^{\infty} \mathfrak{F}_i(jt)e^{-j\omega t}\,dt \tag{9-30b}$$

Each of these is similar to Eq. (9-25a).

No condition has been stated whereby we can know when an arbitrarily given $\mathfrak{F}(j\omega)$ function will be the Fourier transform of some $f(t)$. In view of the symmetry properties mentioned above, we can at least say that if

$$\int_{-\infty}^{\infty} |\mathfrak{F}(j\omega)|\,d\omega$$

converges, then $\mathfrak{F}(j\omega)$ is the Fourier transform of some $f(t)$ function. This gives no information about the case where the inversion integral converges only in the principal-value sense. In that case, if

$$\frac{1}{2\pi} PV \int_{-\infty}^{\infty} \mathfrak{F}(j\omega)e^{j\omega t}\,d\omega$$

converges to a function $f(t)$, this function can then be tested to determine whether or not it has a Fourier transform.

These remarks are of more than academic interest, as we can see by consideration of the following example. In system analysis we often deal with a rectangular pulse defined by

$$f(t) = \begin{cases} 1 & |t| < 1 \\ 0 & |t| > 1 \end{cases}$$

Its Fourier transform is readily found to be

$$\mathfrak{F}(j\omega) = 2\frac{\sin \omega}{\omega}$$

This function is not absolutely integrable. Thus we have an example where a function which is not absolutely integrable is a Fourier transform, emphasizing the fact that absolute integrability is sufficient but not necessary. By invoking the ideas of symmetry, we can also conclude that

$$\frac{\sin t}{t}$$

has a Fourier transform. This function does not satisfy the sufficiency conditions of Theorem 9-1.

9-5. Parseval's Theorem. Consider two functions $f(t)$ and $g(t)$ having the property that the integrals

$$\int_{-\infty}^{\infty} |f(t)|\, dt \quad \text{and} \quad \int_{-\infty}^{\infty} |g(t)|\, dt$$

exist, and let one of these functions be PC and bounded for all t, and the other APC. From these we define a third function

$$r(t) = \int_{-\infty}^{\infty} f(\tau)g(\tau + t)\, d\tau \tag{9-31}$$

Ultimately we shall want the Fourier integral of $r(t)$, but first we investigate convergence of the defining integral. Suppose that $g(t)$ is the bounded PC function. Since it is bounded, for all t and τ we can say that there is a constant M such that

$$|g(\tau + t)| < M$$
and therefore
$$|f(\tau)g(\tau + t)| < M|f(\tau)|$$

for all t. The term on the right is absolutely integrable, and therefore, by Theorem 8-11, it follows that the integral in Eq. (9-31) converges uniformly for all t. If $f(t)$ is the function designated as being PC, we replace $\tau + t$ by u in Eq. (9-31) to give

$$r(t) = \int_{-\infty}^{\infty} f(u - t)g(u)\, du$$

which can be treated by a similar argument.

Since Eq. (9-31) converges uniformly, we can obtain $r(\infty)$ by allowing t to become infinite under the integral. If $g(\infty)$ exists it must be zero, otherwise $g(t)$ would not be absolutely integrable. In that event $r(\infty) = 0$. It can also be shown that $r(t)$ is continuous (see Prob. 9-23).

We shall show that $r(t)$ has a Fourier transform by investigating the integral

$$\int_{-\infty}^{\infty} r(t)e^{-j\omega t}\, dt = \lim_{\substack{A \to \infty \\ B \to \infty}} \int_{-B}^{A} e^{-j\omega t}\, dt \int_{-\infty}^{\infty} f(\tau)g(\tau + t)\, d\tau \qquad (9\text{-}32)$$

By virtue of uniform convergence of the last integral on the right, the order of integration can be changed, giving

$$\int_{-\infty}^{\infty} r(t)e^{-j\omega t}\, dt = \lim_{\substack{A \to \infty \\ B \to \infty}} \int_{-\infty}^{\infty} f(\tau)\, d\tau \int_{-B}^{A} g(\tau + t)e^{-j\omega t}\, dt$$

Now change the variable of integration in the integral with respect to t, by making the substitution $u = \tau + t$, as follows:

$$\int_{-\infty}^{\infty} r(t)e^{-j\omega t}\, dt = \lim_{\substack{A \to \infty \\ B \to \infty}} \int_{-\infty}^{\infty} f(\tau)e^{j\omega \tau}\, d\tau \int_{-B+\tau}^{A+\tau} g(u)e^{-j\omega u}\, du \qquad (9\text{-}33)$$

The integral of a nonnegative function over finite limits is not more than its integral over infinite limits. Therefore, for all values of A, B, and τ, it is true that

$$\int_{-B+\tau}^{A+\tau} |g(u)|\, du \leqq \int_{-\infty}^{\infty} |g(u)|\, du$$

and with this information we can write the following sequence of inequalities for the absolute value of the integrand of the τ integral in Eq. (9-33):

$$\left| f(\tau)e^{j\omega \tau} \int_{-B+\tau}^{A+\tau} g(u)e^{-j\omega u}\, du \right| \leqq |f(\tau)| \int_{-B+\tau}^{A+\tau} |g(u)|\, du \leqq |f(\tau)| \int_{-\infty}^{\infty} |g(u)|\, du$$

The integral on the right is constant, and $f(\tau)$ is absolutely integrable. Therefore, by Theorem 8-11 we can say that the integral on the right of Eq. (9-33) converges uniformly with respect to either A or B. According to Theorem 8-10, it is possible to place the limits shown in Eq. (9-33) inside the τ integral, giving

$$\int_{-\infty}^{\infty} r(t)e^{-j\omega t}\, dt = \int_{-\infty}^{\infty} f(\tau)e^{j\omega \tau}\, d\tau \lim_{\substack{A \to \infty \\ B \to \infty}} \int_{-B+\tau}^{A+\tau} g(u)e^{-j\omega u}\, du$$

$$= \int_{-\infty}^{\infty} f(\tau)e^{j\omega \tau}\, d\tau \int_{-\infty}^{\infty} g(u)e^{-j\omega u}\, du$$

If we let $\mathcal{F}(j\omega)$ and $\mathcal{G}(j\omega)$ represent the respective Fourier transforms of $f(t)$ and $g(t)$, we see that the two integrals on the right above are, respectively, $\mathcal{F}(-j\omega)$ and $\mathcal{G}(j\omega)$. We have now proved the interesting result

$$\int_{-\infty}^{\infty} r(t)e^{-j\omega t}\, dt = \mathcal{F}(-j\omega)\mathcal{G}(j\omega) \qquad (9\text{-}34)$$

The inversion integral can be applied to the right-hand side of Eq. (9-34) to give

$$\int_{-\infty}^{\infty} f(\tau)g(\tau + t)\, d\tau = \frac{1}{2\pi} PV \int_{-\infty}^{\infty} \mathfrak{F}(-j\omega)\mathfrak{G}(j\omega)e^{j\omega t}\, d\omega \quad (9\text{-}35)^*$$

Since $r(t)$ is continuous Eq. (9-35) is valid for all values of t. For the particular value $t = 0$,

$$\int_{-\infty}^{\infty} f(\tau)g(\tau)\, d\tau = \frac{1}{2\pi} PV \int_{-\infty}^{\infty} \mathfrak{F}(-j\omega)\mathfrak{G}(j\omega)\, d\omega$$
$$= \frac{1}{2\pi} PV \int_{-\infty}^{\infty} \mathfrak{F}(j\omega)\mathfrak{G}(-j\omega)\, d\omega \quad (9\text{-}36)$$

This result has a particularly significant physical interpretation if $f(t)$ and $g(t)$ are identical. This function is necessarily PC and bounded, because in the derivation of Eq. (9-36) we established that only one of the two functions could be APC. We also note that $\mathfrak{F}(-j\omega)\mathfrak{F}(j\omega)$ is an even function of ω, and therefore an integral of this function cannot depend upon odd-function properties to make the principal value exist in the absence of ordinary convergence. Accordingly, the principal value called for in Eq. (9-36) is not required. Equation (9-36) now becomes

$$\int_{-\infty}^{\infty} [f(\tau)]^2\, d\tau = \frac{1}{2\pi} \int_{-\infty}^{\infty} \mathfrak{F}(j\omega)\mathfrak{F}(-j\omega)\, d\omega \quad (9\text{-}37)$$

This can be written in another way by recalling that $\mathfrak{F}(-j\omega)$ is the conjugate of $\mathfrak{F}(j\omega)$, so that

$$\mathfrak{F}(j\omega)\mathfrak{F}(-j\omega) = |\mathfrak{F}(j\omega)|^2$$

and finally
$$\int_{-\infty}^{\infty} [f(\tau)]^2\, d\tau = \frac{1}{2\pi} \int_{-\infty}^{\infty} |\mathfrak{F}(j\omega)|^2\, d\omega \quad (9\text{-}38)$$

The result just derived is summarized in the following theorem:

Theorem 9-2. Parseval's Theorem. If $f(t)$ is piecewise continuous, absolutely integrable, and bounded, and if its Fourier transform is designated by $\mathfrak{F}(j\omega)$, then $f(t)$ and $\mathfrak{F}(j\omega)$ are related by the formula

$$\int_{-\infty}^{\infty} [f(t)]^2\, dt = \frac{1}{2\pi} \int_{-\infty}^{\infty} |\mathfrak{F}(j\omega)|^2\, d\omega$$

* Equations (9-34) and (9-35) were derived on the assumption that one of the two functions is PC, while the other may be APC. It can be shown that these equations are still valid if both of these functions are APC, if one of them remains bounded as $|t|$ becomes infinite. However, in that case, $r(t)$ will be APC. In the present discussion we want $r(t)$ to be continuous, and so we adhere to the original conditions on $f(t)$ and $g(t)$. This footnote is intended primarily for later reference, which occurs in Chap. 11.

A physical interpretation of this theorem is possible. If $f(t)$ represents a time-varying physical quantity, under certain conditions the left-hand integral is a measure of the energy transfer. The right-hand integral is a summation of the energies of the differential components of the Fourier spectrum of the function. Engineers are familiar with the Fourier-series counterpart of this, that the power associated with a periodic function equals the sum of the powers associated with its harmonic components.

PROBLEMS

9-1. Determine whether or not the following functions meet the conditions of the Fourier integral theorem:

(a) $f(t) = \begin{cases} 1 & |t| < 1 \\ \dfrac{1}{t^2} & 1 < |t| \end{cases}$

(b) $f(t) = \begin{cases} e^{-t} & t > 0 \\ -e^{t} & t < 0 \end{cases}$

(c) $f(t) = \dfrac{\sin t}{t}$

(d) $f(t) = t^n e^{-t^2}$

(e) $f(t) = \dfrac{\sin^2 t}{t^2}$

(f) $f(t) = \dfrac{\log t}{t^2 + 1}$

9-2. Check the function

$$f(t) = \begin{cases} 0 & |t| > 1 \\ -1 & -1 < t < 0 \\ 1 & 0 < t < 1 \end{cases}$$

in the Fourier integral theorem. That is, find $\mathfrak{F}(j\omega)$, and then recover $f(t)$ from the inversion integral.

9-3. Do Prob. 9-2 for the function

$$f(t) = \begin{cases} 0 & t < 0, t > 2 \\ -1 & 0 < t < 1 \\ 1 & 1 < t < 2 \end{cases}$$

9-4. Do Prob. 9-2 for the function

$$f(t) = \begin{cases} 0 & |t| > \dfrac{\pi}{2} \\ \cos t & |t| < \dfrac{\pi}{2} \end{cases}$$

9-5. Do Prob. 9-2 for the function

$$f(t) = \begin{cases} 0 & |t| > \dfrac{\pi}{2} \\ \cos^2 t & |t| < \dfrac{\pi}{2} \end{cases}$$

9-6. Do Prob. 9-2 for the function

$$f(t) = \begin{cases} 0 & |t| > 1 \\ 1 + t & |t| < 1 \end{cases}$$

9-7. Do Prob. 9-2 for the function

$$f(t) = \frac{\sin t}{t(1 + t^2)}$$

9-8. Do Prob. 9-2 for the function

$$f(t) = \frac{t}{1 + t^4}$$

9-9. For the function

$$f(t) = e^{-at^2}$$

where $a > 0$, show that

$$\mathcal{F}(j\omega) = \sqrt{\frac{\pi}{a}} \, e^{-\omega^2/4a}$$

(HINT: Use the property $d\mathcal{F}/d\omega$ discussed in Sec. 9-3, and then integrate by parts. Also, observe that $\int_0^\infty e^{-x^2} \, dx = \sqrt{\pi}/2$.)

9-10. For the function

$$f(t) = te^{-|t|}$$

show that

$$\mathcal{F}(j\omega) = -j \frac{4\omega}{(1 + \omega^2)^2}$$

(a) By actually performing the integration indicated by the Fourier integral.
(b) By using the result stated in Eq. (9-26).

9-11. For the function

$$f(t) = |t|e^{-|t|}$$

show that

$$\mathcal{F}(j\omega) = \frac{2(1 - \omega^2)}{(1 + \omega^2)^2}$$

and use the inversion integral to recover $f(t)$.

9-12. Prove that, if $f(t)$ meets the conditions of the Fourier integral theorem, then $\mathcal{F}(j\omega)$ approaches zero as ω becomes infinite.

9-13. If $f(t)$ and its first n derivatives are absolutely integrable from minus to plus infinity, and if the nth derivative is APC, show that

$$\int_{-\infty}^{\infty} f^{(n)}(t)e^{-j\omega t} \, dt = (j\omega)^n \mathcal{F}(j\omega)$$

where

$$\mathcal{F}(j\omega) = \int_{-\infty}^{\infty} f(t)e^{-j\omega t} \, dt$$

9-14. Consider the function

$$f(x) = \begin{cases} 0 & |t| > 1 \\ (t^2 - 1)^n & -1 \le t \le 1 \end{cases}$$

where n is even.

(a) Show that this is an even function and that the first $n - 1$ derivatives are continuous.
(b) Obtain the function $\mathcal{F}(j\omega)$ from the Fourier integral.
(c) Check whether or not $(\omega)^n\mathcal{F}(j\omega)$ goes to zero as ω becomes infinite, as would be indicated by the result stated in Probs. 9-12 and 9-13.

9-15. Using the idea presented in Prob. 9-13, and using the functions given in Probs. 9-10 and 9-11,

(a) Obtain the Fourier transforms of

$$f(t) = (1 - t)e^{-|t|} \quad \text{and} \quad f(t) = (1 - |t|)e^{-|t|}$$

(b) Using the $j\omega$ functions obtained in part a, recover the given $f(t)$ functions, through the inversion integral.

9-16. If $f'(t)$ becomes infinite at $t = t_0$, but in such a way that $f(t)$ is APC, show that $f(t)$ satisfies a Lipshitz condition at t_0.

9-17. Show that, if $f(t)$ has a derivative at t_0, then at this point it satisfies the Lipshitz condition of order unity.

9-18. Show that the function

$$f(t) = \sqrt{-\frac{1}{\log |t|}}$$

does not satisfy a Lipshitz condition at $t = 0$.

9-19. Referring to the statement of the Fourier integral theorem, we note that $(\sin t)/t$ does not meet the condition stated in the theorem. Nevertheless, it is found that this function satisfies the integral relationship stated in the theorem. Discuss whether or not the theorem is stated in such a way as to permit this special case.

9-20. If $f(t)$ is made up of the finite sum

$$f(t) = \sum_{n=1}^{\infty} A_n e^{-b_n|t|}$$

where each b_n is real and positive, show that $\mathfrak{F}(j\omega)$ is a rational function having simple poles at $j\omega = b_1$, b_2, etc.

9-21. Use the Parseval relation to evaluate the following integrals:

(a) $\displaystyle\int_{-\infty}^{\infty} \left(\frac{\sin x}{x}\right)^2 dx$
(b) $\displaystyle\int_{-\infty}^{\infty} \frac{\sin^2 t}{t^2(t^2 - \pi^2)^2} dt$

9-22. Use the principles established in Sec. 9-5 to evaluate the integrals

(a) $\displaystyle\int_{-\infty}^{\infty} \frac{\sin x}{x(1 + x^2)} dx$
(b) $\displaystyle\int_{-\infty}^{\infty} \frac{e^{-|x|}}{1 + x^2} dx$

9-23. Prove that the function

$$r(t) = \int_{-\infty}^{\infty} f(\tau)g(\tau + t) \, d\tau$$

is continuous at each value of t under the conditions given in the text.

9-24. Obtain the Fourier transform for

$$f(t) = \frac{\sin bt}{t}$$

CHAPTER 10

THE LAPLACE TRANSFORM

10-1. Introduction. The Fourier integral theorem occupies a central position in the theory of linear integrodifferential equations. This fact is brought out in Chap. 1, and the development given there may be considered motivation for the present chapter.

Three basic tasks lie before us. As is stated in Chap. 9, the Fourier integral theorem is restricted to functions which approach zero at $t = \pm \infty$ fast enough to make the Fourier integral converge. One task is to show that this restriction can be removed. The second task is to extend the Fourier integral theorem to those cases where we want to investigate the response of a linear system to an excitation which commences at $t = 0$. The third task is to develop certain properties of the resulting modified transforms, which are relabeled Laplace transforms. The properties to be investigated are those which make the Laplace transform a useful tool in the solution of linear equations.

Another introductory note is in order. In Chap. 8 several theorems on real integration are enunciated, such as the theorem on integration by parts. In Chap. 10 we shall use these theorems, but frequently the integrand will be a *complex function* of a real variable. A complex function $g(x)$ of a real variable x can always be written

$$g(x) = g_1(x) + jg_2(x)$$

where $g_1(x)$ and $g_2(x)$ are real. Since these various theorems on real integration apply to each of the functions $g_1(x)$ and $g_2(x)$, the respective theorems apply also to $g(x)$.

10-2. The Two-sided Laplace Transform. In Chap. 9 we saw that for certain functions $f(t)$ the Fourier integral

$$\mathfrak{F}(j\omega) = \int_{-\infty}^{\infty} f(t)e^{-j\omega t}\, dt \tag{10-1}$$

exists and yields a transform which is a complex function of the *real* variable ω. Now we allow ω to be complex, but for later convenience it is better to let $j\omega$ be the general complex number s. Then, in a purely

285

formal sense we can replace $j\omega$ by s in Eq. (10-1), giving

$$\mathfrak{F}(s) = \int_{-\infty}^{\infty} f(t)e^{-st}\, dt \tag{10-2}$$

where the function $\mathfrak{F}(s)$ is represented by the integral for any value of s for which the integral converges. $\mathfrak{F}(s)$ is the *two-sided Laplace transform* of $f(t)$, and the defining integral is the *two-sided Laplace integral*. A detailed consideration of convergence of this integral will take some time to develop. Initial insight is gained by writing $s = \sigma + j\omega$, so that Eq. (10-2) becomes

$$\mathfrak{F}(s) = \int_{-\infty}^{\infty} f(t)e^{-\sigma t}e^{-j\omega t}\, dt \tag{10-3}$$

which is the Fourier integral of the function

$$f(t)e^{-\sigma t}$$

From the Fourier integral theory we can say that at least a sufficient condition for the existence of the integral in Eq. (10-3) is that the integral

$$\int_{-\infty}^{\infty} |f(t)|e^{-\sigma t}\, dt$$

shall exist.

We immediately sense that for a given $f(t)$ this integral can exist for certain values of σ and not for others. In particular, it may converge for certain $f(t)$ functions which are not themselves absolutely integrable. This would be the case for the function

$$f(t) = \begin{cases} 1 & t > 0 \\ e^t & t < 0 \end{cases}$$

for which $\quad \displaystyle\int_{-\infty}^{\infty} f(t)e^{-\sigma t}\, dt = \int_{-\infty}^{0} e^{(1-\sigma)t}\, dt + \int_{0}^{\infty} e^{-\sigma t}\, dt$

The first integral on the right converges if $\sigma < 1$, and the second integral converges if $\sigma > 0$. Therefore, the combination converges if $0 < \sigma < 1$, showing that the integral of Eq. (10-2) converges in a vertical strip in the s plane. Later on it will be shown that this is the general situation, that in all cases the integral of Eq. (10-2) converges in a vertical strip. However, this strip may range from the whole plane down to a single vertical line, depending on the nature of $f(t)$. The function

$$f(t) = \frac{\sin bt}{t(t^2 - \pi^2/b^2)}$$

which appears as an example in Chap. 9, is a case where the strip of convergence is reduced to the imaginary axis. Later on we shall find that, if the integral in Eq. (10-2) converges in a strip of finite width, then $\mathfrak{F}(s)$ is an analytic function of s.

We could go on to develop a detailed analysis of the two-sided Laplace transform. However, a better procedure is to recognize that the defining integral can be written in two parts, as follows:

$$\int_{-\infty}^{\infty} f(t)e^{-st}\,dt = \int_{-\infty}^{0} f(t)e^{-st}\,dt + \int_{0}^{\infty} f(t)e^{-st}\,dt$$

$$= \int_{0}^{\infty} f(-t)e^{st}\,dt + \int_{0}^{\infty} f(t)e^{-st}\,dt$$

Therefore, it will be sufficient to study the single integral

$$\int_{0}^{\infty} f(t)e^{-st}\,dt$$

Having done this, with due regard for sign changes, we can apply the results to

$$\int_{-\infty}^{0} f(t)e^{-st}\,dt = \int_{0}^{\infty} f(-t)e^{st}\,dt$$

and in this way we can get the information we want about the two-sided Laplace transform from properties of integrals from 0 to ∞.

10-3. Functions of Exponential Order.   In a series of steps we shall investigate convergence properties of the integral

$$F(s) = \int_{0}^{\infty} f(t)e^{-st}\,dt \tag{10-4}$$

$F(s)$ is called the *one-sided Laplace transform* of $f(t)$, or merely the *Laplace transform*. It is represented by the integral in Eq. (10-4), which we shall call the *Laplace integral*, for all values of s for which the integral converges.

As a first step a new class of functions is defined. Let $f(t)$ be APC, and let it have the further property that there is a real number α_0 such that

$$\lim_{t \to \infty} f(t)e^{-\alpha t} = 0 \qquad \text{when } \alpha > \alpha_0 \tag{10-5}$$

and with the limit not existing when $\alpha < \alpha_0$. A function satisfying this condition is said to be of *exponential order* α_0. Note that Eq. (10-5) is not necessarily satisfied if $\alpha = \alpha_0$. Henceforth, we shall often abbreviate the phrase *exponential order* α_0 by the symbol EO,α_0.

Functions occurring in the solution for time response of stable linear systems are of exponential order 0. Such variables as current and velocity always remain finite, which means that $f(t)$ is bounded. The product of a bounded function by $e^{-\alpha t}$ approaches zero for all $\alpha > 0$. Thus the order for such functions is 0. Other variables like electrical charge and mechanical displacement may increase without limit, but always eventually in proportion to t. However, such a function is also of exponential order 0, as we see by observing that

$$\lim_{t \to \infty} te^{-\alpha t} = 0 \quad \text{when } \alpha > 0$$

In fact, t^n is of exponential order 0. In an unstable system a function may increase as e^{at}, and we see that

$$\lim_{t \to \infty} e^{at} e^{-\alpha t} = 0$$

if $\alpha > a$. Thus, the function e^{at} is of exponential order a.

The order number α_0 will be $-\infty$ for all functions which are identically zero beyond some finite value of t. Thus, we may expect α_0 to lie in the range

$$-\infty \leqq \alpha_0 < \infty \qquad (10\text{-}6)$$

10-4. The Laplace Integral for Functions of Exponential Order. We now consider convergence of the integral in Eq. (10-4) when $f(t)$ is APC and EO,α_0. From Theorem 8-6 we know that

$$\int_0^\infty f(t) e^{-st} \, dt$$

converges if

$$\int_0^\infty |f(t) e^{-st}| \, dt = \int_0^\infty |f(t)| e^{-\sigma t} \, dt \qquad \sigma = \operatorname{Re} s$$

converges. Convergence of the integral on the right will be investigated for σ in the range

$$\alpha_0 < \sigma \qquad (10\text{-}7)$$

For any σ in this range we can pick a number α_2 such that $\alpha_0 < \alpha_2 < \sigma$, and since $f(t)$ is of exponential order α_0, it is known that for any given small positive number ϵ there exists a T_0 such that

$$|f(t)| e^{-\alpha_2 t} < \epsilon \qquad \text{when } t > T_0$$

Therefore,

$$\int_{T_0}^\infty |f(t)| e^{-\sigma t} \, dt = \int_{T_0}^\infty |f(t)| e^{-\alpha_2 t} e^{-(\sigma - \alpha_2)t} \, dt < \epsilon \int_{T_0}^\infty e^{-(\sigma - \alpha_2)t} \, dt \qquad (10\text{-}8)$$

The integral on the right exists, and so we have established absolute and ordinary convergence with respect to the infinite limit in the region $\operatorname{Re} s > \alpha_0$.

For uniform convergence, an integrable function independent of s (or σ) is needed for use in the test given in Theorem 8-11. Let α_1 be a number greater than α_0, and let σ be in the range

$$\alpha_0 < \alpha_1 \leqq \sigma \qquad (10\text{-}9)$$

For any choice of α_1 we can find a number α_2 such that $\alpha_0 < \alpha_2 < \alpha_1$. Relation (10-8) is again valid, by use of the presently defined α_2; and with the introduction of α_1, this inequality can be extended to

$$\int_{T_0}^{\infty} |f(t)| e^{-\sigma t} < \epsilon \int_{T_0}^{\infty} e^{-(\sigma - \alpha_2)t}\, dt < \epsilon \int_{T_0}^{\infty} e^{-(\alpha_1 - \alpha_2)t}\, dt$$

The integral on the right converges and is independent of σ. Therefore, by Theorem 8-11 the Laplace integral converges uniformly for $\mathrm{Re}\, s \geqq \alpha_1 > \alpha_0$.

We have proved the following theorem:

Theorem 10-1.* If $f(t)$ is APC and EO,α_0, then the integral which defines the Laplace transform (the Laplace integral)

$$\int_0^{\infty} f(t) e^{-st}\, dt$$

converges and converges absolutely for

$$\alpha_0 < \mathrm{Re}\, s$$

and converges uniformly with respect to the infinite limit for

$$\alpha_0 < \alpha_1 \leqq \mathrm{Re}\, s$$

10-5. Convergence of the Laplace Integral for the General Case.

Theorem 10-1 tells a great deal about convergence of the Laplace integral for practical functions. In the process, we have gained the important idea of a *half plane* of convergence. For functions which are not necessarily of exponential order, the following slightly different theorem is possible. Assume that $f(t)$ is APC and that for some complex number s_0 the integral

$$\int_0^{\infty} f(t) e^{-s_0 t}\, dt$$

converges. We shall show that it converges for $\mathrm{Re}\, s > \mathrm{Re}\, s_0$. The proof requires an auxiliary function

$$w(t) = \int_t^{\infty} f(\tau) e^{-s_0 \tau}\, d\tau \tag{10-10}$$

This is a continuous function of t, and its derivative is

$$w'(t) = -f(t) e^{-s_0 t}$$

and in terms of this the Laplace integral can be written

$$\int_0^{\infty} f(t) e^{-st}\, dt = \int_0^{\infty} f(t) e^{-s_0 t} e^{-zt}\, dt = -\int_0^{\infty} w'(t) e^{-zt}\, dt \tag{10-11}$$

---

* Obviously the APC condition is for $t \geqq 0$. We omit this qualification in this and following theorems because it would amount to a trivial redundancy, integration being always from 0 to ∞.

where the complex variable s has been replaced by $s = s_0 + z$. The Cauchy principle of convergence will be used to establish conditions for convergence of the integral on the right of Eq. (10-11). Thus, we are to show that corresponding to an arbitrary small $\epsilon > 0$ we can find a number T_0 such that

$$\left| \int_{A'}^{A} f(t)e^{-st}\,dt \right| = \left| \int_{A'}^{A} w'(t)e^{-st}\,dt \right| < \epsilon \qquad (10\text{-}12)$$

when $\qquad\qquad A', A > T_0$

The above integral satisfies the conditions for integration by parts, given in Theorem 8-4, subject to the comments given in Sec. 10-1 relative to e^{-st} being a complex function of the real variable t. Thus,

$$\int_{A'}^{A} w'(t)e^{-st}\,dt = w(t)e^{-st}\Big|_{A'}^{A} + z \int_{A'}^{A} w(t)e^{-st}\,dt$$
$$= -w(A')e^{-zA'} + w(A)e^{-zA} + z \int_{A'}^{A} w(t)e^{-st}\,dt \qquad (10\text{-}13)$$

The absolute value of the right-hand side of Eq. (10-13) is less than the sum of the absolute values of individual terms. Therefore,

$$\left| \int_{A'}^{A} f(t)e^{-st}\,dt \right| < |w(A')|e^{-xA'} + |w(A)|e^{-xA} + |z| \int_{A'}^{A} |w(t)e^{-st}|\,dt \quad (10\text{-}14)$$

where $x = \operatorname{Re} z$. Since the integral

$$\int_{0}^{\infty} f(\tau)e^{-s_0\tau}\,d\tau$$

converges, it follows from the definition of $w(t)$ that, given an arbitrary small $\epsilon' > 0$, we can find a number T_0 such that

$$|w(t)| < \epsilon'$$

when $\qquad\qquad t > T_0$

Thus, if $A', A > T_0$, we have

$$|w(A')|,\ |w(A)| < \epsilon'$$

and if $x > 0$ and if A is the larger of A and A', then relation (10-14) becomes

$$\left| \int_{A'}^{A} f(t)e^{-st}\,dt \right| < \epsilon'\left[e^{-xA'} + e^{-xA} + \frac{|z|}{x}\left(e^{-xA'} - e^{-xA}\right) \right]$$

which in turn is less than

$$\epsilon'\left(2 + \frac{|z|}{x}\right) = \epsilon \qquad (10\text{-}15)$$

This can be the ϵ of relation (10-12).

For any fixed value of z, with $x > 0$, the above quantity in parentheses is finite, and by making ϵ' small enough the whole quantity (ϵ) is arbitrarily small. Since ϵ' determines T_0, we see that

$$\int_0^\infty f(t)e^{-st}\,dt$$

converges when

$$\operatorname{Re} s > \sigma_0$$

where $\sigma_0 = \operatorname{Re} s_0$. This condition on $\operatorname{Re} s$ comes from the condition $x > 0$, since $x = \operatorname{Re} z = \operatorname{Re}(s - s_0)$.

The above discussion provides the range of s for ordinary convergence. This is not the range for uniform convergence, because T_0 is dependent upon $|z|$, through Eq. (10-15) and the dependence of T_0 on ϵ'. In order to get the region of uniform convergence, let θ be the angle of z, and observe that

$$\frac{|z|}{x} = \left| \frac{1}{\cos \theta} \right|$$

when $x > 0$. If θ is restricted to the range

$$|\theta| \leqq \theta' < \frac{\pi}{2}$$

we see that

$$\frac{1}{\cos \theta} \leqq \frac{1}{\cos \theta'}$$

and

$$\epsilon = \epsilon'\left(2 + \frac{|z|}{x}\right) \leqq \epsilon'\left(2 + \frac{1}{\cos \theta'}\right)$$

The quantity on the right is independent of z. Therefore, if ϵ is given, we can now find ϵ', and then T_0. Thus it is established that relationship (10-12) can be satisfied, showing that the Laplace integral converges uniformly in an angular sector $|\operatorname{ang}(s - s_0)| \leqq \theta' < \pi/2$.

The following theorem has been proved:

Theorem 10-2. Let $f(t)$ be an almost piecewise continuous function for which the integral

$$\int_0^\infty f(t)e^{-s_0 t}\,dt$$

converges. Then the Laplace integral

$$\int_0^\infty f(t)e^{-st}\,dt$$

converges when

$$\operatorname{Re} s > \operatorname{Re} s_0$$

Furthermore, convergence of the Laplace integral is uniform with respect to the upper limit when

$$|\operatorname{ang}(s - s_0)| \leqq \theta' < \frac{\pi}{2}$$

Note that we do not get convergence when Re $s = \sigma_0$. This means that, although the integral converges when $s = s_0$, it does not necessarily converge for some other s having the same real part as s_0. A simple example can be given to illustrate this point. Let

$$f(t) = \begin{cases} 0 & 0 \leqq t < 1 \\ \dfrac{1}{t} & 1 \leqq t \end{cases}$$

Then for $s_0 = 0 + j\omega_0$ the integral

$$\int_1^\infty \frac{e^{-j\omega_0 t}}{t}\, dt = \int_1^\infty \frac{\cos \omega_0 t}{t}\, dt - j \int_1^\infty \frac{\sin \omega_0 t}{t}\, dt$$

converges, but it diverges at $s_0 = 0$. This one point on the $j\omega$ axis makes it impossible to say that the integral converges for Re $s \geqq 0$.

Theorems 10-1 and 10-2 are similar to the extent of identifying a half plane of convergence for the Laplace integral. Theorem 10-2 includes the functions of exponential order, which are the sole concern of Theorem 10-1; but Theorem 10-2 also covers certain functions which are not of exponential order. Perhaps this remark would seem to imply that Theorem 10-1 is a special case of Theorem 10-2. There are two reasons why not. In the first place, Theorem 10-1 not only tells us that the Laplace integral converges in a half plane; it also gives a specific number (α_0) for the abscissa of a left-hand boundary of such a half plane. Theorem 10-2 merely states convergence to the right of any point where we happen already to know that the integral converges. In fact, some functions of exponential order exhibit Laplace integral convergence to the left of abscissa α_0, in which case Theorem 10-2 would establish a half-plane boundary to the left of the one given by Theorem 10-1. For example, $\alpha_0 = 0$ for cos (e^t) but there is convergence for Re $s > -1$. In the second place, the regions of uniform convergence are specified differently by the two theorems. Even though a function of exponential order satisfies Theorem 10-2, this theorem tells us only that the Laplace integral converges uniformly in an angular sector of the right half plane. Theorem 10-1 indicates uniform convergence in a less restricted region, namely, a half plane.

Thus we see that the two theorems are essentially different, but are similar to the extent of establishing regular convergence in a half plane. The smallest coordinate defining such a half-plane of convergence is called the *abscissa of convergence* and designated σ_c. The numerical value of σ_c depends on characteristics of $f(t)$ which influence convergence of the integral. In terms of σ_c we now state the region of convergence of the Laplace integral as follows:

$$\text{Re } s > \sigma_c$$

If $f(t)$ is EO,α_0, Theorem 10-1 provides the additional information

$$\sigma_c \leqq \alpha_0$$

We observe that α_0 is obtained directly from the function, through relation (10-5). In any case, to obtain the abscissa of convergence σ_c, it is necessary to determine the set S of values of σ_0 for which the integral converges. The set S may be open or closed at the lower end ($\sigma_0 > 0$ and $\sigma_0 \geqq 0$ being respective examples of S sets which are open and closed at the lower end). In either case, σ_c is the greatest lower bound of set S. Both the sets specified in the above parenthetical example have $\sigma_c = 0$. If the set S is closed at the bottom, the Laplace integral converges at least for some points Re $s = \sigma_c$; and if S is open at the lower end, the Laplace integral converges for no points for which Re $s = \sigma_c$.

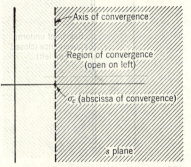

In connection with the question of convergence for Re $s = \sigma_c$, we note that neither theorem establishes whether there will be convergence at points on the vertical line Re $s = \sigma_c$. The integral must be evaluated to determine this. Earlier we saw an example where the Laplace integral converges for Re $s = 0$, except for $s = 0$. The abscissa of convergence in this case

Fig. 10-1. Region of convergence of Laplace integral.

is 0, a fact which could be determined by inspecting the function and determining that it is EO,0. But this would not give information about convergence for Re $s = 0$.

It has been mentioned that the numerical value of σ_c depends on certain characteristics of $f(t)$. A few cases are illustrated in the following table:

$f(t)$		σ_c
e^t		1
$\sin t$		0
1		0
e^{-t}		-1
1	$0 \leqq t < 1$	$-\infty$
0	$1 < t$	

It is often convenient to refer in geometrical terms to the *half plane of convergence* and to the vertical line $s = \sigma_c$, which is its left-hand boundary. This line is called the *axis of convergence*. The ideas relevant to regions of convergence are illustrated in Fig. 10-1.

10-6. Further Ideas about Uniform Convergence. Theorem 10-1 states that for a function of EO,α_0 convergence is uniform for Re $s \geqq \alpha_1 > \alpha_0$.

But for such a function $\sigma_c \leqq \alpha_0$, and so we can obtain a statement of the region of uniform convergence in relation to the abscissa of convergence as follows:

$$\text{Re } s \geqq \alpha_1 > \alpha_0 \geqq \sigma_c \tag{10-16}$$

This is for a function of exponential order. This region is closed on the left, in contrast with the region of ordinary convergence, which is open on the left. For the case of a function having a convergent Laplace integral at some point s_0, but not of exponential order, Theorem 10-2 specifies the region of uniform convergence in relation to the abscissa of convergence. The region is an *angular sector* formed by two lines radiating into the right half plane from a point s' and making an angle $2\theta'$, where $\theta' < 90°$. The

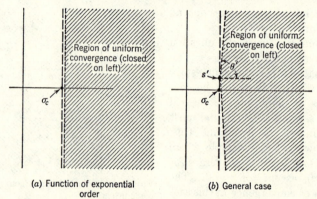

(a) Function of exponential order

(b) General case

Fig. 10-2. Regions of uniform convergence of the Laplace integral. In case b, s' may be to the right of the axis of convergence. Case a is shown for $\sigma_c = \alpha_0$.

region consists of the points on and to the right of these lines. Two cases can occur. If there are any points $\text{Re } s = \sigma_c$ where the integral converges, s' can be one of these. If there are no such points, s' is any point for which $\text{Re } s' > \sigma_c$. Regions of uniform convergence are illustrated in Fig. 10-2.

Up to this point we have been considering convergence with respect to the infinite limit. In addition, since $f(t)$ may be APC, it is necessary to consider convergence with respect to points where $f(t)$ becomes singular. The APC category of functions is defined in such a way that, if t_k is a singular point, the integral

$$\int_{t_k-\delta_1}^{t_k+\delta_2} f(t) \, dt$$

exists, where δ_1 and δ_2 are arbitrary small positive numbers. If σ_1 is any real number, we see that

$$\int_{t_k-\delta_1}^{t_k+\delta_2} |f(t)| e^{|\sigma_1|t}| \, dt \leqq e^{|\sigma_1|(t_k+\delta_2)} \int_{t_k-\delta_1}^{t_k+\delta_2} |f(t)| \, dt$$

also converges. If $\mathrm{Re}\ s \geqq \sigma_1$,

$$|f(t)e^{-st}| = |f(t)|e^{-\mathrm{Re}\ (s)t} \leqq |f(t)|e^{|\sigma_1|t}$$

and so by Theorem 8-11 it follows that

$$\int_{t_k - \delta_1}^{t_k + \delta_2} f(t)e^{-st}\, dt$$

converges uniformly for $\mathrm{Re}\ s \geqq \sigma_1$. Since σ_1 is any number, the Laplace integral converges uniformly with respect to finite singular points of $f(t)$ for s in *any right half plane*.

10-7. Convergence of the Two-sided Laplace Integral. In Sec. 10-2 the two-sided Laplace integral was shown to be the sum of the two integrals

$$\int_{-\infty}^{\infty} f(t)e^{-st} = \int_0^{\infty} f(t)e^{-st}\, dt + \int_0^{\infty} f(-t)e^{st}\, dt \qquad (10\text{-}17)$$

The two-sided Laplace integral converges if each of the two integrals on the right converges. The first of these converges in a right half plane. The second integral is similar but has the exponent $+s$ instead of $-s$. This makes the second integral converge in a right half of the *minus s* plane and thus in a left half of the s plane. The given integral will converge if these two half planes overlap.

To pursue this in more detail, suppose that

$$\int_0^{\infty} f(-t)e^{-st}\, dt$$

has an abscissa of convergence $-\sigma_{c2}$. It converges for

$$\mathrm{Re}\ s > -\sigma_{c2}$$

Then

$$\int_0^{\infty} f(-t)e^{st}\, dt$$

converges for

$$-\mathrm{Re}\ s = \mathrm{Re}\ (-s) > -\sigma_{c2}$$

or

$$\mathrm{Re}\ s < \sigma_{c2}$$

Now suppose that the first integral on the right of Eq. (10-17) has an abscissa of convergence σ_{c1}. Then the two-sided integral will converge in the strip

$$\sigma_{c1} < \mathrm{Re}\ s < \sigma_{c2} \qquad (10\text{-}18)$$

provided that $\sigma_{c1} < \sigma_{c2}$; otherwise there would be no overlap of the two half planes. Equation (10-18) suggests that two abscissas of convergence (σ_{c1} and σ_{c2}) are defined for the two-sided Laplace integral.

All the discussions of the single-sided Laplace integral given in Secs. 10-3 through 10-6 apply to the second integral on the right of Eq. (10-17). It is necessary only to consider $f(-t)$ and to change the sign of s. The abscissa of convergence σ_{c2} can be determined by the conditions of either

Theorem 10-1 or Theorem 10-2.   Theorem 10-1 applies if $f(-t)$ is EO,α_0, where then σ_{c2} equals $-\alpha_0$.   Theorem 10-2 applies if

$$\int_0^\infty f(-t)e^{s_{02}t}\, dt$$

exists, and then σ_{c2} will be the least upper bound of admissible values of Re s_{02}.   Thus, σ_{c2} can be found from properties of $f(-t)$, and then the range of convergence given by relation (10-18) is established.

Some $f(t)$ functions will yield strips of convergence of finite width, and others will not.   The function

$$f(t) = \begin{cases} e^{at} & 0 < t \\ e^{bt} & t < 0 \end{cases}$$

has a convergence strip of finite width, if $a < b$.   Recalling a previous example, $f(t)$ is EO,a and $f(-t)$ is EO,$-b$.   Thus,

$$\sigma_{c1} = a \qquad \sigma_{c2} = b$$

and the range of convergence of the two-sided integral is

$$a < \text{Re } s < b \qquad (10\text{-}19)$$

Examples of various cases of this function are shown in Fig. 10-3.   Note that the left-hand and right-hand abscissas of convergence (σ_{c1} and σ_{c2}, respectively) are influenced by the behavior of $f(t)$ as $t \to +\infty$ and $t \to -\infty$, respectively.

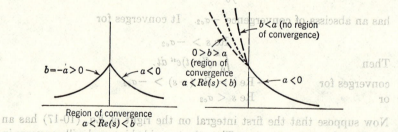

Fig. 10-3. Examples of exponential functions which do and do not have regions of convergence for the two-sided Laplace integral.

Uniform convergence of the two-sided Laplace integral occurs in the sort of region we should expect to get by superimposing two regions like Fig. 10-2, one of them being reversed.   Two cases occur, where $f(t)$ and $f(-t)$ are both of exponential order, and where the two integrals

$$\int_0^\infty f(t)e^{-s_{01}t}\, dt \quad \text{and} \quad \int_0^\infty f(-t)e^{s_{02}t}\, dt$$

converge, where Re $s_{01} <$ Re s_{02}.   Corresponding regions of uniform convergence are shown in Fig. 10-4.

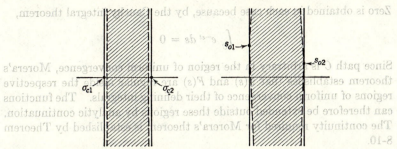

FIG. 10-4. Examples of regions of uniform convergence of the two-sided Laplace transform.

10-8. The One- and Two-sided Laplace Transforms. In Chap. 9 you were introduced to the notion of the Fourier transform as distinct from the Fourier integral. Also, in Secs. 10-2 and 10-3 the Laplace transform is described as the function of s obtained from the Laplace integral. We are now prepared to deal with these transform functions in more detail.

First we summarize the facts known at this point. Under certain conditions on $f(t)$ we know that the integrals in the formulas

$$\mathfrak{F}(s) = \int_{-\infty}^{\infty} f(t)e^{-st}\,dt \tag{10-20}$$

$$F(s) = \int_{0}^{\infty} f(t)e^{-st}\,dt \tag{10-21}$$

have certain regions of convergence. In those regions of convergence, the appropriate integral represents $\mathfrak{F}(s)$ or $F(s)$. However, we are not satisfied with Eqs. (10-20) and (10-21), for several reasons. In the first place, an integral is not always the most convenient type of formula, particularly because it does not clearly put into evidence the properties of the function it defines. Second, it is valid only in a restricted region. If $\mathfrak{F}(s)$ or $F(s)$ is analytic, it will exist outside the range of convergence of its integral representation and can be uniquely determined by analytic continuation.

Let us consider whether or not $\mathfrak{F}(s)$ and $F(s)$ are analytic functions. Assume that each of them has a region of convergence, thereby eliminating the special case where the two-sided Laplace integral converges only on a vertical line. We have shown for the one- and two-sided cases that there is then also a region of uniform convergence in the s plane. Perform a contour integration with respect to s over an arbitrary simple closed curve C in the region of uniform convergence. By virtue of Theorem 8-9, the order of integration may be inverted, and we get

$$\int_{C} \mathfrak{F}(s)\,ds = \int_{-\infty}^{\infty} f(t) \left(\int_{C} e^{-st}\,ds \right) dt = 0$$

$$\int_{C} F(s)\,ds = \int_{0}^{\infty} f(t) \left(\int_{C} e^{-st}\,ds \right) dt = 0$$

Zero is obtained in each case because, by the Cauchy integral theorem,

$$\int_C e^{-st}\, ds = 0$$

Since path C is arbitrary in the region of uniform convergence, Morera's theorem establishes that $\mathfrak{F}(s)$ and $F(s)$ are regular inside the respective regions of uniform convergence of their defining integrals. The functions can therefore be extended outside these regions by analytic continuation. The continuity required for Morera's theorem is established by Theorem 8-10.

With the revelation that $\mathfrak{F}(s)$ and $F(s)$ are analytic functions, we can now see a reason for making a distinction between the Laplace integral and the Laplace transform. Up to this point in the text the words have been applied, respectively, to the right and left sides of the equation

$$F(s) = \int_0^\infty f(t)e^{-st}\, dt \qquad \text{Re } s > \sigma_c$$

but until we were able to show that $F(s)$ exists outside the region Re $s > \sigma_c$ and can be expressed by formulas other than an integral, there was no reason in evidence for making a distinction between the transform and the integral.*

This property of analyticity of $\mathfrak{F}(s)$ provides the necessary vehicle for gaining insight as to why the Laplace integral lends power to the Fourier integral theorem. We now have a precise statement of a condition sufficient to make $\mathfrak{F}(s)$ analytic: the two-sided Laplace integral of $f(t)$ should have a strip of convergence of *finite width*. In Chap. 9 we obtained a result related to this. There we found that $\mathfrak{F}(j\omega)$ is indefinitely differentiable if $t^n f(t)$ has a Fourier integral for all values of n. However, the new result is a stronger one, because analyticity is more restrictive than existence of all derivatives with respect to the real variable ω.

The above discussion has established that $\mathfrak{F}(s)$ and $F(s)$ are regular at all points in the region of convergence of the integral representation. However, they can have singular points outside the region of convergence. Thus, $\mathfrak{F}(s)$ can have singular points to the right and left of the strip of convergence and $F(s)$ can have singular points to the left of the half plane of convergence. The region of convergence extends to the nearest singular point.

10-9. Significance of Analytic Continuation in Evaluating the Laplace Integral. The Laplace transform is defined by the integration process

$$F(s) = \int_0^\infty f(t)e^{-st}\, dt$$

* We now see that Eqs. (10-20) and (10-21) are incomplete as written, since abscissas of convergence are not stipulated. However, in those cases where the specific value of σ_c is not required, the designation Re $s > \sigma_c$ may be omitted without difficulty, so long as its implied existence is recognized.

When it comes to a question of finding explicit transforms, we need to perform this integration by standard methods of integration. However, these methods apply to real integrands, whereas e^{-st} is complex. One way to proceed would be to use

$$e^{-st} = e^{-\sigma t} \cos \omega t - je^{-\sigma t} \sin \omega t$$

which yields two integrals, each with a real integrand. However, this is more complicated than necessary. An easier method, and one which is conceptually more satisfying, is to use the fact, which has been proved, that $F(s)$ is analytic. This means that if we know $F(\sigma)$ we can get $F(s)$ by replacing σ by s.

In most cases this formality is not necessary; seemingly it amounts merely to a change in symbol. But without analytic continuation we could not regard replacing σ by s as a mere formality. Therefore, even though we might manipulate the defining integral as it stands, with complex s as the parameter, the concept of analytic continuation lurks in the background. In particular, those cases leading to multivalued $F(s)$ functions cannot be treated without giving heed in detail to the difference between σ and s. An example is given in Sec. 10-11.

Since it has been shown that the two-sided Laplace transform can be written as the sum of two one-sided transforms, these comments automatically apply also to the two-sided case.

10-10. Linear Combinations of Laplace Transforms. Now that the Laplace transform has been identified as an analytic function, we shall be interested in its various properties. Notation will be simplified by introducing a new symbolism to imply the Laplace transform. If $\mathfrak{F}(s)$ and $F(s)$ are, respectively, the two- and one-sided transforms of $f(t)$, we shall write

$$\mathfrak{F}(s) = \mathcal{L}_2[f(t)] \qquad (10\text{-}22)$$
$$F(s) = \mathcal{L}[f(t)] \qquad (10\text{-}23)$$

The notation using $\mathcal{L}$ and $\mathcal{L}_2$ is a convenience since it implies the pertinent function of t. But you must always regard a symbol like $\mathcal{L}[f(t)]$ as a function of s.

Linear combinations of functions are completely treated if we consider two situations, multiplication by a constant, and the sum of two functions. Let $f(t)$ have a Laplace transform, and let k be a real constant. From the defining integral,

$$\mathcal{L}[kf(t)] = \int_0^\infty kf(t)e^{-st}\,dt = k\int_0^\infty f(t)e^{-st}\,dt$$

Thus it is obvious that

$$\mathcal{L}[kf(t)] = k\mathcal{L}[f(t)] \qquad (10\text{-}24)$$

Now let $g(t)$ be a second function, also Laplace transformable. The two functions $f(t)$ and $g(t)$ will have abscissas of convergence σ_f and σ_g, respectively. If $\sigma_f \neq \sigma_g$, let σ_g be the greater. Then we can write

$$\mathcal{L}[f(t) + g(t)] = \int_0^\infty [f(t) + g(t)]e^{-st}\, dt$$
$$= \int_0^\infty f(t)e^{-st}\, dt + \int_0^\infty g(t)e^{-st}\, dt \qquad \text{Re } s > \sigma_g$$

The condition Re $s > \sigma_g$ is important in the sense that this condition can always be satisfied. The abscissa of convergence of the Laplace integral of the sum of two functions will be the larger of the two abscissas of the individual functions. Thus, if $\mathcal{L}[f(t)]$ and $\mathcal{L}[g(t)]$ each exist, then $\mathcal{L}[f(t) + g(t)]$ also exists and is given by

$$\mathcal{L}[f(t) + g(t)] = \mathcal{L}[f(t)] + \mathcal{L}[g(t)] \qquad (10\text{-}25)$$

The situation is somewhat different for two-sided transforms. It is merely a detail to show that Eq. (10-24) can be modified for $\mathcal{L}_2[f(t)]$, giving

$$\mathcal{L}_2[kf(t)] = k\mathcal{L}_2[f(t)] \qquad (10\text{-}26)$$

but there is no general theorem for addition of two-sided Laplace transforms. The reason can be found by considering the formal expression

$$\int_{-\infty}^\infty f(t)e^{-st}\, dt + \int_{-\infty}^\infty g(t)e^{-st}\, dt = \int_{-\infty}^\infty [f(t) + g(t)]e^{-st}\, dt$$

Even though each of the integrals on the left might converge, the integral on the right exists (and an equality sign can be used) only if there is an overlap of the strips of convergence of the integrals on the left. Thus, if the two-sided Laplace integrals of $f(t)$ and $g(t)$ have overlapping strips of convergence, it is true that $f(t) + g(t)$ has a two-sided Laplace transform given by

$$\mathcal{L}_2[f(t) + g(t)] = \mathcal{L}_2[f(t)] + \mathcal{L}_2[g(t)]$$

If there is no overlapping strip of convergence, the above sum on the right will exist but the left side will not exist. The right side will be the two-sided transform for some function other than $f(t) + g(t)$. Further explanation of this statement is found in Secs. 10-16 and 10-17.

10-11. Laplace Transforms of Some Typical Functions. Later on we shall develop some general properties of Laplace transforms. For the present, some feeling for these functions will be obtained by working out a few examples.

1. $f(t) = e^{at}$. Convergence is for Re $s > a$. We shall follow the plan of Sec. 10-9, first integrating the defining integral with s real, as follows:

$$\int_0^\infty e^{-(\sigma-a)t}\, dt = -\left.\frac{e^{-(\sigma-a)t}}{\sigma - a}\right|_0^\infty = \frac{1}{\sigma - a}$$

Replacing σ by s gives

$$\mathcal{L}(e^{at}) = \frac{1}{s-a} \tag{10-27}$$

2. $f(t) = 1$. Convergence is for Re $s > 0$, and this case can be obtained from the above by taking $a = 0$, giving

$$\mathcal{L}(1) = \frac{1}{s} \tag{10-28}$$

3. $f(t) = \sin bt$, *and* $\cos bt$. Convergence is for Re $s > 0$. Again using s real, we get the following:

$$\int_0^\infty (\sin bt)e^{-\sigma t}\, dt = \frac{e^{-\sigma t}(-\sigma \sin bt - b \cos bt)}{\sigma^2 + b^2} \bigg|_0^\infty$$

$$= \frac{b}{\sigma^2 + b^2}$$

$$\int_0^\infty (\cos bt)e^{-\sigma t}\, dt = \frac{e^{-\sigma t}(-\sigma \cos bt + b \sin bt)}{\sigma^2 + b^2} \bigg|_0^\infty$$

$$= \frac{\sigma}{\sigma^2 + b^2}$$

Again replacing σ by s, we have the results

$$\mathcal{L}(\sin bt) = \frac{b}{s^2 + b^2} \tag{10-29}$$

$$\mathcal{L}(\cos bt) = \frac{s}{s^2 + b^2} \tag{10-30}$$

4. $f(t) = e^{at} \sin bt$, *and* $e^{at} \cos bt$. This is a special case of a general case to be treated later, where a t function, having a known transform $F(s)$, is multiplied by an exponential. For the above three functions, convergence is for Re $s > a$. The required integration is readily performed, being basically the same as case 3, but with $-\sigma$ replaced by $a - \sigma$. Thus, referring to that case, we have

$$\int_0^\infty (\sin bt)e^{(a-\sigma)t}\, dt = \frac{e^{(a-\sigma)t}[(a-\sigma)\sin bt - b\cos bt]}{(\sigma - a)^2 + b^2} \bigg|_0^\infty$$

$$= \frac{b}{(\sigma - a)^2 + b^2}$$

$$\int_0^\infty (\cos bt)e^{(a-\sigma)t}\, dt = \frac{e^{(a-\sigma)t}[(a-\sigma)\cos bt + b\sin bt]}{(\sigma - a)^2 + b^2} \bigg|_0^\infty$$

$$= \frac{\sigma - a}{(\sigma - a)^2 + b^2}$$

In each case the results are dependent upon the fact that $\sigma > a$. Now σ is replaced by s, to yield

$$\mathcal{L}(e^{at} \sin bt) = \frac{b}{(s - a)^2 + b^2} \tag{10-31}$$

$$\mathcal{L}(e^{at} \cos bt) = \frac{s - a}{(s - a)^2 + b^2} \tag{10-32}$$

5. $f(t) = t^n$, *Where n Is an Integer.*   Convergence is for Re $s > 0$.   The defining formula can be integrated by parts, giving

$$\int_0^\infty t^n e^{-\sigma t} \, dt = -\left. \frac{t^n e^{-\sigma t}}{\sigma} \right|_0^\infty + \frac{n}{\sigma} \int_0^\infty t^{n-1} e^{-\sigma t} \, dt$$

The integral on the right exists, and the lower limit can be used in the first term, if $n \geqq 1$.   Since $\sigma > 0$, the exponent in the first term goes to zero as t goes to infinity.   Thus we have the formula

$$\mathcal{L}(t^n) = \frac{n}{s} \mathcal{L}(t^{n-1}) \tag{10-33}$$

Case 2 is the same as the present case, if $n = 0$.   Thus, Eq. (10-33) leads by induction to the sequence

$$\mathcal{L}(t^0) = \frac{1}{s}$$

$$\mathcal{L}(t) = \frac{1}{s^2}$$

$$\mathcal{L}(t^2) = \frac{2}{s^3} \tag{10-34}$$

$$\cdots \cdots \cdots$$

$$\mathcal{L}(t^n) = \frac{n!}{s^{n+1}}$$

6. $f(t) = 1/\sqrt{t}$.   This function is APC by virtue of its behavior near the singularity at $t = 0$.   However, it is also EO,0, and so its Laplace integral converges for Re $s > 0$.   Here we have a case toward which the comments of Sec. 10-9 were directed.   As we shall see, it is essential that s shall be replaced by σ.   In the integral

$$\int_0^\infty \frac{1}{\sqrt{t}} e^{-\sigma t} \, dt$$

replace σt by x^2, giving

$$\frac{2}{\sqrt{\sigma}} \int_0^\infty e^{-x^2} \, dx = \frac{\sqrt{\pi}}{\sqrt{\sigma}} \tag{10-35}$$

This integral is well known and can be found in tables of improper integrals. In writing $\sqrt{t}$ and $\sqrt{\sigma}$ in the above formulas we understand a single branch of the $t^{\frac{1}{2}}$ or $\sigma^{\frac{1}{2}}$ functions. That is to say, a minus sign is not admitted. Now, when we analytically continue $\sqrt{\sigma}$, we go into only *one sheet* of the Riemann surface of the function $s^{\frac{1}{2}}$, the sheet in which the points of $\sqrt{\sigma}$ are located. Let us then use the symbol $\sqrt{s}$ to imply this *single-valued* branch of the function $s^{\frac{1}{2}}$. Accordingly, the result is

$$\mathcal{L}\left(\frac{1}{\sqrt{t}}\right) = \sqrt{\frac{\pi}{s}} \qquad (10\text{-}36)$$

By first considering the Laplace integral for the real number σ we are able to arrive at the correct choice of the two possible values of $s^{\frac{1}{2}}$.

If this case had been treated throughout with the variable s retained, the formal manipulations of variable change would have led to the factor $1/\sqrt{s}$. However, we would not then have a clear meaning for $\sqrt{s}$. Notation is merely a symbolism for ideas; symbols which are not properly defined have no meaning, a point which is illustrated by this example.

7. $f(t) = \sqrt{t^k}$, $k \geqq 1$ *and Odd Integer*. This function gives convergence for Re $s > 0$. As in case 5, integration by parts yields a general recurrence formula, as follows:

$$\int_0^\infty \sqrt{t^k}\, e^{-\sigma t}\, dt = -\left. \frac{\sqrt{t^k}\, e^{-\sigma t}}{\sigma}\right|_0^\infty + \frac{k}{2\sigma}\int_0^\infty \sqrt{t^{k-2}}\, e^{-\sigma t}\, dt$$

If $k \geqq 1$, the lower limit can be used in the first term on the right and the integral exists. The result can be stated as

$$\mathcal{L}(\sqrt{t^k}) = \frac{k}{2s}\mathcal{L}(\sqrt{t^{k-2}}) \qquad k \geqq 1 \text{ and odd} \qquad (10\text{-}37)$$

In similarity with Eqs. (10-34), this yields a sequence of formulas, starting with $\sqrt{t^{-1}}$, which is obtained from case 6. Thus,

$$\mathcal{L}\left(\frac{1}{\sqrt{t}}\right) = \frac{\sqrt{\pi}}{\sqrt{s}}$$

$$\mathcal{L}(\sqrt{t}) = \frac{\sqrt{\pi}}{2\sqrt{s^3}}$$

$$\mathcal{L}(\sqrt{t^3}) = \frac{3\sqrt{\pi}}{4\sqrt{s^5}} \qquad (10\text{-}38)$$

$$\cdots \cdots \cdots \cdots \cdots \cdots$$

$$\mathcal{L}(\sqrt{t^k}) = \frac{(k+1)!\,\sqrt{\pi}}{2^{(k+1)}\left(\dfrac{k+1}{2}\right)!\,\sqrt{s^{(k+2)}}}$$

In this general formula the root of a power of s is always interpreted to be on that sheet of the Riemann surface on which are found the values of $\sqrt{\sigma^{k+2}}$.

8. $f(t) = $ *Pulse of Unit Height and Duration* T. This function is defined as

$$f(t) = \begin{cases} 1 & 0 < t < T \\ 0 & T < t \end{cases}$$

and it has the Laplace transform

$$\int_0^T e^{-\sigma t}\, dt = \frac{1 - e^{-\sigma T}}{\sigma}$$

$$F(s) = \frac{1 - e^{-sT}}{s} \tag{10-39}$$

9. $f(t) = $ *Triangular Pulse of Duration* T. The function is

$$f(t) = \begin{cases} \dfrac{2}{T}\, t & 0 < t < \dfrac{T}{2} \\ 2 - \dfrac{2}{T}\, t & \dfrac{T}{2} < t < T \\ 0 & T < t \end{cases}$$

The Laplace integral, for real s, is

$$\frac{2}{T} \int_0^{T/2} te^{-\sigma t}\, dt + \int_{T/2}^T \left(2 - \frac{2}{T}\, t\right) e^{-\sigma t}\, dt = \frac{2}{\sigma^2 T}\left(1 - 2e^{-\sigma T/2} + e^{-\sigma T}\right)$$

and the transform is

$$F(s) = \frac{2}{T}\frac{1 - 2e^{-sT/2} + e^{-sT}}{s^2} \tag{10-40}$$

10. $f(t) = $ *Sinusoidal Pulse*. This is the function

$$f(t) = \begin{cases} \sin bt & 0 < t < \dfrac{\pi}{b} \\ 0 & \dfrac{\pi}{b} < t \end{cases}$$

The Laplace integral is

$$\int_0^{\pi/b} \sin bt e^{-\sigma t}\, dt = \frac{b(1 + e^{-\sigma \pi/b})}{\sigma^2 + b^2}$$

and the transform is

$$F(s) = \frac{b(1 + e^{-s\pi/b})}{s^2 + b^2} \tag{10-41}$$

The denominator of this function is zero at $s = \pm jb$, but $e^{\pm j\pi} = -1$, showing that the numerator is zero also. $F(s)$ has no poles and is an entire function. In each of the last three cases $f(t)$ is identically zero beyond a certain value of t, and in each case $F(s)$ has no finite singular points. Later on it will be shown that this is a general property of transforms of functions which eventually become identically zero.

From these examples we can construct Table 10-1, a short table of Laplace transforms.

TABLE 10-1

	$f(t)$	$F(s)$
1.	1	$\dfrac{1}{s}$
2.	e^{at}	$\dfrac{1}{s - a}$
3.	$\sin bt$	$\dfrac{b}{s^2 + b^2}$
4.	$\cos bt$	$\dfrac{s}{s^2 + b^2}$
5.	(a) $e^{at} \sin bt$	$\dfrac{b}{(s - a)^2 + b^2}$
	(b) $e^{at} \cos bt$	$\dfrac{s - a}{(s - a)^2 + b^2}$
6.	$t^n \quad n \geqq 0$	$\dfrac{n!}{s^{n+1}}$
7.	$\sqrt{t^k} \quad k \geqq -1$, and odd	$\dfrac{(k + 1)! \sqrt{\pi}}{2^{k+1} \left(\dfrac{k + 1}{2}\right)! \sqrt{s^{k+2}}}$
8.	$\begin{array}{ll} 1 & 0 < t < T \\ 0 & T < t \end{array}$	$\dfrac{1 - e^{-sT}}{s}$
9.	$\begin{array}{ll} \dfrac{2}{T}t & 0 < t < \dfrac{T}{2} \\ 2 - \dfrac{2}{T}t & \dfrac{T}{2} < t < T \\ 0 & T < t \end{array}$	$\dfrac{2}{T}\dfrac{1 - 2e^{-sT/2} + e^{-sT}}{s^2}$
10.	$\begin{array}{ll} \sin bt & 0 < t < \dfrac{\pi}{b} \\ 0 & \dfrac{\pi}{b} < t \end{array}$	$b\dfrac{1 + e^{-s\pi/b}}{s^2 + b^2}$

In conjunction with the combinational properties described in Sec. 10-10, this table provides the transforms for many particular functions. These examples show that $F(s)$ is rational if $f(t)$ is a polynomial or a sum of exponentials. We conjecture that a product of a polynomial by an exponential might also yield a rational $F(s)$. When the square root appears, we do not get a rational function. Another observation is that in each case $F(s)$ approaches zero as $|s|$ becomes infinite, but at different rates. We also note that, if $f(t)$ ultimately becomes identically zero, its transform is an entire function. These ideas, based here on only a few examples, are offered in order to awaken your curiosity about possible relationships between properties of $F(s)$ and $f(t)$.

Another interesting question arises. In Table 10-1 we find $F(s)$ functions tabulated for given $f(t)$ functions. The $F(s)$ functions have been uniquely determined for these $f(t)$ functions. The question arises: Is each of the $f(t)$ functions shown in the table the only t function which will give the $F(s)$ function appearing opposite it in the table? This is an important question because the solution of a practical problem usually presents a known $F(s)$, from which $f(t)$ must be found. A table such as this is therefore more useful if it can be used to find $f(t)$ from $F(s)$. But this is possible only if there is a unique $f(t)$ for each $F(s)$. In Sec. 10-17 we show that $F(s)$ uniquely determines $f(t)$ for $t > 0$. Meanwhile, in the following four sections we shall answer the above questions about properties of $F(s)$ and shall establish some other properties which are not so obviously anticipated by Table 10-1.

10-12. Elementary Properties of $F(s)$. In the previous section we obtained some empirical information about the properties of $\mathcal{L}[f(t)]$. Now we shall derive an assortment of different general properties of $F(s)$, which are valid for a variety of conditions on $f(t)$. The presentation is in terms of a sequence of theorems and proofs.

Theorem 10-3. If $f(t)$ is a real function of t and if $F(s) = \mathcal{L}[f(t)]$ is single-valued, then $F(s)$ is a real function of s.

PROOF. It is necessary only to look at the defining integral

$$F(s) = \int_0^\infty f(t)e^{-st}\,dt$$

In this integral t is real, and so $f(t)$ is real. Also, when $s = \sigma > \sigma_c$ the integrand is real and therefore the integral is real. This establishes that $F(s)$ is real on the real axis to the right of point σ_c. Also, if $F(s)$ is single-valued, by virtue of Theorem 7-2 it can be said that $F(s)$ is a real analytic function.

Theorem 10-4. If $f(t)$ is an APC function having a Laplace integral which converges at s_0, and if $F(s)$ is the transform, then

$$\lim_{|s|\to\infty} F(s) = 0$$

uniformly in the region

$$|\text{ang}\,(s - s_0)| \leqq \theta' < \frac{\pi}{2}$$

PROOF. From Theorem 10-2 it is known that the Laplace integral converges uniformly in the region specified above. Therefore, if we are given an arbitrary small positive number ϵ, we can find a number A_1 which is independent of s such that

$$\left| \int_{A_1}^{\infty} f(t)e^{-st}\, dt \right| < \frac{\epsilon}{3}$$

when

$$|\text{ang}\,(s - s_0)| \leqq \theta' < \frac{\pi}{2}$$

Also, since the function is integrable, another number $A_2 < A_1$ can be found such that

$$\left| \int_{0}^{A_2} f(t)e^{-st}\, dt \right| < \frac{\epsilon}{3}$$

Finally, for the remaining interval of integration we have

$$\left| \int_{A_2}^{A_1} f(t)e^{-st}\, dt \right| \leqq e^{-\sigma A_2} \int_{A_2}^{A_1} |f(t)|\, dt \qquad \sigma > 0$$

The integral on the right is a constant, and therefore, A_1 and A_2 having been found, it is possible to find a number σ' such that

$$e^{-\sigma A_2} \int_{A_2}^{A_1} |f(t)|\, dt < \frac{\epsilon}{3}$$

when

$$\sigma > \sigma' > 0$$

Now, as $|s|$ approaches infinity in the region

$$|\text{ang}\,(s - s_0)| \leqq \theta' < \frac{\pi}{2}$$

we recognize from Fig. 10-5 that ultimately $\text{Re}\, s > \sigma'$ when $|s - s_0|$ is

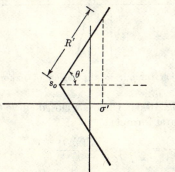

Fig. 10-5. A graphical illustration of the fact that a number R' can be found such that $\text{Re}\, s > \sigma'$ when $R > R'$, in the sector $|\text{ang}\,(s - s_0)| \leqq \theta' < \pi/2$.

greater than some number R'. Therefore, we can write

$$\left| \int_{0}^{\infty} \right| \leqq \left| \int_{0}^{A_2} \right| + \left| \int_{A_2}^{A_1} \right| + \left| \int_{A_1}^{\infty} \right| < \epsilon$$

when

$$|s - s_0| > R'$$

for all angles in the region

$$|\text{ang } (s - s_0)| \leqq \theta' < \frac{\pi}{2}$$

Since R' does not depend on θ', but only on ϵ, it is established that $F(s)$ approaches zero uniformly in the angular sector specified.

This theorem yields some useful information about the behavior of $F(s)$ at infinity. However, you will observe that behavior along the j axis is not predicted. This omission is necessary because in some cases $F(s)$ will have an essential singularity at infinity.

Theorem 10-5. If $F(s)$ is the Laplace transform of an almost piecewise continuous function $f(t)$, then at all regular points of $F(s)$ its nth derivative is

$$\frac{d^n F(s)}{ds^n} = \int_0^\infty (-t)^n f(t) e^{-st} \, dt \qquad \text{Re } s > \sigma_c$$
$$= \mathcal{L}[(-t)^n f(t)] \tag{10-42}$$

PROOF. From Theorem 10-2 it is known that there is a point s_0 where the Laplace integral converges and that convergence is uniform in an

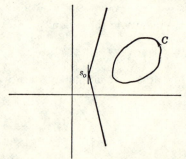

FIG. 10-6. An integration contour in the region of uniform convergence of the Laplace integral.

angular sector like Fig. 10-6. Choose a closed curve C in this region, and use the Cauchy integral formula

$$\frac{d^n F(s)}{ds^n} = \frac{n!}{2\pi j} \int_C \frac{F(z)}{(z - s)^{n+1}} \, dz$$
$$= \frac{n!}{2\pi j} \int_C \frac{dz}{(z - s)^{n+1}} \int_0^\infty f(t) e^{-zt} \, dt$$

But since C lies in a region of uniform convergence, by Theorem 8-9 we can interchange the order of integration, giving

$$\frac{d^n F(s)}{ds^n} = \frac{n!}{2\pi j} \int_0^\infty f(t) \, dt \int_C \frac{e^{-zt}}{(z - s)^{n+1}} \, dz$$

In the above expression the contour integral can be evaluated by finding the residue of the integrand at the pole $z = s$, which is obtained from Eq. (5-80) as follows:

$$\text{Residue } \frac{e^{-zt}}{(z - s)^{n+1}} = \frac{1}{n!} \frac{d^n(e^{-zt})}{dz^n}\bigg|_{z=s} = \frac{(-t)^n e^{-st}}{n!}$$

Therefore, we get

$$\frac{n!}{2\pi j} \int_C \frac{e^{-zt}}{(z - s)^{n+1}} dz = (-t)^n e^{-st}$$

proving the theorem.

Theorem 10-6. If $f(t)$ is APC and is identically zero for $t > T_0$, where T_0 is any positive number, then $F(s)$ is an entire function.

PROOF. In this case the Laplace integral reduces to an integral with finite limits, namely,

$$F(s) = \int_0^{T_0} f(t)e^{-st} dt$$

Since the limits are finite and the APC condition ensures integrability, this integral exists for all finite s and we may say that its abscissa of convergence is $-\infty$. From Sec. 10-8 we know that the Laplace integral is regular to the right of the axis of convergence, in the finite plane in this case.

If you will refer to Table 10-1, you will find illustrations of Theorems 10-3 through 10-6. Many more general properties of $F(s)$ can be derived. Some of the simpler ones are given in the next section, and others are developed in later chapters. The intention here is to present only the relatively simple properties that can be proved relatively easily.

10-13. The Shifting Theorems. Suppose that we have the transform $F(s)$ of a transformable function $f(t)$ and then consider the transform of the function

$$e^{-at}f(t)$$

where a is real or complex. If σ_c is the abscissa of convergence for $f(t)$, then the integral

$$\int_0^\infty e^{-at}f(t)e^{-st} dt = \int_0^\infty f(t)e^{-(s+a)t} dt$$

converges for $\text{Re } s > \sigma_c - \text{Re } a$. However, this integral is an expression for $F(s + a)$, showing that, if the Laplace transform is known for any function, the transform of that function multiplied by an exponential can immediately be obtained by a simple change in variable. There is a "shift," or translation, in the s variable. Thus we have the general result

$$\mathcal{L}[e^{-at}f(t)] = F(s + a) \tag{10-43}$$

where

$$F(s) = \mathcal{L}[f(t)]$$

For the next case again assume that the function $f(t)$ has a transform $F(s)$, and consider a shift in the t variable. From $f(t)$ a new function is found by changing the variable to $t - T$, where T is a positive constant, while stipulating that the new function shall be zero for $t < T$. With the aid of the notation for the unit step,

$$u(t) = \begin{cases} 0 & t < 0 \\ 1 & t > 0 \end{cases} \tag{10-44}$$

this function can be written

$$f(t - T)u(t - T)$$

In the graphical interpretation, this transformation shifts the graph an amount T in the positive t direction. The transform is represented by the integral

$$\int_0^\infty f(t - T)u(t - T)e^{-st}\, dt = \int_T^\infty f(t - T)e^{-st}\, dt$$

The factor $u(t - T)$ is dropped in the second integral because the lower limit has been changed to T. Now change the variable of integration to $t' = t - T$, giving

$$\mathcal{L}[f(t - T)u(t - T)] = e^{-sT} \int_0^\infty f(t')e^{-st'}\, dt'$$
$$= e^{-sT}F(s) \tag{10-45}$$

These results are summarized in the following two theorems:

Theorem 10-7.   If $f(t)$ has a Laplace transform $F(s)$, then the Laplace transform of $e^{-at}f(t)$ is $F(s + a)$, where a is real or complex.

Theorem 10-8.   If $f(t)$ has a Laplace transform $F(s)$, then the Laplace transform of $f(t - T)u(t - T)$ is $e^{-sT}F(s)$, where T is real and positive.

Examples of applications of these two theorems are found in Table 10-1. To cite one, we can find the transform of $\cos bt$ by knowing the transform of $f(t) = 1$, which is $1/s$. Thus,

$$\cos bt = \frac{e^{jbt} + e^{-jbt}}{2}$$

$$\mathcal{L}(\cos bt) = \frac{1}{2}\left(\frac{1}{s - jb} + \frac{1}{s + jb}\right) = \frac{s}{s^2 + b^2}$$

10-14. Laplace Transform of the Derivative of $f(t)$. We shall now obtain a relationship between the Laplace transform of the derivative of a function and the Laplace transform of the function itself. We assume that $f(t)$ is continuous for $t > 0$ and has a limit at $t = 0$ and that its derivative is APC. We also require $f(t)$ to be of exponential order α_0. Consider the Laplace integral, written with $s = \sigma$, as follows:

$$\int_0^\infty f'(t)e^{-\sigma t}\,dt = \lim_{A\to\infty}\int_0^A f'(t)e^{-\sigma t}\,dt$$

The integral on the right meets the conditions for integrating by parts (Theorem 8-4), namely, continuity of $f(t)$ and $e^{-\sigma t}$. Thus, we get

$$\int_0^A f'(t)e^{-\sigma t}\,dt = \lim_{\epsilon\to 0} f(t)e^{-\sigma t}\Big|_\epsilon^A + \sigma\int_0^A f(t)e^{-\sigma t}\,dt$$
$$= f(A)e^{-\sigma A} - f(0+) + \sigma\int_0^A f(t)e^{-\sigma t}\,dt \quad (10\text{-}46)$$

It is necessary to write

$$\lim_{\epsilon\to 0} f(\epsilon) = f(0+)$$

in order to say precisely what we mean, because quite often $f(t)$ is discontinuous at $t = 0$ and $f(0)$ may be undefined or defined as some value other than $f(0+)$, usually $f(0+)/2$. This is the only discontinuity permitted in the theorem we are developing. A similar theorem, where $f(t)$ is allowed other points of discontinuity, is considered in Chap. 12.

We recall that $f(t)$ is EO,α_0. Therefore, by relation (10-5),

$$\lim_{A\to\infty} f(t)e^{-\sigma A} = 0$$

when $\sigma > \alpha_0$. Also,

$$\lim_{A\to\infty}\int_0^A f(t)e^{-\sigma t}\,dt$$

converges to $F(\sigma)$, for $\sigma > \alpha_0$, by Theorem 10-1. Thus, for $\sigma > \alpha_0$, Eq. (10-46) yields

$$\int_0^\infty f'(t)e^{-\sigma t}\,dt = \sigma F(\sigma) - f(0+)$$

or
$$\mathcal{L}[f'(t)] = sF(s) - f(0+) \quad (10\text{-}47)$$

This proof shows that the abscissa of convergence for the Laplace integral of the derivative function is at most no larger than α_0. We also note that $f(t)$ must be of exponential order but that $f'(t)$ is not necessarily of exponential order. For example,

$$f(t) = \cos e^{t^2}$$

is EO,0, but its derivative

$$f'(t) = -2te^{t^2} \sin e^{t^2}$$

is not of exponential order. But by the proof just completed we know that it has a Laplace integral with zero for its abscissa of convergence.

This proof can be applied repeatedly to successive derivatives, as long as the *next-to-the-last* derivative is EO,α_0. The result is stated as follows:

Theorem 10-9.   Let $f(t)$ and its derivatives of orders up to and including order $n - 1$ be continuous for $t > 0$, with limits existing at $t = 0$, and of exponential order. Then the Laplace transform of $f^{(n)}(t)$ is

$$\mathcal{L}[f^{(n)}(t)] = s^n F(s) - s^{n-1}f(0+) - s^{n-2}f^{(1)}(0+) - \cdots - f^{(n-1)}(0+)$$

$$(10\text{-}48)$$

Table 10-1 provides illustrations of this theorem. For example,

$$\mathcal{L}(t) = \frac{1}{s^2}$$

and, from Theorem 10-9,

$$\mathcal{L}(1) = s\frac{1}{s^2} - 0 = \frac{1}{s}$$

in agreement with the table. As another example, we can get $\mathcal{L}(\sin bt)$ from $\mathcal{L}(\cos bt)$ as follows:

$$\mathcal{L}(\cos bt) = \frac{s}{s^2 + b^2}$$

$$\mathcal{L}(-b \sin bt) = \frac{s^2}{s^2 + b^2} - 1 = \frac{-b^2}{s^2 + b^2}$$

$$\mathcal{L}(\sin bt) = \frac{b}{s^2 + b^2}$$

10-15. Laplace Transform of the Integral of a Function.   Let $f(t)$ be an APC function having a Laplace transform and an abscissa of convergence σ_c. This function is not necessarily of exponential order. Now define the function

$$g(t) = \int_0^t f(\tau)\, d\tau$$

and investigate whether or not it has a Laplace transform $G(s)$. We begin by defining another function

$$A(T,\sigma) = \int_0^T e^{-\sigma t} \int_0^t f(\tau)\, d\tau\, dt \qquad (10\text{-}49)$$

If the limit of the above exists, as T goes to infinity, this limit will be $G(\sigma)$. Routine integration by parts yields

$$\sigma A(T,\sigma) = -e^{-\sigma T} \int_0^T f(\tau)\, d\tau + \int_0^T e^{-\sigma t} f(t)\, dt$$

From Eq. (10-49) it is apparent that

$$\frac{\partial A(T,\sigma)}{\partial T} = e^{-\sigma T} \int_0^T f(\tau)\, d\tau$$

and so, in combination with the previous equation, we get

$$\sigma A(T,\sigma) + \frac{\partial A(T,\sigma)}{\partial T} = \int_0^T e^{-\sigma t} f(t)\, dt \qquad (10\text{-}50)$$

Observe that the left-hand side of Eq. (10-50) can also be written

$$\sigma A(T,\sigma) + \frac{\partial A(T,\sigma)}{\partial T} = \frac{1}{e^{\sigma T}} \frac{\partial}{\partial T}[e^{\sigma T} A(T,\sigma)] = \sigma \frac{(\partial/\partial T)[e^{\sigma T} A(T,\sigma)]}{(\partial/\partial T)e^{\sigma T}} \qquad (10\text{-}51)$$

and, referring to Eq. (10-50), we see that, if $\sigma > \sigma_c$,

$$\lim_{T\to\infty} \left[\sigma A(T,\sigma) + \frac{\partial A(T,\sigma)}{\partial T} \right] = F(\sigma)$$

is obtained for the limit of the left-hand side of Eq. (10-51). An expression for the limit of the right-hand side of Eq. (10-51) can be obtained, if $\sigma > 0$, by application of the Lhopital rule, giving

$$\lim_{T\to\infty} \frac{(\partial/\partial T)[e^{\sigma T} A(T,\sigma)]}{(\partial/\partial T)e^{\sigma T}} = \lim_{T\to\infty} \frac{e^{\sigma T} A(T,\sigma)}{e^{\sigma T}} = \lim_{T\to\infty} A(T,\sigma)$$

The condition $\sigma > 0$ is necessary to ensure that $e^{\sigma T}$ will approach infinity. In order to ensure the existence of limits for both the right and left sides of Eq. (10-51), it is necessary to have $\sigma > \sigma_1$, where σ_1 is 0 or σ_c, whichever is greater. Accordingly, we have shown that

$$\lim_{T\to\infty} A(T,\sigma) = \frac{1}{\sigma} F(\sigma) \qquad \sigma > \sigma_1$$

In the original integral notation, this is

$$\int_0^\infty e^{-\sigma t} \int_0^t f(\tau)\, d\tau\, dt = \frac{1}{\sigma} \int_0^\infty e^{-\sigma t} f(t)\, dt \qquad \sigma > \sigma_1$$

By virtue of Theorem 10-2 the same formula can now be written with σ replaced by s, subject to the restriction Re $s > \sigma_1$. Thus, the function

$$\int_0^t f(\tau)\, d\tau$$

has a Laplace integral which converges for Re $s > \sigma_1$ and a Laplace transform given by

$$\mathcal{L}\left[\int_0^t f(\tau)\,d\tau\right] = \frac{1}{s}\,\mathcal{L}[f(t)]$$

It is of interest to observe how the abscissa of convergence σ_1 is related to σ_c, the abscissa of convergence of $f(t)$. We found that if $\sigma_c < 0$ then $\sigma_1 = 0$, but if $\sigma_c > 0$ then $\sigma_1 = \sigma_c$. To illustrate, consider the example

$$f(t) = e^{-bt} \qquad b > 0$$

Its abscissa of convergence is $-b$, but the integral of this function

$$\int_0^t e^{-b\tau}\,d\tau = \frac{1 - e^{-bt}}{b}$$

has zero as its abscissa of convergence. We have now proved the following theorem:

Theorem 10-10. If $f(t)$ is APC and has a Laplace transform, then the function $\int_0^t f(\tau)\,d\tau$ has a Laplace transform given by

$$\mathcal{L}\left[\int_0^t f(\tau)\,d\tau\right] = \frac{\mathcal{L}[f(t)]}{s} \tag{10-52}$$

As an example of an application of this theorem, we can get $\mathcal{L}(\sin bt)$ from $\mathcal{L}(\cos bt)$ by recognizing that

$$\int_0^t \cos bx\,dx = \frac{\sin bt}{b}$$

From Theorem 10-10 we immediately have

$$\mathcal{L}\left(\frac{\sin bt}{b}\right) = \frac{1}{s^2 + b^2}$$

$$\mathcal{L}(\sin bt) = \frac{b}{s^2 + b^2}$$

A theorem similar to Theorem 10-10, but in which $f(t)$ is of exponential order, has a much simpler proof. Its proof is left to you as an exercise.

10-16. Initial- and Final-value Theorems. The derivation of the formula for the transform of a derivative provides two theorems which are sometimes useful in analysis. Let $f(t)$ be continuous for $t > 0$, with a limit at $t = 0$, of exponential order, and with a derivative $f'(t)$ which is APC. These are the conditions for Theorem 10-9, and accordingly the transform of $f'(t)$ exists, and Eq. (10-47) applies. By Theorem 10-4

$$\lim_{\sigma \to \infty} \mathcal{L}[f'(t)] = 0$$

and therefore it follows from Eq. (10-47) that

$$\lim_{\sigma \to \infty} \sigma F(\sigma) = f(0+) \qquad (10\text{-}53)$$

We shall now see that the value approached by $f(t)$ as t becomes infinite can also be determined from $F(s)$. Assume the same conditions as before on $f(t)$ and $f'(t)$, with the additional stipulation that the Laplace integral

$$\int_0^\infty f'(t) e^{-st} \, dt$$

shall converge for $s = 0$. From Theorem 10-2 it follows that convergence of this integral is uniform for s real and nonnegative, and so we can take the limit as $\sigma \to 0$ inside the integral, as follows:

$$\lim_{\sigma \to 0} \int_0^\infty f'(t) e^{-\sigma t} \, dt = \int_0^\infty f'(t) \, dt = f(\infty) - f(0+)$$

Now if you will refer to the equation preceding Eq. (10-47) and take the limit indicated above, the result is

$$f(\infty) - f(0+) = \lim_{\sigma \to 0} \sigma F(\sigma) - f(0+)$$

or

$$\lim_{\sigma \to 0} \sigma F(\sigma) = f(\infty) \qquad (10\text{-}54)$$

These results are derived here for a restricted class of functions. The requirement that $f(t)$ shall be continuous would perhaps seem to provide an unreasonable limitation. In Chap. 12 we shall review this topic again and shall find that some of the restrictions at present placed on $f(t)$ can be removed. However, the results given above are as general as we are able to prove with the theory presented up to this point. Pending the treatment of the more general conditions, a formal statement of these results is omitted here. It should be mentioned, however, that Eqs. (10-53) and (10-54) state the results of two theorems which are known, respectively, as the *initial-value theorem* and the *final-value theorem*.

10-17. Nonuniqueness of Function Pairs for the Two-sided Laplace Transform. In Sec. 10-2 it is shown that the two-sided Laplace transform can be defined as the sum of two one-sided transforms. To reiterate, if $\mathfrak{F}(s) = \mathcal{L}_2[f(t)]$, $F_1(s) = \mathcal{L}[f(t)]$, and $F_2(s) = \mathcal{L}[f(-t)]$, then

$$\mathfrak{F}(s) = F_1(s) + F_2(-s) \qquad (10\text{-}55)$$

In this way we see that a table of one-sided transforms can also be used to obtain two-sided transforms. But the table cannot be used to obtain $f(t)$ if $\mathfrak{F}(s)$ is known. In order to determine $f(t)$, knowledge of

$\mathfrak{F}(s)$ must be supplemented with directions as to how it is to be split up into $F_1(s)$ and $F_2(-s)$, or some equivalent information.

In order to clarify these ideas, assume that two functions $f_g(t)$ and $f_h(t)$ are given, with converging single-sided Laplace integrals as follows:

$$F_g(s) = \int_0^\infty f_g(t)e^{-st}\,dt \qquad \text{Re } s > \sigma_g$$

and $$F_h(-s) = \int_0^\infty f_h(-t)e^{st}\,dt \qquad \text{Re } s < \sigma_h$$

Also, choose a number b in the range $\sigma_g < b < \sigma_h$, and define the two additional functions

$$f_a(t) = \begin{cases} f_g(t) + e^{bt} & t > 0 \\ f_h(t) & t < 0 \end{cases}$$

and $$f_b(t) = \begin{cases} f_g(t) & t > 0 \\ f_n(t) - e^{bt} & t < 0 \end{cases}$$

We shall show that $\mathcal{L}_2[f_a(t)] = \mathcal{L}_2[f_b(t)]$. In addition to $F_g(s)$ and $F_h(-s)$

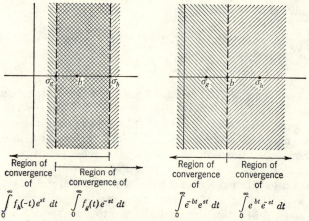

Fig. 10-7. Regions of convergence of various integrals used in defining $F_1(s)$ and $F_i(-s)$, in the equation $\mathfrak{F}(s) = F_1(s) + F_2(-s)$.

defined above, we also have

$$\frac{1}{s - b} = \int_0^\infty e^{bt}e^{-st}\,dt \qquad \text{Re } s > b$$

$$\frac{1}{-s + b} = \int_0^\infty e^{-bt}e^{st}\,dt \qquad \text{Re } s < b$$

Regions of convergence of the four pertinent defining integrals are shown in Fig. 10-7. We then get

$$\mathcal{L}_2[f_a(t)] = \left[F_g(s) + \frac{1}{s-b}\right] + F_h(-s)$$

$$\mathcal{L}_2[f_b(t)] = F_g(s) + \left[F_h(-s) - \frac{1}{-s+b}\right]$$

where the quantities in brackets are respectively $F_1(s)$ or $F_2(-s)$ in the notation of Eq. (10-55). Obviously, $\mathcal{L}_2[f_a(t)] = \mathcal{L}_2[f_b(t)]$, although $f_a(t) \neq f_b(t)$.

We see that shifting the term e^{bt} from the region of positive t to the region of negative t causes the term $1/(s-b)$ in the transform to change its association from $F_1(s)$ to $F_2(-s)$. The region of convergence for the defining integral of this term changes from Re $s > b$ to Re $s < b$, as

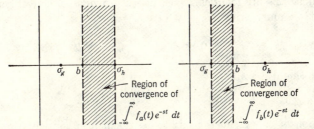

FIG. 10-8. Regions of convergence of functions having identical two-sided transforms.

indicated in Fig. 10-7. As a result, the regions of convergence of the two defining integrals

$$\mathcal{L}_2[f_a(t)] = \int_{-\infty}^{\infty} f_a(t)e^{-st}\, dt \qquad \text{and} \qquad \mathcal{L}_2[f_b(t)] = \int_{-\infty}^{\infty} f_b(t)e^{-st}\, dt$$

are different, as indicated in Fig. 10-8.

As a specific example, let

$$f_a(t) = \begin{cases} e^{-2t} - e^{-t} & t > 0 \\ 0 & t < 0 \end{cases}$$

$$f_b(t) = \begin{cases} e^{-2t} & t > 0 \\ e^{-t} & t < 0 \end{cases}$$

For $f_a(t)$ we have

$$F_1(s) = \int_0^{\infty} (e^{-2t} - e^{-t})\, dt \qquad \text{Re } s > -1$$

$$= \frac{1}{s+2} - \frac{1}{s+1}$$

and $\qquad F_2(-s) = 0$

Similarly, for $f_b(t)$,

$$F_1(s) = \int_0^\infty e^{-2t}\, dt \qquad \text{Re } s > -2$$

$$= \frac{1}{s+2}$$

and

$$F_2(-s) = \int_0^\infty e^t e^{st}\, dt \qquad \text{Re } s < -1$$

$$= -\frac{1}{s+1}$$

In each case $F_1(s) + F_2(-s)$ is the same.

The important observation is that $f_a(t)$ and $f_b(t)$ have different regions of convergence of the defining integrals; but after analytic continuation their transforms are identical. Because the regions of convergence are different, we must write

$$\int_{-\infty}^\infty f_a(t) e^{-st}\, dt \neq \int_{-\infty}^\infty f_b(t) e^{-st}\, dt$$

while we can also write

$$\mathfrak{F}_a(s) = \mathfrak{F}_b(s)$$

The region of convergence needs to be taken into consideration if we are to get a unique $f(t)$ if $\mathfrak{F}(s)$ is given. We get more insight into this question after we develop a specific process for getting $f(t)$ from $F(s)$ or $\mathfrak{F}(s)$.

10-18. The Inversion Formula. As is pointed out in Eq. (10-3), the two-sided Laplace transform

$$\mathfrak{F}(s) = \int_{-\infty}^\infty f(t) e^{-st}\, dt$$

is identical with the Fourier transform

$$\mathfrak{F}(\sigma + j\omega) = \int_{-\infty}^\infty f(t) e^{-\sigma t} e^{-j\omega t}\, dt \qquad \sigma_{c1} < \sigma < \sigma_{c2}$$

if we regard σ as fixed. This is a function of ω only. Of course, the fixed σ must lie in the strip of convergence for the particular function $f(t)$. Since the above is really a Fourier integral, we can use the inversion formula of the Fourier integral theorem

$$f(t) e^{-\sigma t} = \frac{1}{2\pi} PV \int_{-\infty}^\infty \mathfrak{F}(\sigma + j\omega) e^{j\omega t}\, d\omega \qquad (10\text{-}56)$$

or

$$f(t) = \frac{1}{2\pi} PV \int_{-\infty}^\infty \mathfrak{F}(\sigma + j\omega) e^{(\sigma + j\omega)t}\, d\omega \qquad (10\text{-}57)$$

The symbol $\sigma + j\omega$ which appears in two places in the above integral can be replaced by s, provided that we stipulate that s *shall lie on the vertical line with abscissa σ.* Then the above integral can be described as a contour integral along the contour shown in Fig. 10-9. The principal-value feature is retained by directing that the contour shall always extend from $\sigma - jR$ to $\sigma + jR$ while R approaches infinity. This contour occurs repeatedly and will be called the *Bromwich contour* (named for a mathematician who did much work in this field). The Bromwich contour will carry the abbreviated designation Br.

On this contour

$$ds = j\, d\omega$$

and so the real integral of Eq. (10-57) becomes

$$f(t) = \frac{1}{2\pi j} \int_{\text{Br}} \mathfrak{F}(s) e^{st}\, ds \qquad (10\text{-}58)$$

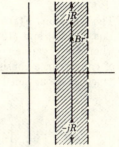

Fig. 10-9. Definition of the Bromwich contour.

The Br contour must lie in the original strip of convergence. One of the reasons for being concerned about regions of convergence is to establish the location of the Br contour. In Sec. 10-8 it was pointed out that $\mathfrak{F}(s)$ has singular points outside the strip of convergence of its defining integral. Therefore, if Br is moved outside this strip, we should expect a change in the integral of Eq. (10-58), because the contour will then have passed over one or more singular points.

This requirement, that the region of convergence of the Laplace integral of $f(t)$ must be known before we can locate the Br contour, is a point of importance. If only $\mathfrak{F}(s)$ were given, we would not be able to locate this region, and therefore $f(t)$ could not be obtained, for lack of knowledge of where to put the Br contour. However, if path Br is specified, $f(t)$ can be obtained from $\mathfrak{F}(s)$ by Eq. (10-58). As a consequence, we can say that $\mathfrak{F}(s)$ determines $f(t)$ uniquely only if by some independent means we know where Br should be located. This information is not automatically provided by the properties of $\mathfrak{F}(s)$. The ambiguity is the same one described at the end of Sec. 10-17, where it was pointed out that $F_1(s)$ and $F_2(-s)$ are not uniquely determined by $\mathfrak{F}(s)$. A change in Br would correspond to a change in $F_1(s)$ and $F_2(-s)$.

We cannot carry the above discussion to completion until the corresponding theory is developed for the single-sided transform, which we now do. Then the relation of contour position to $F_1(s)$ and $F_2(-s)$ of Eq. (10-55) can be clarified. Therefore, we proceed by considering the single-sided case. Let us compare $\mathcal{L}_2[f(t)]$ and $\mathcal{L}[f(t)]$, where $f(t)$ is the same function in both cases and is defined for all t. Upon recalling the defining integrals for the one- and two-sided transforms, it is evident that

$$F(s) = \mathcal{L}[f(t)] = \mathcal{L}_2[f(t)u(t)] \qquad (10\text{-}59)$$

Having related $F(s)$ to a two-sided transform, we can use the inversion integral for the latter, given by Eq. (10-58); and we have

$$f(t)u(t) = \frac{1}{2\pi j} \int_{\text{Br}} F(s)e^{st}\,ds \qquad (10\text{-}60)$$

The functions $F(s)$ and $\mathfrak{F}(s)$ are different, and so the integrations of Eqs. (10-58) and (10-60) yield different results, the difference occurring when t is negative. The two-sided transform $\mathcal{L}_2[f(t)u(t)]$ has a "strip" of convergence consisting of a right half plane, the half plane of convergence of a one-sided transform. Thus, in Eq. (10-60) the Br contour can be any vertical line in the half plane of convergence, as indicated in Fig. 10-10.

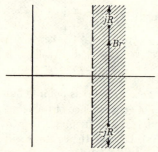

With this background we can turn to the question of whether or not a single-sided transform function uniquely determines a t function. Suppose that $F(s)$ is given. Equation (10-60) provides $f(t)u(t)$ uniquely. In other words, $F(s)$ determines $f(t)$ uniquely for $t > 0$ and gives zero for $t < 0$. In appraising these comments we recognize the role played by the Br contour. The important fact is that from the location of singularities of the

FIG. 10-10. The Br contour which yields zero $f(t)$ for negative t.

given $F(s)$ we know Br must be *to the right* of all singular points, and therefore Eq. (10-60) gives a unique function $f(t)u(t)$.

The fact that $F(s)$ does uniquely determine $f(t)u(t)$ means that we can adopt a notation to indicate that relationship. The notation $F(s) = \mathcal{L}[f(t)]$ has been introduced to symbolize that $f(t)$ determines $F(s)$ uniquely. Now we are going in the other direction, and so it is convenient to use the notation

$$f(t)u(t) = \mathcal{L}^{-1}[F(s)] \qquad (10\text{-}61)$$

to represent Eq. (10-60). The function $\mathcal{L}^{-1}[F(s)]$ is often called the *inverse transform* of $F(s)$.

In the two-sided transform, $\mathfrak{F}(s)$ is uniquely related to $f(t)$ by the formula

$$\mathfrak{F}(s) = \int_{-\infty}^{\infty} f(t)e^{-st}\,dt$$

and this fact is implied when we write $\mathfrak{F}(s) = \mathcal{L}_2[f(t)]$, but, as pointed out earlier, a notation similar to Eq. (10-61) cannot be employed, because of lack of uniqueness.

The numerical example given in Sec. 10-17 provides a case in point. There we obtained

$$\mathcal{L}_2[f_a(t)] = \mathcal{L}_2\begin{pmatrix} e^{-2t} - e^{-t} & t > 0 \\ 0 & t < 0 \end{pmatrix} = \frac{1}{s+2} - \frac{1}{s+1}$$

where convergence is for Re $s > -1$, and

$$\mathcal{L}_2[f_b(t)] = \mathcal{L}_2\begin{pmatrix} e^{-2t} & t > 0 \\ e^{-t} & t < 0 \end{pmatrix} = \frac{1}{s+2} - \frac{1}{s+1}$$

where convergence is for $-2 < $ Re $s < -1$. To these we add a third case,

$$\mathcal{L}_2[f_c(t)] = \mathcal{L}_2\begin{pmatrix} 0 & t > 0 \\ e^{-t} - e^{-2t} & t < 0 \end{pmatrix} = \frac{1}{s+2} - \frac{1}{s+1}$$

where convergence is for Re $s < -2$. You can work out this third case for yourself. Each of the s functions is the same and can be labeled $\mathcal{F}(s)$. The pole locations of $\mathcal{F}(s)$, and the various related Bromwich contours to give the three possible $f(t)$ functions, are shown in Fig. 10-11. From the specified regions of convergence, we see that contours Br_a, Br_b, and Br_c, respectively, yield $f_a(t)$, $f_b(t)$, and $f_c(t)$ when used in Eq. (10-58), and where

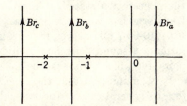

FIG. 10-11. Various Br contours which yield different $f(t)$ functions for a given $\mathcal{F}(s)$ function.

$$\mathcal{F}(s) = \frac{1}{s+2} - \frac{1}{s+1} = -\frac{1}{(s+1)(s+2)}$$

Thus, indirectly we have shown that

$$\text{Case } a \quad -\frac{1}{2\pi j}\int_{Br_a} \frac{e^{st}\,ds}{(s+1)(s+2)} = \begin{cases} e^{-2t} - e^{-t} & t > 0 \\ 0 & t < 0 \end{cases}$$

$$\text{Case } b \quad -\frac{1}{2\pi j}\int_{Br_b} \frac{e^{st}\,ds}{(s+1)(s+2)} = \begin{cases} e^{-2t} & t > 0 \\ e^{-t} & t < 0 \end{cases}$$

$$\text{Case } c \quad -\frac{1}{2\pi j}\int_{Br_c} \frac{e^{st}\,ds}{(s+1)(s+2)} = \begin{cases} 0 & t > 0 \\ e^{-t} - e^{-2t} & t < 0 \end{cases}$$

Now if we adopt the notation of Eq. (10-55) and write

$$\mathcal{F}(s) = F_1(s) + F_2(-s)$$

where $F_1(s) = \mathcal{L}[f(t)]$ and $F_2(s) = \mathcal{L}[f(-t)]$, we see that $F_1(s)$ and $F_2(-s)$ are the following, for the three cases:

$$F_1(s) = \frac{1}{s+2} - \frac{1}{s+1} \qquad F_2(s) = 0 \qquad\qquad \text{for case } a$$

$$F_1(s) = \frac{1}{s+2} \qquad\qquad F_2(-s) = -\frac{1}{s+1} \qquad \text{for case } b$$

$$F_1(s) = 0 \qquad\qquad\qquad F_2(-s) = \frac{1}{s+2} - \frac{1}{s+1} \qquad \text{for case } c$$

The present discussion is somewhat hampered by a lack of a direct means of evaluating the above three integrals. We got the answers only by knowing regions of convergence for the various t functions and by knowing that to recover any one of these t functions by Eq. (10-58) requires a Br contour in that region. This is not a satisfactory state of affairs, and it will be corrected in Sec. 10-19, where a method is developed for evaluating these integrals directly.

Perhaps the main accomplishment of this section has been to confirm the expectation stated at the end of Sec. 10-11, that the single-sided transform has a unique inverse. We have a formula for it, namely, Eq. (10-60), but as yet no means of evaluating it. In obtaining this formula a sound basis has been established for the concept of function pairs, for which tables can be constructed.

It is emphasized that the concept of paired functions does not apply for the general case of the two-sided Laplace transform, for the reasons cited above. However, function pairs for the two-sided case can be constructed in a restricted sense, by considering only those $f(t)$ functions which have a prescribed strip of convergence. Then we know that the Br path of the inversion integral must be in this strip, and Eq. (10-58) becomes unique. This is what happens in the case of the Fourier transform, which uniquely determines $f(t)$. The Fourier transform is a special case of the two-sided Laplace transform for which the strip of convergence spans the imaginary axis and for which the Br contour is actually the imaginary axis. Thus, it is possible to construct a table pairing functions with their Fourier transforms.

10-19. Evaluation of the Inversion Formula. We have yet to consider in detail what kinds of functions we get for $F(s)$ and $\mathfrak{F}(s)$. The answer to this has bearing on methods which will be applicable in evaluating the inversion integrals

$$\frac{1}{2\pi j} \int_{\text{Br}} \mathfrak{F}(s) e^{st} \, ds \qquad \text{and} \qquad \frac{1}{2\pi j} \int_{\text{Br}} F(s) e^{st} \, ds$$

which appear in Eqs. (10-58) and (10-60), respectively.

The form of these integrands suggests the possibility of using Jordan's lemma. However, one adjustment is needed. Jordan's lemma is stated with reference to an integration along the $j\omega$ axis, rather than a Bromwich contour. Suppose that the Br contour is at abscissa σ_1. In the integrals in question we can change the variable of integration from s to z, so that the new contour will be the imaginary axis, as required in Jordan's lemma. Thus, let

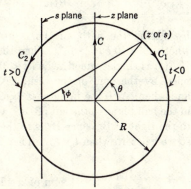

$$z = s - \sigma_1 \qquad (10\text{-}62)$$

so that points $s = \sigma_1 + j\omega$ on contour Br transform to $z = jy$. We shall consider only the inversion integral for $F(s)$, the case of $\mathfrak{F}(s)$ being essentially the same. Let C be a contour on the jy axis, from $-jR$ to jR. It is the same as the contour C of Figs. 8-8 and 8-10. As R approaches infinity,

FIG. 10-12. Change in variable in the inversion integral, to meet the conditions of Jordan's lemma.

this contour becomes the special case of a Br contour which coincides with the j axis. Equation (10-60) now becomes

$$f(t)u(t) = \frac{e^{\sigma_1 t}}{2\pi j} \lim_{R \to \infty} \int_C F(z + \sigma_1)e^{zt}\,dz \qquad (10\text{-}63)$$

The details of this variable change are given in Fig. 10-12.

Jordan's lemma applies to the integral of Eq. (10-63) for two cases, as follows: If

$$\lim_{|z| \to \infty} F(z + \sigma_1) = 0 \qquad \text{uniformly for } -\frac{\pi}{2} \leqq \theta \leqq \frac{\pi}{2} \qquad (10\text{-}64)$$

and $t < 0$, then the integral over C_1 approaches zero as R approaches infinity. Similarly, if

$$\lim_{|z| \to \infty} F(z + \sigma_1) = 0 \qquad \text{uniformly for } \frac{\pi}{2} \leqq \theta \leqq \frac{3\pi}{2} \qquad (10\text{-}65)$$

and $t > 0$, then the integral over C_2 approaches zero as R approaches infinity. These conditions must be satisfied, respectively, for $t < 0$ and $t > 0$. In many practical cases $F(s)$ meets the stronger condition

$$\lim_{|s| \to \infty} F(s) = 0 \qquad \text{uniformly for all } \phi \qquad (10\text{-}66)$$

which of course is sufficient for both conditions (10-64) and (10-65).

We have reached an important milepost. If conditions (10-64) and (10-65) are satisfied, or if condition (10-66) is satisfied, we can evaluate the integral of Eq. (10-63) or its equivalent,

$$f(t)u(t) = \frac{1}{2\pi j} \int_{\text{Br}} F(s)e^{st}\, ds \qquad (10\text{-}67)$$

by closing the path on the right or left, depending, respectively, on whether $t < 0$ or $t > 0$.* Then the calculus of residues comes into play. If $F(s)$ is single-valued, when $t < 0$ the integral is equal to $-2\pi j$ multiplied by the sum of the residues in the half plane to the right of Br, and when $t > 0$ the integral is $2\pi j$ multiplied by the sum of the residues in the half plane to the left of Br. Of course, we know that $F(s)$ is regular to the right of Br, since Br is to the right of the axis of convergence, and so we get zero when $t < 0$. Therefore, Eq. (10-67) reduces to

$$f(t)u(t) = \begin{cases} 0 & t < 0 \\ \text{sum of residues to left of Br} & t > 0 \end{cases} \qquad (10\text{-}68)$$

If $F(s)$ is multivalued, the paths C_1 and C_2 must of course be appropriately modified to avoid crossing branch cuts, in some manner such as that described in Sec. 8-12.

The two-sided transform $\mathcal{F}(s)$ is enough different to warrant summarizing results in this case. If $\mathcal{F}(s)$ satisfies conditions (10-64) and (10-65), or condition (10-66), and if $\mathcal{F}(s)$ is single-valued, then the calculus of residues is used again. However, now path Br lies in a vertical strip, with singular points of $\mathcal{F}(s)$ to the right of this strip as well as to the left. Therefore, a nonzero result is obtained when $t < 0$, and we can summarize:

$$f(t) = \begin{cases} - \text{ sum of residues to right of Br} & t < 0 \\ \text{sum of residues to left of Br} & t > 0 \end{cases} \qquad (10\text{-}69)$$

In this equation we have confirmation of the statement made in Sec. 10-17, that $f(t)$ is not uniquely determined if $\mathcal{F}(s)$ is known. There are singularities to the right and left of Br, and Br can be moved into a different strip of regularity, thereby changing one or more poles from one side of Br to the other.

10-20. Evaluating the Residues (The Heaviside Expansion Theorem). On the basis of the previous section, we can now derive one of the classical theorems in Laplace transform theory. This result first appeared in engineering literature in the operational calculus of Oliver Heaviside.

In Eqs. (10-68) and (10-69) we are directed to find residues of the function $F(s)e^{st}$ or $\mathcal{F}(s)e^{st}$. Consider the former, and assume that $F(s)$ is

* If $F(s)$ includes the factor e^{-sT}, then these paths apply respectively when $t < T$ and $t > T$.

a rational function (ratio of polynomials) with the degree of the denominator at least one greater than the numerator. This condition ensures that condition (10-66) will be satisfied. $F(s)$ could be one of the functions tabulated in Table 10-1, for example.

In Chap. 5 we considered the representation of a ratio of polynomials in a partial-fraction expansion. There will be poles at $s_1, s_2, \ldots, s_M$ of orders $N_1, N_2, \ldots, N_M$. The expansion will look like

$$F(s) = \sum_{k=1}^{M} \sum_{n=1}^{N_k} \frac{a_{n,k}}{(s - s_k)^n} \tag{10-70}$$

All we need do, then, is to find the residue for the typical term

$$\frac{a_{n,k}e^{st}}{(s - s_k)^n}$$

This term itself has a Laurent expansion, for which we want the coefficient of $(s - s_k)^{-1}$, the residue. This problem is worked out in Chap. 5, and the answer is given by Eq. (5-80) as

$$\begin{aligned}
\text{Residue} &= \frac{a_{n,k}}{(n - 1)!} \frac{d^{n-1}}{ds^{n-1}} (e^{st}) \bigg|_{s=s_k} \\
&= \frac{a_{n,k}}{(n - 1)!} t^{n-1} e^{s_k t}
\end{aligned} \tag{10-71}$$

Thus, when $F(s)$ is a rational function, its inverse transform will be a sum of terms like Eq. (10-71). Note that there is a certain amount of work in arriving at these terms which is not shown here, the finding of the coefficients $a_{n,k}$ from the partial-fraction expansion of $F(s)$. The combination of Eqs. (10-70) and (10-71) is an expression of the *Heaviside expansion theorem.*

Although Eq. (10-71) represents all possible cases when $F(s)$ is rational, there is an important subclassification when s_k is complex. It can be shown that for physical problems $F(s)$ is real. Assume that this is the case, and recall, from Chap. 7, that there must be a conjugate pole at $s_{k+1} = \bar{s}_k$. The coefficients in Laurent expansions about these two poles will be conjugates, giving

$$a_{n,k+1} = \bar{a}_{n,k} \tag{10-72}$$

Consequently, the partial-fraction expansion of $F(s)$ will include the two terms

$$\left[\frac{a_{n,k}}{(s - s_k)^n} + \frac{\bar{a}_{n,k}}{(s - \bar{s}_k)^n} \right] e^{st}$$

In similarity with Eq. (10-71) we get

$$\text{Sum of two residues} = \frac{t^{n-1}}{(n-1)!}\left(a_{n,k}e^{s_k t} + \bar{a}_{n,k}e^{\bar{s}_k t}\right) \qquad (10\text{-}73)$$

This is reduced to a more useful form by writing

$$a_{n,k} = A_{n,k}e^{j\alpha_{n,k}}$$
$$s_k = \sigma_k + j\omega_k$$

with the result

$$\text{Sum of two residues} = \frac{2A_{n,k}t^{n-1}e^{\sigma_k t}}{(n-1)!}\cos\left(\omega_k t + \alpha_{n,k}\right) \qquad (10\text{-}74)$$

You should not necessarily think of Eqs. (10-71) and (10-74) as formulas to be remembered and applied in a "crank-turning" manner. They are important because they show that all systems for which $F(s)$ is a real rational function lead to solutions in t which are in the form of either of these two equations.

The forms of Eqs. (10-73) and (10-74) are the *natural modes* of the classical theory of linear differential equations, and the quantities s_k are the *characteristic values*. Although we are not yet considering the Laplace transform method of solving equations, it is very important to recognize that Eqs. (10-71) and (10-74) include specific directions for finding the numerical coefficients. The fact that the inverse of Eq. (10-70) leads to terms like Eqs. (10-71) and (10-74) is the substance of the Heaviside expansion theorem.

Equations (10-71) and (10-74) have been established as the general terms obtained when the inversion integral is applied to a rational function $F(s)$ of degree in the numerator less than the degree of the denominator. Furthermore, the number of such terms will be finite, and hence their sum will be an entire function which is also of exponential order. These observations can be formalized as follows:

Theorem 10-11.   If $F(s)$ is a rational function for which the degree of the numerator is less than the degree of the denominator, then the inverse transform function $f(t) = \mathcal{L}^{-1}[F(s)]$ exists, is an entire function consisting of a finite sum of products of powers of t by exponentials,* and is of exponential order α_0, where α_0 is the largest real part of all the poles of $F(s)$.

10-21. Evaluating the Inversion Integral When $F(s)$ Is Multivalued. As an example of a function $F(s)$ which is multivalued we shall consider the simple case

$$f(t) = \frac{1}{2\pi j}\int_{\text{Br}} \frac{e^{st}}{\sqrt{s}}\,ds \qquad (10\text{-}75)$$

---

* The sine and cosine functions are classified as exponentials in this statement.

where we understand $\sqrt{s}$ to mean values of $s^{1/2}$ in the same sheet of the
Riemann surface as $\sqrt{\sigma}$. The
function $1/\sqrt{s}$ approaches zero
uniformly with respect to ang s
as $|s|$ becomes infinite, and so
Jordan's lemma is applicable. We
still get a result which is identically zero when $t < 0$. For $t > 0$,
we close the path by a circle to
the left, bypassing the branch cut
in the manner shown in Fig. 10-13.
No singularities are enclosed by
the closed curve consisting of
$C + C_2 + C'$, where C_2 is a semicircle from which the infinitesimal
gap at the branch cut has been
omitted. Thus

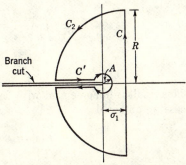

FIG. 10-13. Modification of the integration path when the integrand of the inversion integral has a first-order branch point
at the origin.

$$\lim_{R \to \infty} \left(\int_C \frac{e^{st}}{\sqrt{s}} ds + \int_{C_2} \frac{e^{st}}{\sqrt{s}} ds + \int_{C'} \frac{e^{st}}{\sqrt{s}} ds \right) = 0 \qquad (10\text{-}76)$$

The second integral approaches zero, and so we evaluate the first integral
by evaluating the third. That is, from Eq. (10-76) we have

$$\frac{1}{2\pi j} \int_{\text{Br}} \frac{e^{st}}{\sqrt{s}} ds = - \lim_{R \to \infty} \frac{1}{2\pi j} \int_{C'} \frac{e^{st}}{\sqrt{s}} \qquad (10\text{-}77)$$

This is like Example 3 of Sec. 8-12. Path C' is replaced by a pair of
straight parallel lines plus a small circle of radius A. On the straight
parts of C', $\sqrt{s} = j\sqrt{\rho}$ and $-j\sqrt{\rho}$, respectively, above and below the
branch cut. On circle A, $\sqrt{s} = \sqrt{A}\, e^{j\phi/2}$, where $-\pi < \phi \leqq \pi$. Also,
on each of the straight lines $ds = -d\rho$. Thus, we have

$$\int_{C'} \frac{e^{st}}{\sqrt{s}} ds = -j \int_{R-\sigma_1}^{A} \frac{e^{-\rho t}}{\sqrt{\rho}} (-d\rho) + j \int_A^{R-\sigma_1} \frac{e^{-\rho t}}{\sqrt{\rho}} (-d\rho)$$
$$+ j \frac{A}{\sqrt{A}} \int_{\pi}^{-\pi} \frac{e^{(A \cos \phi)t} e^{j(A \sin \phi)t} e^{j\phi}}{e^{j\phi/2}} d\phi$$

We let A go to zero and R approach infinity and note that the first
two integrals on the right combine into a single integral. The last
integral on the right approaches zero as A goes to zero, in similarity
with Example 3 of Sec. 8-12. Thus, from the above and Eq. (10-77),
we have

$$\frac{1}{2\pi j} \int_{\text{Br}} \frac{e^{st}}{\sqrt{s}} = \frac{1}{\pi} \int_0^{\infty} \frac{e^{-\rho t}}{\sqrt{\rho}} d\rho$$

This integral was previously met in Example 6 of Sec. 10-11. It is evaluated in a routine manner by substituting $\rho t = x^2$, to give

$$\frac{1}{2\pi j} \int_{\text{Br}} \frac{e^{st}}{\sqrt{s}} = \frac{2}{\pi \sqrt{t}} \int_0^\infty e^{-x^2}\, dx$$

$$= \frac{1}{\sqrt{\pi t}} \tag{10-78}$$

This result is in agreement with Example 6 of Sec. 10-11. In the earlier case we obtained

$$F(s) = \mathcal{L}\left(\frac{1}{\sqrt{t}}\right) = \sqrt{\frac{\pi}{s}}$$

Now we have obtained the inverse,

$$f(t) = \mathcal{L}^{-1}\left(\frac{1}{\sqrt{s}}\right) = \frac{1}{\sqrt{\pi t}}$$

Thus, we see that a table of transforms can be evaluated in either way, by finding $\mathcal{L}[f(t)]$ or $\mathcal{L}^{-1}[F(s)]$. It is interesting to observe a significant difference between these two procedures. $\mathcal{L}[f(t)]$ requires a real integration, whereas $\mathcal{L}^{-1}[F(s)]$ leads to an integration in the complex plane. In the latter case the introduction of Jordan's lemma is a natural extension because we are already thinking about integration in a complex plane. But in Chap. 8 we showed how Jordan's lemma can be used to evaluate real integrals, and so we naturally ask whether this powerful method might not be applicable to the integration in $\mathcal{L}[f(t)]$. In some cases it would be applicable, but there are many where it would not. There may be two reasons: $f(t)$ does not necessarily approach zero as t becomes infinite, and $f(t)$ is not necessarily capable of an analytic continuation. The latter would be the case, for example, if $f(t)$ should consist of a sequence of sections of straight lines.

PROBLEMS

10-1. Determine which of the following functions are of exponential order, and, for those which are, determine α_0. Also, determine whether or not α_0 is included in the set of values of α for which

$$\lim_{t \to \infty} f(t)e^{-\alpha t} = 0$$

(a) $f(t) = \dfrac{1}{\sqrt{t}}$

(b) $f(t) = e^{\sqrt{t^3}}$

(c) $f(t) = \sin t^2$

(d) $f(t) = e^{3t}$

(e) $f(t) = t^n$

(f) $f(t) = \sqrt{|\tan t|}$

10-2. Do Prob. 10-1 for the functions

(a) $f(t) = \dfrac{\sin t}{t}$ $\qquad\qquad\qquad$ (b) $f(t) = \sin^2 t$

(c) $f(t) = \dfrac{1}{\sqrt{t^3}}$ $\qquad\qquad\qquad$ (d) $f(t) = \log t$

(e) $f(t) = \dfrac{\log t}{\sqrt{t^3}}$ $\qquad\qquad\qquad$ (f) $f(t) = e^{t \log t}$

10-3. In Sec. 10-4 it is established that the single-sided Laplace integral converges for Re $s > \alpha_0$, rather than Re $s \geqq \alpha_0$. Is the absence of an equality sign due to the fact that for some functions α_0 is not included in the set of α numbers for which $\lim\limits_{t \to \infty} f(t)e^{-\alpha t} = 0$, or is it because

$$\int_0^\infty f(t)e^{-\alpha_0 t}\, dt$$

might not exist even if $\lim\limits_{t \to \infty} f(t)e^{-\alpha_0 t} = 0$?

10-4. Let $f(t)$ be a complex function of the real variable t. A Laplace integral can be defined for such a function, and assume that it converges at a point s_0. Determine a region of convergence for such a function.

10-5. Let $f(t)$ be a real function of t, meaning that $f(t)$ is real when t is real. Now suppose that t is allowed to become complex, using the notation

$$t = re^{j\theta}$$

and define a generalized Laplace integral

$$\int_C f(t)e^{-st}\, dt$$

where C is a straight line in the t plane radiating at angle θ_0 and extending to infinity. If this integral converges at s_0, determine a region of convergence.

10-6. State and justify the regions of convergence and uniform convergence of the one-sided Laplace integral, for the following functions:

(a) $f(t) = e^{-t^2}$ $\qquad\qquad\qquad$ (b) $f(t) = \dfrac{\sin t}{t}$

(c) $f(t) = \sin t^2$ $\qquad\qquad\qquad$ (d) $f(t) = e^{\sin t}$

(e) $f(t) = \log t$ $\qquad\qquad\qquad$ (f) $f(t) = t^n$

10-7. Suppose that $f(t)$ is known to be EO,α_0. Prove that $e^{at}f(t)$, where a is real, is EO,$\alpha_0 + a$.

10-8. Let a function $f(t)$ be defined as follows:

$$f(t) = (-1)^n n \qquad \log \log n < t < \log \log (n + 1)$$

where n represents a succession of integers. Determine whether or not this function is of exponential order, and determine the region of convergence of its Laplace integral.

10-9. Determine the strip of convergence of the two-sided Laplace integral, for the following:

(a) $f(t) = e^{-|t|}$ $\qquad\qquad$ (b) $f(t) = \dfrac{\sin t}{t}$ $\qquad\qquad$ (c) $f(t) = e^{-t^2}$

10-10. In forming the Laplace integral, it is not necessary to define $f(t)$ as identically zero for $t < 0$ or to introduce the factor $u(t)$. Explain why it is nevertheless customary to do so, explaining what conceptual simplification is thereby attained.

10-11. Explain the difference between the notations

$$F(s) = \int_0^\infty f(t)e^{-st}\, dt \qquad \text{Re } s > \sigma_c$$

and

$$F(s) = \int_0^\infty f(t)e^{-st}\, dt$$

explaining which is correct. Also, is it your understanding that $\mathcal{L}[f(t)]$ is a symbol for $F(s)$ or for the integral?

10-12. What is the significance in knowing that the Laplace integral converges in a right half plane, if it converges at all, in determining the properties of $F(s)$?

10-13. Give an example to show that, if $f(t)$ and $g(t)$ have Laplace transforms, it is not necessarily true that $f(t)g(t)$ has a Laplace transform. State and prove conditions on $f(t)$ and $g(t)$ which will ensure that $f(t)g(t)$ will have a transform.

10-14. For each of the following cases, determine the strips of convergence for the two-sided Laplace transforms of $f(t)$ and $g(t)$, and determine whether or not $f(t) + g(t)$ has a two-sided transform:

(a) $f(t) = e^{-|t|}$
  $g(t) = \dfrac{\sin t}{t}$

(b) $f(t) = \begin{cases} e^{-t} & t > 0 \\ e^{-0.5t} & t < 0 \end{cases}$
  $g(t) = e^{-0.5|t|}$

(c) $f(t) = \dfrac{1}{1 + t^2}$
  $g(t) = \begin{cases} e^t & t > 0 \\ e^{2t} & t < 0 \end{cases}$

10-15. Using trigonometric identities and evaluating the Laplace integral, find the Laplace transforms of the following functions:

(a) $f(t) = \sin at \cos bt$
(c) $f(t) = \sin^2 bt$
(b) $f(t) = \cos^2 bt$
(d) $f(t) = t^2 e^{at}$

10-16. A sinusoidal wave train of $2N$ cycles duration is described by the function

$$f(t) = \begin{cases} \sin \omega_0 t & |t| < \dfrac{2\pi N}{\omega_0} \\ 0 & |t| > \dfrac{2\pi N}{\omega_0} \end{cases}$$

(a) Find the two-sided Laplace transform of this function.

(b) Now suppose that the same wave train occurs completely after $t = 0$, being translated to the right to occur in the interval $0 < t < 4\pi N/\omega_0$. Outside this interval the function is zero. What is the one-sided transform of this function? Compare it with the answer to part a.

10-17. Obtain a formula for the Laplace transform of the binomial

$$f(t) = (t^2 - 1)^n$$

10-18. Using the shifting theorem appropriately and any appropriate information in Table 10-1, obtain the Laplace transforms of the following:

(a) $f(t) = \begin{cases} t & 0 < t < 1 \\ 1 & 1 < t < 2 \\ 3 - t & 2 < t < 3 \\ 0 & 3 < t \end{cases}$

(b) $f(t) = \begin{cases} 1 - e^{-t} & 0 < t < 1 \\ \dfrac{e-1}{e} e^{-t} & 1 < t \end{cases}$

(c) $f(t) = \begin{cases} \sin bt & 0 < t < \dfrac{\pi}{b} \\ 0 & \dfrac{\pi}{b} < t < \dfrac{2\pi}{b} \\ \sin bt & \dfrac{2\pi}{b} < t < \dfrac{3\pi}{b} \\ 0 & \dfrac{3\pi}{b} < t \end{cases}$

(d) $f(t) = \begin{cases} t & 0 < t < 1 \\ -1 + t & 1 < t < 2 \\ -2 + t & 2 < t < 3 \\ -3 + t & 3 < t < 4 \\ 0 & 4 < t \end{cases}$

10-19. Discuss the difference between the Laplace integral and the Laplace transform. Answer such questions as: Which has an abscissa of convergence? Which is an analytic function? Which has singular points? Suppose that you were given a Laplace transform, what would you do to get the corresponding Laplace integral? Is there any difference in your answer to the last question if it applies to the two-sided transform rather than to the single-sided transform?

10-20. Let $F(s)$ be the Laplace transform of $f(t)$, and define

$$F_0(s) = \int_0^T f(t)e^{-st}\, dt$$

Show that

$$\mathcal{L}[f(t + T)] = e^{sT}[F(s) - F_0(s)]$$

Test this result on the function $f(t) = \sin(t + \alpha)$.

10-21. Find the Laplace transform of

$$f(t) = (1 + e^{\alpha t})^n$$

10-22. Use appropriate theorems to find the Laplace transforms of the following (information in Table 10-1 may be used):

(a) $f(t) = \sin bt \cos bt$ (b) $f(t) = t^2 \cos bt$

(c) $f(t) = t^2 \sin^2 bt$ (d) $f(t) = (1 - e^{-\alpha t})^2 (1 + t)^2$

10-23. Prove that

$$f(t) = e^{t + e^t} \sin e^{e^t}$$

has a Laplace transform.

10-24. Beginning with the formula for the nth-order Legendre polynomial

$$P_n(t) = \frac{1}{2^n n!} \frac{d^n}{dt^n}(t^2 - 1)^n$$

show that

$$\mathcal{L}[P_n(t)] = \frac{1}{2^n} \sum_{k=\left[\frac{n+1}{2}\right]}^{n} \frac{(-1)^{k+n}(2k)!}{k!(n-k)! s^{2k-n+1}}$$

where $[(n+1)/2]$ is the greatest integer in $(n+1)/2$. What is the region of convergence of the Laplace integral?

10-25. Prove that, if $\mathcal{L}[f(t)] = F(s)$ and if $f(t)/t$ has a transform $G(s)$, then

$$G(s) = \int_s^\infty F(z)\, dz$$

where the path of integration lies in the half plane of convergence.

10-26. Show that

(a) $\mathcal{L}\left(\dfrac{1 + 2bt}{\sqrt{t}}\, e^{bt}\right) = \dfrac{\sqrt{\pi}\, s}{\sqrt{(s + b)^3}}$

(b) $\mathcal{L}\left[\dfrac{1}{\sqrt{t^3}}\, (e^{bt} - e^{at})\right] = 2\sqrt{\pi}\,(\sqrt{s - a} - \sqrt{s - b})$

by starting with $\mathcal{L}(1/\sqrt{t})$ and with the aid of suitable theorems.

10-27. Starting with the formula for the gamma function

$$\Gamma(x) = \int_0^\infty e^{-t} t^{x-1}\, dt$$

show that

$$\mathcal{L}(\log t) = \frac{\Gamma'(1) - \log s}{s}$$

and determine the region of convergence of the Laplace integral.   (HINT: Recall that $dt^x/dx = t^x \log t$.)

10-28. Prove that

$$\mathcal{L}\left(\frac{1 - e^{-at}}{t}\right) = \log\left(1 + \frac{a}{s}\right)$$

(a) Using Theorem 10-5.

(b) Using the property stated in Prob. 10-25.

What is the region of convergence of the Laplace integral?

10-29. From the known transform

$$\mathcal{L}(\sin bt) = \frac{b}{s^2 + b^2}$$

show that

$$\mathcal{L}\left(\frac{\sin bt}{t}\right) = \tan^{-1}\frac{b}{s}$$

and determine the region of convergence of the Laplace integral.

10-30. Consider the Laplace transform of $(\sin bt)/t$ given in Prob. 10-29.   With due consideration of multivaluedness, consider $F_1(s)$ and $F_1(-s)$ as s approaches the $j\omega$ axis respectively from the right and left half planes, and observe whether the sum of these is the same as the Fourier transform given in Prob. 9-24.

10-31. Show that, for real $a > 0$,

$$\mathcal{L}(e^{-at^2}) = \frac{1}{2}\sqrt{\frac{\pi}{a}}\, e^{s^2/4a}\left[1 - \operatorname{erf}\left(\frac{s}{2\sqrt{a}}\right)\right]$$

where

$$\operatorname{erf}(x) = \frac{2}{\sqrt{\pi}}\int_0^x e^{-y^2}\, dy$$

What is the region of convergence of the Laplace integral?   [HINT: Obtain the integral representation for $dF(s)/ds$, and integrate by parts, thereby obtaining a differential equation.]

10-32. Consider the Laplace transform of e^{-at^2} given in Prob. 10-31. From this result, obtain the two-sided Laplace transform of this function, and specify the strip of convergence of the two-sided Laplace integral. Compare the result with the Fourier transform given in Prob. 9-9.

10-33. Prove that

$$\mathcal{L}\left(\frac{\cos b \sqrt{t}}{\sqrt{t}}\right) = \sqrt{\frac{\pi}{s}}\, e^{-b^2/4s}$$

and determine the region of convergence of the Laplace integral. [HINT: We have

$$\int_0^\infty \frac{\cos b \sqrt{t}}{\sqrt{t}}\, e^{-\sigma t}\, dt = \int_{-\infty}^\infty e^{-\sigma u^2} e^{jbu}\, du$$

and this is recognized as the Fourier integral $\mathcal{F}(-jb)$ of $e^{-\sigma u^2}$, which is treated in Prob. 9-9.]

10-34. From the transform given in Prob. 10-33, show that

$$\mathcal{L}(\sin b \sqrt{t}) = \frac{b}{2} \sqrt{\frac{\pi}{s^3}}\, e^{-b^2/4s}$$

What is the region of convergence of the Laplace integral?

10-35. Obtain the Laplace transform

$$\mathcal{L}[J_0(at)] = \frac{1}{\sqrt{a^2 + s^2}}$$

where $J_0(at)$ is the zeroth-order Bessel function of the first kind, given by

$$J_0(at) = \frac{1}{\pi}\int_0^\pi \cos(at \cos \phi)\, d\phi$$

What is the region of convergence of the Laplace integral? Do this by forming the Laplace integral, justifying an interchange of order of integration, giving an integral in terms of $\cos^2 \phi$. The variable change

$$z = \frac{1 - \cos \phi}{1 + \cos \phi}$$

will yield an integral from 0 to ∞ in the variable z, which can then be regarded as complex. With due regard to properties of a Riemann surface and the calculus of residues, the required answer is obtained.

10-36. Starting with the Laplace transform of e^{-t^2}, as given in Prob. 10-31, and recalling that the error function is

$$\text{erf } t = \frac{2}{\sqrt{\pi}}\int_0^t e^{-x^2}\, dx$$

obtain the transform of erf (at), and determine the region of convergence of the Laplace integral for this function.

10-37. Starting with the Laplace transform of e^{-t^2}, given in Prob. 10-31, derive the Laplace transform

$$\mathcal{L}\left(\frac{e^{-a\sqrt{t}}}{\sqrt{t}}\right) = \sqrt{\frac{\pi}{s}}\, e^{a^2/4s}\left(1 - \text{erf}\frac{a}{2\sqrt{s}}\right)$$

and specify the region of convergence of the Laplace integral.

10-38. Referring to the information given in Prob. 10-31, show that

$$\mathcal{L}(e^{-a\sqrt{t}}) = \frac{a}{2} \sqrt{\frac{\pi}{s^3}} e^{a^2/4s} \left(\operatorname{erf} \frac{a}{2\sqrt{s}} - 1 \right) + \frac{1}{s}$$

and specify the region of convergence of the Laplace integral.

10-39. Referring to the information given in Prob. 10-29, show that

$$\mathcal{L}[\operatorname{Si}(at)] = \frac{1}{s} \tan^{-1} \frac{b}{s}$$

where

$$\operatorname{Si} x = \int_0^x \frac{\sin y}{y} \, dy$$

What is the region of convergence of the Laplace integral?

10-40. Evaluate the inversion integral, to show that

$$\frac{1}{2\pi j} \int_{\text{Br}} \frac{e^{st}}{\sqrt{a^2 + s^2}} \, ds = \begin{cases} 0 & t < 0 \\ J_0(at) & t > 0 \end{cases}$$

See Prob. 10-35 for the formula for $J_0(at)$.

10-41. This is a generalization of Prob. 10-40. The formula

$$J_n(at) = \frac{(-j)^n}{\pi} \int_0^\pi e^{jat \cos \phi} \cos n\phi \, d\phi$$

gives the nth-order Bessel function of the first kind. Evaluate the inversion integral to show that, for $n \geq 0$,

$$\mathcal{L}[J_n(at)] = \frac{(\sqrt{s^2 + a^2} - s^2)^n}{a^n \sqrt{s^2 + a^2}}$$

10-42. By use of the inversion integral, show that

$$\mathcal{L}[J_0(a\sqrt{t})] = \frac{1}{s} e^{-a^2/4s}$$

Refer to Prob. 10-35 for the definition of $J_0(x)$.

10-43. By use of the inversion integral, show that

$$\mathcal{L}\left(\frac{1}{\sqrt{t^3}} e^{-a^2/4t} \right) = \frac{2\sqrt{\pi}}{a} e^{-a\sqrt{s}}$$

where $a > 0$. Also, from this result, show that

$$\mathcal{L}\left(1 - \operatorname{erf} \frac{a}{2\sqrt{t}} \right) = \frac{1}{s} e^{-a\sqrt{s}}$$

10-44. The function

$$F(s) = \frac{2s \cosh as + 2 \sinh as}{s^2 - 1}$$

can be a two-sided transform, but not a single-sided transform. Explain how this is known. Also, using three Br paths, to the right of both singular points, between singular points, and to the left of both singular points, obtain three functions for which the above is the two-sided Laplace transform.

10-45. Obtain inverse transforms for the following:

(a) $F(s) = \dfrac{6s^2 + s - 1}{s^3 + s}$ (b) $F(s) = \dfrac{4s^2 + 16s + 16}{s^3 + 5s^2 + 9s + 5}$

(c) $F(s) = \dfrac{3s^3 + 8s^2 + 9s + 4}{s^4 + 5s^3 + 9s^2 + 7s + 2}$

10-46. If $F(s) = \mathcal{L}[f(t)]$, prove that

$$\mathcal{L}[f(at)] = \frac{1}{a} F\left(\frac{s}{a}\right)$$

(a) From the defining integral.
(b) From the inversion integral.

CHAPTER 11

CONVOLUTION THEOREMS

11-1. Introduction. In the application of the Laplace transform to the solution of the equations of linear systems one encounters products like $F(s)G(s)$, where each of these two functions is the Laplace transform of a function of t. You can refer to Chap. 1, in which a preview is given of how this comes about. In general at least one of these functions appearing in the product is a so-called "system function," the steady-state sinusoidal response function analytically continued into the s plane. The other function in the product is often the Laplace transform of the driving function. However, it sometimes occurs that each of the factors is a system function, as when two systems are combined in such a way that the over-all system function is the product of two individual system functions. This is the case, for example, when two networks are cascaded under conditions where the second network does not load the first. This situation arises in the design of corrective networks, where one of the factors is the system function of a given system and the other is the function for the corrective network.

If one arrives at the point of having a function $F(s)G(s)$ for which the inverse transform is wanted, the various techniques of finding an inverse are of course available. However, here we are not interested in solutions for specific problems; rather, we are interested in finding out how the inverse

$$\mathcal{L}^{-1}[F(s)G(s)]$$

is related to the individual inverse functions

$$f(t) = \mathcal{L}^{-1}[F(s)] \quad \text{and} \quad g(t) = \mathcal{L}^{-1}[G(s)]$$

By having such a relationship it is sometimes possible to arrive at general theorems and properties of systems without the need for specific solutions.

In dealing with systems which are essentially linear but which have isolated nonlinear operations, like modulation and demodulation, we can have products like $f(t)g(t)$. In this connection we are interested in knowing the relationship between the Laplace transform

$$\mathcal{L}[f(t)g(t)]$$

and the individual Laplace transforms

$$F(s) = \mathcal{L}[f(t)] \quad \text{and} \quad G(s) = \mathcal{L}[g(t)]$$

It would require a wide digression into the entire field of applications of the Laplace transform to provide much more motivation than this for the subject at hand. It is hoped that this brief introduction will serve to show that there is some reason to consider these two questions.

11-2. Convolution in the t Plane (Fourier Transform). In Sec. 9-5 you will find a development showing that, if $f(t)$ and $g(t)$ are two functions, one of which is PC and bounded and the other APC, and both of which are absolutely integrable from $-\infty$ to ∞, then the function

$$r(t) = \int_{-\infty}^{\infty} f(\tau)g(t + \tau) \, d\tau \tag{11-1}$$

has a Fourier integral which converges to

$$\int_{-\infty}^{\infty} r(t)e^{-j\omega t} \, dt = \mathfrak{F}(-j\omega)\mathcal{G}(j\omega) \tag{11-2}$$

where

$$\mathfrak{F}(j\omega) = \int_{-\infty}^{\infty} f(t)e^{-j\omega t} \, dt \quad \mathcal{G}(j\omega) = \int_{-\infty}^{\infty} g(t)e^{-j\omega t} \, dt \tag{11-3}$$

These results are obtained specifically from Eq. (9-34) and the various defining expressions for $r(t)$, $\mathfrak{F}(j\omega)$, and $\mathcal{G}(j\omega)$. Furthermore, in the footnote on page 281, it is pointed out that the above is still true if $f(t)$ and $g(t)$ are both APC, if one of them remains bounded as $|t|$ becomes infinite.

Equation (11-2) serves as the starting point for finding the inverse of $\mathfrak{F}(j\omega)\mathcal{G}(j\omega)$. The similarity with the right side of Eq. (11-2) is at once apparent, but a sign change is needed in the argument of $\mathfrak{F}(-i\omega)$.

Instead of Eq. (11-1) consider the function

$$w(t) = \int_{-\infty}^{\infty} f(-\tau)g(t + \tau) \, d\tau \tag{11-4}$$

In doing so we note that $f(-t)$ also satisfies the Fourier integral theorem and has a Fourier transform obtained by the following sequence of steps:

$$\int_{-\infty}^{\infty} f(-t)e^{-j\omega t} \, dt = \int_{-\infty}^{\infty} f(t)e^{j\omega t} \, dt = \mathfrak{F}(-j\omega) \tag{11-5}$$

We can also say that $f(-t)$ satisfies the same conditions as $f(t)$, and therefore the steps used in Sec. 9-5 to show that $r(t)$ exists are also applicable to show that $w(t)$ exists. Now we write Eq. (11-2), with $w(t)$ replacing $r(t)$; and from Eq. (11-5) it is evident that $\mathfrak{F}(-j\omega)$ replaces $\mathfrak{F}(j\omega)$. Therefore, we get

$$\mathfrak{F}(j\omega)\mathcal{G}(j\omega) = \int_{-\infty}^{\infty} w(t)e^{-j\omega t} \, dt \tag{11-6}$$

showing that the product $\mathcal{F}(j\omega)\mathcal{G}(j\omega)$ is the Fourier transform of $w(t)$. It also follows that

$$w(t) = \frac{1}{2\pi} PV \int_{-\infty}^{\infty} \mathcal{F}(j\omega)\mathcal{G}(j\omega)e^{j\omega t}\,d\omega \qquad (11\text{-}7)$$

The formula for $w(t)$ in Eq. (11-4) can be put into a more convenient form by changing the variable of integration from τ to $-\tau$, with the result

$$w(t) = \int_{-\infty}^{\infty} f(\tau)g(t-\tau)\,d\tau \qquad (11\text{-}8)$$

The integral in Eq. (11-8) is called a *convolution integral*. Later on other types of convolution integrals will be defined. We now state this result formally as the following theorem:

Theorem 11-1. Let $f(t)$ and $g(t)$ each be APC, while one of them is bounded as $|t|$ becomes infinite and has at most a finite number of infinite discontinuities, and let both be absolutely integrable from $-\infty$ to ∞, with respective Fourier transforms $\mathcal{F}(j\omega)$ and $\mathcal{G}(j\omega)$. Then,

$$w(t) = \int_{-\infty}^{\infty} f(\tau)g(t-\tau)\,d\tau$$

is an APC function of t, and is continuous if either $f(t)$ or $g(t)$ is PC, and its Fourier integral converges to

$$\mathcal{F}(j\omega)\mathcal{G}(j\omega)$$

11-3. Convolution in the t Plane (Two-sided Laplace Transform). Let $f(t)$ and $g(t)$ be APC functions for which the integrals

$$\int_{-\infty}^{\infty} |f(t)|e^{-\sigma t}\,dt \qquad \text{and} \qquad \int_{-\infty}^{\infty} |g(t)|e^{-\sigma t}\,dt$$

have strips of convergence which overlap in a strip designated by

$$\sigma_a < \text{Re } s < \sigma_b$$

Also assume that there is a number σ_1 between σ_a and σ_b such that $|g(t)|e^{-\sigma_1 t}$ is bounded as $|t|$ becomes infinite, and is integrable over the infinite interval. Note that

$$\left| \int_{-\infty}^{\infty} f(t)e^{-st}\,dt \right| \leqq \int_{-\infty}^{\infty} |f(t)e^{-st}|\,dt = \int_{-\infty}^{\infty} |f(t)|e^{-\sigma t}\,dt$$

and since it is stipulated that the integral on the right shall converge at least in the strip $\sigma_a < \sigma < \sigma_b$, it is concluded that the integral

$$\int_{-\infty}^{\infty} f(t)e^{-st}\,dt$$

must certainly converge in this same strip. A similar statement applies to the corresponding integral of $g(t)$. Therefore, we can define a pair of two-sided Laplace transforms as follows,

$$\mathfrak{F}(s) = \int_{-\infty}^{\infty} f(t)e^{-st}\, dt \qquad \mathfrak{G}(s) = \int_{-\infty}^{\infty} g(t)e^{-st}\, dt \qquad (11\text{-}9)$$

with assurance that each defining integral will converge in the strip

$$\sigma_a < \sigma < \sigma_b \qquad (11\text{-}10)$$

Using the previously defined number σ_1, we write the above transforms as explicit functions of ω, as follows:

$$\mathfrak{F}(\sigma_1 + j\omega) = \int_{-\infty}^{\infty} f(t)e^{-\sigma_1 t}e^{-j\omega t}\, dt \qquad \mathfrak{G}(\sigma_1 + j\omega) = \int_{-\infty}^{\infty} g(t)e^{-\sigma_1 t}e^{-j\omega t}\, dt$$
$$(11\text{-}11)$$

These are, respectively, the Fourier transforms of

$$f(t)e^{-\sigma_1 t} \qquad \text{and} \qquad g(t)e^{-\sigma_1 t}$$

Furthermore, because σ_1 lies in the common strip of convergence of the two-sided Laplace integrals of $|f(t)|$ and $|g(t)|$, it follows that

$$|f(t)|e^{-\sigma_1 t} \qquad \text{and} \qquad |g(t)|e^{-\sigma_1 t}$$

are integrable over the infinite interval. It is now evident that $f(t)e^{-\sigma_1 t}$ and $g(t)e^{-\sigma_1 t}$ satisfy the conditions specified for $f(t)$ and $g(t)$ in Theorem 11-1 and that they can be, respectively, substituted in that theorem. The conclusion is that the integral

$$\int_{-\infty}^{\infty} f(\tau)e^{-\sigma_1 \tau}g(t - \tau)e^{-\sigma_1 t}e^{\sigma_1 \tau}\, d\tau = e^{-\sigma_1 t}\int_{-\infty}^{\infty} f(\tau)g(t - \tau)\, d\tau \qquad (11\text{-}12)$$

converges and has the Fourier transform

$$\mathfrak{F}(\sigma_1 + j\omega)\mathfrak{G}(\sigma_1 + j\omega)$$

The integral on the right of Eq. (11-12) has previously been defined as $w(t)$. Thus, we can write explicitly

$$\mathfrak{F}(\sigma_1 + j\omega)\mathfrak{G}(\sigma_1 + j\omega) = \int_{-\infty}^{\infty} w(t)e^{-\sigma_1 t}e^{-j\omega t}\, dt \qquad (11\text{-}13)$$

The function on the left can be analytically continued, and it is regular in the vertical strip between σ_a and σ_b. Thus, the integral on the right converges in this strip, and we have

$$\mathfrak{F}(s)\mathfrak{G}(s) = \int_{-\infty}^{\infty} w(t)e^{-st}\, dt \qquad \sigma_a < \mathrm{Re}\ s < \sigma_b \qquad (11\text{-}14)$$

We also know that $w(t)$ can be obtained from $\mathfrak{F}(s)\mathfrak{G}(s)$, from the inversion integral.

These results are now stated in the form of a theorem, as follows:

Theorem 11-2.   Let $f(t)$ and $g(t)$ each be APC, and let $|f(t)|$ and $|g(t)|$ have convergent two-sided Laplace integrals, both converging in a strip

$$\sigma_a < \operatorname{Re} s < \sigma_b$$

Furthermore, let

$$\mathfrak{F}(s) = \mathfrak{L}_2[f(t)] \qquad \text{and} \qquad \mathfrak{G}(s) = \mathfrak{L}_2[g(t)]$$

and assume there is a number σ_1 in the above range such that $|g(t)|e^{-\sigma_1 t}$ is bounded as $|t|$ becomes infinite.   It is then true that

$$w(t) = \int_{-\infty}^{\infty} f(\tau)g(t-\tau)\,d\tau$$

is an APC function of t and is continuous if $g(t)$ is PC, and that

$$\mathfrak{F}(s)\mathfrak{G}(s) = \int_{-\infty}^{\infty} w(t)e^{-st}\,dt$$

where convergence is in the strip $\sigma_a < \operatorname{Re} s < \sigma_b$; and, finally,

$$w(t) = \frac{1}{2\pi j} \int_{\text{Br}} \mathfrak{F}(s)\mathfrak{G}(s)e^{st}\,ds$$

where the path Br lies in the strip specified.

If $f(t)$ and $g(t)$ have converging Fourier integrals, there is no difference between Theorems 11-1 and 11-2, except for the difference in notation. Then the Fourier transforms are the same as the two-sided Laplace transforms.   However, Theorem 11-2 is more general, since it applies to a larger class of functions than does Theorem 11-1.   It must be noted, however, that even in Theorem, 11-2 $f(t)$ and $g(t)$ are still somewhat restricted, to the extent that their two-sided Laplace integrals of their absolute values must have overlapping strips of convergence.

This restriction on $f(t)$ and $g(t)$ is important, as will be shown by the following example: Consider the two cases shown in Fig. 11-1.   For case *a*

$$f(t) = u(t) \qquad \text{and} \qquad \mathfrak{F}(s) = \frac{1}{s}$$

$$g(t) = 2u(-t) \qquad \text{and} \qquad \mathfrak{G}(s) = -\frac{2}{s}$$

and therefore

$$\mathfrak{F}(s)\mathfrak{G}(s) = -\frac{2}{s^2}$$

which is a perfectly good analytic function having two possible inverse functions, depending on whether the Br path would be to the right or to the left of the origin. However, in this case, the formal expression

$$w(t) = 2 \int_{-\infty}^{\infty} u(\tau)u(-t+\tau) \, d\tau = 2 \int_{0}^{\infty} u(\tau - t) \, d\tau$$

leads to a divergent integral. This becomes clearly evident when we recognize the properties of $u(t)$, giving

$$2 \int_{0}^{\infty} d\tau \qquad t < 0 \qquad\qquad 2 \int_{t}^{\infty} d\tau \qquad 0 < t$$

when we attempt to evaluate the integral. Neither of the above integrals exists, and so $w(t)$ does not exist. Theorem 11-2 has failed in this case,

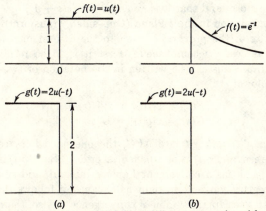

Fig. 11-1. Examples of functions which have different properties with respect to the conditions of Theorem 11-2. (a) A case which does not satisfy the theorem; (b) a case which does satisfy the theorem.

and an inspection of the original functions shows why. The two-sided Laplace integral of $|f(t)|$ converges to the right of the $j\omega$ axis, and the corresponding integral of $|g(t)|$ converges to the left of the $j\omega$ axis. There is no overlapping strip of convergence, as required in the theorem.

In case b, $g(t)$ is the same as before, and

$$f(t) = e^{-t}u(t)$$

The two-sided Laplace integral of $|f(t)|$ now converges to the right of the point -1, and so the two integrals have a common strip of convergence in the range

$$-1 < \operatorname{Re} s < 0$$

In this case

$$f(t) = e^{-t}u(t) \qquad \text{and} \qquad \mathfrak{F}(s) = \frac{1}{s+1}$$

and $\mathcal{G}(s)$ is the same as before, giving the product

$$\mathcal{F}(s)\mathcal{G}(s) = -\frac{2}{s(s+1)}$$

Now $w(t) = 2\int_0^\infty e^{-\tau}u(\tau - t)\,d\tau = \begin{cases} 2\int_0^\infty e^{-\tau}\,d\tau & t < 0 \\ 2\int_t^\infty e^{-\tau}\,d\tau & 0 < t \end{cases}$

and the result is

$$w(t) = \begin{cases} 2 & t < 0 \\ 2e^{-t} & 0 < t \end{cases}$$

A simple check shows that this function does indeed have a two-sided Laplace transform and that it is given by $-2/s(s+1)$.

11-4. Convolution in the t Plane (One-sided Transform). Extension of the results of the previous section to the case of one-sided Laplace transforms is easily accomplished. First it is to be recalled that the one-sided transforms can be written in the notation of two-sided transforms as follows:

$$\begin{aligned} F(s) &= \mathcal{L}[f(t)] = \mathcal{L}_2[f(t)u(t)] \\ G(s) &= \mathcal{L}[g(t)] = \mathcal{L}_2[g(t)u(t)] \end{aligned} \tag{11-15}$$

Assume that $f(t)$ and $g(t)$ are APC, the one-sided Laplace integral of $|f(t)|$ converges, and $g(t)$ is of exponential order. Since $g(t)$ is of exponential order, $|g(t)|$ has a convergent Laplace integral, and since one-sided Laplace integrals converge in right half planes, it follows that there will always be an overlapping region of convergence. Also, there will always be a number σ_1 in this region for which $|g(t)|e^{-\sigma_1 t}$ approaches zero as t becomes infinite. The conditions of Theorem 11-2 are thus satisfied by $f(t)u(t)$ and $g(t)u(t)$, and so we know that

$$w(t) = \int_{-\infty}^\infty f(\tau)u(\tau)g(t-\tau)u(t-\tau)\,d\tau \tag{11-16}$$

is APC and that this function has a two-sided Laplace transform given by $F(s)G(s)$.

Now observe that

$$\begin{aligned} u(\tau) &= 0 & \tau < 0 \\ u(t - \tau) &= 0 & t < \tau \end{aligned}$$

Therefore, the integral in Eq. (11-16) reduces to

$$w(t) = \begin{cases} \int_0^t f(\tau)g(t-\tau)\,d\tau & 0 < t \\ 0 & t < 0 \end{cases} \tag{11-17}$$

Since $w(t)$ is zero for negative t, the above statement that $F(s)G(s)$ is the two-sided Laplace transform of $w(t)$ can be revised to state that $F(s)G(s)$ is the one-sided Laplace transform of $w(t)$.

The results presented above are formally stated as follows:

Theorem 11-3. Let $f(t)$ and $g(t)$ each be APC, where $|f(t)|$ has a convergent one-sided Laplace integral and $g(t)$ is of exponential order. If we use the notation

$$F(s) = \mathcal{L}[f(t)] \qquad G(s) = \mathcal{L}[g(t)]$$

then the integral

$$h(t) = \int_0^t f(\tau)g(t - \tau)\, d\tau \qquad (11\text{-}18)$$

is an APC function of t, is continuous if $g(t)$ is PC, and has a single-sided Laplace transform given by $F(s)G(s)$.

The integral on the right of Eq. (11-18) is also called a *convolution integral*. It is perhaps more frequently met in practice than the one given in Eq. (11-8). Your understanding of the convolution integral will be enhanced if you will pick some simple $f(t)$ and $g(t)$ functions and make graphs of the integrand for various fixed values of t.

It is possible that you may have come to regard the one- and two-sided Laplace transforms as quite similar, and if so it may be surprising to find a significant difference in these two convolution integrals, namely, the appearance of t in the limit of one and not of the other. It is possible to get a degree of correlation between this difference and the severity of restrictions on $f(t)$ and $g(t)$. For the case of Theorem 11-3 it has been pointed out that, if $|f(t)|$ and $|g(t)|$ have convergent one-sided Laplace integrals, an overlapping region of convergence is automatically obtained. On the other hand, in the case of Theorem 11-2, convergence of the appropriate integrals does not in itself ensure an overlapping region of convergence. Thus, $f(t)$ and $g(t)$ are more severely restricted in Theorem 11-2, on the basis of condition of convergence of their Laplace integrals.

11-5. Convolution in the s Plane (One-sided Transform). In view of similarities between the Laplace integral and the inversion integral, we might expect to have a theorem relating the Laplace transform of the product of two functions with a convolution integral of their transforms. If we were interested in analytic t functions, it would be a simple matter to establish such a theorem by reference to the previous results. However, many practical applications involve functions of t which are not analytic. We want at least to admit the piecewise continuous functions of exponential order. An adequate treatment of s-plane convolution for functions of this generality would require an unreasonably extensive digression into further theory of improper integrals, and particularly the Fourier integral. In view of these realities, we shall not attempt here to establish conditions and a rigorous proof of a theorem on s-plane convolution. Instead, we shall develop the subject in a purely formal way, to establish the forms of the formulas, and to locate the pertinent contours of integration.

We begin by considering two functions $f(t)$ and $g(t)$ which have, respectively, single-sided transforms $F(s)$ and $G(s)$. In the single-sided Laplace integral of the product, namely,

$$\int_0^\infty f(t)g(t)e^{-st}\,dt$$

we replace $g(t)$ by its inversion integral

$$g(t) = \frac{1}{2\pi j} \int_{Br_1} G(z)e^{zt}\,dz \tag{11-19}$$

where Br_1 has an abscissa x_1 in the z plane. If σ_g is the abscissa of convergence of $g(t)$, then we know that $x_1 > \sigma_g$. We now have

$$\int_0^\infty f(t)g(t)e^{-st}\,dt = \frac{1}{2\pi j} \int_0^\infty f(t)\,dt \int_{Br_1} G(z)e^{(z-s)t}\,dz \tag{11-20}$$

An interchange of order of integration is anticipated on the right. This would be justifiable if $G(z)$ had an absolutely convergent integral over the Br_1 path. However, we do not have this property in general, and so we proceed to make this interchange without justification, with the result

$$\int_0^\infty f(t)g(t)e^{-st}\,dt = \frac{1}{2\pi j} \int_{Br_1} G(z)\,dz \int_0^\infty f(t)e^{-(s-z)t}\,dt \tag{11-21}$$

For the integral on the right of Eq. (11-21) we have

$$F(s - z) = \int_0^\infty f(t)e^{-(s-z)t}\,dt \qquad \text{Re } (s - z) > \sigma_f \tag{11-22}$$

where σ_f is the abscissa of convergence of $f(t)$. Since Re $z = x_1$ for the z-plane integration, it follows that if

$$\text{Re } s > \sigma_f + x_1 \tag{11-23}$$

then

$$\mathcal{L}[f(t)g(t)] = \frac{1}{2\pi j} \int_{Br_1} G(z)F(s - z)\,dz \tag{11-24}$$

Now observe that if

$$\text{Re } s > \sigma_f + \sigma_g \tag{11-25}$$

a value of $x_1 > \sigma_g$ can always be found so that relation (11-23) is satisfied. Thus, the abscissa of convergence of $f(t)g(t)$ is $\sigma_f + \sigma_g$.

The situation is portrayed graphically in Fig. 11-2 for the case where both σ_f and σ_g are positive. Normally the contour integral can be evaluated by employing Theorem 8-13 or 8-14 and the calculus of residues, in the manner described in Chap. 10 for the inversion integral. In this connection it is useful to observe where (in the z plane) we may expect to find the poles of the integrand in Eq. (11-24). From the properties of the single-sided transform we have the following information:

Poles of $F(s - z)$ are in the region Re $(s - z) < \sigma_f$
Poles of $G(z)$ are in the region Re $z < \sigma_g$

The first of these conditions, in conjunction with Eq. (11-23), yields

$$\text{Re } z > \text{Re } s - \sigma_f > x_1$$

and since $\sigma_g < x_1$, the second condition becomes

$$\text{Re } z < x_1$$

Thus, the poles of $F(s - z)$ are to the right of Br_1, and the poles of $G(z)$ are to the left of Br_1. It is emphasized that poles of $F(s - z)$ are specified in the z plane (not the s plane) because integration is in the z plane. We make the interesting observation that poles of $F(s - z)$ are functions of s but that poles of $G(z)$ are fixed in relation to s.

If the Br_1 contour is closed to the right with a large semicircle C_1, the integral is $-2\pi j$ times the sum of residues of $F(s - z)$. If a large semicircle C_2 to the left is used, the integral is $2\pi j$ times the sum of residues at poles of $G(z)$. In some cases closure is permissible to the left or right, and in other cases only one is permitted. The properties at infinity of the $F(s - z)$ and $G(z)$ functions determine which semicircle should be used. The closure semicircle must be on the side where the integrand approaches zero.

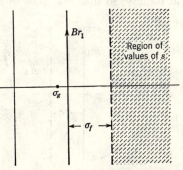

Fig. 11-2. The Br_1 contour and allowed range of s, in relation to the abscissas of convergence of the two functions in the s-plane convolution integral. (Shown for positive values of σ_f and σ_g.)

The shifting theorem offers an instructive vehicle to illustrate s-plane convolution. Let $g(t) = e^{-at}$, which has the transform

$$G(z) = \frac{1}{z + a}$$

Then Eq. (11-24) becomes

$$\mathcal{L}[e^{-at}f(t)] = \frac{1}{2\pi j} \int_{Br_1} \frac{1}{z + a} F(s - z) \, dz \qquad (11\text{-}26)$$

The integrand will always behave appropriately in the left-hand z plane for the application of Theorem 8-13 or 8-14 to contour C_2. The pertinent geometry is given in Fig. 11-3. The residue of the integrand at pole $z = -a$ is

$$F(s - z)\Big|_{z=-a} = F(s + a)$$

which is in agreement with the expected result.

As a second example, suppose that $g(t)$ is the "gate function,"

$$g(t) = \begin{cases} 0 & 0 < t < T_1 \\ 1 & T_1 < t < T_2 \\ 0 & T_2 < t \end{cases}$$

having the transform

$$G(z) = \frac{e^{-zT_1} - e^{-zT_2}}{z}$$

Furthermore, suppose that $f(t)$ has a transform which is rational and zero at infinity, so that $F(s)$ goes to zero at least as fast as $1/s$ as s goes to

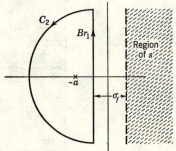

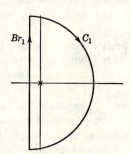

FIG. 11-3. The Br_1 contour and contour of closure for deriving the shifting theorem from s-plane convolution, shown for $\sigma_f > 0$.

FIG. 11-4. The Br_1 contour and contour of closure for obtaining the Laplace transform of t multiplied by a gate function.

infinity in any direction. To be specific, let $f(t) = t$. Then the integrand of

$$\frac{1}{2\pi j} \int_{Br_1} \frac{e^{-zT_1} - e^{-zT_2}}{z} \left(\frac{1}{s - z}\right)^2 dz$$

approaches zero properly in the right half plane, and we use contour C_1, as in Fig. 11-4.

The pole to the right of Br_1 is at $z = s$, and we have

$$\text{Residue} = \frac{d}{dz}\left(\frac{e^{-zT_1} - e^{-zT_2}}{z}\right)_{z=s}$$

$$= \left(\frac{-T_1 e^{-zT_1} + T_2 e^{-zT_2}}{z} - \frac{e^{-zT_1} - e^{-zT_2}}{z^2}\right)_{z=s}$$

USE RESPONSE OF ENTIRE INTEGRAND!

In this case, owing to the negative sense in which the contour is traced, we change the sign and get

$$\mathcal{L}[tg(t)] = \left(\frac{T_1}{s} + \frac{1}{s^2}\right)e^{-sT_1} - \left(\frac{T_2}{s} + \frac{1}{s^2}\right)e^{-sT_2} \qquad (11\text{-}27)$$

Since a valid proof is lacking, we do not state the results of this section as a formal theorem. However, the results are summarized by saying that, if $f(t)$ and $g(t)$ have respective Laplace transforms $F(s)$ and $G(s)$, then at least in some cases

$$\mathcal{L}[f(t)g(t)] = \frac{1}{2\pi j}\int_{\mathrm{Br}_1} G(z)F(s-z)\,dz \qquad (11\text{-}28a)$$

$$\mathcal{L}[f(t)g(t)] = \frac{1}{2\pi j}\int_{\mathrm{Br}_1} F(z)G(s-z)\,dz \qquad (11\text{-}28b)$$

for

$$\mathrm{Re}\ s > \sigma_f + \sigma_g$$

and where Br_1 is a Bromwich path to the right of all singular points of $G(z)$ or to the right of all singular points of $F(z)$ for Eqs. (11-28a) and (11-28b), respectively. The second form in Eq. (11-28) is obtained by interchanging the roles of $f(t)$ and $g(t)$ in the derivation. Furthermore, it was found that the abscissa of convergence of each integral in Eqs. (11-28) is the sum of the abscissas of convergence of $f(t)$ and $g(t)$.

11-6. Application of Convolution in the s Plane to Amplitude Modulation. The process of amplitude modulation can be analyzed readily by the s-plane convolution and is illustrative of an essentially nonlinear process that can be handled by the Laplace transform.

Let the carrier signal be represented by

$$g(t) = \cos bt$$

and let $f(t)$ be the modulation function, with modulation factor m. The modulated signal is then

$$r(t) = [1 + mf(t)]\cos bt \qquad (11\text{-}29)$$

The following transforms:

$$\mathcal{L}(\cos bt) = \frac{s}{s^2 + b^2} = \frac{1}{2}\left(\frac{1}{s - jb} + \frac{1}{s + jb}\right)$$

$$\mathcal{L}[1 + mf(t)] = \frac{1}{s} + mF(s)$$

are used in the convolution formula to give

$$\mathcal{L}[r(t)] = \frac{1}{4\pi j}\int_{\mathrm{Br}_1}\left(\frac{1}{z - jb} + \frac{1}{z + jb}\right)\left[\frac{1}{s - z} + mF(s - z)\right]dz \qquad (11\text{-}30)$$

The integral is evaluated by summing residues at poles to the left of Br_1, in similarity with the treatment of Eq. (11-26). The residues at the two poles are readily obtained, giving the result

$$\mathcal{L}[r(t)] = \frac{1}{2}\left[\frac{1}{s - jb} + \frac{1}{s + jb} + mF(s - jb) + mF(s + jb)\right]$$

$$= \frac{s}{s^2 + b^2} + \frac{m}{2}[F(s - jb) + F(s + jb)] \qquad (11\text{-}31)$$

Here we see three terms, the transforms, respectively, of the carrier and two translations of the transform of $f(t)$.

Now let this modulated signal be demodulated in a square-law detector followed by a low-pass filter having a response function $H(s)$. Convolution in the s plane is used again, to obtain the transform of the squared function. The transform of the squared signal is

$$\frac{1}{8\pi j}\int_{Br_1}\left[\frac{1}{z - jb} + \frac{1}{z + jb} + mG(z)\right]\left[\frac{1}{s - z - jb} + \frac{1}{s - z + jb}\right.$$

$$\left. + mG(s - z)\right]dz = I_1 + I_2 + I_3 + I_4$$

where $\qquad\qquad G(s) = F(s - jb) + F(s + jb)$

and where

$$I_1 = \frac{1}{8\pi j}\int_{Br_1}\left(\frac{1}{z - jb} + \frac{1}{z + jb}\right)\left(\frac{1}{s - z - jb} + \frac{1}{s - z + jb}\right)dz$$

$$I_2 = \frac{1}{8\pi j}\int_{Br_1}\left(\frac{1}{z - jb} + \frac{1}{z + jb}\right)mG(s - z)\,dz$$

$$I_3 = \frac{1}{8\pi j}\int_{Br_1}mG(z)\left(\frac{1}{s - z - jb} + \frac{1}{s - z + jb}\right)dz$$

$$I_4 = \frac{m^2}{8\pi j}\int_{Br_1}G(z)G(s - z)\,dz$$

Integrals I_1 and I_2 are evaluated by evaluating residues to the left of Br_1, and I_3 is evaluated by using residues at poles to the right of Br_1, giving

$$I_1 = \frac{1}{4}\left(\frac{1}{s - j2b} + \frac{1}{s} + \frac{1}{s} + \frac{1}{s + j2b}\right)$$

$$= \frac{1}{2}\left(\frac{1}{s} + \frac{s}{s^2 + 4b^2}\right)$$

$$I_2 = I_3 = \frac{m}{4}[K(s) + 2F(s)]$$

where $\qquad\qquad K(s) = F(s - j2b) + F(s + j2b)$

From the definition of $G(s)$ it is evident that $\mathcal{L}^{-1}[G(s)] = 2f(t)\cos bt$ and I_4 is therefore recognized as the transform of

$$m^2[f(t)]^2\cos^2 bt = m^2[f(t)]^2\frac{1 + \cos 2bt}{2}$$

Thus, the output of a square-law detector followed by a filter of characteristic $H(s)$ is described by the transform function

$$\frac{H(s)}{2}\left(\frac{1}{s} + \frac{s}{s^2 + 4b^2} + 2m[F(s) + K(s)] + m^2\mathcal{L}\{[f(t)]^2(1 + \cos 2bt)\}\right)$$

The filter will normally be a low-pass filter with cutoff considerably below frequency b. Therefore, as an approximation, the transform of the output is

$$\frac{1}{2}\left\{\frac{1}{s} + 2mF(s) + m^2\mathcal{L}[f(t)]^2\right\}$$

and the inversion to give a function of t yields

$$\frac{1}{2}\left\{1 + 2mf(t) + m^2[f(t)]^2\right\}$$

For linear detection, we take the square root of this, giving

$$\frac{1}{\sqrt{2}}\left[1 + mf(t)\right]$$

This differs from the original only by the factor $1/\sqrt{2}$, which appears because this is the case for rms detection.

11-7. Convolution in the s Plane (Two-sided Transform). The arguments given in Sec. 11-5 are applicable to the two-sided transform. The only difference is in regions of convergence and locations of the Bromwich contours. We assume that two functions $f(t)$ and $g(t)$ have the respective two-sided transforms $\mathfrak{F}(s)$ and $\mathcal{G}(s)$. Furthermore, let the two-sided Laplace integrals converge in the following regions:

$$\sigma_{f1} < \text{Re } s < \sigma_{f2} \quad \text{for } f(t)$$
$$\sigma_{g1} < \text{Re } s < \sigma_{g2} \quad \text{for } g(t)$$

The two-sided Laplace integral of the product, namely,

$$\int_{-\infty}^{\infty} f(t)g(t)e^{-st}\,dt$$

is treated by replacing $g(t)$ by the inversion formula

$$g(t) = \frac{1}{2\pi j}\int_{\text{Br}_2} \mathcal{G}(z)e^{zt}\,dz$$

where x_2, the abscissa of contour Br_2, is in the range

$$\sigma_{g1} < x_2 < \sigma_{g2}$$

Again, without justification, we interchange the order of integration and obtain

$$\frac{1}{2\pi j} \int_{-\infty}^{\infty} f(t) \, dt \int_{\text{Br}_3} \mathcal{G}(z)e^{-(s-z)t} \, dz = \frac{1}{2\pi j} \int_{\text{Br}_3} \mathcal{G}(z) \, dz \int_{-\infty}^{\infty} f(t)e^{-(s-z)t} \, dt \tag{11-32}$$

The last integral on the right converges for

$$\sigma_{f1} < \text{Re } (s - z) < \sigma_{f2} \tag{11-33}$$

or

$$\text{Re } z + \sigma_{f1} < \text{Re } s < \text{Re } z + \sigma_{f2}$$

But, since

$$\text{Re } z = x_2$$

it follows that in Eq. (11-32) s is restricted to the range

$$x_2 + \sigma_{f1} < \text{Re } s < x_2 + \sigma_{f1} \tag{11-34}$$

If s is in the range

$$\sigma_{g1} + \sigma_{f1} < \text{Re } s < \sigma_{g2} + \sigma_{f2} \tag{11-35}$$

a value of x_2 can always be found in the range

$$\sigma_{g1} < x_2 < \sigma_{g2}$$

such that relation (11-34) can be satisfied. Therefore, we conclude that relation (11-35) gives the strip of convergence of the two-sided Laplace integral of $f(t)g(t)$. A similar result is obtained if the roles of $f(t)$ and $g(t)$ are interchanged. Accordingly, we have

$$\mathcal{L}_2[f(t)g(t)] = \frac{1}{2\pi j} \int_{\text{Br}_2} \mathcal{G}(z)\mathcal{F}(s - z) \, dz \tag{11-36}$$

or

$$\mathcal{L}_2[f(t)g(t)] = \frac{1}{2\pi j} \int_{\text{Br}_3} \mathcal{F}(z)\mathcal{G}(s - z) \, dz \tag{11-37}$$

where

$$\sigma_{g1} + \sigma_{f1} < \text{Re } s < \sigma_{g2} + \sigma_{f2}$$

Br_2 and Br_3 lie in the respective convergence strips of $g(t)$ and $f(t)$.

Evaluating the integrals in Eqs. (11-36) and (11-37) is slightly more complicated than in the single-sided case because each factor in the integrand will have poles on both sides of the Br contour. Therefore, no matter which side is chosen to close the contour, poles of both factors in the integrand must be considered.

PROBLEMS

11-1. You are given

$$f(t) = e^{-0.5|t|}$$

and the following three $g(t)$ functions:

(a) $g(t) = \begin{cases} e^{-2t} & t > 0 \\ e^{-t} & t < 0 \end{cases}$ (b) $g(t) = \begin{cases} e^{t} & t > 0 \\ e^{2t} & t < 0 \end{cases}$

(c) $g(t) = e^{-|t|}$

In each case determine whether or not Theorem 10-1 is applicable, and where it is, check the validity of the theorem. If the theorem is not applicable, explain why.

11-2. Do Prob. 11-1 for the case

$$f(t) = \frac{\sin t}{t}$$

11-3. Considering the function

$$f(t) = \begin{cases} e^{-3t} & t > 0 \\ e^{-t} & t < 0 \end{cases}$$

which of the $g(t)$ functions given in Prob. 11-1 can be used in conjunction with the above, in Theorem 11-2? If the theorem applies, check it; if it does not apply, explain why.

11-4. Let the two functions

$$f(t) = \begin{cases} 0 & t < -1 \\ 1 & -1 < t < 1 \\ 0 & 1 < t \end{cases} \qquad g(t) = e^{-|t|}$$

be given.

(a) Find $\phi(t) = \int_{-\infty}^{\infty} f(\tau)g(t - \tau)\, d\tau$.

(b) For this example, show that $\mathcal{L}_2[\phi(t)] = \mathfrak{F}(s)\mathfrak{G}(s)$.

11-5. The function

$$f(t) = \begin{cases} t & 0 < t < 1 \\ t^2 & 1 < t \end{cases}$$

can be regarded as the product of two functions

$$f_a(t) = t \qquad f_b(t) = (t - 1)u(t - 1) + 1$$

Use the appropriate convolution integral to get $\mathcal{L}[f(t)]$.

11-6. The two functions

$$f(t) = \begin{cases} 0 & t < 0 \\ 1 & 0 < t < 1 \\ 0 & 1 < t \end{cases} \qquad g(t) = \begin{cases} 0 & t < 0 \\ 1 & 0 < t < 2 \\ 0 & 2 < t \end{cases}$$

are given.

(a) Find

$$\phi(t) = \int_0^t f(\tau)g(t - \tau)\, d\tau$$

(b) For this example show that

$$\mathcal{L}[\phi(t)] = F(s)G(s)$$

11-7. Show that the t-plane convolution integral can be written

$$\int_0^t f(\tau)g(t - \tau)\, d\tau = t \int_0^1 f[t(1 - w)]g(tw)\, dw$$

$$= t \int_0^1 f(tw)g[t(1 - w)]\, dw$$

Check this on the case $f(t) = e^t$, $g(t) = \cos t$.

11-8. Assume that the transform

$$\mathcal{L}[J_0(at)] = \frac{1}{\sqrt{a^2 + s^2}}$$

is known, where J_0 is the zeroth-order Bessel function of the first kind, given by

$$J_0(x) = \frac{1}{\pi} \int_0^\pi \cos(x \cos \phi) \, d\phi$$

By writing $\sqrt{a^2 + s^2} = \sqrt{s + ja} \, \sqrt{s - ja}$, use the appropriate convolution integral to show that this is the correct transform for $J_0(at)$.

11-9. Using the functions

$$f(t) = \frac{1}{\sqrt{t}} \qquad g(t) = \frac{1}{\sqrt{t}}$$

we recognize that each has $\sqrt{\pi/s}$ as its Laplace transform and therefore that the t-plane convolution of f and g should have π/s as its Laplace transform. Check whether or not this is true.

11-10. Given the two Laplace transforms

$$\mathcal{L}\left(\frac{1}{\sqrt{t}} e^{-a/t}\right) = \sqrt{\frac{\pi}{s}} \, e^{-2\sqrt{as}} \qquad \mathcal{L}\left(\frac{1}{\sqrt{t}}\right) = \sqrt{\frac{\pi}{s}}$$

use convolution to prove that

$$\mathcal{L}\left(1 - \operatorname{erf} \sqrt{\frac{a}{t}}\right) = \frac{1}{s} e^{-2\sqrt{as}}$$

11-11. Use s-plane convolution to obtain the Laplace transform of $\sin at \cos bt$.

11-12. Use the appropriate convolution integral to prove that

$$\mathcal{L}^{-1}\left[\frac{a - b}{(s - a)(s - b)}\right] = e^{at} - e^{bt}$$

11-13. Use the appropriate convolution integral to get the inverse transform of

$$\frac{1 - e^{-s}}{s^2 + s}$$

11-14. Use s-plane convolution to prove the theorems

(a) $\mathcal{L}[e^{-at}f(t)] = F(s + a)$ (b) $\mathcal{L}[tf(t)] = -\dfrac{dF(s)}{ds}$

11-15. Use the appropriate convolution theorem to obtain the following:

(a) $\mathcal{L}\left(e^{-at} \displaystyle\int_0^{\sqrt{at}} e^{x^2} \, dx\right) = \dfrac{\sqrt{a\pi}}{2 \sqrt{s} \, (s + a)}$

(b) $\mathcal{L}\left(\sqrt{at} - e^{-at} \displaystyle\int_0^{\sqrt{at}} e^{x^2} \, dx\right) = \dfrac{\sqrt{a^3\pi}}{2 \sqrt{s^3} \, (s + a)}$

11-16. Use t-plane convolution to prove that

$$\mathcal{L}^{-1}\left[\frac{F(s)}{s}\right] = \int_0^t f(\tau) \, d\tau$$

where $F(s) = \mathcal{L}[f(t)]$. Why cannot convolution be used to obtain the inverse of $sF(s)$?

CHAPTER 12

FURTHER PROPERTIES OF THE LAPLACE TRANSFORM

12-1. Introduction. There are at least two kinds of "operational calculus" which lead to formulas identical with, or similar to, the Laplace transform. The best known of these methods is the operational calculus invented by Oliver Heaviside. Therefore, since the Laplace integral is not the only formula which will yield a unique $F(s)$ corresponding to a given $f(t)$, the serious student will naturally ask whether or not the Laplace integral is preferable to the other possible formulations and, if so, why.

Several advantages of the Laplace integral formulation have already been mentioned in this book, at least by implication. The close relationship with the Fourier integral theorem, from which an inversion formula for $f(t)$ can be established, is one of these reasons. Another reason is that the Laplace theory puts into evidence a generalized viewpoint and interpretation of the steady-state-system response function, as a function of a complex variable, making it possible conclusively to establish certain properties of these functions. The Laplace transform theory is the catalyst which provides the bridge between the system equations and the powerful methods of the theory of complex variables.

In Chap. 10 it has been shown that $F(s)$ must have certain properties (analyticity, for example), and we found that these properties are established by relying on the Laplace integral for the definition of $F(s)$. Properties such as this can be inferred by the other methods of arriving at $F(s)$, but possibly less conclusively and less easily. In this chapter, as we develop further properties of $F(s)$, you will come to a fuller understanding of why the Laplace transform is so important.

12-2. Behavior of $F(s)$ at Infinity. In Chap. 10 it was demonstrated that, if $f(t)$ is APC, then its transform $F(s)$ approaches zero uniformly as $|s|$ becomes infinite in an angular sector in the right half plane described by

$$|\text{ang } (s - s_0)| \leqq \theta' < \frac{\pi}{2}$$

where s_0 is any point where the Laplace integral converges. We shall now consider two special cases where $f(t)$ is more restricted and for which it is possible to derive stronger conditions on the behavior of $F(s)$ at infinity.

First we treat the case where $f(t)$ is identically zero for t greater than some number T. From Sec. 10-12 we recall that the transform of such a function is an entire function. It is singular only at infinity, and now we shall learn something about its behavior at this singularity. Specifically we shall look at

$$\lim_{|s| \to \infty} \int_0^T f(t)e^{-st}\, dt$$

where $|s|$ approaches infinity along the ray defined by

$$s = \rho e^{j\phi}$$

with ϕ held constant. The above integral bears a similarity with the integral in Theorem 8-12, but the exponential replaces the trigonometric function. Use will be made of this similarity and we shall rely heavily on the proof given for Theorem 8-12; and so you should refer to that proof for details. A function $g(t)$ will be defined as a staircase approximation for $f(t)$. Assuming $f(t)$ is piecewise continuous, we have

$$g(t) = \begin{cases} A_0 & 0 = t_0 < t < t_1 \\ A_1 & t_1 < t < t_2 \\ \cdots\cdots\cdots\cdots\cdots\cdots \\ A_{N-1} & t_{N-1} < t < T_N = T \end{cases}$$

such that

$$\int_0^T |f(t) - g(t)|\, dt < \frac{\epsilon}{2}$$

Now, if Re $s \geqq 0$, it is true that

$$|e^{-st}| \leqq 1$$

and so, for Re $s \geqq 0$, we have

$$\left| \int_0^T f(t)e^{-st}\, dt \right| - \left| \int_0^T g(t)e^{-st}\, dt \right| \leqq \left| \int_0^T [f(t) - g(t)]e^{-st}\, dt \right|$$

$$\leqq \int_0^T |f(t) - g(t)|\, dt < \frac{\epsilon}{2}$$

which then yields

$$\left| \int_0^T f(t)e^{-st}\, dt \right| < \frac{\epsilon}{2} + \left| \int_0^T g(t)e^{-st}\, dt \right|$$

Recalling the definition of $g(t)$, we can evaluate the integral on the right as follows,

$$\left| \int_0^T g(t)e^{-st}\, dt \right| = \left| \sum_{k=0}^{N-1} A_k \int_{t_k}^{t_{k+1}} e^{-st}\, dt \right|$$

$$= \left| \sum_{k=0}^{N-1} - A_k \frac{e^{-st_{k+1}} - e^{-st_k}}{s} \right| < \frac{2NM}{\rho}$$

where M is an upper bound of the set of numbers A_k, and Re $s \geqq 0$. Since the above relations are true for Re $s \geqq 0$, we can also say that they are true for

$$|\phi| \leqq \frac{\pi}{2}$$

Thus, if

$$\rho > \frac{4NM}{\epsilon}$$

we conclude that

$$\left| \int_0^T f(t)e^{-st}\, dt \right| < \epsilon$$

In view of the independence of ϕ, in the range specified, it follows that as ρ becomes infinite the integral approaches zero uniformly with respect to ϕ. Now suppose that

$$s - s_0 = Re^{j\theta}$$

where $s_0 = \sigma_0 + j\omega_0$ is any complex constant. We can write

$$\int_0^T f(t)e^{-st}\, dt = \int_0^T f(t)e^{-s_0 t}e^{-(s-s_0)t}\, dt = \int_0^T [f(t)e^{-\sigma_0 t} \cos \omega_0 t]e^{-(s-s_0)t}\, dt$$
$$- j \int_0^T [f(t)e^{-\sigma_0 t} \sin \omega_0 t]e^{-(s-s_0)t}\, dt$$

and each of the integrals on the right meets the conditions of the proof just given, if $s - s_0$ is replaced by s. Thus, we can use the previous development to conclude that each of these integrals approaches zero as R becomes infinite. By following the outline given in Sec. 8-10 we can extend these results to the case where $f(t)$ is APC, rather than PC as we assumed here. We have established the following theorem:

Theorem 12-1. Let $f(t)$ be a function which is APC and which is identically zero for t greater than some number T. Then, the Laplace transform of $f(t)$ approaches zero uniformly as s becomes infinite in a right half plane,

$$|\text{ang } (s - s_0)| \leqq \frac{\pi}{2}$$

This theorem shows that Jordan's lemma will always be applicable in evaluating the inversion integral of functions of this type. You will recall that this statement cannot be made for the general class of Laplace transformable functions.

For the second case, assume that $f(t)$ is piecewise continuous and of exponential order α_0. This condition is sufficient to ensure that

$$\int_0^\infty f(t)e^{-\sigma t}\,dt$$

converges for Re $s > \alpha_0$. With the observation that the Laplace integral can be written

$$\int_0^\infty f(t)e^{-st}\,dt = \int_0^\infty f(t)e^{-\sigma t}\cos \omega t\,dt - j\int_0^\infty f(t)e^{-\sigma t}\sin \omega t\,dt$$

we see that Theorem 8-12 applies to each of the integrals on the right. Therefore, it follows that the Laplace transform of such a function approaches zero as s goes to infinity along a vertical line to the right of α_0. Referring to Theorem 10-4, we recall that uniform approach to zero in the region

$$|\text{ang }(s - s_0)| \leqq \theta' < \frac{\pi}{2}$$

implies approach to zero (but not uniformly) in the region

$$|\text{ang }(s - s_0)| < \frac{\pi}{2}$$

But, in view of the present proof that $F(s)$ also approaches zero along a vertical line through s_0, we can add an equality sign to the above, giving the following theorem:

Theorem 12-2. If $f(t)$ is PC and EO,α_0, then the Laplace transform of $f(t)$ has the property

$$\lim_{|s|\to\infty} F(s) = 0$$

for

$$|\text{ang }(s - s_0)| \leqq \frac{\pi}{2}$$

and where Re $s_0 > \alpha_0$.

The following comments are offered in appraisal of this theorem. Comparison shows that Theorems 12-1 and 12-2 both deal with behavior as s approaches infinity along a radial line, making an angle less than or equal to $\pi/2$ with the real axis. There are two important differences, however. In Theorem 12-1 the line can radiate from any point in the s plane, and $F(s)$ approaches zero *uniformly* with respect to its angle in a right half plane. In Theorem 12-2 $F(s)$ approaches zero in a similar sector, but the apex must be in the region of convergence of the Laplace integral; and there is no proof that zero is approached uniformly. Thus,

Theorem 12-2 is not strong enough to ensure applicability of Jordan's lemma to the inversion formula.

12-3. Functions of Exponential Type. You will recall from Chap. 5 that an entire function is an analytic function having no singularities in the finite plane. Any Taylor series of such a function has an infinite radius of convergence. Let t be the real part of a complex variable $w = t + ju$, and let $f(w)$ be an entire function. In addition, assume that it is possible to find numbers M and γ such that for all w

$$|f(w)| \leqq M e^{\gamma|w|} \tag{12-1}$$

where M and γ are real numbers greater than zero. Many functions fit into this category; functions such as $\sin bt$, e^{at}, any polynomial in t are typical. An entire function satisfying condition (12-1) is said to be of *exponential type*. This designation is not to be confused with exponential order. Functions of exponential order are not necessarily entire functions and, indeed, need not even be defined for complex values of the variable. We note that the function e^{-t^2} is of exponential order, but not of exponential type, because $e^{-(ju)^2} = e^{u^2}$ cannot be dominated by the exponential in Eq. (12-1). Also, any function having discontinuities, or discontinuous derivatives, cannot be of exponential type, because such a function is not analytic.

In all but the trivial case $f(t) \equiv 0$, the set of values for which relation (12-1) is satisfied will have a greatest lower bound γ_0 which can never be negative. This number may or may not be a member of the set of γ's for which relation (12-1) is true. We shall call γ_0 the *order* of $f(t)$.

Since $f(w)$ is an entire function, it possesses all derivatives and we shall now show that each derivative is of exponential type, of the same order γ_0 as $f(t)$. The Cauchy integral formula for the nth derivative is

$$f^{(n)}(w) = \frac{n!}{2\pi j} \int_C \frac{f(z)}{(z - w)^{n+1}} \, dz \tag{12-2}$$

where C is a circle of radius R centered at $z = w$. The change of variable

$$z = w + Re^{j\theta} \qquad dz = jRe^{j\theta} \, d\theta$$

leads to

$$f^{(n)}(w) = \frac{n!}{2\pi j} \int_0^{2\pi} \frac{f(w + Re^{j\theta})e^{-jn\theta}}{R^n} \, d\theta$$

But since $f(w)$ is of exponential type of order γ_0,

$$|f(w + Re^{j\theta})| < M e^{\gamma|w + Re^{j\theta}|} < M e^{\gamma R} e^{\gamma|w|} \qquad \gamma > \gamma_0$$

and consequently

$$|f^{(n)}(w)| < \left(\frac{n! M e^{\gamma R}}{R^n} \right) e^{\gamma|w|} \qquad \gamma > \gamma_0$$

The quantity in parentheses is independent of w, and so we conclude that $f^{(n)}(w)$ is of exponential type and of order γ_0. A more convenient form for the above, for later use, is obtained by choosing $R = n/\gamma$, which can be done because R, the radius of the circle used in the Cauchy integral formula, is arbitrary. Thus,

$$|f^{(n)}(w)| < \frac{Mn!e^n\gamma^n}{n^n}\, e^{\gamma|w|} \qquad \gamma > \gamma_0 \qquad (12\text{-}3)$$

Next we show that a function of exponential *type*, and all its derivatives, are of exponential *order*. This can easily be shown by observing that if

$$|f(w)| \leqq Me^{\gamma|w|}$$
$$|f^{(n)}(w)| \leqq M_n e^{\gamma|w|}$$

then for $w = t > 0$ (w real) we have

$$|f(t)|e^{-\alpha t} \leqq Me^{-(\alpha-\gamma)t}$$
$$|f^{(n)}(t)|e^{-\alpha t} \leqq M_n e^{-(\alpha-\gamma)t} \qquad (12\text{-}4)$$

For each of the above, t can be made sufficiently large to make the right-hand side arbitrarily small, if the condition

$$\alpha > \gamma$$

is satisfied. Thus, $f(t)$ and $f^{(n)}(t)$ are of exponential order α_0, where in each case $\alpha_0 \leqq \gamma_0$. Also, $f(t)$ and all its derivatives are continuous. These conditions satisfy Theorem 10-9. Furthermore, Eq. (10-48) can be written by using the Laplace integral formulation for the transforms, if $\mathrm{Re}\ s > \gamma$. The latter restriction arises from conditions (12-4) and Theorem 10-1. The $+$ is not needed on $f(0+)$, etc. Thus, we can write

$$\int_0^\infty f(t)e^{-st}\,dt = \frac{f(0)}{s} + \frac{f'(0)}{s^2} + \frac{f''(0)}{s^3} + \cdots + \frac{1}{s^n}\int_0^\infty f^{(n)}(t)e^{-st}\,dt$$
$$\mathrm{Re}\ s > \gamma \quad (12\text{-}5)$$

which carries the implication that the Laplace transform of a function of exponential type might possibly be represented by a series in negative powers of s. This proves to be the case, as we now find by investigating the remainder term

$$A_n = \frac{1}{s^n}\int_0^\infty f^{(n)}(t)e^{-st}\,dt \qquad (12\text{-}6)$$

as n approaches infinity. Relation (12-3) makes possible the following appraisal:

$$|A_n| < \frac{Mn!e^n\gamma^n}{|s|^n n^n}\int_0^\infty e^{-(\sigma-\gamma)t}\,dt = \frac{Mn!e^n\gamma^n}{|s|^n n^n(\sigma-\gamma)} = B_n \qquad \sigma > \gamma \quad (12\text{-}7)$$

In order to show that B_n approaches zero, we start with the ratio

$$\frac{B_{n+1}}{B_n} = \frac{(n+1)!e^{n+1}\gamma^{n+1}}{|s|^{n+1}(n+1)^{n+1}} \frac{|s|^n n^n}{n!e^n\gamma^n} = \frac{e\gamma}{(1+1/n)^n|s|}$$

and recall the definition of e, namely,

$$e = \lim_{n\to\infty} \left(1 + \frac{1}{n}\right)^n$$

It is now evident that

$$\lim_{n\to\infty} \frac{B_{n+1}}{B_n} = \frac{\gamma}{|s|} \qquad \sigma > \gamma \qquad (12\text{-}8)$$

The condition $|s| > \gamma$ is satisfied when $\sigma > \gamma$, and therefore this limit is less than 1. We have just completed the ratio test for the series

$$B_0 + B_1 + \cdots$$

showing that convergence occurs for $\sigma > \gamma$. But a series can converge only if the terms approach zero, and since $|A_n| < B_n$, we have proved

$$\lim_{n\to\infty} |A_n| = 0 \qquad \text{if } \sigma > \gamma \qquad (12\text{-}9)$$

It follows that Eq. (12-5) can be replaced by the series

$$\int_0^\infty f(t)e^{-st}\,dt = \sum_{n=0}^\infty \frac{f^{(n)}(0)}{s^{n+1}} \qquad \sigma > \gamma$$

If a power series in positive powers of $1/s$ converges at some point $1/\sigma$, it will converge in the circular region

$$\frac{1}{|s|} \leqq \frac{1}{\sigma} < \frac{1}{\gamma}$$

and therefore the above series converges in the region

$$|s| > \gamma > \gamma_0$$

It is sufficient to designate this region by

$$|s| > \gamma_0 \qquad (12\text{-}10)$$

Thus, although the Laplace integral has a half plane of convergence, the series expression for the transform converges in the region defined above and accordingly we get the result

$$F(s) = \sum_{n=0}^\infty \frac{f^{(n)}(0)}{s^{n+1}} \qquad |s| > \gamma_0 \qquad (12\text{-}11)$$

Next we observe that the Taylor expansion of $f(t)$ is

$$f(t) = \sum_{n=0}^{\infty} \frac{f^{(n)}(0)}{n!} t^n \tag{12-12}$$

Thus, in Eqs. (12-11) and (12-12) we have an indirect relationship between $f(t)$ and its Laplace transform, in terms of correspondence of coefficients in series expansions for the two functions. We also get two specific items of information about $F(s)$. First, the series in Eq. (12-11) is regular outside the circle of radius γ_0. Therefore, all singular points of $F(s)$ are on or inside the circle $|s| = \gamma_0$. Second, Eq. (12-11) shows that $F(s)$ has a zero at infinity and that the order of this zero is one greater than the order of the lowest derivative of $f(t)$ which is not zero at the origin. In general terms, this shows that, the more smoothly $f(t)$ starts out from the origin, the higher will be the order of the zero of $F(s)$ at infinity. The existence of at least a first-order zero at infinity guarantees that Jordan's lemma is applicable to $\mathcal{L}[f(t)]$ for all functions of exponential type.

These ideas are illustrated by the following three cases:

$$\mathcal{L}(e^{bt}) = \frac{1}{s - b} \qquad \mathcal{L}(e^{-bt}) = \frac{1}{s + b} \qquad \mathcal{L}(\sin bt) = \frac{b}{s^2 + b^2}$$

Each of the above t functions is of exponential type, with $\gamma_0 = b$. Thus, for each case we should expect singularities to be in or on a circle of radius b. The above formulas show that this is indeed the case: the poles are located on the circle $|s| = b$.

These examples provide insight into the difference between the conditions of exponential *type* and of exponential *order*. As far as the condition of exponential order is concerned, each of these functions is of different order, the orders being respectively b, $-b$, and 0. However, the same γ_0 is obtained in each case. Each of these transforms has poles inside the same circle, but the abscissas of convergence of the various Laplace integrals are different. Observe that a function of exponential type is of exponential order but that the converse is not true.

The important features of the preceding discussion are collected and stated as follows:

Theorem 12-3. If $f(w)$ is a function of exponential type of order γ_0, which means that, if $\gamma > \gamma_0$,

$$|f(w)| < Me^{\gamma|w|}$$

for all complex values of w, then for the Laplace transform

$$F(s) = \mathcal{L}[f(t)]$$

we can say that:

1. $F(s)$ has a zero at infinity, of order one greater than the order of the lowest-order nonzero derivative at $t = 0$.

2. Singularities of $F(s)$ lie inside or on the circle $|s| = \gamma_0$.

Now suppose that $F(s)$ is given by the series

$$F(s) = \sum_{n=0}^{\infty} \frac{a_n}{s^{n+1}} \qquad (12\text{-}13)$$

which converges for $|s| > \gamma_0$. The series therefore converges uniformly in the region $|s| \geq \gamma' > \gamma_0$. In Sec. 10-18 it is demonstrated that for $t > 0$ an integration over the infinite path Br is equivalent to integration around a closed path enclosing all singular points of $F(s)$. Since all singularities of $F(s)$ are inside a circle of radius γ_0, it follows that if Cr is a circle of radius $R > \gamma_0$, centered at the origin, then, for $t > 0$,

$$f(t) = \frac{1}{2\pi j} \int_{\text{Cr}} F(s) e^{st} \, ds = \sum_{n=0}^{\infty} \frac{1}{2\pi j} \int_{\text{Cr}} \frac{a_n e^{st}}{s^{n+1}} \, ds$$

where uniform convergence is used to justify term-by-term integration. Each term on the right is $\mathcal{L}^{-1}(a_n/s^{n+1}) = a_n t^n/n!$. Therefore,

$$f(t) = \sum_{n=0}^{\infty} \frac{a_n t^n}{n!} \qquad (12\text{-}14)$$

Since the series in Eq. (12-13) converges, outside a circle of radius γ_0,

$$\varlimsup_{n \to \infty} \sqrt[n]{|a_n|} = \gamma_0$$

and this means that, corresponding to a small positive number ϵ, there is a number N such that when $n > N$

$$\sqrt[n]{|a_n|} - \gamma_0 < \epsilon$$

or
$$|a_n| < (\gamma_0 + \epsilon)^n \qquad (12\text{-}15)$$

since $\gamma_0 > 0$. For each $n \leq N$ we can define a number M_n by the relation

$$|a_n| = M_n (\gamma_0)^n$$

There will be one value of M_n which is the largest of this finite number of terms. Let M be this largest M_n or unity, whichever is larger. Accordingly,

$$|a_n| < M(\gamma_0)^n < M(\gamma_0 + \epsilon)^n \qquad 0 \leq n \leq N$$

Relation (12-15) is also true if M is included as a factor on the right, because $M \geqq 1$. Thus, we have

$$|a_n| < M(\gamma_0 + \epsilon)^n \qquad \text{all } n \tag{12-16}$$

and

$$\left| \sum_{n=0}^{\infty} \frac{a_n}{n!} t^n \right| \leqq \sum_{n=0}^{\infty} \left| \frac{a_n}{n!} \right| |t|^n < M \sum_{n=0}^{\infty} \frac{[(\gamma_0 + \epsilon)|t|]^n}{n!}$$

But the series on the right is the exponential function of argument $(\gamma_0 + \epsilon)|t|$, which converges for all t. Therefore, the series in Eq. (12-14) also converges, and hence $f(t)$ is an entire function. Also, the above can be written

$$f(t) < Me^{(\gamma_0 + \epsilon)|t|} \tag{12-17}$$

showing that $f(t)$ is of exponential type and that γ can be any number greater than γ_0. This leads to the following theorem:

Theorem 12-4.   If

$$F(s) = \sum_{n=0}^{\infty} \frac{a_n}{s^{n+1}}$$

converges for $|s| > \gamma_0$, then $\mathcal{L}^{-1}[F(s)]$ is of exponential type and there is a number M such that

$$|f(t)| < Me^{\gamma|t|} \qquad \text{all } t$$

where $\gamma > \gamma_0$.

12-4. A Special Class of Piecewise Continuous Functions.   Many functions of importance in engineering are made up of continuous sections, each of which is a portion of a function consisting of sums of terms like $t^n e^{at}$. The following two examples illustrate this type of function:

Case a

$$f(t) = \begin{cases} 1 & 0 < t < 1 \\ e^{-(t-1)} & 1 < t < 2 \\ 3 - t & 2 < t < 3 \\ \sin \dfrac{\pi}{3} t & 3 < t < 6 \\ 1 & 6 < t \end{cases}$$

Case b

$$f(t) = \begin{cases} 1 & 2n\pi < t < (2n + 1)\pi \\ \sin t & (2n + 1)\pi < t < (2n + 2)\pi \end{cases}$$

where n is an integer.* Case a is illustrated in Fig. 12-1.

In case a there are a finite number of sections, and in case b the number of sections is infinite. The function may or may not be continuous

---

* In this classification sine and cosine are regarded as sums of two exponentials.

at a point of transition from one function to another. This kind of function will be said to be *piecewise continuous with sections of exponentials times polynomials*, abbreviated PCSEP.

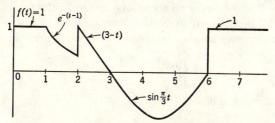

FIG. 12-1. Example of a function of PCSEP type.

The next objective is to study the properties of the Laplace transform of PCSEP functions. As a start, let us inspect the transform of case *a*. From information available in Chap. 10, the transform can be written by inspection, as follows:

$$F(s) = \frac{1}{s}(1 - e^{-s}) + \frac{1}{s+1}\left(e^{-s} - \frac{1}{e}e^{-2s}\right) + \left(\frac{1}{s} - \frac{1}{s^2}\right)e^{-2s} + \frac{1}{s^2}e^{-3s}$$
$$- \frac{3\pi}{9s^2 + \pi^2}(e^{-3s} + e^{-6s}) + \frac{e^{-6s}}{s}$$

This transform consists of a sum of terms of the form

$$F_k(s)e^{-sT_k}$$

where each $F_k(s)$ is a rational function and T_k corresponds to the successive values of t where $f(t)$ changes from one functional form to another. Observe that a new term like the above appears for each transition, whether or not the function is continuous at the point of transition. We shall now show that the observations made for this particular case are generally valid for the PCSEP functions.

Before proceeding with the general treatment, it is to be emphasized that this class of functions is considerably more restricted than the PC functions. For example, the functions

$$f(t) = \begin{cases} \sqrt{t} & 0 < t < 1 \\ 1 & 1 < t \end{cases}$$

and

$$f(t) = \begin{cases} 1 & 0 < t < 1 \\ \dfrac{1}{\sin{(\pi t/4)}} & 1 < t < 2 \\ t & 2 < t \end{cases}$$

are piecewise continuous and of exponential order; but each one includes a section of a function [respectively, $\sqrt{t}$ and $1/\sin{(\pi t/4)}$] which is not of the proper type.

The general case can be written

$$f(t) = f_k(t) \qquad T_k < t < T_{k+1} \tag{12-18}$$

where $T_0 = 0$ and k takes on successive integral values. This notation implies that k becomes infinite. If $f(t)$ should have only a finite number (N) of discontinuous points, the appropriate notation would be

$$f(t) = \begin{cases} f_k(t) & T_k < t < T_{k+1} \\ f_N(t) & T_N < t \end{cases} \tag{12-19}$$

The Laplace integral of Eq. (12-18) is

$$\int_0^\infty f(t)e^{-st}\,dt = \sum_{k=0}^\infty \int_{T_k}^{T_{k+1}} f_k(t)e^{-st}\,dt \tag{12-20}$$

and if $f(t)$ is in the form of Eqs. (12-19) we get

$$\int_0^\infty f(t)e^{-st}\,dt = \sum_{k=0}^{N-1} \int_{T_k}^{T_{k+1}} f_k(t)e^{-st}\,dt + \int_{T_N}^\infty f_N(t)e^{-st}\,dt \tag{12-21}$$

In either case the general term of the series can be changed to the following, with σ used in place of s for later convenience:

$$\int_{T_k}^{T_{k+1}} f_k(t)e^{-\sigma t}\,dt$$

$$= \int_{T_k}^\infty f_k(t)e^{-\sigma t}\,dt - \int_{T_{k+1}}^\infty f_k(t)e^{-\sigma t}\,dt$$

$$= e^{-\sigma T_k}\int_0^\infty f_k(t+T_k)e^{-\sigma t}\,dt - e^{-\sigma T_{k+1}}\int_0^\infty f_k(t+T_{k+1})e^{-\sigma t}\,dt \tag{12-22}$$

This step is possible because each $f_k(t)$ is defined for all t. Therefore, each of the Laplace integrals in Eq. (12-22) exists. In order to see how to proceed in substituting Eq. (12-22) into Eq. (12-20), it is convenient to adopt the shorthand notation

$$\begin{aligned} A_k &= \int_0^\infty f_k(t+T_k)e^{-\sigma t}\,dt \\ B_k &= \int_0^\infty f_k(t+T_{k+1})e^{-\sigma t}\,dt \end{aligned} \tag{12-23}$$

Then Eq. (12-20) can be written

$$\int_0^\infty f(t)e^{-\sigma t}\,dt = (A_0 - B_0 e^{-\sigma T_1}) + (A_1 e^{-\sigma T_1} - B_1 e^{-\sigma T_2})$$
$$+ (A_2 e^{-\sigma T_2} - B_2 e^{-\sigma T_3}) + \cdots$$

where the parentheses indicate the individual terms for the summation as written in Eq. (12-20). If this series converges absolutely, the terms can be rearranged in the more convenient form

$$\int_0^\infty f(t)e^{-\sigma t}\,dt = A_0 + (A_1 - B_0)e^{-\sigma T_1} + (A_2 - B_1)e^{-\sigma T_2} + \cdots$$

Next it will be shown that the series of Eq. (12-20) will converge absolutely if $f(t)$ is EO,α_0. We recall the definition

$$\lim_{t \to \infty} f(t)e^{-\alpha t} = 0 \qquad \alpha > \alpha_0$$

and this can be interpreted to mean that if a small number ϵ is given a number T' can be found such that

$$|f(t)| < \epsilon e^{\alpha t}$$

when $t > T'$

Having chosen ϵ and found T', choose a number N' such that $T_{N'} > T'$. Then, for $k > N'$

$$\left| \int_{T_k}^{T_{k+1}} f(t)e^{-\sigma t}\,dt \right| < \epsilon \int_{T_k}^{T_{k+1}} e^{-(\sigma-\alpha)t}\,dt$$

and if $n > N'$ and $\sigma > \alpha$,

$$\sum_{k=n}^\infty \left| \int_{T_k}^{T_{k+1}} f(t)e^{-\sigma t}\,dt \right| < \epsilon \sum_{k=n}^\infty \int_{T_k}^{T_{k+1}} e^{-(\sigma-\alpha)t}\,dt = \epsilon \int_{T_n}^\infty e^{-(\sigma-\alpha)t}\,dt < \frac{\epsilon}{\sigma - \alpha}$$

Now, if $\sigma \geq \alpha' > \alpha$, the above gives

$$\sum_{k=n}^\infty \left| \int_{T_k}^{T_{k+1}} f(t)e^{-\sigma t}\,dt \right| < \frac{\epsilon}{\alpha' - \alpha} = \epsilon' \qquad n > N'$$

Since ϵ was arbitrarily chosen, ϵ' is also arbitrary and so it follows that the series converges uniformly for Re $s \geq \alpha' > \alpha_0$. Absolute convergence shows that the rearrangement suggested for the series in Eq. (12-20) is valid, and accordingly we assume that $f(t)$ is EO,α_0 and write

$$\int_0^\infty f(t)e^{-\sigma t}\,dt = A_0 - \sum_{k=1}^\infty (A_k - B_{k-1})e^{-\sigma T_k} \qquad \sigma > \alpha_0 \qquad (12\text{-}24)$$

Each integral in Eqs. (12-23) has a half plane of convergence, and so we can regard A_k and B_k each as functions of s which are in reality the respective Laplace transforms of $f_k(t + T_k)$ and $f_k(t + T_{k+1})$. Accordingly, let us define

$$\psi_k(s) = \begin{cases} \mathcal{L}[f_0(t)] & k = 0 \\ \mathcal{L}[f_k(t + T_k) - f_{k-1}(t + T_k)] & k \neq 0 \end{cases} \qquad (12\text{-}25)$$

giving, finally,

$$F(s) = \sum_{k=0}^{\infty} \psi_k(s)^{-sT_k} \qquad \text{Re } s > \alpha_0 \tag{12-26}$$

The case where the number of transition points is a finite number N, represented by Eq. (12-21), is simpler. In that case it is unnecessary to be concerned about convergence of a series. The only formal change is in the summation limit, and we have

$$F(s) = \sum_{k=0}^{N} \psi_k(s) e^{-sT_k} \tag{12-27}$$

Equations (12-26) and (12-27) have a unique form. As expected from the earlier example, each transition point produces a factor like e^{-sT_k}. Furthermore, each $\psi_k(s)$ is the Laplace transform of a combination of terms like $t^n e^{at}$ and is therefore known to be a rational function in s.

The case where each section of $f(t)$ is a constant,

$$f_k(t) = D_k$$

is of particular interest. From Eq. (12-25) we then have

$$\psi_0(s) = \frac{D_0}{s}$$

$$\psi_k(s) = \frac{D_k - D_{k-1}}{s} \qquad k \geq 1$$

and Eq. (12-26) gives

$$F(s) = \frac{D_0}{s} + \sum_{k=1}^{\infty} \frac{D_k - D_{k-1}}{s} e^{-sT_k} \tag{12-28}$$

You should be aware in this development of the importance of having $f(t)$ of exponential order, in addition to being PCSEP in type. The latter condition does not ensure the former. For example, the successive $f_k(t)$ functions can have multiplying constants that increase so rapidly with increasing k that $f(t)$ might not be of exponential order.

As a theorem, we can now state these results as follows:

Theorem 12-5.   Let $f(t)$ be made up of sections of functions of combinations of terms like $t^n e^{at}$, and let $f(t)$ be EO,α_0. Furthermore, let T_k represent the point where $f(t)$ experiences a transition from section $f_{k-1}(t)$ to section $f_k(t)$, taking $k = 0$ as the origin. Then, $f(t)$ has a Laplace transform in the form

$$F(s) = \sum_{k=0}^{\infty} \psi_k(s) e^{-sT_k} \qquad \text{Re } s > \alpha_0$$

where each $\psi_k(s)$ is a rational function given by

$$\psi_k(s) = \begin{cases} \mathcal{L}[f_0(t)] & k = 0 \\ \mathcal{L}[f_k(t + T_k) - f_{k-1}(t + T_k)] & k \geq 1 \end{cases}$$

The series expression for $F(s)$ converges uniformly for Re $s \geq \alpha' > \alpha_0$.

Corollary. If section k of a function meeting the conditions of Theorem 12-5 is a constant D_k, then the $\psi_k(s)$ functions in that theorem are given by

$$\psi_0(s) = \frac{D_0}{s}$$

$$\psi_k(s) = \frac{D_k - D_{k-1}}{s} \qquad k \geq 1$$

In the above theorem we do not have any particular requirement on the spacing of the transition points T_k. Of course, the results presented here include the case where the spacing is uniform and where each section is a repetition of the previous one; i.e., the periodic functions are included. However, this theorem gives somewhat less than the maximum information about the transform of a periodic function. A detailed consideration of the periodic case is given in Chap. 15.

12-5. Laplace Transform of the Derivative of a Piecewise Continuous Function of Exponential Order. In Sec. 10-14, when considering the Laplace transform of the derivative of a function, we considered the function to be continuous. This is in agreement with the usual practice of requiring a function to be continuous if we are to discuss its derivative. Of course, situations like point T_1 in Fig. 12-2 are admitted, where right- and left-handed derivatives exist at the point, making the derivative approach different values from the two sides.

Now suppose that $f(t)$ has isolated points of discontinuity, like T_2 and T_3. No derivative can be defined at these points, but the function can be differentiated at all neighboring points. Points T_1 and T_2 differ to the extent that the derivative (in the one-sided sense) is double-valued at T_1 and does not exist at T_2. Thus, it is not unreasonable to talk about the "derivative," when a function has isolated discontinuities, *if we understand that the derivative function is undefined at the points of discontinuity.* In this sense we shall define the derivative of a PC function and shall use the symbol $f'(t)$ to represent it.

Having defined $f'(t)$ in the manner described above, we now form the function

$$f_0(t) = \int_0^t f'(\tau) \, d\tau \tag{12-29}$$

This function is not affected by the fact that $f'(t)$ is undefined at isolated points, because these points form a set of measure 0 and, by Theorem 8-5, we understand that the value of an integral is not affected if the integrand is undefined over such a set. An integral is a continuous function of its upper limit. Thus $f_0(t)$ is continuous and has the same derivative as $f(t)$ at points where $f'(t)$ is defined. Figure 12-2 shows an example of $f_0(t)$ and also the staircase function $f(t) - f_0(t)$.

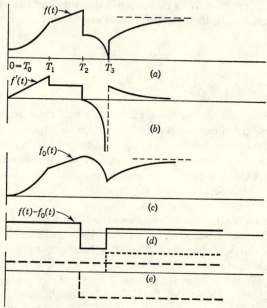

Fig. 12-2. Resolution of a discontinuous function into continuous and discontinuous components.

The fact that $f(t)$ is of exponential order does not ensure that $f_0(t)$, and also $f(t) - f_0(t)$, will be of exponential order. An illustration can be derived from the function

$$e^{t^2}$$

which is not of exponential order. At values

$$t_n = \sqrt{\log n} \qquad n \text{ integer}$$

the function has the value n, and so if a downward jump of unity is introduced at each t_n, a function like Fig. 12-3 (which is never greater than 1) will be obtained. If Fig. 12-3 is labeled $f(t)$, then Eq. (12-29) will yield

$$f_0(t) = e^{t^2} - 1$$

Whereas $f(t)$ is of exponential order 0, $f_0(t)$ is not of exponential order.

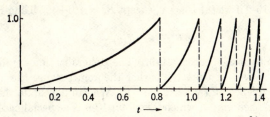

FIG. 12-3. Example of a function of exponential order for which $\int_0^t f'(\tau)\,d\tau$ is not of exponential order.

In view of the above comments, it is necessary also to require $f_0(t)$ to be of exponential order; and then $f(t) - f_0(t)$ will also be of exponential order. We can adopt the notation of the corollary of Theorem 12-5 and write

$$f(t) - f_0(t) = D_k \qquad T_k < t < T_{k+1} \tag{12-30}$$

and this function can also be written as a series of displaced steps, as illustrated in Fig. 12-2e, giving

$$f(t) - f_0(t) = \sum_{k=0}^{\infty} B_k u(t - T_k) \tag{12-31}$$

where
$$\begin{aligned} B_0 &= D_0 \\ B_k &= D_k - D_{k-1} \qquad k \geqq 1 \end{aligned} \tag{12-32}$$

The corollary of Theorem 12-5 is applicable to Eq. (12-30), and, using the notation of Eq. (12-32),

$$F(s) - F_0(s) = \sum_{k=0}^{\infty} \frac{B_k}{s} e^{-sT_k} \tag{12-33}$$

Note that this result would have been obtained by a formal term-by-term transformation of the series in Eq. (12-31). However, Theorem 12-5 provides justification of this step.

Equation (12-29) fits the conditions of Theorem 10-10, and therefore

$$\mathcal{L}[f_0(t)] = \frac{\mathcal{L}[f'(t)]}{s} \tag{12-34}$$

As usual, let $F(s) = \mathcal{L}[f(t)]$. Then Eq. (12-33) becomes

$$\mathcal{L}[f'(t)] = sF(s) - \sum_{k=0}^{\infty} B_k e^{-sT_k} \tag{12-35}$$

This result is a generalization of Theorem 10-9. In that earlier theorem, $f(t)$ is continuous. Thus, the present result should reduce to Theorem 10-9 if we set $B_k = 0$, except $B_0 = f(0+)$. Then Eq. (12-35) gives

$$\mathcal{L}[f'(t)] = sF(s) - B_0$$

in agreement with Theorem 10-9. The generalization implied in Eq. (12-35) is not evident from the earlier theorem because the exponential e^0 is not in evidence on the term $f(0+)$. We can now state the following more general theorem, of which Theorem 10-9 is a special case:

Theorem 12-6.   Let $f(t)$ be a piecewise continuous function of exponential order, and let the discontinuous function

$$f(t) - \int_0^t f'(\tau)\, d\tau = \sum_{k=0}^{\infty} B_k u(t - T_k)$$

also be of exponential order α_0, it being understood that $f'(t)$ is undefined at points where $f(t)$ is discontinuous.   Then, if $F(s) = \mathcal{L}[f(t)]$, the Laplace transform of $f'(t)$ exists and is given by

$$\mathcal{L}[f'(t)] = sF(s) - \sum_{k=0}^{\infty} B_k e^{-sT_k} \qquad \text{Re } s > \alpha_0$$

Equation (12-35) can also be written

$$F(s) = \frac{\mathcal{L}[f'(t)]}{s} + \sum_{k=0}^{\infty} \frac{B_k}{s} e^{-sT_k} \tag{12-36}$$

which bears a certain resemblance to Theorem 12-5.   However, in Eq. (12-36) we are told that exponential terms like

$$\frac{B_k}{s} e^{-sT_k}$$

must appear, but from this equation we know nothing about the form of $\mathcal{L}[f'(t)]$, which can include nonrational functions of s.   In Theorem 12-5, the function is PCSEP, but no such stipulation is required for Eqs. (12-35) and (12-36).

12-6. Approximation of $f(t)$ by Polynomials.   It is sometimes inconvenient to find an exact Laplace transform of a given function; or sometimes it is impossible, as when the function is specified graphically or as a tabulated set of numbers.   This fact lends pertinence to the commonly used method of approximating $f(t)$ by a piecewise continuous function made up of sections of polynomials.   We shall now consider the Laplace transform of such a function.   Assume that each section is an nth-degree polynomial.   All derivatives of $f(t)$ will be defined in the sense given in Theorem 12-6; and each derivative, up to and including the nth, can be discontinuous at the ends of the sections (points T_k). However, the $(n + 1)$st derivative will be zero, because each polynomial is of degree n. Equation (12-35) will be applied to the successive derivatives, until the $(n + 1)$st is reached. The following notation will be used:

$$f_0(t) = \int_0^t f'(\tau)\, d\tau$$
$$f_0'(t) = \int_0^t f''(\tau)\, d\tau$$
$$\dots\dots\dots\dots\dots \tag{12-37}$$
$$f^{(n)}(t) = \int_0^t f^{(n-1)}(\tau)\, d\tau$$

and

$$f(t) - f_0(t) = \sum_{k=0}^{\infty} B_{0k} u(t - T_k)$$
$$f'(t) - f_0'(t) = \sum_{k=0}^{\infty} B_{1k} u(t - T_k) \tag{12-38}$$
$$\dots\dots\dots\dots\dots\dots$$
$$f^{(n)}(t) - f_0^{(n)}(t) = \sum_{k=0}^{\infty} B_{nk} u(t - T_k)$$

each of which is assumed to be EO,α_0. Equation (12-36) then gives

$$F(s) = \frac{\mathcal{L}[f'(t)]}{s} + \sum_{k=0}^{\infty} \frac{B_{0k}}{s} e^{-sT_k}$$

$$\mathcal{L}[f'(t)] = \frac{\mathcal{L}[f''(t)]}{s} + \sum_{k=0}^{\infty} \frac{B_{1k}}{s} e^{-sT_k}$$

$$\dots\dots\dots\dots\dots\dots\dots$$

$$\mathcal{L}[f^{(n)}(t)] = \frac{\mathcal{L}[f^{(n-1)}(t)]}{s} + \sum_{k=0}^{\infty} \frac{B_{(n-1)k}}{s} e^{-sT_k}$$

$$0 = \frac{\mathcal{L}[f^{(n)}(t)]}{s} + \sum_{k=0}^{\infty} \frac{B_{nk}}{s} e^{-sT_k}$$

The zero on the left of the last equation is, of course, the transform of the $(n + 1)$st derivative, which is zero. By substituting successive formulas for transforms of derivatives we get, for Re $s > \alpha_0$,

$$F(s) = \sum_{k=0}^{\infty} \left(\frac{B_{0k}}{s} + \frac{B_{1k}}{s^2} + \frac{B_{2k}}{s^3} + \cdots + \frac{B_{nk}}{s^{n+1}} \right) e^{-sT_k} \tag{12-39}$$

The interesting point about this result is that the transform is completely determined by the discontinuities in the derivatives of the polynomials. The coefficients of the polynomials do not enter explicitly. However, this is not surprising, if we think of each polynomial section being a Taylor series, since the coefficients of the Taylor series are indeed determined by the function and its derivatives at a point.

This same result could have been obtained from Theorem 12-5. That theorem can also be useful if $f(t)$ is sectionally approximated by functions other than polynomials—exponentials, for example. Details of special cases can be worked out as exercises.

12-7. Initial- and Final-value Theorems. In Sec. 10-16 we introduced the relationship between limits approached by $sF(s)$ and the initial and final values of $f(t)$. In that earlier development $f(t)$ was required to be continuous, but on the strength of Sec. 12-5 it is now possible to relax the conditions on $f(t)$. Accordingly, we shall assume that $f(t)$ is piecewise continuous and of exponential order and that

$$f(t) - \int_0^t f'(\tau)\, d\tau = f(0+) + \sum_{k=1}^{\infty} B_k u(t - T_k) \qquad (12\text{-}40)$$

is also of exponential order. The derivative $f'(t)$ is defined arbitrarily to be zero wherever $f(t)$ is discontinuous. From Theorem 12-6 we can write

$$\sigma F(\sigma) = f(0+) + \mathcal{L}[f'(t)] + \sum_{k=1}^{\infty} B_k e^{-\sigma T_k} \qquad (12\text{-}41)$$

Theorem 10-4 provides the information

$$\lim_{\sigma \to \infty} \mathcal{L}[f'(t)] = 0$$

Furthermore, the series

$$\sum_{k=1}^{\infty} \frac{B_k}{s} e^{-sT_k}$$

is a special case of Theorem 12-5, and therefore it exhibits uniform convergence for Re $s \geqq \alpha' > \alpha_0$. The factor $1/s$ does not affect convergence, and therefore the series in Eq. (12-41) converges uniformly in the same region; and the limit as $\sigma \to \infty$ can be taken prior to summation, giving

$$\lim_{\sigma \to \infty} \sum_{k=1}^{\infty} B_k e^{-\sigma T_k} = 0$$

Returning to Eq. (12-41), we have the result

$$\lim_{\sigma \to \infty} \sigma F(\sigma) = f(0+) \tag{12-42}$$

It is concluded that the appearance of discontinuities in $f(t)$ has no effect on the initial-value theorem.

For the next development, directed toward finding the limit of $\sigma F(\sigma)$ as σ goes to zero, Eq. (12-41) is the starting point. Assume that the integral

$$\int_0^\infty f'(t)\, dt$$

exists and that the series

$$\sum_{k=1}^\infty |B_k|$$

converges. The above integral is the Laplace integral of $f'(t)$, with $s = 0$, and therefore from Theorem 10-2 it can be said that

$$\int_0^\infty f'(t) e^{-\sigma t}\, dt$$

converges uniformly for $\sigma \geqq 0$. Accordingly, the limit as σ goes to zero can be taken under the integral, giving

$$\lim_{\sigma \to 0} \int_0^\infty f'(t) e^{-\sigma t}\, dt = \int_0^\infty f'(t)\, dt \tag{12-43}$$

Furthermore, since the infinite sum of $|B_k|$ converges,

$$\sum_{k=1}^\infty B_k e^{-\sigma T_k}$$

converges uniformly for $\sigma \geqq 0$ and therefore

$$\lim_{\sigma \to 0} \sum_{k=1}^\infty B_k^{-\sigma T_k} = \sum_{k=1}^\infty B_k \tag{12-44}$$

Equation (12-41) now yields

$$\lim_{\sigma \to 0} \sigma F(\sigma) = f(0+) + \int_0^\infty f'(t)\, dt + \sum_{k=1}^\infty B_k \tag{12-45}$$

Recalling Eq. (12-29), we recognize the above integral as $f_0(\infty)$, the value of $f(t)$ at infinity with the discontinuous jumps omitted. However, the terms

$$f(0+) + \sum_{k=1}^\infty B_k$$

constitute the contributions of these jumps to $f(t)$, at $t = \infty$, and thus we have

$$\lim_{\sigma \to 0} \sigma F(\sigma) = f(\infty) \tag{12-46}$$

Again we find that the final-value theorem can be stated without change, if the function $f(t)$ has discontinuities. In summary, these results are stated as the following pair of theorems:

Theorem 12-7. Initial-value Theorem. Let $f(t)$ be piecewise continuous, define $f'(t)$ as the derivative of $f(t)$ at points of continuity, and define $f'(t)$ as zero, where $f(t)$ is discontinuous. Each of the functions $f(t)$ and

$$f(t) - \int_0^t f'(\tau)\, d\tau$$

shall be of exponential order. Then,

$$\lim_{\sigma \to \infty} \sigma F(\sigma) = f(0+)$$

where $F(s) = \mathcal{L}[f(t)]$.

Theorem 12-8. Final-value Theorem. Let $f(t)$ be a piecewise continuous function of exponential order, having a derivative $f'(t)$ in the sense defined in Theorem 12-7. Let the integral

$$\int_0^\infty f'(t)\, dt$$

exist, and let the series

$$\sum_{k=1}^\infty |B_k|$$

converge, the B_k terms of this series being defined by

$$f(t) - \int_0^t f'(\tau)\, d\tau = \sum_{k=0}^\infty B_k u(t - T_k)$$

where T_k represents the points of discontinuity of $f(t)$. Then, if $F(s) = \mathcal{L}[f(t)]$,

$$\lim_{\sigma \to 0} \sigma F(\sigma) = f(\infty)$$

12-8. Conditions Sufficient to Make $F(s)$ a Laplace Transform. Previous sections of this chapter have developed some *necessary* conditions that $F(s)$ must satisfy if it is to be a Laplace transform. Now we shall deal briefly with some conditions on $F(s)$ which are *sufficient* to ensure that it is the transform of some function. Conditions which are both necessary and sufficient will not be found.

From Chap. 10, and indirectly from the Fourier integral theorem, it is known that a given function $F(s)$ is the Laplace transform of some function $f(t)$ if $F(s)$ is regular for Re $s > \sigma_c$ and if the integral

$$\int_{Br} F(s)e^{st}\, dt$$

exists for all t. Path Br is to the right of some abscissa σ_c. Existence of this integral is a sufficient condition, but not very enlightening. Conditions that can be checked by inspection are more useful. However, this integral does provide one simple sufficient condition, arrived at by recalling that the above can be written

$$PV \int_{-\infty}^{\infty} F(\sigma + j\omega)e^{\sigma t}e^{j\omega t}\, d\omega \qquad \sigma > \sigma_c$$

Now, by recalling Theorem 8-6, it follows that this integral exists if

$$PV \int_{-\infty}^{\infty} |F(\sigma + j\omega)|\, d\omega \qquad \sigma > \sigma_c$$

exists, and, finally, the last integral exists if

$$\lim_{|\omega| \to \infty} |\omega|\, |F(\sigma + j\omega)| = 0 \qquad \sigma > \sigma_c \qquad (12\text{-}47)$$

Theorem 10-11 provides another useful sufficient condition, namely, that $F(s)$ shall be a rational function for which the degree of the denominator is greater than the numerator.

Now suppose that $\psi(s)$ is a Laplace transform. From Theorem 10-8 we know, if $T \geq 0$, that

$$F(s) = \psi(s)e^{-sT}$$

is a Laplace transform. Finally, by virtue of the convolution theorem, we can say that the product of two transform functions of functions satisfying the conditions of the convolution theorem (Theorem 11-3) is itself a transform function. We collect and state these results in the following theorem:

Theorem 12-9. A function $F(s)$ is the Laplace transform of some function $f(t)$ if any one of the following conditions is satisfied:

1. If a number σ_c exists such that $F(s)$ is regular for Re $s > \sigma_c$ and

$$\lim_{|\omega| \to \infty} |\omega|\, |F(\sigma + j\omega)| = 0 \qquad \sigma > \sigma_c$$

2. If $F(s)$ is a rational function of s for which the degree of the denominator is greater than the numerator

3. If $\psi(s)$ is a Laplace transform and $F(s)$ is in the form

$$F(s) = \psi(s)e^{-sT}$$

where $T \geq 0$

4. If $F(s) = G(s)H(s)$, where $G(s)$ and $H(s)$ are each transforms of functions satisfying the conditions of Theorem 11-3

12-9. Relationships between Properties of $f(t)$ and $F(s)$. Theorems 12-7 and 12-8 give a small amount of information about $f(t)$, at $t = 0$ and $t = \infty$; and Theorem 10-11 has already been cited to show that certain properties of $f(t)$ are related to identifiable properties of $F(s)$, namely, that if $F(s)$ is a rational function with degree of denominator greater than the numerator, $f(t)$ will be a combination of terms like $t^n e^{at}$. Depending on how detailed one wants to become, many other properties of $f(t)$ can be related to properties of $F(s)$. In this section we shall deal with one particular case, of importance in the following chapter. We shall prove the following theorem:

Theorem 12-10. Let $F(s)$ be of the form

$$F(s) = G(s)H(s)$$

where $H(s)$ is the Laplace transform of $h(t)$, a piecewise continuous function of exponential order α_0, and where $G(s)$ is a rational function for which the degree of the numerator is no greater than the degree of the denominator. Then:

1. $\mathcal{L}^{-1}[F(s)]$ exists and is piecewise continuous and of exponential order.
2. $\mathcal{L}^{-1}[F(s)]$ exists and is continuous for $t > 0$ if either of the following is true:
   a. The denominator of $G(s)$ is greater in degree than the numerator.
   b. $h(t)$ is continuous for $t > 0$.

PROOF. Since the numerator of $G(s)$ is no higher in degree than the denominator, we can write

$$G(s) = A + B(s) \tag{12-48}$$

where the numerator of $B(s)$ is of lower degree than the denominator and A is a constant. By Theorem 10-11, $B(s)$ has an inverse transform $b(t)$ which is continuous and of exponential order α_1. Now

$$F(s) = AH(s) + B(s)H(s) \tag{12-49}$$

and since $H(s)$ is a Laplace transform, $AH(s)$ obviously has the inverse $Ah(t)$. Also, $B(s)$ and $H(s)$ have inverses which meet the conditions of the convolution theorem (Theorem 11-3), and therefore $B(s)H(s)$ is the transform of

$$\int_0^t b(\tau)h(t - \tau)\, d\tau$$

But $h(t)$ and $b(t)$ are both the exponential order, respectively, of orders

α_0 and α_1. Also, each is bounded for finite t, and so each is dominated by an exponential, as follows:

$$|h(t)| < M_h e^{\alpha_h t} \qquad \alpha_h > \alpha_0$$
$$|b(t)| < M_b e^{\alpha_b t} \qquad \alpha_b > \alpha_1$$

making possible the estimate

$$\left| \int_0^t b(\tau) h(t - \tau) \, d\tau \right| < M_h M_b e^{\alpha_h t} \int_0^t e^{(\alpha_b - \alpha_h)\tau} \, d\tau = \frac{M_h M_b}{\alpha_b - \alpha_h} (e^{\alpha_b t} - e^{\alpha_h t})$$

The expression on the right can be dominated by an exponential (there is no loss of generality if we assume that $\alpha_h \neq \alpha_b$), and consequently $\mathcal{L}^{-1}[B(s)H(s)]$ is of exponential order. By now we have shown that each term on the right side of Eq. (12-49) has an inverse which is of exponential order. The sum of two functions of exponential order is of exponential order, and so we have proved that

$$\mathcal{L}^{-1}[F(s)]$$

exists and is of exponential order. This completes the proof of part 1 of the theorem.

Part 2 is proved by first observing that, even though $h(t)$ may have points of discontinuity,

$$\mathcal{L}^{-1}[B(s)H(s)] = \int_0^t b(\tau) h(t - \tau) \, d\tau$$

is a continuous function of t. Thus, when referring to Eq. (12-49), $\mathcal{L}^{-1}[F(s)]$ is continuous if $\mathcal{L}^{-1}[AH(s)]$ is continuous. This leads directly to the two cases listed under (2) in the theorem, namely:

a. If the degree of the denominator of $G(s)$ is greater than the degree of the numerator, A will be zero and then $\mathcal{L}^{-1}[AH(s)]$ is identically zero.

b. If $h(t)$ is continuous, obviously

$$\mathcal{L}^{-1}[AH(s)] = Ah(t)$$

is continuous. Thus part 2 has been proved.

Another useful theorem can be proved for functions which can be written $G(s)H(s)$:

Theorem 12-11. If $F(s)$ is of the form

$$F(s) = G(s)H(s)$$

where $G(s)$ is a rational function for which the degree of the numerator is n greater than the degree of the denominator, and if $\mathcal{L}^{-1}[H(s)]$ has $n - 1$ continuous derivatives, all of which are zero at $t = 0$, and if the nth derivative is piecewise continuous and of exponential order, then $\mathcal{L}^{-1}[F(s)]$ exists and is piecewise continuous.

PROOF.   We write

$$F(s) = \frac{G(s)}{s^n} s^n H(s)$$

and observe from Theorem 10-9 that $\mathcal{L}^{-1}[s^n H(s)]$ exists and is piecewise continuous.   The functions $G(s)/s^n$ and $s^n H(s)$, respectively, meet the conditions on $G(s)$ and $H(s)$ in Theorem 12-10.   Therefore, from that theorem, we conclude that $\mathcal{L}^{-1}[F(s)]$ exists and is piecewise continuous.

From Theorem 12-11 we can immediately derive the following corollary, by allowing $G(s)$ to be the single term s^n.   Thus, we have:

Corollary.   If $\mathcal{L}^{-1}[H(s)]$ has $n - 1$ continuous derivatives, all of which are zero at $t = 0$, and if the nth derivative is piecewise continuous and of exponential order, then $\mathcal{L}^{-1}[s^n H(s)]$ exists and is piecewise continuous.

PROBLEMS

12-1. For each of the following special cases establish that the Laplace transform is an entire function, by actually finding the transform:

(a) $f(t) = \begin{cases} t^n & 0 < t < T \\ 0 & T < t \end{cases}$ (b) $f(t) = \begin{cases} e^{at} & 0 < t < T \\ 0 & T < t \end{cases}$

(c) $f(t) = \begin{cases} t^n e^{at} & 0 < t < T \\ 0 & T < t \end{cases}$

12-2. Test the validity of Theorem 12-2 on the following cases:

(a) $f(t) = \begin{cases} 1 & 0 < t < 1 \\ 2 & 1 < t \end{cases}$ (b) $f(t) = (-1)^n e^n \quad n < t < n + 1$

12-3. Prove that if $f(t)$ is continuous and of exponential order, with a PC derivative, then $F(s) = \mathcal{L}[f(t)]$ has the property

$$\lim_{|s| \to \infty} F(s) = 0$$

uniformly with respect to ang s, in a right half plane.

12-4. Of the following functions, which are of exponential order, but not of exponential type, and which are of exponential type?   In each case, give the order.

(a) $f(t) = |\sin t|$ (b) $f(t) = \sin^2 t$
(c) $f(t) = \cos \sqrt{t}$ (d) $f(t) = (-1)^n \quad n < t < n + 1$
(e) $f(t) = e^{-t^2}$ (f) $f(t) = e^{\sin t}$

12-5. Using Eqs. (12-11) and (12-12), show that

$$\mathcal{L}[f'(t)] = sF(s) - f(0+)$$

for the cases to which these equations apply.

12-6. Using Eqs. (12-11) and (12-12), show that

$$\mathcal{L}\left[\int_0^t f(\tau)\ d\tau\right] = \frac{F(s)}{s}$$

for the cases to which these equations apply.

12-7. Using Eqs. (12-11) and (12-12), show that

$$\mathcal{L}\left[(-t)^n f(t)\right] = \frac{d^n}{ds^n} F(s)$$

for the cases to which these equations apply.

12-8. Obtain a power-series expansion for the function whose transform is

$$F(s) = \sin\frac{1}{s}$$

12-9. Let a function be specified at the discrete points 0, T, $2T$, etc., at which the values of the function are, respectively, y_0, y_1, y_2, etc. Let the function be interpolated by holding it constant in the interval $nT < t < (n+1)T$, at the value y_n. Show that the Laplace transform of this interpolated function is

$$F(s) = \frac{y_0}{s} + \frac{1}{s}\sum_{n=1}^{\infty}(y_n - y_{n-1})e^{-nTs}$$

12-10. Let a function be specified at the discrete points 0, T, $2T$, etc., at which the values of the function are, respectively, y_0, y_1, y_2, etc. Let the function be defined for other values of t by using a straight-line interpolation between each pair of defined points. Show that the Laplace transform of the interpolated function is

$$F(s) = \frac{y_0}{s} + \frac{y_1 - y_0}{Ts^2} + \frac{1}{Ts^2}\sum_{n=1}^{\infty}(y_{n+1} - 2y_n + y_{n-1})e^{-nTs}$$

12-11. Let a function be specified at the discrete points 0, T, $2T$, etc., at which the values of the function are, respectively, y_0, y_1, y_2, etc. Let the function be defined at other values of t by an interpolation in which each successive triplet of points (y_0, y_1, y_2), (y_2, y_3, y_4), etc., is connected by a parabolic arc. Show that the Laplace transform of this interpolated function is

$$F(s) = \frac{y_0}{s} - \frac{1}{2Ts^2}\sum_{n=1}^{\infty}(y_{2n+2} - 4y_{2n+1} + 6y_{2n} - 4y_{2n-1} + y_{2n-2})e^{-2nTs}$$

$$+ \frac{1}{T^2s^3}\sum_{n=1}^{\infty}(2y_{2n+2} - 2y_{2n+1} - y_{2n} + 2y_{2n-1} - y_{2n-2})e^{-2nTs}$$

12-12. For each of the functions

(a) $f(t) = n(-1)^n$ $n < t < n + 1$

(b) $f(t) = t(-1)^n$ $n < t < n + 1$

(c) $f(t) = \begin{cases} 0 & 2n < t < 2n + 1 \\ \dfrac{1}{t} & 2n + 1 < t < 2n + 2 \end{cases}$

obtain the derived function

$$f(t) - \int_0^t f'(\tau)\, d\tau$$

12-13. Check the validity of Theorem 12-6 for each of the functions given in Prob. 12-12.

12-14. Find the Laplace transform of each of the functions

(a) $f(t) = n^2(-1)^n \qquad \sqrt{n} < t < \sqrt{n+1}$

(b) $f(t) = n^2(-1)^n \qquad n^2 < t < (n+1)^2$

where n is the succession of integers, in each case.

12-15. Of the following functions, which are you sure have an inverse transform, and which are you sure have a continuous inverse? Explain your answers, and obtain your answers without actually finding the inverse functions.

(a) $\dfrac{\sin^2 s}{s^2}$

(b) $\dfrac{1 - 2e^{-s} + e^{-2s}}{s}$

(c) $\dfrac{1 - 2e^{-s} + e^{-2s}}{s^2}$

(d) $\dfrac{1 + se^{-sT}}{s^2 + as + b}$

(e) $\dfrac{1}{\sqrt{a + bs + s^2}}$

(f) $s \tan^{-1}\dfrac{1}{s}$

12-16. Assume that a given function $f(t)$, defined for all t, has a two-sided Laplace transform which converges in a strip $\sigma_{c_1} < \text{Re } s < \sigma_{c_2}$. State and prove a theorem relating the two-sided transform $\mathcal{L}_2[f(t)]$ with the two-sided transform of its derivative $\mathcal{L}_2[df(t)/dt]$. State conditions used in the derivation, with particular emphasis on permissible range of the strip of convergence.

12-17. Assume that a given function $f(t)$, defined for all t, has a two-sided Laplace transform which converges in a strip $\sigma_{c_1} < \text{Re } s < \sigma_{c_2}$. State and prove a theorem relating the two-sided transform $\mathcal{L}_2[f(t)]$ with the two-sided transform of its integral $\mathcal{L}_2\left[\int_{-\infty}^t f(\tau)\, d\tau\right]$. State conditions used in the derivation, with particular emphasis on permissible range of the strip of convergence.

12-18. Justify the correctness of finding $\mathcal{L}^{-1}[1/(s+1)]$ by applying the inversion integral to the individual terms of the expansion

$$\frac{1}{s+1} = \frac{1}{s} - \frac{1}{s^2} + \frac{1}{s^3} \cdots$$

and check whether or not the result is correct.

CHAPTER 13

SOLUTION OF ORDINARY LINEAR EQUATIONS
WITH CONSTANT COEFFICIENTS

13-1. Introduction. An outline of how the Laplace transform fits into the theory of linear equations is found in Chap. 1. There the emphasis is on motivation, and the treatment is necessarily nonrigorous. In the present chapter it is assumed that you have assimilated the essential concepts of the theory of functions of a complex variable and that the theory and properties of the Laplace transform are familiar.

Most of the material in Chaps. 10 and 11 is relevant to the present chapter. Chapter 12 will also be helpful as background, but specific reference is made only to Secs. 12-8 and 12-9. If you omitted Chap. 12, you should at least study these two sections before proceeding with the present chapter. It might also be helpful for you to reread Chap. 1 at this time.

13-2. Existence of a Laplace Transform Solution for a Second-order Equation. The simple second-order integrodifferential equation

$$a \frac{dy}{dt} + by + c + d \int_0^t y(\tau) \, d\tau = x(t) \qquad (13\text{-}1)$$

is used as the starting point, but now we shall inspect each step more critically than before. First, it is stipulated that $x(t)$ shall be PC and of exponential order. The equality sign in Eq. (13-1) is really an identity, and so, since $x(t)$ has a transform, so also does the left side; and they are equal. Therefore, we immediately have

$$\mathcal{L}\left[a \frac{dy}{dt} + by + c + d \int_0^t y(\tau) \, d\tau \right] = X(s) \qquad (13\text{-}2)$$

where $X(s) = \mathcal{L}[x(t)]$. Since Laplace transforms are unique, any $y(t)$ which is a solution of Eq. (13-2) will also be a solution of Eq. (13-1). However, Eq. (13-2) is of no help in solving for $y(t)$ unless it can be converted to

$$\mathcal{L}\left(a \frac{dy}{dt} \right) + \mathcal{L}(by) + \frac{c}{s} + \mathcal{L}\left[d \int_0^t y(\tau) \, d\tau \right] = X(s) \qquad (13\text{-}3)$$

and so we must investigate whether or not this step can be permitted.

The question before us may be clarified by recalling that, if $f_1(t)$ and $f_2(t)$ are two functions having Laplace transforms, then if $f(t) = f_1(t) + f_2(t)$,

$$\mathcal{L}[f(t)] = \mathcal{L}[f_1(t)] + \mathcal{L}[f_2(t)]$$

However, the converse is not true: if only $f(t)$ is given, then we cannot arbitrarily write $f(t)$ as the sum of two functions and expect the above relation to hold. As an example, consider

$$\frac{t - \sin t}{t^2} = \frac{1}{t} - \frac{\sin t}{t^2}$$

The left side has a transform, but neither function on the right has a transform.

Now return to Eq. (13-2), and *assume*, on a trial basis, that $y(t)$ is continuous and of exponential order. These conditions are sufficient to ensure the existence of

$$Y(s) = \mathcal{L}[y(t)]$$

and to allow Theorems 10-9 and 10-10 to be applied as follows:

$$\mathcal{L}\left(\frac{dy}{dt}\right) = sY(s) - y(0+)$$

$$\mathcal{L}\left[\int_0^t y(\tau)\, d\tau\right] = \frac{Y(s)}{s} \tag{13-4}$$

Thus, if the assumed condition on $y(t)$ is true, each of the transforms on the left of Eq. (13-3) exists and this equation is identical with Eq. (13-2). Then, the solution of Eq. (13-3) is the Laplace transform of the solution of Eq. (13-1). Furthermore, if the notation $Y(s)$ is used, Eq. (13-3) reduces to

$$\left(as + b + \frac{d}{s}\right) Y(s) = X(s) + ay(0+) - \frac{c}{s} \tag{13-5}$$

Equation (13-5) yields the explicit solution

$$Y(s) = \frac{s}{as^2 + bs + d}\left[X(s) + \frac{say(0+) - c}{s}\right] \tag{13-6}$$

Now recall Theorem 12-10, from which it is evident that

$$\mathcal{L}^{-1}\left[\frac{s}{as^2 + bs + d} X(s)\right]$$

is continuous and of exponential order. Also, Theorem 10-11 (or direct observation of the inverse) is available to show that

$$\mathcal{L}^{-1}\left[\frac{say(0+) - c}{as^2 + bs + a}\right]$$

is also continuous and of exponential order. Consequently, if $Y(s)$ is given by Eq. (13-6), its inverse

$$y(t) = \mathcal{L}^{-1}[Y(s)] \tag{13-7}$$

is continuous and of exponential order. Therefore, Eq. (13-5) is equivalent to Eq. (13-1), and since $Y(s)$ is a solution of Eq. (13-5), it follows that $y(t)$ as given by Eq. (13-7) is a solution of the original equation.

Two special cases are of interest. Suppose that the derivative term is missing, giving

$$by + c + d \int_0^t y(\tau)\, d\tau = x(t) \tag{13-8}$$

as the equation to be solved. Since the Laplace transform of a derivative is not needed, it is not necessary for $y(t)$ to be continuous. Assume merely that it has a transform $Y(s)$. Then, by Theorem 10-10,

$$\int_0^t y(\tau)\, d\tau = \frac{Y(s)}{s}$$

and, by virtue of the same reasoning as before, we get

$$\left(b + \frac{d}{s}\right) Y(s) = X(s) - \frac{c}{s} \tag{13-9}$$

which has the solution

$$Y(s) = \frac{s}{bs + d}\left[X(s) - \frac{c}{s}\right] \tag{13-10}$$

This function meets one of the sufficiency conditions for being a Laplace transform, as stated in Theorem 12-9, and therefore the sufficient condition for Eq. (13-9) to be equivalent to Eq. (13-8) is met. It follows that $y(t)$ is the solution of Eq. (13-8).

Now turn to the case

$$d \int_0^t y(\tau)\, d\tau = x(t) \tag{13-11}$$

and proceed as before. Assuming that $y(t)$ has a transform $Y(s)$, we get

$$d\, \frac{Y(s)}{s} = X(s) \tag{13-12}$$

from which

$$Y(s) = \frac{s}{d} X(s)$$

is obtained as the tentative transform of the solution. However, although $Y(s)$ is certainly a function of s, we have no assurance that it is a Laplace transform. It meets none of the conditions specified in Theorem 12-9. In fact, if $x(t)$ is the unit step, $X(s) = 1/s$ and $Y(s) = 1/d$, which is not a Laplace transform. Theorem 12-11 shows that $Y(s)$ can be a transform if $x(0) = 0$ and $x(t)$ is continuous, with a derivative of exponential order. Thus, Eq. (13-11) has a solution if $x(t)$ is $E0,\alpha_0$, continuous at all points and is zero at the origin. The same conclusion could have been reached without recourse to the Laplace transform. It is impossible to define a function such that its integral is a *discontinuous* function of its upper limit, a condition which $y(t)$ would have to satisfy if $x(t)$ were to be discontinuous.

In much of the engineering literature an "impulse function" $\delta(t)$ is "defined" which supposedly does have a discontinuous integral. In the present section we are ruling out such "functions." They are not really functions at all, but certain manipulations can be performed with them as if they were functions. Chapter 14 is devoted to this subject, and so we avoid a digression at this time. These comments are intended to allay any concern you may have if what you read here seems to be in conflict with notions you may already have. In this section we are adhering to the pure concepts of the Laplace transform, and under these conditions Eq. (13-11) has no solution if $x(t)$ is discontinuous.

13-3. Solution of Simultaneous Equations. Suppose that we have n simultaneous integrodifferential equations in the n unknown functions $y_1(t), y_2(t), \ldots, y_n(t)$, as follows:

$$a_{11}\frac{dy_1}{dt} + b_{11}y_1 + d_{11}\int_0^t y_1(\tau)\,d\tau + \cdots + a_{1n}\frac{dy_n}{dt} + b_{1n}y_n$$

$$+ d_{1n}\int_0^t y_n(\tau)\,d\tau + \beta_1 = x_1(t)$$

$$\cdots\cdots\cdots\cdots\cdots\cdots\cdots\cdots\cdots\cdots\cdots\cdots \quad (13\text{-}13)$$

$$a_{n1}\frac{dy_1}{dt} + b_{n1}y_1 + d_{n1}\int_0^t y_1(\tau)\,d\tau + \cdots + a_{nn}\frac{dy_n}{dt} + b_{nn}y_n$$

$$+ d_{nn}\int_0^t y_n(\tau)\,d\tau + \beta_n = x_n(t)$$

This system of equations is typical of an electric network or possibly a mechanical system. We assume that each driving function $x_1(t), \ldots, x_n(t)$ is PC and of exponential order. The constants $\beta_1, \ldots, \beta_n$ take the place of the constant d of Eq. (13-1), representing such quantities as initial displacements or initial charges or flux linkages.

The procedure for solving these is the same as for the single equation. Assume that each unknown function $y_1(t), \ldots, y_n(t)$ is continuous and of exponential order. Then we have

$$Y_1(s) = \mathcal{L}[y_1(t)] \quad sY_1(s) - y_1(0+) = \mathcal{L}\left(\frac{dy_1}{dt}\right) \quad \frac{Y_1(s)}{s} = \mathcal{L}\left[\int_0^t y_1(\tau)\,d\tau\right]$$

$$\cdots\cdots\cdots\cdots\cdots$$

$$Y_n(s) = \mathcal{L}[y_n(t)] \quad sY_n(s) - y_n(0+) = \mathcal{L}\left(\frac{dy_n}{dt}\right) \quad \frac{Y_n(s)}{s} = \mathcal{L}\left[\int_0^t y_n(\tau)\,d\tau\right]$$

If we take the Laplace transform of each side of Eqs. (13-13) and then use the fact that each of the above transforms exists, we can have a sum of transforms on the left equal to a transform of the corresponding $x(t)$. It is too cumbersome to write this all out, but we see that the kth term in the jth equation is

$$\left(a_{jk}s + b_{jk} + \frac{d_{jk}}{s}\right) Y_j(s) - a_{jk}y_j(0+)$$

Accordingly, we adopt the simplifying notation

$$M_{jk}(s) = a_{jk}s + b_{jk} + \frac{d_{jk}}{s} \tag{13-14}$$

so that the transform equations can be written in matrix form:

$$\left\| \begin{matrix} M_{11}(s) & \cdots & M_{1n}(s) \\ \cdots & \cdots & \cdots \\ M_{n1}(s) & \cdots & M_{nn}(s) \end{matrix} \right\| \left\| \begin{matrix} Y_1(s) \\ \cdots \\ Y_n(s) \end{matrix} \right\| = \left\| \begin{matrix} X_1(s) \\ \cdots \\ X_n(s) \end{matrix} \right\| - \frac{1}{s} \left\| \begin{matrix} \beta_1 \\ \cdots \\ \beta_n \end{matrix} \right\|$$

$$+ \left\| \begin{matrix} a_{11} & \cdots & a_{1n} \\ \cdots & \cdots & \cdots \\ a_{n1} & \cdots & a_{nn} \end{matrix} \right\| \left\| \begin{matrix} y_1(0+) \\ \cdots \\ y_n(0+) \end{matrix} \right\| \tag{13-15}$$

where $X_1(s), \ldots, X_n(s)$ are the Laplace transforms of $x_1(t), \ldots, x_n(t)$. Equation (13-15) can be further simplified, as follows:

$$\|M(s)\| \, \|Y(s)\| = \|X(s)\| - \frac{1}{s} \|\beta\| + \|a\| \, \|y(0+)\| \tag{13-16}$$

The definition of each matrix in Eq. (13-16) is apparent by comparing with the expanded matrices in Eq. (13-15).

Each matrix on the right of Eq. (13-16) is known, from the specified information about the system. Therefore, let us combine these into a single known matrix

$$\|E(s)\| = \|X(s)\| - \frac{1}{s} \|\beta\| + \|a\| \, \|y(0+)\| \tag{13-17}$$

to give

$$\|M(s)\| \, \|Y(s)\| = \|E(s)\| \tag{13-18}$$

as the matrix formulation of a system of algebraic equations in the variable s. Matrix notation having been used to arrive at this point, perhaps it is helpful to look at the set of equations themselves, namely,

$$M_{11}(s)Y_1(s) + \cdots + M_{1n}(s)Y_n(s) = E_1(s)$$
$$\cdots\cdots\cdots\cdots\cdots\cdots\cdots\cdots\cdots\cdots\cdots \tag{13-19}$$
$$M_{n1}(s)Y_1(s) + \cdots + M_{nn}(s)Y_n(s) = E_n(s)$$

As in the case of a single equation, if each function $y_1(t)$, . . . , $y_n(t)$ is continuous and of exponential order, the solutions of Eqs. (13-19) are the Laplace transforms of the solutions of Eqs. (13-13). That is, if the set of functions in the matrix $\|Y(s)\|$ is a solution of Eqs. (13-19), then the set of functions in the matrix

$$\left\| \begin{matrix} y_1(t) \\ \cdots \\ y_n(t) \end{matrix} \right\| = \left\| \begin{matrix} \mathcal{L}^{-1}[Y_1(s)] \\ \cdots\cdots\cdots \\ \mathcal{L}^{-1}[Y_n(s)] \end{matrix} \right\| \tag{13-20}$$

will be solutions of Eqs. (13-13). A simpler notation for Eq. (13-20) is

$$\|y(t)\| = \mathcal{L}^{-1}[\|Y(s)\|] \tag{13-21}$$

where the symbol $\mathcal{L}^{-1}$ outside the matrix is defined as the matrix of the inverse transform functions given in Eq. (13-20).

Now we proceed to obtain the solutions of Eqs. (13-19) and then to determine whether each of the functions in the matrix of Eq. (13-20) exists, is continuous, and is of exponential order, as assumed. One way to solve Eqs. (13-19) is to define the determinant

$$D(s) = \begin{vmatrix} M_{11} & \cdots & M_{1n} \\ \cdots\cdots\cdots\cdots \\ M_{n1} & \cdots & M_{nn} \end{vmatrix} \tag{13-22}$$

and the set of cofactors

$$D_{jk}(s) = (-1)^{j+k} \text{ [minor of } D \text{ obtained by omitting row } j \text{ and column } k] \tag{13-23}$$

Then, by Cramer's rule, if $D(s)$ is not identically zero,

$$Y_1(s) = \frac{1}{D}[D_{11}E_1(s) + \cdots + D_{n1}E_n(s)]$$
$$\cdots\cdots\cdots\cdots\cdots\cdots\cdots\cdots\cdots \tag{13-24}$$
$$Y_n(s) = \frac{1}{D}[D_{1n}E_1(s) + \cdots + D_{nn}E_n(s)]$$

This result can also be written in matrix notation,

$$\|Y(s)\| = \|D(s)\|^{-1}\|E(s)\| \tag{13-25}$$

where

$$\|D(s)\|^{-1} = \frac{1}{D} \left\| \begin{matrix} D_{11} & \cdots & D_{n1} \\ \cdots\cdots\cdots\cdots \\ D_{1n} & \cdots & D_{nn} \end{matrix} \right\|$$

is the *inverse* of $\|D(s)\|$. The condition that $D(s)$ shall not be identically zero means that the n functions $Y_1(s) \cdots Y_n(s)$ shall be independent. Such will be the case for equations appropriately chosen to represent a physical problem.

Let us now look at the expression for $Y_j(s)$, which can be written

$$Y_j(s) = \sum_{k=1}^{n} \frac{D_{kj}(s)}{D(s)} E_k(s) \tag{13-26}$$

The determinant $D(s)$ is made up of a sum of n products of factors like

$$as + b + \frac{d}{s}$$

and therefore it is a rational function of the form

$$D(s) = \frac{\text{polynomial of degree } 2n}{s^n} \tag{13-27}$$

Similarly, observing that each minor has a product of $n - 1$ factors like the above,

$$D_{kj}(s) = \frac{\text{polynomial of degree } 2n - 2}{s^{n-1}} \tag{13-28}$$

Therefore, if $D_{kj}(s)/D(s)$ is multiplied in numerator and denominator by s^n we can write it as a rational function

$$\frac{D_{kj}(s)}{D(s)} = \left(\frac{\text{polynomial of degree } 2n - 1}{\text{polynomial of degree } 2n} \right)_{kj} \tag{13-29}$$

The subscript kj is retained as a reminder that these polynomials are different for each k and j. Owing to the special form of $D(s)$ and $D_{kj}(s)$, having only a power of s in each denominator, their ratio is a rational function, even though each one is not a polynomial.

The degrees of numerator and denominator in Eq. (13-29) apply only to the general case where all a, b, and d coefficients in Eqs. (13-13) are non-zero. Otherwise, the degrees of the polynomials specified in Eqs. (13-27) and (13-28) can be lower than indicated; and no general statement can be made about the degree of numerator and denominator of their ratio.

In the general case represented by Eq. (13-29) we see that each term in Eq. (13-26) consists of the product of $E_k(s)$ by a rational function in which the numerator is less in degree than the denominator. The $E_k(s)$ functions are defined by Eq. (13-17), which shows that each one is made up as follows:

$$E_k(s) = X_k(s) + \frac{[a_{k1}y_1(0+) + \cdots + a_{kn}y_n(0+)]s - \beta_k}{s} \tag{13-30}$$

Now recall that $x_k(t)$ is piecewise continuous and of exponential order. Therefore, by Theorem 12-10,

$$\frac{D_{kj}(s)}{D(s)} X_k(s)$$

is the Laplace transform of a continuous function of exponential order. Also, by Theorem 10-11, the same statement can be made about the function

$$\frac{D_{kj}(s)\{[a_{k1}y_1(0+) + \cdots + a_{kn}y_n(0+)]s - \beta_k\}}{sD(s)}$$

Thus, each of the functions

$$\frac{D_{kj}(s)}{D(s)} E_k(s)$$

is a Laplace transform, and

$$\mathcal{L}^{-1}\left[\frac{D_{kj}(s)}{D(s)} E_k(s)\right]$$

is continuous and of exponential order. Now we can write

$$y_j(t) = \mathcal{L}^{-1}[Y_j(s)]$$

$$= \sum_{k=1}^{n} \mathcal{L}^{-1}\left[\frac{D_{kj}(s)}{D(s)} E_k(s)\right] \qquad (13\text{-}31)$$

and we draw the conclusion that $y_j(t)$ is continuous and of exponential order. These are the conditions upon which we based the conclusion that Eqs. (13-18) and (13-19) could be derived from the original equations, and since $\|Y(s)\|$, as given by Eq. (13-25), is a solution of Eq. (13-18), so also is

$$\|y(t)\| = \mathcal{L}^{-1}[\|Y(s)\|]$$

a solution of Eqs. (13-13).

As in the case of the single equation, degenerate cases can occur in which some of the a, b, and c coefficients are zero, and $D_{kj}(s)/D(s)$ is not zero at infinity (the degree of the denominator is not greater than the numerator). Then, whether or not a solution exists depends on the nature of $\|E(s)\|$, as indicated by Theorem 12-11. A solution will not exist for any physical situation requiring an integral to be a discontinuous function of its upper limit. Whenever the original equations have a solution, it can be obtained by the Laplace transform method.

13-4. The Natural Response. Let us now analyze the general solution given by Eq. (13-31). Assume that each zero of $D(s)$ is a regular point of each $E_k(s)$. Since $D(s)$ has a finite number of zeros, we can separate out the sum of the principal parts of the Laurent expansions about each of these poles, giving

$$\sum_{k=1}^{n} \frac{D_{kj}(s)}{D(s)} E_k(s) = G_j(s) + H_j(s) \qquad (13\text{-}32)$$

where $G_j(s)$ is the sum of these principal parts. It follows that $H_j(s)$ is an analytic function and that it is regular at each of the zeros of $D(s)$. $G_j(s)$ is a rational function with denominator higher in degree than the numerator, and therefore $G_j(s)$ is a transform. Also, the left-hand side of Eq. (13-32) is a transform, and therefore $H_j(s)$ is also a transform function, and the solution can be written

$$y_j(t) = g_j(t) + h_j(t) \tag{13-33}$$

where
$$g_j(t) = \mathcal{L}^{-1}[G_j(s)]$$
$$h_j(t) = \mathcal{L}^{-1}[H_j(s)] \tag{13-34}$$

Of course, this is not the only possible way to break $y_j(t)$ into separate terms; but these components have particular significance, which will now be discussed. Since $G_j(s)$ is a rational function, from Sec. 10-20 we know precisely what form $g_j(t)$ will have. A real pole of order p at $s = \sigma_r$ will contribute a term like

$$(A_1 + A_2 t + \cdots + A_p t^{p-1})e^{\sigma_r t}$$

and a pair of complex-conjugate poles, each of order p (one of them being $s = \sigma_r + j\omega_r$), will contribute

$$[B_1 \cos (\omega_r t + \alpha_1) + B_2 t \cos (\omega_r t + \alpha_2) + \cdots$$
$$+ B_p t^{p-1} \cos (\omega_r t + \alpha_p)]e^{\sigma_r t}$$

where the A's, B's, and α's are constants. Each of the terms like

$$t^m e^{\sigma_r t} \qquad \text{and} \qquad t^m \cos (\omega_r t + \alpha_{m+1})e^{\sigma_r t}$$

is a *natural mode* of the system.

The numbers σ_r and $\sigma_r + j\omega_r$ represent the various roots of the equation

$$D(s) = 0 \tag{13-35}$$

This equation is called the *characteristic equation* of the system, and the roots are called the *characteristic values*. The characteristic values are so named because they are the essential parameters which determine the characteristics of the natural modes. The complete function $g_j(t)$, consisting of a sum of all the possible terms like the above, is called the *natural response* of the system. Sometimes it is referred to as the *transient response*, but the latter terminology leaves something to be desired because the natural response does not always die out with increasing t. If any characteristic value has a positive real part, the corresponding natural mode will increase without limits; and the natural response is not a transient.

The form of the natural response is determined by the characteristic values, and therefore by the system parameters, exclusive of the driving functions. However, the A's, B's, and α's appearing in the expressions for the natural modes are obtained from the coefficients in the principal parts of the Laurent expansions of

$$\sum_{k=1}^{n} \frac{D_{kj}(s)}{D(s)} E_k(s)$$

about the zeros of $D(s)$. Therefore, these A, B, and α parameters are influenced by the driving function. However, these affect only the amplitudes and initial angles of the various terms, not their general form.

13-5. Stability. The comments of the previous section introduce the question of stability. A *stable system* is defined as one whose natural response goes to zero as t increases; its natural response is a transient. Thus, we see that the criterion for stability is that the characteristic values shall have nonpositive real parts. The transitional condition, where a characteristic value is a pure imaginary, separates the stable condition from the unstable. It is customary to define this case as stable.

From the foregoing discussion it is obvious that, if the characteristic values of a system are known, then from observation of their real parts it can be determined whether or not the system is stable. In many cases the main interest is in whether the system is stable, rather than in an explicit determination of the natural response. Several methods are available for determining whether or not the zeros of $D(s)$ are in the left half plane, without actually finding their values. Three methods are commonly used for this purpose. The *Routh algorithm** is a method which uses an algorithm of the coefficients of $D(s)$. The *Nyquist criterion* accomplishes the same result by a complex-plane plot of the complex function $D(j\omega)$. A brief description of the Nyquist criterion is given in Sec. 7-7. The third method is the *root-locus* method, briefly described in Sec. 6-10. These three methods vary in the kind of information required, and one or the other may be appropriate in different situations.

13-6. The Forced Response. Returning to Eq. (13-33), we now consider the $h_j(t)$ portion of the response. Its transform $H_j(s)$ is regular at the poles of $G_j(s)$. However, $H_j(s)$ may have singular points of its own, which are determined by the driving functions $E_k(s)$. Accordingly, since $H_j(s)$ is influenced by the driving functions, $h_j(t)$ is called the *forced response*.

Very little general discussion of the forced response is possible without specific knowledge of the driving function. It is apparent that the initial

---

* E. A. Guillemin, "Mathematics of Circuit Analysis," p. 395, John Wiley & Sons, Inc., New York, 1949.

conditions of the system, the β's and various values of $y_j(0+)$, effectively become part of the driving function. The forced response is determined by $E(s)$, and it is recalled from Eq. (13-17) that $E(s)$ includes all terms due to initial conditions.

An important degenerate case is worthy of a brief discussion. It is entirely possible for

$$\sum_{k=1}^{n} D_{kj} E_k(s)$$

to have one or more poles coincident with zeros of $D(s)$. This condition was ruled out in the earlier discussion, but now it will be considered briefly. The routine process described for forming $G_j(s)$ and $H_j(s)$ will now introduce some extraneous terms in $G_j(s)$, because one or more of the poles will be of higher multiplicity than if all the poles are due to zeros of $D(s)$. To illustrate, suppose that s_r is an mth-order zero of $D(s)$, and suppose that

$$\sum_{k=1}^{n} D_{kj} E_k(s)$$

has a pole of order p at s_r. The principal part of the expansion of

$$\sum_{k=1}^{n} \frac{D_{kj}(s)}{D(s)} E_k(s)$$

about this pole is of the form

$$\frac{a_{-1}}{s - s_r} + \frac{a_{-2}}{(s - s_r)^2} + \cdots + \frac{a_{-m}}{(s - s_r)^m} + \cdots + \frac{a_{-(m+p)}}{(s - s_r)^{m+p}}$$

The first m terms in this expansion can be assigned to $G_j(s)$ and the remainder to $H_j(s)$. In this way it is still possible to differentiate between a natural response and a forced response. The situation just described is a generalization of the condition of *resonance*.

13-7. Illustrative Examples. Some of the important ideas relevant to the discussion of the response functions can be illustrated by a few simple examples. In each of these, a single equation is used in order to avoid obscuring the important points in a welter of detail. In each case the system will be assumed to have zero stored energy at $t = 0$.

Example 13-1

$$\frac{dy}{dt} + y + \int_0^t y(\tau)\, d\tau = e^{-t} \qquad t > 0$$

In transformed functions this becomes

$$\left(s + 1 + \frac{1}{s}\right) Y(s) = \frac{1}{s+1}$$

$$Y(s) = \frac{s}{s^2 + s + 1} \frac{1}{s+1}$$

The function $D(s)$ in this case is $(s^2 + s + 1)/s$, and its zeros are

$$s = \tfrac{1}{2}(-1 \pm j\sqrt{3})$$

To find $G(s)$, we want the sum of the principal parts due to these two poles. Routine manipulation yields

$$G(s) = \frac{1}{j\sqrt{3}} \left(\frac{-\tfrac{1}{2} + j\sqrt{3/2}}{s + \tfrac{1}{2} + j\sqrt{3/2}} + \frac{\tfrac{1}{2} + j\sqrt{3/2}}{s + \tfrac{1}{2} - j\sqrt{3/2}}\right)$$

$$= \frac{s + \tfrac{1}{2}}{(s + \tfrac{1}{2})^2 + \tfrac{3}{4}} + \frac{1}{\sqrt{3}} \frac{\sqrt{3/2}}{(s + \tfrac{1}{2})^2 + \tfrac{3}{4}}$$

Referring to Table 10-1 for the inverse of the above, we get

$$g(t) = e^{-t/2}\left(\cos\frac{\sqrt{3}\,t}{2} + \frac{1}{\sqrt{3}}\sin\frac{\sqrt{3}\,t}{2}\right)$$

as the natural response of the system. $H(s)$ is obtained most readily by evaluating the principal part of $Y(s)$ at the pole -1; or it could be obtained by subtracting $G(s)$ from $Y(s)$. The result is

$$H(s) = -\frac{1}{s+1}$$

and

$$h(t) = -e^{-t}$$

This is the forced response. The complete response is, of course,

$$y(t) = e^{-t/2}\left(\cos\frac{\sqrt{3}\,t}{2} + \frac{1}{\sqrt{3}}\sin\frac{\sqrt{3}\,t}{2}\right) - e^{-t}$$

The natural response is a damped sinusoid, and the forced response is a decaying exponential.

Example 13-2

$$y + \int_0^t y(\tau)\,d\tau = \sin t \qquad t > 0$$

$$\left(1 + \frac{1}{s}\right) Y(s) = \frac{1}{s^2 + 1}$$

$$Y(s) = \frac{s}{s+1} \frac{1}{s^2 + 1}$$

In this case $G(s)$ has a single term

$$G(s) = -\frac{\frac{1}{2}}{s + 1}$$

and
$$g(t) = -\frac{1}{2}e^{-t}$$

is the natural response. $H(s)$ is easily obtained as follows:

$$H(s) = Y(s) - G(s)$$
$$= \frac{s + \frac{1}{2}(s^2 + 1)}{(s + 1)(s^2 + 1)} = \frac{1}{2}\frac{(s + 1)^2}{(s + 1)(s^2 + 1)} = \frac{1}{2}\frac{s + 1}{s^2 + 1}$$

Therefore, the forced response is

$$h(t) = \frac{1}{2}\cos t + \frac{1}{2}\sin t$$

and the complete response is

$$y(t) = \frac{1}{2}(-e^{-t} + \cos t + \sin t)$$

In this example the natural response is a decaying exponential, and the forced response is a constant-amplitude sinusoid.

Example 13-3

$$y + \int_0^t y(\tau)\, d\tau = e^{-t} \qquad t > 0$$
$$\left(1 + \frac{1}{s}\right) Y(s) = \frac{1}{s + 1}$$
$$Y(s) = \frac{s}{(s + 1)^2}$$

In this case $D(s) = 1 + 1/s = (1 + s)/s$. The second degree in the denominator of $Y(s)$ arises because $\mathcal{L}(e^{-t})$ contributes a like factor. The expansion of $Y(s)$ is

$$Y(s) = \frac{1}{s + 1} - \frac{1}{(s + 1)^2}$$

and we attribute only the first term to $G(s)$. Thus,

$$G(s) = \frac{1}{s + 1} \qquad H(s) = -\frac{1}{(s + 1)^2}$$

From Table 10-1 we obtain

$$g(t) = e^{-t} \qquad h(t) = -te^{-t}$$

and so the complete response is

$$y(t) = (1 - t)e^{-t}$$

Example 13-4

$$\frac{dy}{dt} + y = 1 \qquad t > 0$$

$$(s + 1)Y(s) = \frac{1}{s}$$

$$Y(s) = \frac{1}{s + 1}\frac{1}{s}$$

Following the usual procedure, we readily obtain

$$G(s) = -\frac{1}{s + 1} \qquad H(s) = \frac{1}{s}$$

and
$$g(t) = -e^{-t} \qquad h(t) = 1$$

are the natural and forced responses, respectively. The complete response

$$y(t) = 1 - e^{-t}$$

is quite familiar, being, among other things, the current in a series resistance-inductance circuit with a constant voltage suddenly applied.

Example 13-5.   The previous example is very simple and is included mainly to set the stage for the following:

$$\frac{dy}{dt} + y = \begin{cases} 1 & 0 < t < 1 \\ 0 & 1 < t \end{cases}$$

In this case the Laplace transform of the driving function is the transcendental function

$$\frac{1 - e^{-s}}{s}$$

and so we have

$$Y(s) = \frac{1}{s + 1}\frac{1 - e^{-s}}{s}$$

The principal part of the Laurent expansion about point $s = -1$ is

$$G(s) = \frac{e - 1}{s + 1}$$

and therefore
$$H(s) = \frac{1 - e^{-s}}{s(s + 1)} - \frac{e - 1}{s + 1}$$

$$= \frac{1}{s} - \frac{e}{s + 1} - e^{-s}\left(\frac{1}{s} - \frac{1}{s + 1}\right)$$

From these results, for the natural and forced response, we respectively get

$$g(t) = (e - 1)e^{-t}$$
$$h(t) = (1 - e^{-(t-1)})u(t) - (1 - e^{-(t-1)})u(t - 1)$$

where the $u(t - 1)$ appears because of the factor e^{-s} in the transform. The above notation somewhat obscures the real nature of $h(t)$, which can also be written

$$h(t) = \begin{cases} 1 - e^{-(t-1)} & 0 < t < 1 \\ 0 & 1 < t \end{cases}$$

Thus, $$y(t) = \begin{cases} (e - 1)e^{-t} + 1 - e^{-(t-1)} & 0 < t < 1 \\ (e - 1)e^{-t} & 1 < t \end{cases}$$

is an alternative formula for the response.

We can make several comments about this example. In the first place, as pointed out in the previous section, the natural response is a continuous exponential, even if the forcing function is discontinuous. Here we find a forced response with a discontinuous derivative. Second, the complete solution is not necessarily written most conveniently by separating the natural and forced responses. To illustrate, the above solution can be put in the form

$$\begin{array}{ll} 1 - e^{-t} & 0 < t < 1 \\ 1 - e^{-t} - (1 - e^{-(t-1)}) & 1 < t \end{array}$$

which puts into evidence a response $1 - e^{-t}$ consisting of a natural response $-e^{-t}$ and a forced response 1, valid for $0 < t < 1$. Over this interval the response is no different from that in Example 13-4. However, when $t > 1$, the solution appears as the sum of two responses, the continuing response due to a positive unit step occurring at $t = 0$, and the other due to a negative unit step occurring at $t = 1$. In this form, there appears to be a new natural response commencing at $t = 1$. However, this formulation is not the one we want here; it is mentioned only because this form is often the one presented in elementary treatments of the subject. Graphs of the natural and forced responses for this example are shown in Fig. 13-1.

13-8. Solution for the Integral Function. Momentarily let us return to the single equation used in the example of Sec. 13-2, namely,

$$a\frac{dy}{dt} + by + c + d \int_0^t y(\tau) \, d\tau = x(t)$$

As previously mentioned, in physical systems y is usually some such quantity as current, voltage, velocity, etc. The variable

$$w(t) = \int_0^t y(\tau) \, d\tau \tag{13-36}$$

has, in these respective cases, physical significance as charge, flux linkage, or displacement. These quantities are sometimes required as part of a solution. There are many ways to obtain $w(t)$. First, if $y(t)$ is

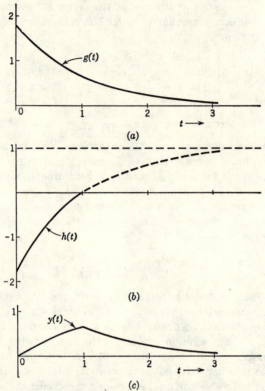

Fig. 13-1. Example of natural and forced response, for the case where the driving function is discontinuous. (a) Natural response; (b) forced response; (c) complete response.

found, Eq. (13-36) can be used to find $w(t)$. Also, in terms of the transform functions it is true that

$$W(s) = \mathcal{L}[w(t)] = \frac{Y(s)}{s}$$

and thus from Eq. (13-6)

$$W(s) = \frac{1}{as^2 + bs + d}\left[X(s) + \frac{say(0+) - c}{s} \right] \qquad (13\text{-}37)$$

from which $w(t)$ is obtained as the inverse transform.

A slightly different approach is to introduce the variable $w(t)$ in the original equation, giving

$$a\,\frac{d^2w}{dt^2} + b\,\frac{dw}{dt} + dw = x(t) - c \qquad (13\text{-}38)$$

If $w(t)$ and $w'(t)$ are both assumed to be continuous and of exponential order,

$$
\begin{aligned}
\mathcal{L}[w'(t)] &= sW(s) - w(0+) \\
&= sW(s) \\
\mathcal{L}[w''(t)] &= s^2W(s) - w'(0+) \\
&= s^2W(s) - y(0+)
\end{aligned}
\tag{13-39}
$$

Equation (13-36) is applied here to provide the information

$$
w(0+) = \int_0^0 y(\tau)\, d\tau = 0
$$

$$
w'(t) = \frac{d}{dt}\int_0^t y(\tau)\, d\tau = y(t)
$$

which is used to get the second form given for each transform. Thus, the transformed equation corresponding to Eq. (13-38) is

$$
(as^2 + bs + d)W(s) = X(s) + ay(0+) - \frac{c}{s}
$$

and this obviously has a solution given by Eq. (13-37). A routine check shows that the conditions assumed for $w(t)$ and $w'(t)$ are satisfied if $x(t)$ is PC and of exponential order.

From this brief development it is seen that it makes little difference whether we solve for $y(t)$ or $w(t)$. An exactly similar development could be given for the set of equations of Sec. 13-4. To write this out in detail would contribute little more than an increase in complexity.

We conclude that an equation carrying a first derivative and an integral is essentially the same as one in which there are second and first derivatives. Accordingly, both are designated as *second-order* equations. To get from one form to another requires a change in variable. In this connection it should be noted that we avoid the formality

$$
\frac{d}{dt}\left[a\frac{dy}{dt} + by + c + d\int_0^t y(\tau)\, d\tau \right] = \frac{dx(t)}{dt}
$$

$$
a\frac{d^2y}{dt^2} + b\frac{dy}{dt} + dy = x'(t)
$$

which is sometimes used to remove the integral. In a general treatment this process is to be avoided because $x'(t)$ would not exist at any point of discontinuity of $x(t)$; and we do not want to be restricted to continuous driving functions.

13-9. Sinusoidal Steady-state Response. The case where the driving function is a suddenly applied sinusoid is of both historical and practical importance. Consider the single equation

$$
a\frac{dy}{dt} + by + d\int_0^t y(\tau)\, d\tau = A\cos(\beta t + \beta)
\tag{13-40}
$$

To expedite taking the Laplace transform, it is convenient to use

$$A \cos (\beta t + \alpha) = A(\cos \alpha \cos \beta t - \sin \alpha \sin \beta t) \qquad (13\text{-}41)$$

so that from Table 10-1 we have

$$\mathcal{L}[A \cos (\beta t + \alpha)] = A \frac{s \cos \alpha - \beta \sin \alpha}{s^2 + \beta^2} \qquad (13\text{-}42)$$

and therefore

$$Y(s) = A \frac{s \cos \alpha - \beta \sin \alpha}{Z(s)(s^2 + \beta^2)} \qquad (13\text{-}43)$$

where

$$Z(s) = as + b + \frac{d}{s} \qquad (13\text{-}44)$$

To find the forced response, we need the sum of the principal parts of the Laurent expansions about poles $+j\beta$ and $-j\beta$. The principal parts for these poles are found, by routine methods, to be

$$\frac{A}{2} \frac{\cos \alpha + j \sin \alpha}{Z(j\beta)(s - j\beta)} \qquad \text{and} \qquad \frac{A}{2} \frac{\cos \alpha - j \sin \alpha}{Z(-j\beta)(s + j\beta)}$$

Note that $Z(s)$ is a real function, because the coefficients a, b, and d are real, and therefore

$$Z(-j\beta) = \overline{Z(j\beta)}$$

and so if we write $Z(j\beta)$ in terms of a magnitude $|Z(j\beta)|$ and an angle $\theta(j\beta)$, we have

$$Z(j\beta) = |Z|e^{j\theta} \qquad \text{and} \qquad Z(-j\beta) = |Z|e^{-j\theta}$$

For the sum of these principal parts we can now write

$$\frac{A}{2|Z|} \left(\frac{e^{j(\alpha-\theta)}}{s - j\beta} + \frac{e^{-j(\alpha-\theta)}}{s + j\beta} \right) = \frac{A}{Z} \frac{s \cos (\alpha - \theta) - \beta \sin (\alpha - \theta)}{s^2 + \beta^2}$$

The forced response is the inverse of this transform, namely,

$$\frac{A}{|Z|} [\cos (\alpha - \theta) \cos \beta t - \sin (\alpha - \theta) \sin \beta t]$$

$$= \frac{A}{|Z|} \cos (\beta t + \alpha - \theta) \qquad (13\text{-}45)$$

This is the familiar solution of "steady-state" circuit theory, which, presumably, you could have written down without recourse to Laplace transform theory. A similar result would be obtained for the case of a set of simultaneous equations. Perhaps this presentation is helpful in establishing an understanding of $Z(s)$, as defined in Laplace transform theory, and $Z(j\omega)$, as defined in the conventional analysis of the sinusoidal steady state. Both are the same function, but with different arguments. The natural response will not be computed for this example because it is not relevant to the point under discussion.

13-10. Immittance Functions. Let us return to the general equations presented in Sec. 13-3. Assume that $x_k(t)$ is the only driving function, and let $y_k(t)$ be the response of the system component acted upon by $x_k(t)$. (The words *system component* are used here because of the desire not to be committed to a particular type of system. Figure 13-2 shows a network loop, a network node, and a mechanical linkage as typical examples of what we mean by a system component.) In agreement with usual network terminology, the forcing function is regarded as being a *source* which maintains the prescribed function under all conditions.

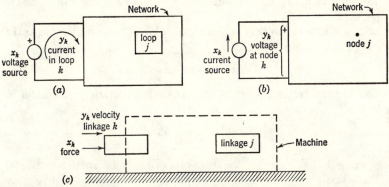

Fig. 13-2. Examples of physical interpretations of system variables for definition of an immittance function.

It is also assumed that the system is initially without stored energy; in the notation of Eqs. (13-13), each β and $y_j(0+)$ is zero. Under these prescribed circumstances, Eq. (13-15) becomes

$$\begin{Vmatrix} M_{11}(s) & \cdots & M_{1n}(s) \\ \cdots & \cdots & \cdots \\ M_{n1}(s) & \cdots & M_{nn}(s) \end{Vmatrix} \begin{Vmatrix} Y_1(s) \\ \cdots \\ Y_n(s) \end{Vmatrix} = \begin{Vmatrix} 0 \\ \cdots \\ X_k(s) \\ \cdots \\ 0 \end{Vmatrix} \qquad (13\text{-}46)$$

Following the notation of Eq. (13-23), we can write the solutions for the transforms of two response functions,

$$Y_k(s) = \frac{D_{kk}(s)}{D(s)} X_k(s) \qquad (13\text{-}47a)$$

$$Y_j(s) = \frac{D_{kj}(s)}{D(s)} X_k(s) \qquad (13\text{-}47b)$$

where $Y_j(s)$ represents any response other than $Y_k(s)$.

The quantities

$$W_{kk}(s) = \frac{D_{kk}(s)}{D(s)}$$

$$W_{kj}(s) = \frac{D_{kj}(s)}{D(s)} \qquad (13\text{-}48)$$

appearing in Eqs. (13-47) are called, respectively, the *self-immittance* presented by the system to source k and the *transfer immittance* from source k to response j. These immittance functions can be defined only when there is a single driving function. The term *immittance* is used here because we are not committed as to what physical quantities are represented by the x and y functions. In network theory, if x is voltage and y is current (Fig. 13-2a), then W_{kk} and W_{jk} are admittances. If the roles of x and y are interchanged (Fig. 13-2b), the functions are impedances.

13-11. Which Is the Driving Function? When $x_k(t)$ was treated as the driving function, we obtained

$$Y_k(s) = W_{kk}(s)X_k(s)$$

as the Laplace transform of the response. It is certainly also valid to write

$$X_k(s) = \frac{1}{W_{kk}(s)} Y_k(s)$$

suggesting that $Y_k(s)$ is then the transform of a driving function and $X_k(s)$ is the transform of the response. This is indeed the case, subject to the comment that $X_k(s)$ must be a transform. This question arises when $W_{kk}(s)$ has a zero at infinity because then $Y_k(s)$ must have a zero of sufficiently high order at infinity to more than cancel the pole due to $1/W_{kk}(s)$. It should furthermore be stated that the so-called driving function, whether it be $X_k(s)$ or $Y_k(s)$, does not need to be due to a so-called source (voltage or current source, for example). Further comments relevant to this point are given at the end of Sec. 13-12.

Of course, we have now suggested that $1/W_{kk}(s)$ is also an immittance function. If $W_{kk}(s)$ is an impedance, $1/W_{kk}(s)$ is an admittance, or vice versa. Thus, impedance and admittance can *each* be determined from *each* of the above equations. In the language of network analysis, this is to say that impedance and admittance can each be found from either a loop analysis or a node analysis.

13-12. Combination of Immittance Functions. Immittance functions are convenient because they often permit the analysis of a complex system by breaking it down into subsystems (called branches, in network analysis) each of which can be analyzed independently. The only requirement is that each subsystem shall interact with the remainder of the system at only one point. For example, suppose that we have two systems, one with response functions $y_1(t), \ldots, y_{k-1}(t)$ and the other with response functions $y_{k+1}(t), \ldots, y_n(t)$. This choice of consecutive numbering from one system to the other is arbitrarily made, to simplify notation. The number k is purposely omitted. Now let the driving function $x_k(t)$ act simultaneously on both systems, in such a way that it acts on component $k - 1$ in the first system and on component

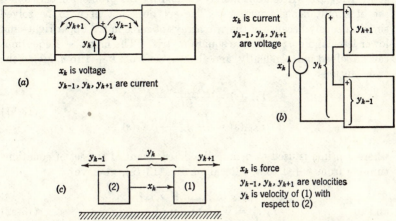

FIG. 13-3. Physical situations which can be described by the addition of immittance functions. (a) Addition of admittances; (b) addition of impedances; (c) mechanical system.

$k + 1$ in the second system. Furthermore, suppose that the systems are connected in such a way that y_{k-1} adds to y_{k+1} to give an additional response variable

$$y_k(t) = y_{k-1}(t) + y_{k+1}(t) \tag{13-49}$$

Illustrations are given in Fig. 13-3.

The combined system is described by the following matrix equation,

$$
\left\|
\begin{array}{ccccccccc}
M_{11} & \cdots & M_{1,k-1} & 0 & \cdots & & \cdots & & 0 \\
\cdots & \cdots & \cdots & \cdots & \cdots & \cdots & \cdots & \cdots & \cdots \\
M_{k-1,1} & \cdots & M_{k-1,k-1} & 0 & \cdots & & & \cdots & 0 \\
0 & \cdots & -1 & 1 & -1 & & 0 & \cdots & 0 \\
0 & \cdots & & \cdots & 0 & M_{k+1,k+1} & \cdots & \cdots & M_{k+1,n} \\
\cdots & \cdots & \cdots & \cdots & \cdots & \cdots & \cdots & \cdots & \cdots \\
0 & \cdots & & \cdots & 0 & M_{k+1,n} & \cdots & \cdots & M_{nn}
\end{array}
\right\|
\times
\left\|
\begin{array}{c}
Y_1 \\
\cdots \\
Y_{k-1} \\
Y_k \\
Y_{k+1} \\
\cdots \\
Y_n
\end{array}
\right\|
=
\left\|
\begin{array}{c}
0 \\
\cdots \\
X_k \\
0 \\
X_k \\
\cdots \\
0
\end{array}
\right\|
\tag{13-50}
$$

where the elements $-1, 1, -1$ represent Eq. (13-49). This set of equations could be solved for $Y_k(s)$, but it is not necessary to treat the whole set. They are written out merely to identify the problem in terms of a set of simultaneous equations. The set does not need to be solved simultaneously because of the two arrays of zeros in the upper right- and lower left-hand corners of the square matrix. The first $k - 1$ equations can be solved independently, and also the set from $k + 1$ to n, giving the two solutions

$$Y_{k-1}(s) = \frac{D_{k-1,k-1}(s)}{D(s)} X_k(s)$$

$$Y_{k+1}(s) = \frac{D'_{k+1,k+1}(s)}{D'(s)} X_k(s) \qquad (13\text{-}51)$$

where a prime is used to denote the determinant of the set of equations running from $k + 1$ to n. Recalling Eq. (13-49), we have

$$Y_k = Y_{k-1} + Y_{k+1}$$

and so

$$Y_k(s) = W_{kk}(s)X_k(s) \qquad (13\text{-}52)$$

where

$$W_{kk}(s) = \frac{D_{k-1,k-1}(s)}{D(s)} + \frac{D'_{k+1,k+1}(s)}{D'(s)} \qquad (13\text{-}53)$$

is the immittance function for the combined system. Each ratio of determinants in Eq. (13-53) is an immittance function of one of the subsystems. When systems are connected together in the manner described, we have shown that their immittance functions add to give the immittance function of the combination.

We stress two points which have already been made: First, the subsystems must have definable immittance functions (zero initial energy and no sources). Second, the interaction of the two systems must be of such a form as to cause two response functions to add, but with no other interaction (so that the over-all system matrix will have appropriately placed sets of zeros).

You will recognize here a generalization of two principles of common usage in network theory. When $x(t)$ represents voltage and $y(t)$ current, we have shown that the parallel connection of two-port networks leads to addition of their admittance functions. In the reverse situation, if $x(t)$ is current and $y(t)$ is voltage, our results indicate the familiar fact that when one-port networks are connected in series their impedance functions add, to give the impedance function for the combination.

Figure 13-3 implies that the two networks in parallel are driven by a specified voltage source $x_k(t)$ or that the two networks in series are driven by a current source $x_k(t)$. In reality, the parallel combination could be considered as driven by a current source or the series combination by a voltage source. If either of the combinations shown in Fig. 13-3 should be a part of a larger network, the actual sources may be absent at the points indicated in Fig. 13-3, as long as there is a source somewhere within the larger network. As an example, consider the networks shown in Fig. 13-4. In Fig. 13-4a $x_0(t)$ represents either a voltage source or a current source, and a voltage $x_1(t)$ will exist and may be considered the driving function for the parallel combination within the dashed rectangle.

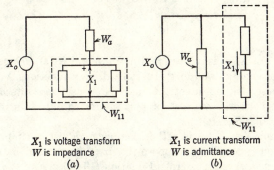

X_1 is voltage transform
W is impedance
(a)

X_1 is current transform
W is admittance
(b)

FIG. 13-4. Simple configurations of immittances.

If W_{11} and W_a are regarded as impedances, the transform of $x_1(t)$ is given by

$$X_1(s) = \frac{W_{11}(s)}{W_a(s) + W_{11}(s)} X_0(s)$$

if $x_0(t)$ is a voltage and by

$$X_1(s) = W_{11}(s)X_0(s)$$

if $x_0(t)$ is a current. Similar comments apply to Fig. 13-4b, if W_{11} is regarded as an admittance.

13-13. Helmholtz Theorem. In the discussion of immittance functions it is carefully stipulated that the system should have zero stored energy at $t = 0$ and that there should be a single driving function. Now we approach the more general situation, where these conditions are not fulfilled. Again, Eqs. (13-13) are taken as the system equations and the transform solution for $Y_j(s)$, given by Eq. (13-26), is the starting point. For notational simplicity, take $j = 1$. Then Eq. (13-26) becomes

$$Y_1(s) = \frac{D_{11}(s)}{D(s)} E_1(s) + \sum_{k=2}^{n} \frac{D_{k1}(s)}{D(s)} E_k(s) \qquad (13\text{-}54)$$

Equation (13-17), defining matrix $\|E(s)\|$, yields the specific formula

$$E_k(s) = X_k(s) - \frac{1}{s}\beta_k + \sum_{v=1}^{n} a_{kv}y_v(0+) \qquad (13\text{-}55)$$

In terms of this, Eq. (13-54) becomes

$$Y_1(s) = W_{11}(s)[X_1(s) + K_1(s)] \tag{13-56}$$

where

$$K_1(s) = \sum_{k=2}^{n} \frac{D_{k1}(s)}{D_{11}(s)} X_k(s) - \frac{1}{s} \sum_{k=1}^{n} \left[\frac{D_{k1}(s)}{D_{11}(s)} \beta_k - \sum_{v=1}^{n} a_{kv} y_v(0+) \right] \tag{13-57}$$

and

$$W_{11}(s) = \frac{D_{11}(s)}{D(s)} \tag{13-58}$$

The function $K_1(s)$ is independent of $Y_1(s)$ and $X_1(s)$, being a function of the initial energies and of the various driving functions $X_k(s)$, where $k \neq 1$. These driving functions may be considered internal to the system, in so far as we are at present interested only in a relationship between $Y_1(s)$ and $X_1(s)$.

The formula for $K_1(s)$ is rather complicated, but a simple physical interpretation is possible. Suppose that $x_1(t)$ is replaced by a function $f(t)$ such that $y_1(t)$ is then zero. Equation (13-56) gives

$$K_1(s) = -\mathcal{L}[f(t)]$$

In other words, $K_1(s)$ is minus the transform of the driving function that gives zero response. In practice, $K_1(s)$ is usually found by modifying the system, to force $y_1(t)$ to be zero rather than to solve the original system with a modified driving function. This modified system is then solved with $x_1(t)$ temporarily regarded as a response variable. In network theory, if x represents voltage and y current, an open circuit forces $y_1(t)$ to be zero; and $\mathcal{L}^{-1}[K_1(s)]$ is the negative of the voltage across the open circuit, the same reference polarity being used as for $x_1(t)$. Similarly, if x represents current and y represents voltage, a short circuit forces y_1 to be zero; and then $K_1(s)$ is the transform of the negative of the current in this short circuit, the current reference direction being kept unchanged. In a mechanical system, if y_1 is a velocity, the system can be modified by clamping linkage 1 and then solving the modified system for the force on that linkage. The factor $W_{11}(s)$ in Eq. (13-56) is, of course, the same immittance function defined in Sec. 13-10.

Equation (13-56) is a mathematical statement of the *Helmholtz theorem*. It is an extension of the immittance concept to the case where "internal sources" and initial energies are not zero. In similarity with Eq. (13-47a), Eq. (13-56) gives a relationship between the transforms of two driving-point quantities, such as $X_1(s)$ and $Y_1(s)$.

In network theory, when x is voltage and y is current, the Helmholtz theorem commonly goes under the name of Thévenin's theorem. It is then customary to draw an equivalent circuit, called the Thévenin equivalent, as shown in Fig. 13-5. In this case $K_1(s)$ is the transform of an *equivalent voltage source*, and $W_{11}(s)$ is an impedance function. If a network is analyzed by regarding x as current and y as voltage, the Helmholtz theorem is known as Norton's theorem and the equivalent network described by Eq. (13-56) is shown in Fig. 13-6. $K_1(s)$ is the Laplace transform of an *equivalent current source*, and $W_{11}(s)$ is an admittance function.

These equivalent circuits, or the corresponding equivalent systems in nonelectrical systems, can then be combined with other subsystems in the manner of combining immittances described in Sec. 13-12. There is one significant difference, however. When Helmholtz equivalent systems are combined, the new system will still have a distribution of driving functions representing equivalent sources, one for each Helmholtz equivalent.

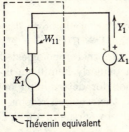

Thévenin equivalent

X_1 and K_1 are voltage transforms
Y_1 is current transform

Fig. 13-5. Circuit of the Thévenin example of the Helmholtz theorem.

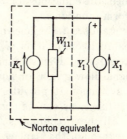

Norton equivalent

X_1 and K_1 are current transforms
Y_1 is voltage transform

Fig. 13-6. Circuit of the Norton example of the Helmholtz theorem.

However, the number of driving functions will, in general, be less than in the original system.

13-14. Appraisal of the Immittance Concept and the Helmholtz Theorem. The Laplace transform does not offer the only technique for solving linear equations, and historically it is not the oldest method. This chapter presents in outline some of the reasons why the Laplace transform has gained wide acceptance for solving engineering problems.

One way to describe the advantage of the Laplace transform is to say that it clearly shows how the integrodifferential equations are reduced to algebraic equations. The algebraic formulation is then exploited further by the definition of immittance functions and Helmholtz equivalents. The advantages accruing from these steps arise particularly because of the existence of the well-developed technique of d-c and sinusoidal steady-state network analysis. With a simple change of variable the immittance functions of sinusoidal analysis become the immittance functions of transform theory. In fact, transform immittances combine in exactly the same way as resistances and conductances of the resistance network. We are also permitted, in the Laplace method, a simple extension of the Helmholtz-Thévenin and Helmholtz-Norton theorems. Except in trivial cases, it is not evident from the original integrodifferential equations how these equivalent systems should be constructed.

From previous studies of network theory you should be aware of why it is such an advantage to be able to define immittance functions and Helmholtz equivalents. The answer is that certain intuitive and formalized topological methods can be used in many cases actually to bypass writing the integrodifferential equations of the complete system. It is usually possible, from an inspection of the network graph, to recognize certain immittances that can be combined and network sections that can be replaced by Helmholtz equivalents. A simpler network is then obtained, the equations for which can be written directly in terms of the transform functions.

The standard techniques of network theory are well known, the series and parallel combination of immittances, the T = Π (or Y = Δ) transformations, and loop and node analyses. Another topological device, the signal flow graph, or flow graph, is particularly helpful in nonelectrical systems. It provides another method for determining how the various functions can be combined to yield the required response function. All these techniques are possible only by virtue of the functions defined through the Laplace transform.

The objective of this chapter is to give the broad concepts of how the Laplace transform serves in the solution of linear equations. The algebraic and topological aspects, as applied to networks, are a very well developed subject, and for further details on that subject you are referred to the many texts on network theory.

13-15. The System Function. Many physical systems described by linear equations are of a particular type in which there are one input port and one output port. In network terminology such a system is called a *two-terminal pair* or a *two-port network*. A rotating shaft driven at one end and loaded at the other is a mechanical two-port system.

In our general notation, let $x_1(t)$ be the input function, and let $y_2(t)$ be the output function. There is no initial stored energy. Then, if we define

$$X_1(s) = \mathcal{L}[x_1(t)]$$
$$Y_2(s) = \mathcal{L}[y_2(t)]$$

the function

$$H(s) = \frac{Y_2(s)}{X_1(s)} \tag{13-59}$$

is called the *system function*. By implication, $x_1(t)$ will be Laplace transformable, and since $y_2(t)$ must also have a transform, it follows that

$H(s)$ must behave appropriately at infinity. For example, if $X_1(s) = 1/s$, $H(s)$ cannot have a pole at infinity.

The system function carries all the information we need to know about the response function $y_2(t)$. In particular, the poles of $H(s)$ are the characteristic values of the natural response.

PROBLEMS

13-1. Referring to the circuit of Fig. P 13-1, the capacitor has an initial charge q_0 and the inductor an initial current i_0. For this specific case, write out the $\|M(s)\|$

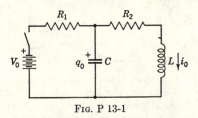

FIG. P 13-1

and $\|E(s)\|$ matrices, and carry out the details leading to a solution for the inductor current, in the form of Eq. (13-23). Determine the natural response of the system, again regarding inductor current as the output.

13-2. Do Prob. 13-1, regarding the voltage across the inductor as the output.

13-3. Do Prob. 13-1, regarding the capacitor current as the output.

13-4. Do Prob. 13-1, regarding the voltage across the capacitor as the output.

13-5. In Fig. P 13-5, up to time $t = 0$, a flywheel of moment of inertia I_0 is rotating freely on shaft s_1, with angular velocity ω_0, while shaft S_1 and all other members are

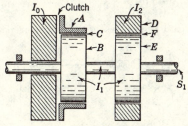

FIG. P 13-5

rotating with angular velocity ω_{10}. At time $t = 0$ the flywheel is clamped rigidly to ring A, which has an ideal viscous frictional coupling to B, through surface C. Structure DEF is a similar frictionally coupled element, ring D being coupled to E by an ideal viscous frictional force at surface F. At each frictional surface let the friction be described by a constant f, the ratio of frictional torque to relative angular velocity. This constant is the same for both surfaces. Assume that ring A has negligible moment of inertia. Obtain a transform solution for the angular velocity ω_1 of shaft S_1. Also obtain the forced and natural responses of the system.

13-6. Figure P 13-6 shows a coupling between two shafts S_0 and S_2. The symbols I_1 and I_2 are moments of inertia of the indicated members, and each spring is described by a constant k (the ratio of torque to relative angular displacement between the ends of the spring). There is assumed to be an ideal viscous frictional force at surface A, such that the frictional torque is given by a frictional constant f multiplied by the relative angular velocity between the two members. Assuming that equilibrium

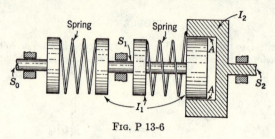

FIG. P 13-6

conditions prevail at $t = 0$, let shaft S_0 be given a sudden angular velocity ω_0 at $t = 0$. Obtain a transform solution for the angular velocity ω_2 of shaft 2. Determine the natural and forced responses of the system. In your solution show the characteristic equation and assign symbols to represent its roots, giving the solution in terms of these symbols.

13-7. For the circuit consisting of R and L in series, with a sinusoidal source

$$v = V_0 \sin (\omega_0 t + \alpha)$$

switched on at $t = 0$, show that the complete response (current in the circuit) approaches the usual steady-state response.

13-8. For the circuit of Fig. P 13-8, the driving voltage V_1 is given by

$$V_1(t) = \begin{cases} 1 & 0 < t < 1 \\ 0 & 1 < t \end{cases}$$

Obtain the forced and natural responses for the output function V_2.

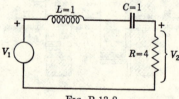

FIG. P 13-8

13-9. Do Prob. 13-8 for the case $R = 2$.

13-10. Do Prob. 13-8 for the function

$$(t) = \begin{cases} \sin t & 0 < t < \pi \\ 0 & \pi < t \end{cases}$$

13-11. Do Prob. 13-8 for the case $R = 2$, using the driving function specified in Prob. 13-10.

13-12. Referring to Fig. P 13-5, what is the immittance function presented to the flywheel, when it is connected to ring A, if immittance is defined as the ratio of transforms of torque to angular velocity. It is assumed that no external load is connected to shaft S_1.

13-13. In the system specified in Fig. 13-5 obtain the Helmholtz equivalent system parameters which could be used if this system were to be connected to another system by shaft S_1. The output functions are ω_1 and T_1, respectively, the angular velocity and torque of shaft S_1.

13-14. In the system specified in Fig. P 13-6, obtain the Helmholtz equivalent system parameters which could be used if this system were to be connected to another system, by shaft S_2. The output functions are ω_2 and T_2, respectively, the angular velocity and torque of shaft S_2.

13-15. In Fig. P 13-1, suppose that there is a pair of output terminals at the ends of the inductance. Assuming that there are an initial current i_0 in the inductance and an initial charge q_0 on the capacitor, and assuming the excitation given in the figure, obtain the parameters of the Thévenin form of the Helmholtz equivalent of this system.

13-16. Using the specifications given in Prob. 13-15, find the parameters of the Norton form of the Helmholtz equivalent.

13-17. In Fig. P 13-1 suppose that there is a pair of output terminals across resistor R_2. Assuming that there are an initial current i_0 in the inductance and an initial charge q_0 on the capacitor and assuming the excitation given in the figure, obtain the parameters of the Thévenin form of the Helmholtz equivalent.

13-18. Using the specifications given in Prob. 13-17, find the parameters of the Norton form of the Helmholtz equivalent.

13-19. Assuming that a system is without stored energy at $t = -\infty$, develop a theory using the two-sided Laplace transform, for finding the response $r(t)$ to a driving function $f(t)$, for the integrodifferential equation

$$a\frac{dr}{dt} + br + c\int_{-\infty}^{t} r(\tau)\,d\tau = f(t)$$

Refer to Probs. 12-16 and 12-17 for further information.

13-20. Referring to Prob. 13-19, let $a = c = 1$ and $b = 4$, and use two-sided Laplace transform theory to solve for $r(t)$ in the following two cases:

(a) $f(t) = \begin{cases} 1 & t > 0 \\ e^t & t < 0 \end{cases}$ (b) $f(t) = e^{-|t|}$

13-21. For a driving function $f(t) = \sqrt{t}$, obtain the response $r(t)$ for the following two systems:

(a) $H(s) = \dfrac{s}{s+1}$ (b) $H(s) = \dfrac{1}{s+1}$

(HINT: See Prob. 11-15 for related transforms.)

13-22. By virtue of the relation $\mathcal{L}[(-t)^n f(t)] = d^n F(s)/ds^n$, certain differential equations involving variable coefficients can be solved. As an example, obtain the solution of the zeroth-order Bessel equation

$$t\frac{d^2 f(t)}{dt^2} + \frac{df(t)}{dt} + tf(t) = 0$$

(HINT: See Prob. 10-35 for related information.)

CHAPTER 14

IMPULSE FUNCTIONS

14-1. Introduction. There are many physical systems in which a driving function has a very short duration and where the response is required only after the driving pulse has died out. Impact between colliding bodies and the response of a ballistic galvanometer are cases in point.

A similar situation also occurs in certain cases where the independent variable is a spatial coordinate. A simple beam with a concentrated weight is an example. In any practical case the weight must cover a finite length of the beam. However, many answers can be obtained, such as reactions at the support and a good approximation of the amount of bending, if the actually distributed force due to the weight is replaced by an equivalent concentrated force at a mathematical point. This conceptual model would not, however, accurately predict the bending directly under the weight. Another example occurs in electrostatic-field theory, where it is recognized that in certain situations charged bodies of finite size can be replaced by point charges.

These cases all share the property that a response is to be computed which will be valid for values of the independent variable outside the range of that variable over which the driving function is not zero. In the notation of the Laplace transform of a function of t, this function of short duration T, and which we shall call $f_T(t)$, can be characterized by a *strength*

$$P = \int_0^T f_T(t)\, dt \tag{14-1}$$

We shall find, within certain limitations, that the response of the system is a function of P and does not depend upon detailed properties of $f_T(t)$. A pulse for which P is *unity* is called a *unit pulse*.

14-2. Examples of an Impulse Response. Consider the circuit of Fig. 14-1, driven by the voltage pulse

$$v_0(t) = \begin{cases} \dfrac{1}{T} & 0 < t < T \\ 0 & T < t \end{cases} \tag{14-2}$$

We assume that the capacitance has zero charge at $t = 0$ and regard

410

the voltage $v_c(t)$ across the capacitor as the response function. From routine analysis, the following relation in transform functions is obtained:

$$V_c(s) = \frac{1}{RCs + 1} \frac{1 - e^{-sT}}{sT}$$

$$= \frac{1}{T}\left(\frac{1}{s} - \frac{1}{s + 1/RC}\right)(1 - e^{-sT}) \quad (14\text{-}3)$$

The inverse transform is readily found to be

$$v_c(t) = \frac{1}{T}[(1 - e^{-t/RC})u(t) - (1 - e^{-(t-T)/RC})u(t - T)] \quad (14\text{-}4)$$

If $t > T$, the above reduces to

$$v_c(t) = \left(\frac{e^{T/RC} - 1}{T/RC}\right)\frac{e^{-t/RC}}{RC} \quad T < t \quad (14\text{-}5)$$

As T approaches zero, the quantity in parentheses approaches 1, so that

$$\lim_{T \to 0} v_c(t) = \frac{1}{RC} e^{-t/RC} \quad (14\text{-}6)$$

The fact that the response function approaches a unique limit is not surprising, since the pulse area

$$\int_0^T v_0(t)\, dt = 1 \quad (14\text{-}7)$$

is constant as the width approaches zero. In the language of Eq. (14-1), the pulse has unit strength.

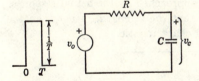

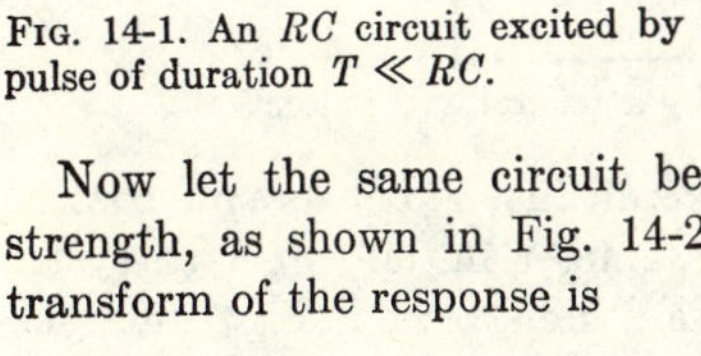

FIG. 14-1. An RC circuit excited by a pulse of duration $T \ll RC$.

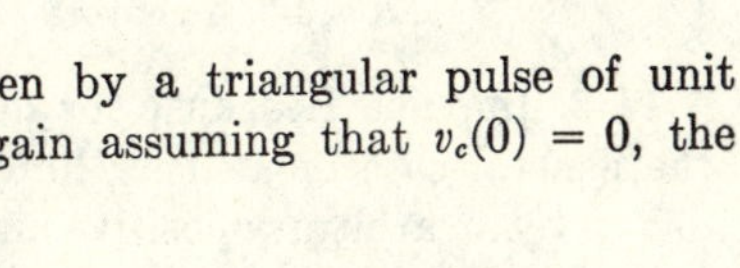

FIG. 14-2. A triangular unit pulse.

Now let the same circuit be driven by a triangular pulse of unit strength, as shown in Fig. 14-2. Again assuming that $v_c(0) = 0$, the transform of the response is

$$V_c(s) = \frac{4}{RCs + 1} \frac{1 - 2e^{-sT/2} + e^{-sT}}{T^2 s^2}$$

$$= \frac{4}{T^2}\left(\frac{1}{s^2} - \frac{RC}{s} + \frac{RC}{s + 1/RC}\right)(1 - 2e^{-sT/2} + e^{-sT}) \quad (14\text{-}8)$$

We are interested only in the response when $T < t$, which is readily seen to be

$$
\begin{aligned}
v_c(t) &= \frac{4}{T^2}\left[t + RC(e^{-t/RC} - 1)\right] \\
&\quad - \frac{8}{T^2}\left[\left(t - \frac{T}{2}\right) + RC(e^{-(t-T/2)/RC} - 1)\right] \\
&\quad + \frac{4}{T^2}\left[(t - T) + RC(e^{-(t-T)/RC} - 1)\right] \\
&= \left(\frac{e^{T/2RC} - 1}{T/2RC}\right)^2 \frac{e^{-t/RC}}{RC} \qquad T < t
\end{aligned}
\tag{14-9}
$$

It is seen that the quantity in parentheses approaches unity as $T \to 0$, and Eq. (14-6) is obtained in the limit.

It being recognized that T will remain finite in any practical situation, it is useful to estimate how small T must be for the actual response to differ negligibly from Eq. (14-6). Considering the rectangular pulse, we see that the correction factor in Eq. (14-5) is

$$
\begin{aligned}
\frac{e^{T/RC} - 1}{T/RC} &= \frac{1 + T/RC + \frac{1}{2}(T/RC)^2 + \cdots - 1}{T/RC} \\
&= 1 + \frac{T}{2RC} + \cdots
\end{aligned}
\tag{14-10}
$$

which shows that, if $T/2RC \ll 1$, the per unit error is approximately $T/2RC$. Thus, for example, if

$$
T < 0.02RC
$$

Eq. (14-6) gives the exact response (for $T < t$) within about 1 per cent.

For these two examples the limits approached by the actual responses for $T < t$ are the same, illustrating that the strength, rather than the shape of the pulse, is the determining characteristic.

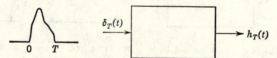

Fig. 14-3. A network excited by a general unit pulse.

14-3. Impulse Response for the General Case. Let us now consider the general case of a linear lumped-parameter initially relaxed system driven by an arbitrary positive unit pulse, as illustrated in Fig. 14-3. The pulse function, designated by $\delta_T(t)$, is one of a family of pulses which are zero outside the variable interval T and which satisfy the conditions

$$
\int_0^T \delta_T(t)\, dt = 1 \qquad \text{and} \qquad \lim_{T \to 0} \int_0^T \delta_T(t)\, dt = 1
\tag{14-11}
$$

$H(s)$ is the system function, and $h_T(t)$ is the response.* Since the system has lumped parameters, it is described by a set of ordinary integro-differential equations and $H(s)$ is therefore a rational function of s. We also temporarily make the assumption that $H(s)$ is of lower degree in the numerator than in the denominator. Under these conditions (see Theorem 10-11) we know that $H(s)$ has a continuous inverse transform

$$h(t) = \mathcal{L}^{-1}[H(s)] \tag{14-12}$$

In transform functions, the system is described by

$$H_T(s) = H(s)\Delta_T(s) \tag{14-13}$$

where $H_T(s)$ and $\Delta_T(s)$ are the respective transforms of $h_T(t)$ and $\delta_T(t)$.

The product in Eq. (14-13) suggests writing the convolution formula

$$h_T(t) = \int_0^T \delta_T(\tau)h(t - \tau)\,d\tau \qquad T < t \tag{14-14}$$

The upper limit of integration is T because $\delta_T(t)$ is identically zero for $T < t$. For simplicity, we shall assume that $\delta_T(t)$ is nonnegative for all t, and it is known that $h(t)$ is continuous. One form of the mean-value theorem for integrals† is used to give

$$h_T(t) = h(t - \lambda) \int_0^T \delta_T(\tau)\,d\tau \qquad T < t$$

or, in view of Eqs. (14-11),

$$h_T(t) = h(t - \lambda) \qquad T < t \tag{14-15}$$

where $0 \leqq \lambda \leqq T$ for all t. Since $h(t)$ is continuous, we can now write

$$\lim_{T \to 0} h_T(t) = h(t) \tag{14-16}$$

* $H(s)$ is the system function defined in Sec. 13-15.
† The mean-value theorem used here is slightly different from that usually stated in elementary texts. The required theorem is readily proved, as follows: Let $u(x)$ be continuous, $a \leqq x \leqq b$, and let $v(x)$ be nonnegative for all x in the interval. Being continuous, $u(x)$ will have a lower bound m and an upper bound M. The quantities $u(x) - m$, $M - u(x)$, and $v(x)$ are all nonnegative. Therefore

$$\int_a^b u(x)v(x)\,dx - m \int_a^b v(x)\,dx = \int_a^b [u(x) - m]v(x)\,dx \geqq 0$$

$$M \int_a^b v(x)\,dx - \int_a^b u(x)v(x)\,dx = \int_a^b [M - u(x)]v(x) \geqq 0$$

and therefore $\quad m \int_a^b v(x)\,dx \leqq \int_a^b u(x)v(x)\,dx \leqq M \int_a^b v(x)\,dx$

In view of this inequality,

$$\int_a^b u(x)v(x)\,dx = u(x') \int_a^b v(x)\,dx$$

where $u(x')$ has a value between m and M. Being continuous, $u(x)$ takes on all values between m and M in the interval of integration. Therefore, we are sure that $a \leqq x' \leqq b$.

In words, we state this result by saying that the inverse transform of the system function is the limit approached by the response to a short pulse of unit area, as the duration of the pulse approaches zero. This general result is confirmed by the example in Sec. 14-2. The limit approached by the response was found to be

$$\frac{1}{RC} e^{-t/RC}$$

and its transform is

$$\frac{1}{RCs + 1}$$

which is indeed the system function $V_c(s)/V_0(s)$.

Let us turn again to the practical situation, in which T can never reach zero. The argument on the right of Eq. (14-15) differs from t by at most the amount T. This fact gives the approximate formula

$$\text{Per unit error} < T \left| \frac{h'(t)}{h(t)} \right| = \left| T \frac{d}{dt} [\log h(t)] \right| \tag{14-17}$$

Applied to the example of Sec. 14-2, this estimate yields T/RC, rather than the previously obtained value of $T/2RC$. The factor of 2 by which these two estimates differ is not significant, since we are obtaining only an order-of-magnitude appraisal. Equation (14-17) is significant because it provides an estimate which is independent of the parameters of a specific system. In general, the result shows that the maximum rate of change of $h(t)$ determines how small T must be for the actual response to differ from $h(t)$ by a negligible amount.

In order to simplify the derivation, we assumed that the pulse was a nonnegative function. However, this restriction is not necessary, as can be seen by applying the mean-value theorem separately to each interval of t over which the function does not change sign. Of course, the condition that its integral from 0 to T shall be unity must be retained.

When the unit-pulse excitation is short enough in duration for $h(t)$ to be indistinguishable from the actual response, such a pulse is called a *unit impulse* and $h(t)$ is called the *impulse response* of the system. The unit-impulse function is not uniquely defined; the acceptable duration T depends on the system function. In practice, a pulse which is short enough in duration to act as an impulse for one system might be too extended to act as an impulse for another system.

The main conclusions reached in this section can be summarized by the following theorem:

Theorem 14-1.* If a linear lumped system, with a transmission function $H(s)$ which has a zero at infinity, is excited by a unit pulse, the limit

---

* A similar theorem can be stated for distributed systems, for which $H(s)$ is a transcendental function of the proper form to have an inverse transform.

approached by the response, as the duration of the pulse approaches zero, is

$$h(t) = \mathcal{L}^{-1}[H(s)]$$

14-4. Impulsive Response. In the previous section we were careful to specify $H(s)$ as a rational function with a zero at infinity, thereby ensuring existence of $\mathcal{L}^{-1}[H(s)]$. We now consider a system for which $H(s)$ is rational but with an nth-order pole at infinity, in which case $\mathcal{L}^{-1}[H(s)]$ does not exist. Clearly, the development of the preceding section does not apply. However, if $\delta_T(t)$ is defined so as to have n continuous derivatives, all of which are zero at $t = 0$, and a piecewise continuous $(n + 1)$st derivative, the following sequence of transforms will exist:

$$\mathcal{L}[\delta_T(t)] = \Delta_T(s)$$
$$\mathcal{L}[\delta_T'(t)] = s\Delta_T(s)$$
$$\cdots \cdots \cdots \cdots$$
$$\mathcal{L}[\delta_T^{(n+1)}(t)] = s^{n+1}\Delta_T(s)$$

Then we can write

$$H_T(s) = H(s)\Delta_T(s)$$
$$= \frac{H(s)}{s^{n+1}}[s^{n+1}\Delta_T(s)]$$

The factor $H(s)/s^{n+1}$ has an inverse transform, and we have shown that the factor in brackets has an inverse. Accordingly, by Theorem 12-10 it is known that $\mathcal{L}^{-1}[H_T(s)]$ exists and can be designated by $h_T(t)$.*

Now let $H(s)$ be written as a partial-fraction expansion (see Sec. 5-15), with the polynomial part written explicitly, giving

$$H(s) = G(s) + A_0 + A_1 s + \cdots + A_n s^n \qquad (14\text{-}18)$$

In the above, $G(s)$ is the sum of the principal parts at the finite poles and therefore has an inverse transform

$$g(t) = \mathcal{L}^{-1}[G(s)] \qquad (14\text{-}19)$$

Now it is possible to write

$$H_T(s) = G(s)\Delta_T(s) + (A_0 + A_1 s + \cdots + A_n s^n)\Delta_T(s) \qquad (14\text{-}20)$$

The first term on the right is the product of two functions, each of which has an inverse transform. Accordingly, the convolution theorem applies, giving

$$\mathcal{L}^{-1}[G(s)\Delta_T(s)] = \begin{cases} \int_0^t g(t - \tau)\delta_T(\tau)\,d\tau & 0 \leq t < T \\ \int_0^T g(t - \tau)\delta_T(\tau)\,d\tau & T \leq t \end{cases} \qquad (14\text{-}21)$$

---

* Theorem 12-11 could be used to yield this result directly.

Furthermore, in view of the existence of n derivatives of $\delta_T(t)$, it follows that

$$\mathcal{L}^{-1}[A_0 \Delta_T(s)] = A_0 \delta_T(t)$$
$$\mathcal{L}^{-1}[A_1 s \Delta_T(s)] = A_1 \delta_T'(t)$$
$$\cdots \cdots \cdots \cdots \cdots \cdots \qquad (14\text{-}22)$$
$$\mathcal{L}^{-1}[A_n s^n \Delta_T(s)] = A_n \delta_T^{(n)}(t)$$

If T is very small, so that there is no significant difference between $g(t - \lambda)$ and $g(t)$, where $0 \leqq \lambda \leqq T$, we can use the mean-value theorem given in Sec. 14-3 to write the approximate equalities

$$\int_0^t g(t - \tau)\delta_T(\tau)\, d\tau = g(t) \int_0^t \delta_T(\tau)\, d\tau = \alpha(t)g(t) \qquad 0 \leqq t < T$$

$$\int_0^T g(t - \tau)\delta_T(\tau)\, d\tau = g(t) \qquad\qquad\qquad t \geqq T \qquad (14\text{-}23)$$

where $$\alpha(t) = \int_0^t \delta_T(\tau)\, d\tau$$

is positive and lies in the interval $0 \leqq \alpha(t) \leqq 1$. Thus, approximately,

$$h_T(t) = \begin{cases} \alpha(t)g(t) + A_0\delta_T(t) + A_1\delta_T'(t) + \cdots + A_n\delta_T^{(n)}(t) \\ \qquad\qquad\qquad\qquad\qquad\qquad\qquad 0 \leqq t < T \quad (14\text{-}24) \\ g(t) \qquad\qquad\qquad\qquad\qquad\qquad\quad T \leqq t \end{cases}$$

The approximation can be as good as we like, by making T sufficiently small. Furthermore, $\delta_T(t)$ and its derivatives become increasingly large, for $0 < t < T$, as T is made small, and so $|\alpha(t)g(t)|$, which remains finite and less than $|g(t)|$, eventually becomes negligible. There is therefore the possibility of saying that in the interval $0 \leqq t < T$ there is no significant change if we replace $\alpha(t)g(t)$ by $g(t)$, thereby allowing $h_T(t)$ to be represented by the single approximate formula

$$h_T(t) = g(t) + A_0\delta_T(t) + A_1\delta_T'(t) + \cdots + A_n\delta_T^{(n)}(t) \quad (14\text{-}25)$$

for $0 < t$. The one point $t = 0$ is excluded, since $g(0)$ is not necessarily zero, whereas $\alpha(t)g(t)$ must be zero at $t = 0$ because $\alpha(0) = 0$.

It is convenient to continue using the notation of the previous section, employing the symbols $\delta(t)$, $\delta'(t)$, etc., to signify $\delta_T(t)$, $\delta_T'(t)$, etc., when T is small enough for Eq. (14-25) to be valid. We shall also drop the subscript on $h_T(t)$, with a similar interpretation. Accordingly, Eq. (14-25) is simplified to

$$h(t) = g(t) + A_0\delta(t) + A_1\delta'(t) + \cdots + A_n\delta^{(n)}(t) \qquad 0 < t \quad (14\text{-}26)$$

as the approximate response of a system to a short unit impulse.

In Eq. (14-17), the symbol $h(t)$ implies the limit of $h_T(t)$ as T goes to zero. However, in the present case, such a limit does not exist, because $\delta_T(t)$, $\delta_T'(t)$, etc., do not approach limits. Thus, we emphasize that in

Eq. (14-26) T is very small, but finite, and that the meaning of the words *very small* depends on the nature of $g(t)$.

When $\delta_T(t)$ is short enough in duration to allow it to be written $\delta(t)$, it becomes a unit impulse. The functions $\delta_T'(t)$, $\delta_T''(t)$, etc., are called, respectively, the *unit doublet, unit triplet*, etc. It is customary, although not necessary, to define $\delta_T(t)$ in such a way that $\delta_T'(t)$ changes sign only once, $\delta_T''(t)$ only twice, etc. Examples are shown in Fig. 14-4 for the case where $\delta_T'(t)$ is piecewise continuous but for which $\delta_T'''(t)$ does not exist. The original pulse can always be made smooth enough at $t = 0$ and $t = T$ to allow any required number of derivatives to be continuous.

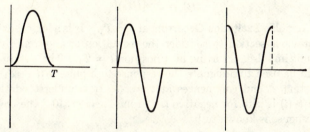

Fig. 14-4. Example of a unit pulse having first and second derivatives which, respectively, yield the unit doublet and the unit triplet.

The fact that $H(s)$ does not have an inverse transform, unless it is zero at infinity, makes it necessary to separate out $G(s)$ and ultimately to arrive at Eq. (14-26) instead of Eq. (14-17). However, a symbolic notation has been invented which allows a similar notation to be used for both cases. This is accomplished by *defining* the symbol $\mathcal{L}^{-1}[H(s)]$ to mean

$$\mathcal{L}^{-1}[H(s)] = \mathcal{L}^{-1}[G(s)] + A_0\delta(t) + A_1\delta'(t) + \cdots + A_n\delta^{(n)}(t) \quad (14\text{-}27)$$

Use of the symbol $\mathcal{L}^{-1}$ might be questioned; but it is reasonable, since the above reduces to a true inverse transform in the event that all A's are zero.

Since Eq. (14-27) should apply if all terms but one are zero, this definition is consistent with the following definitions of symbolic transforms and their inverses:

$$\begin{array}{ll} \mathcal{L}^{-1}(1) = \delta(t) & 1 = \mathcal{L}[\delta(t)] \\ \mathcal{L}^{-1}(s) = \delta'(t) & s = \mathcal{L}[\delta'(t)] \\ \mathcal{L}^{-1}(s^2) = \delta''(t) & s^2 = \mathcal{L}[\delta''(t)] \end{array} \quad (14\text{-}28)$$

$\cdots \cdots \cdots \cdots \quad \cdots \cdots \cdots \cdots$

With the aid of this notation, it is now possible to state a theorem which is more general than Theorem 14-1, as follows:

Theorem 14-2.   Let a linear system be represented by a rational function $H(s)$ having a pole of order n at infinity, and let the function be written

$$H(s) = G(s) + A_0 + A_1s + \cdots + A_ns$$

where $G(s)$ is zero at infinity.   If $\mathcal{L}^{-1}[H(s)]$ is defined as

$$\mathcal{L}^{-1}[H(s)] = \mathcal{L}^{-1}[G(s)] + A_0\delta(t) + A_1\delta'(t) + \cdots + A_n\delta^{(n)}(t)$$

then the response of the system to a unit pulse which is short enough in duration to be considered a unit impulse is given approximately by

$$h(t) = \mathcal{L}^{-1}[H(s)]$$

14-5. Impulse Excitation Occurring at $t = T_1$.   It is a logical extension of the previous section to consider the excitation of a system by a displaced unit impulse, occurring at a position $t = T_1$.   The notation $h_T(t)$ will again be used to designate the response to a pulse $\delta_T(t - T_1)$ which has duration T and commences at $t = T_1$.   It is understood that the function $\delta_T(t)$ is zero for negative t.   From Theorem 10-8 the transform of the response is known to be

$$H_T(s) = H(s)\Delta_T(s)e^{-sT_1} \tag{14-29}$$

If $H(s)$ is given by Eq. (14-18), we have

$$H_T(s) = G(s)\Delta_T(s)e^{-sT_1} + (A_0 + A_1s + \cdots + A_ns^n)\Delta_T(s)e^{-sT_1} \tag{14-30}$$

By arguments involving the convolution theorem, in similarity with the development of Eq. (14-21), it is found that when T is very small we can write

$$\mathcal{L}^{-1}[G(s)\Delta_T(s)e^{-sT_1}] = g(t - T_1)u(t - T_1)$$

and, approximately,

$$\mathcal{L}^{-1}[A_0\Delta_T(s)e^{-sT_1}] = A_0\delta_T(t - T_1)$$

etc., and so when T is sufficiently small to write $h(t)$ in place of $h_T(t)$ we get

$$h(t) = g(t - T_1)u(t - T_1) + A_0\delta(t - T_1) + A_1\delta'(t - T_1) + \cdots \\ + A_n\delta^{(n)}(t - T_1) \qquad t \neq T_1 \tag{14-31}$$

This can be written, symbolically,

$$h(t) = \mathcal{L}^{-1}[H(s)e^{-sT_1}] \qquad t \neq T_1$$

where, by definition,

$$\mathcal{L}^{-1}[H(s)e^{-sT_1}] = \mathcal{L}^{-1}[G(s)e^{-sT_1}] + A_0\delta(t - T_1) + A_1\delta'(t - T_1) \\ + \cdots + A_n\delta^{(n)}(t - T_1) \tag{14-32}$$

This last definition amounts to an extension of the symbolic notation of Eq. (14-27), to which it reduces when $T_1 = 0$. Also, Eq. (14-32) is consistent with the following definitions:

$$\mathcal{L}^{-1}(e^{-sT_1}) = \delta(t - T_1) \qquad \mathcal{L}[\delta(t - T_1)] = e^{-sT_1}$$
$$\mathcal{L}^{-1}(se^{-sT_1}) = \delta'(t - T_1) \qquad \mathcal{L}[\delta'(t - T_1)] = se^{-sT_1} \qquad (14\text{-}33)$$
$$\mathcal{L}^{-1}(s^2 e^{-sT_1}) = \delta''(t - T_1) \qquad \mathcal{L}[\delta''(t - T_1)] = s^2 e^{-sT_1}$$

In concluding this and the previous section it is to be emphasized that we are using a symbolic notation which looks like Laplace transform notation, and which is derived from the Laplace transform theory. However, the functions 1, s, s^2, etc., are not Laplace transforms and hence have no inverse transforms. When applied in these cases, the $\mathcal{L}^{-1}$ symbol should be regarded purely as a symbolism for the definitions given in Eqs. (14-28) and (14-33).

14-6. Generalization of the "Laplace Transform" of the Derivative. The symbolic notation defined in the last two sections permits a generalized treatment of the derivative, extending the earlier results presented in Theorems 10-9 and 12-6. First we shall generalize Theorem 10-9.

Consider a continuous function $f(t)$ of exponential order and having a derivative for $t > 0$ which approaches a limit as t approaches zero; and let the value of $f(0+)$ be different from zero. Let this function be approximated by a continuous function $f_T(t)$, which is identical with $f(t)$, for $t > T$, but which is zero at $t = 0$. An example is shown in Fig. 14-5. The symbol $f'(t)$ will be used to designate the derivative of $f(t)$ for $t > 0$. The modified function can be written with the aid of a $\delta_T(t)$ function, as follows:*

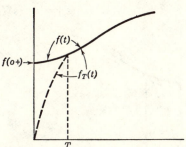

FIG. 14-5. Approximation at a discontinuity by a continuous function.

$$f_T(t) = \int_0^t f'(\tau)\, d\tau + f(0+) \int_0^t \delta_T(\tau)\, d\tau \qquad (14\text{-}34)$$

The derivative is

$$f_T'(t) = f'(t) + f(0+)\delta_T(t) \qquad (14\text{-}35)$$

Although this formula is valid for all values of T, $\delta_T(t)$ will be regarded as a unit impulse in any specific case for which $f'(t)$ is negligible compared with $f(0+)\delta_T(t)$ in the interval $0 < t < T$. This provides a criterion for estimating under what conditions $\delta_T(t)$ can be replaced by $\delta(t)$. To

---

* It is assumed that $f_T'(t)$ has the same sign throughout the interval $0 \leqq t \leqq T$.

drop the T subscript on $f'_T(t)$ would make the symbol indistinguishable from $f'(t)$ as used on the right of Eq. (14-35). Accordingly, $f'_\delta(t)$ is used to designate the derivative when T is sufficiently small. Equation (14-35) now becomes

$$f'_\delta(t) = f'(t) + f(0+)\delta(t) \qquad (14\text{-}36)$$

and from Theorem 10-9 we have

$$f'(t) = \mathcal{L}^{-1}[sF(s) - f(0+)] \qquad (14\text{-}37)$$

and finally
$$f'_\delta(t) = \mathcal{L}^{-1}[sF(s) - f(0+)] + f(0+)\delta(t) \qquad (14\text{-}38)$$

It being recognized that the first term on the right is a true inverse transform, it is convenient to write

$$sF(s) = [sF(s) - f(0+)] + f(0+) \qquad (14\text{-}39)$$

and to recognize this as similar to Eq. (14-18), where

$$A_1 = A_2 = \cdots = A_n = 0$$

Then, by following the definition in Eqs. (14-28), it is consistent to write the symbolic equation

$$\mathcal{L}^{-1}[sF(s)] = \mathcal{L}^{-1}[sF(s) - f(0+)] + f(0+)\delta(t) \qquad (14\text{-}40)$$

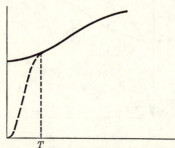

in terms of which Eq. (14-38) becomes

$$f'_\delta(t) = \mathcal{L}^{-1}[sF(s)] \qquad 0 < t \quad (14\text{-}41)$$

This process can be extended to higher-order derivatives. If we are interested in the nth derivative, $f(t)$ and all its derivatives must be modified close to the origin, giving a sequence of functions $f_T(t)$, $f'_T(t)$, . . . , $f_T^{(n-1)}(t)$ each of which is zero at $t = 0$. This requirement places a restriction on the modification of $f(t)$, as compared with the previous case (see

FIG. 14-6. Approximation at a discontinuity by a function having a continuous first derivative.

Fig. 14-6 for the case where the second derivative is considered). An analysis similar to the previous case then yields the symbolic result

$$f_\delta^{(n)}(t) = \mathcal{L}^{-1}[s^nF(s)] \qquad 0 < t$$

where $\mathcal{L}^{-1}[s^nF(s)]$ is defined as

$$\mathcal{L}^{-1}[s^nF(s)] = \mathcal{L}^{-1}[s^nF(s) - s^{n-1}f(0+) - \cdots - f^{(n-1)}(0+)]$$
$$+ f(0+)\delta^{(n-1)}(t) + \cdots + f^{(n-1)}(0+)\delta(t) \quad (14\text{-}42)$$

Of course, this definition is consistent with Eqs. (14-28). In this case, omission of the subscript T implies that T is small enough to ensure negligibility of $f^{(k)}(0+)$ compared with $f^{(k+1)}(0+)\delta^{(k+1)}(t)$ in the small interval $0 \leq t < T$, for each k in the range $0 \leq k \leq n$.

Now we shall generalize Theorem 12-6, which deals with the case where $f(t)$ has discontinuities at T_1, T_2, etc. It will be sufficiently general to consider a single discontinuity, at T_1. Also, since the case where $f(0+) \neq 0$ has been covered, we now assume $f(0+) = 0$. The derivative symbol is used in the sense defined in Theorem 12-6. That is, $f'(t)$ is undefined at T_1. It is assumed that $f'(t)$ exists for $t \neq T_1$ and has finite limits (not necessarily the same) as T_1 is approached from right or left. In similarity with the previous case, a modified function $f_T(t)$ is defined, in which the jump at T_1 is replaced by a continuous curve (with slope of constant sign) extending over a short interval T.

In similarity with Eq. (14-34), we can write

$$f_T(t) = \int_0^t f'(\tau)\, d\tau + B_1 \int_0^t \delta_T(\tau - T_1)\, d\tau \qquad (14\text{-}43)$$

where $\qquad B_1 = f(T_1+) - f(T_1-)$

is the jump at the discontinuity. When T is sufficiently small, the derivative is

$$f_\delta'(t) = f'(t) + B_1\delta(t - T_1) \qquad (14\text{-}44)$$

and from Theorem 12-6

$$f'(t) = \mathcal{L}^{-1}[sF(s) - B_1 e^{-sT_1}]$$

Thus, Eq. (14-44) can be written

$$f_\delta'(t) = \mathcal{L}^{-1}[sF(s) - B_1 e^{-sT_1}] + B_1\delta(t - T_1) \qquad t \neq T_1 \quad (14\text{-}45)$$

Now we write

$$sF(s) = [sF(s) - B_1 e^{-sT_1}] + B_1 e^{-sT_1} \qquad (14\text{-}46)$$

The quantity in brackets has an inverse transform, and so this is similar to Eq. (14-32). Thus, using the symbolic notation of Eqs. (14-33), we have

$$\mathcal{L}^{-1}[sF(s)] = \mathcal{L}^{-1}[sF(s) - B_1 e^{-sT_1}] + B_1\delta(t - T_1) \qquad (14\text{-}47)$$

which allows us again to write

$$f_\delta'(t) = \mathcal{L}^{-1}[sF(s)] \qquad t \neq T_1$$

This process can be extended to higher-order derivatives. As a theorem, we can now state the following:

Theorem 14-3. If a function $f(t)$ and $n - 1$ derivatives are piecewise continuous and of exponential order (the derivatives being considered in the sense defined in Theorem 12-6), and if $f_\delta'(t)$, $f_\delta''(t)$, etc., are the derivatives of approximating continuous functions, where the interval over which each actual discontinuity is approximated by a continuous function is made arbitrarily small, then the symbolic Laplace transform of $f_\delta^{(n)}(t)$ is related to $F(s) = \mathfrak{L}[f(t)]$ by

$$\mathfrak{L}[f_\delta^{(n)}(t)] = s^n F(s) \tag{14-48}$$

Thus, in the sense that the derivative is now defined, the formula for the "Laplace transform" is simplified. This result is of minor value in problem solving, but it has some conceptual value, particularly in the following section. To illustrate how these ideas would apply to a practical problem, consider the equation

$$L\frac{di}{dt} + Ri = V$$

where V is constant, subject to the condition $i(0+) = i_0$. Using the modified current $i_T(t)$, we have

$$i_T'(t) = i'(t) + i_0 \delta_T(t)$$

and so the original equation can also be written

$$L i_T'(t) - L i_0 \delta_T(t) + Ri = V$$

By making T small, the difference between i and i_T becomes negligible for $t > 0$, and the subscript T can be dropped, giving

$$L i_\delta'(t) + Ri(t) = V + L i_0 \delta(t)$$

$\mathfrak{L}[i_\delta'(t)] = sI(s)$, where $I(s) = \mathfrak{L}[i(t)]$, and from Eqs. (14-28) we have

$$\mathfrak{L}[\delta(t)] = 1$$

all in symbolic notation. Therefore,

$$I(s)(Ls + R) = V + L i_0$$

and thus

$$I(s) = \frac{V + L i_0}{R + Ls}$$

is the transform of the solution. This same result would have been obtained by using Theorem 10-9, apparently with less effort than by the present method. However, the example does show that it is possible to regard the initial value i_0 as contributing an impulse excitation of strength i_0 and then to disregard i_0 in writing the Laplace transform of the derivative.

14-7. Response to the Derivative and Integral of an Excitation. The main purpose of introducing the generalized treatment in the previous section was to make it possible to develop a rather interesting and simple relationship which otherwise could be stated only with encumbering qualifications. We consider a lumped linear initially relaxed system, with an excitation $f(t)$ and response $r(t)$. Assume that $f(t)$ and $r(t)$ have Laplace transforms, respectively designated by $F(s)$ and $R(s)$. Then,

$$R(s) = H(s)F(s) \qquad (14\text{-}49)$$

Now suppose that the driving function is replaced by the nth derivative of $f(t)$, where the derivative is interpreted as $f_\delta^{(n)}(t)$, as defined in Sec. 14-6. The transform of the derivative is $s^n F(s)$, and so the response is

$$\mathcal{L}^{-1}[H(s)s^n F(s)] = \mathcal{L}^{-1}[s^n H(s)F(s)] \qquad (14\text{-}50)^*$$

However, the right-hand side is $r_\delta^{(n)}(t)$. This result includes the special case where the nth derivative of $r(t)$ exists in the ordinary sense.

Of course, it is true that the response to the integral of a certain excitation function will be the integral of the response to the excitation. That is, after n integrations of the excitation, the transform of the response is

$$H(s)\frac{F(s)}{s^n} = \frac{H(s)F(s)}{s^n} \qquad (14\text{-}51)$$

No difficulty of existence of transforms is experienced in the case of integral relationships, because each integration increases the degree of the denominator by 1. These results are now summarized as the following theorem:

Theorem 14-4. If a system function $H(s)$ is rational, and if when initially relaxed the response $r(t)$ is obtained from a driving function $f(t)$, then if the driving function is changed to the nth derivative $f_\delta^{(n)}(t)$, the response will change to the nth derivative $r_\delta^{(n)}(t)$, these derivatives being in the sense of the approximations obtained when continuous approximations are used over arbitrarily small intervals at all discontinuities. Also, if the function

$$\int_0^{t_{(n)}} f(t)\, dt = \int_0^t \cdots \int_0^{t_2} \int_0^{t_1} f(t_0)\, dt_0\, dt_1 \cdots dt_{n-1}$$

acts as an excitation, the response is

$$\int_0^{t_{(n)}} r(t)$$

where the symbol (n) implies the iterated process of integration indicated above.

It is interesting to note that even though $f(t)$ might be discontinuous, causing its derivative to have impulsive components, it is not necessary

---

* This equation is written with the knowledge that we are permitted to commute the factors of a transform function. We need consider only the definition of $\mathcal{L}[\delta(t)]$ to realize that this interchange is proper for symbolic as well as true transforms.

that these impulsive components appear in the response. If $s^n H(s)F(s)$ is zero at infinity, there will be no impulsive component in the response.

A particularly useful application of Theorem 14-4 relates the impulse response of a system to the response to the unit step. The derivative of the unit step, in the sense used here, is

$$u_\delta'(t) = \delta(t)$$

and so if $S(t)$ is the response to $u(t)$, it follows from Theorem 14-4 that the impulse response is

$$h(t) = \frac{dS(t)}{dt} \tag{14-52}$$

It is left to you as an exercise to try this out on simple examples.

14-8. The Singularity Functions. The group of functions $\delta_T(t)$, $\delta_T'(t)$, $\delta_T''(\delta)$, etc., have been defined and used in the previous work, showing how it is possible, in a specific problem, to make T small enough so that the response of a system to one of these functions is substantially independent of T. When T is small enough to satisfy this condition, these functions are designated by $\delta(t)$, $\delta'(t)$, $\delta''(t)$, etc., and have been called the unit impulse, the unit doublet, etc. The doublet and triplet are also called *second-order* and *third-order* impulses, respectively. These functions are members of the family of functions called *singularity functions*. Additional members of the family are obtained by successive integration. For example,

$$\delta_T{}^{(-1)}(t) = \int_0^t \delta_T(\tau)\, d\tau$$

$$\delta_T{}^{(-2)}(t) = \int_0^t \delta_T{}^{(-1)}(\tau)\, d\tau \tag{14-53}$$

. . . .

are additional functions, which are designated by $\delta^{(-1)}(t)$, $\delta^{(-2)}(t)$, etc., when T is sufficiently small. Although the functions $\delta_T(t)$, $\delta_T'(t)$, etc., do not approach limits as T goes to zero, it is noted that the *negative-order* singularity functions do approach limits. In fact

$$\lim_{T \to 0} \delta_T{}^{(-1)}(t) = \delta^{(-1)}(t) = u(t) \tag{14-54}$$

is identical with the unit-step function. Also

$$\lim_{T \to 0} \delta_T{}^{(-2)}(t) = \delta^{(-2)}(t) = \begin{cases} 0 & t < 0 \\ t & t > 0 \end{cases}$$

is known as the *unit-ramp* function.

In view of the existence of the limits as T goes to zero, the negative-order singularity functions are precisely definable true functions, as compared with the impulse "functions," which are not true functions because for them limits do not exist. However, the singularity function of order $-k$ is singular to the extent that the kth derivative does not exist at $t = 0$. Of course, singularity functions of all orders can be defined so as to be singular at any prescribed value of t.

Theorem 14-4 applies to the negative-order singularity functions. As an example, suppose that we begin with the known current response of an initially relaxed RL circuit to a unit impulse of voltage, namely,

$$i = \mathcal{L}^{-1}\left(\frac{1}{R + Ls}\right) = \frac{1}{L}\,e^{-Rt/L}$$

The response to a unit step is

$$\frac{1}{L}\int_0^t e^{-R\tau/L}\,d\tau = -\frac{1}{R}\,e^{-R\tau/L}\bigg|_0^t = \frac{1}{R}\,(1 - e^{-Rt/L})$$

and the response to a unit ramp is

$$\frac{1}{R}\int_0^t (1 - e^{-R\tau/L})\,d\tau = \frac{1}{R}\left(\tau + \frac{L}{R}\,e^{-R\tau/L}\right)\bigg|_0^t$$

$$= -\frac{L}{R^2} + \frac{t}{R} + \frac{L}{R^2}\,e^{-Rt/L}$$

14-9. Interchangeability of Order of Differentiation and Integration.

An interesting interpretation of the positive-order singularity functions is obtained when we consider the difference between the results obtained when a function is differentiated and then integrated and those when it is integrated and then differentiated. These operations are noncommutative, in general. We shall consider here only the first-order impulse.

In the notation we have been using, if $f(t)$ is integrable and differentiable for $t > 0$, differentiating and then integrating yields

$$\int_0^t f'(\tau)\,d\tau = f(t) - f(0+) \tag{14-55}$$

However, integrating first, and then differentiating, yields the different result

$$\frac{d}{dt}\int_0^t f(\tau)\,d\tau = f(t) \tag{14-56}$$

This situation can be extended to the case where $f(t)$ has a discontinuity at T_1 and $f'(t)$ is defined in the sense given in Theorem 12-6. Then, for $t > T_1$,

$$\int_0^t f'(\tau)\,d\tau = \int_0^{T_1} f'(\tau)\,d\tau + \int_{T_1}^t f'(\tau)\,d\tau$$
$$= f(t) + f(T_1-) - f(T_1+) - f(0) \tag{14-57}$$

whereas Eq. (14-56) still holds for the inverted order. If we differentiate first and then integrate, we get a continuous function which is zero at $t = 0$. The constants in Eqs. (14-55) and (14-57) subtract out the amounts of the discontinuous jumps in $f(t)$.

From the above, it is evident that noncommutability of differentiation and integration is due to $f(0+)$ not necessarily being zero and to discontinuities of $f(t)$. Accordingly, it is evident that the operations are commutable for a continuous function if it is zero for $t = 0$. The approximation for $f(t)$ previously described and labeled $f_T(t)$ meets this condition. Thus,

$$\int_0^t f_T'(\tau) \, d\tau = \frac{d}{dt} \int_0^t f_T(\tau) \, d\tau \qquad (14\text{-}58)$$

for all approximations of $f(t)$ for which T is finite. A symbolic notation can now be introduced whereby T is considered to be small enough to write $f(t)$ on the right of Eq. (14-58) and $f_\delta'(t)$ on the left, giving

$$\int_0^t f_\delta'(\tau) \, d\tau = \frac{d}{dt} \int_0^t f(\tau) \, d\tau \qquad (14\text{-}59)$$

This result is consistent with the symbolic property of the impulse function whereby

$$\int_0^t \delta(\tau) \, d\tau = \begin{cases} 1 & t > 0 \\ 0 & t \leq 0 \end{cases}$$

Referring to Eqs. (14-36) and (14-44), we have

$$f_\delta'(t) = f'(t) + f(0+)\delta(t) + [f(T_1+) - f(T_1-)]\delta(t - T_1)$$

and its integral, for $0 < t < T_1$, is

$$\int_0^t f_\delta'(\tau) \, d\tau = f(t) - f(0+) + f(0+) = f(t)$$

For $T_1 < t$, the integral is

$$\int_0^t f_\delta'(\tau) \, d\tau = f(t) - f(0+) + f(0+) - [f(T_1+) - f(T_1-)]$$
$$+ [f(T_1+) - f(T_1-)]$$
$$= f(t)$$

In both cases, $f(t)$ agrees with the known correct function for the right-hand side of Eq. (14-59).

14-10. Integrands with Impulsive Factors. A unit pulse has been defined to have the property

$$\int_0^T \delta_T(\tau) \, d\tau = 1$$

and $\delta_T(t) = 0$ when $t < 0$ and when $t > T$. An integral like

$$\int_0^t \delta_T(\tau)f(\tau)\, d\tau$$

sometimes occurs in problems arising from physical situations. Since $\delta_T(t)$ is zero for $T < t$, we immediately have

$$\int_0^t \delta_T(\tau)f(\tau)\, d\tau = \int_0^T \delta_T(\tau)f(\tau)\, d\tau \qquad T < t \qquad (14\text{-}60)$$

Except for differences in symbols, the right-hand side of Eq. (14-60) is like the integral in Eq. (14-14). Relying on the discussion of the latter, as given in Sec. 14-3, it is evident that, if $\delta_T(t)$ is nonnegative, and if $f(t)$ is continuous in the interval $0 \leqq t \leqq T$,

$$\int_0^t \delta_T(\tau)f(\tau)\, d\tau = f(\lambda) \int_0^T \delta_T(\tau)\, d\tau = f(\lambda) \qquad T < t$$

where $0 \leqq \lambda \leqq T$. If T is so small that $f(t)$ does not change appreciably between 0 and T, $\delta_T(t)$ can be called an impulse function and written $\delta(t)$ and $f(\lambda)$ can be replaced by $f(0)$ as an approximate equivalent, giving

$$\int_0^t \delta(\tau)f(\tau)\, d\tau = f(0) \qquad t > 0 \qquad (14\text{-}61)$$

Similarly, if $\delta_T(t - T_1)$ is defined with T small enough so that $f(t)$ does not change appreciably between T_1 and $T_1 + T$, we have the general case

$$\int_0^t \delta(t - T_1)f(\tau)\, d\tau = f(T_1) \qquad T_1 < t \qquad (14\text{-}62)$$

Corresponding formulas can be written for the higher-order impulse functions. Taking the second-order case as an example, we assume that $\delta_T(t)$ has a piecewise continuous derivative $\delta_T'(t)$ and consider

$$\int_0^t \delta_T'(\tau)f(\tau)\, d\tau = \int_0^T \delta_T'(\tau)f(\tau)\, d\tau \qquad T < t$$

According to Theorem 8-4, the integral on the right can be integrated by parts, giving

$$\int_0^T \delta_T'(\tau)f(\tau)\, d\tau = \delta_T(T)f(T) - \delta_T(0)f(0) - \int_0^T \delta_T(\tau)f'(\tau)\, d\tau$$

However, $\delta_T(t)$ must be continuous in order to be piecewise differentiable, and $\delta_T(t) = 0$ for $t < 0$ and $t > T$. Therefore $\delta_T(T) = \delta_T(0) = 0$. Furthermore, if $f'(\tau)$ is PC, the principles applied to Eq. (14-60) apply, giving

$$\int_0^T \delta_T(\tau)f'(\tau)\, d\tau = -f'(\lambda) \qquad 0 \leqq \lambda \leqq T \qquad (14\text{-}63)$$

If $f'(t)$ is approximately constant between 0 and T, $\delta'_T(\tau)$ can be replaced by the second-order impulse $\delta'(t)$, and $f'(\lambda)$ can be replaced by $f'(0)$, giving the approximate result

$$\int_0^t \delta'(\tau)f(\tau)\,d\tau = -f'(0) \qquad t > 0 \qquad (14\text{-}64)$$

Repeated application of this analysis yields a result for the general nth-order impulse occurring at the general point T_1, leading to the following theorem:

Theorem 14-5.   If the nth derivative of $f(t)$ is piecewise continuous, and if $\delta^{(n)}(t)$ is an nth-order impulse defined as $\delta_T{}^{(n)}(t)$ when T is small enough so that $f^{(n)}(t)$ is approximately constant in the range

$$T_1 \leqq t \leqq T_1 + T$$

then approximately

$$\int_0^t \delta^{(n)}(\tau - T_1)f(\tau)\,d\tau = (-1)^n f^{(n)}(T_1) \qquad (14\text{-}65)$$

14-11. Convolution Extended to Impulse Functions.   In Theorem 12-3 it is stated that under certain conditions

$$\mathcal{L}^{-1}[F(s)G(s)] = \int_0^t f(\tau)g(t - \tau)\,d\tau$$
$$= \int_0^t f(t - \tau)g(\tau)\,d\tau$$

where the integral on the right is called a convolution integral.   With the aid of ideas presented in Sec. 14-10, it is possible symbolically to extend the convolution theorem to certain cases where $F(s)$ and/or $G(s)$ do not have inverse transforms.   Suppose, for example, that in the symbolic sense we are interested in the inverse transform of the product

$$[F(s) + Ae^{-sT_1}][G(s) + Be^{-sT_2}]$$

where $T_1 = T_2$ is a possibility, and where $F(s)G(s)$ meets conditions sufficient to have an inverse.   Then, by using the definition of an inverse given in Eq. (14-32), the inverse of the above is found to be

$$\mathcal{L}^{-1}[F(s)G(s)] + \mathcal{L}^{-1}[AG(s)e^{-sT_1} + BF(s)e^{-sT_2}] + AB\mathcal{L}^{-1}(e^{-s(T_1+T_2)})$$
$$= \int_0^t f(\tau)g(t - \tau)\,d\tau + Ag(t - T_1)u(t - T_1) + Bf(t - T_2)u(t - T_2)$$
$$+ AB\delta(t - T_1 - T_2) \qquad (14\text{-}66)$$

Next we shall apply the principles established in Sec. 14-10 to the convolution integral, to establish that the same result is obtained.   The integral in question is obtained by using the following individual relationships:

$$\mathcal{L}^{-1}[F(s) + Ae^{-sT_1}] = f(t) + A\delta_a(t - T_1)$$
$$\mathcal{L}^{-1}[G(s) + Be^{-sT_2}] = g(t) + B\delta_b(t - T_2) \qquad (14\text{-}67)$$

The impulse functions would normally be written without designating finite widths by the a and b subscripts. However, these are included as a reminder that the widths must be small enough so that $f(t)$ and $g(t)$ do not vary appreciably as t varies over the respective intervals a and b.

The convolution integral of the above functions gives the formal expression

$$\int_0^t [f(\tau) + A\delta_a(\tau - T_1)][g(t - \tau) + B\delta_b(t - \tau - T_2)]\, d\tau$$

$$= \int_0^t f(\tau)g(t - \tau)\, d\tau + A \int_0^t g(t - \tau)\delta_a(\tau - T_1)\, d\tau$$

$$+ B \int_0^t f(\tau)\delta_b(t - \tau - T_2)\, d\tau + AB \int_0^t \delta_a(\tau - T_1)\delta_b(t - \tau - T_2)\, d\tau$$

$$(14\text{-}68)$$

The first integral on the right includes no impulse function, and so no further discussion of this term is required. The second and third integrals are similar. When a and b are sufficiently small, the subscripts are omitted, giving

$$A \int_0^t g(t - \tau)\delta(\tau - T_1)\, d\tau = \begin{cases} 0 & t < T_1 \\ Ag(t - T_1) & T_1 < t \end{cases}$$
$$= Ag(t - T_1)u(t - T_1) \qquad (14\text{-}69a)$$

$$B \int_0^t f(\tau)\delta(t - \tau - T_2)\, d\tau = \begin{cases} 0 & t < T_2 \\ Bf(t - T_2) & T_2 < t \end{cases}$$
$$= Bf(t - T_2)u(t - T_2) \qquad (14\text{-}69b)$$

The last integral on the right of Eq. (14-68) presents a different situation, because Theorem 14-5 does not provide for a product of two impulse functions in the integrand. In view of the fact that $\delta_a(t)$ is zero except in the interval $0 < t < a$ and $\delta_b(t)$ is zero except in the interval $0 < t < b$, it follows that the integral

$$\int_0^t \delta_a(\tau - T_1)\delta_b(t - \tau - T_2)\, d\tau$$

is zero for $t < T_1$; and, for $t > T_1 + a$, the integral takes on fixed limits as follows:

$$\int_{T_1}^{T_1+a} \delta_a(\tau - T_1)\delta_b(t - \tau - T_2)\, d\tau$$

The second factor of the integrand is zero for $t < \tau + T_2$, and the integration limits put τ in the range $T_1 < \tau < T_1 + a$. Accordingly, we see that the integral is zero for $t < T_1 + T_2$. The same factor is also zero for $t > T_2 + b + \tau$, and since τ has $T_1 + a$ as its maximum value, the integral is zero for $t > T_1 + T_2 + a + b$. Thus the integral in question is a function of t, with the property of being identically zero for $t < T_1 + T_2$ and for $t > T_1 + T_2 + a + b$. It is a pulse of duration

$a + b$. We have yet to show that its integral is unity, making it a unit pulse.

We proceed by showing that the integral from 0 to ∞ is unity, recognizing that the integral over this range is the same as the integral over the pulse width. Since a and b are finite, the following Laplace transforms,

$$\Delta_a(s) = \mathcal{L}[\delta_a(t)]$$
$$\Delta_b(s) = \mathcal{L}[\delta_b(t)] \tag{14-70}$$

exist. From Theorem 11-3 for the convolution integral and Theorem 10-10 for the transform of an integrated function, we have

$$\mathcal{L}\left\{\int_0^t \left[\int_0^\lambda \delta_a(\tau - T_1)\delta_b(\lambda - \tau - T_2)\,d\tau\right] d\lambda\right\} = \frac{\Delta_a(s)\Delta_b(s)}{s} \tag{14-71}$$

The final-value theorem can be used on the function on the right, to yield the value of the integral from 0 to ∞. We get

$$\lim_{\sigma \to 0} \frac{\sigma \Delta_a(\sigma)\Delta_b(\sigma)}{\sigma} = \lim_{\sigma \to 0}\left[\frac{\sigma\Delta_a(\sigma)}{\sigma}\frac{\sigma\Delta_b(\sigma)}{\sigma}\right]$$
$$= \int_0^\infty \delta_a(\tau)\,d\tau \int_0^\infty \delta_b(\tau)\,d\tau = 1 \tag{14-72}$$

Thus, it is proved that the integral in question is a unit pulse, of width $a + b$.

Earlier, we placed restrictions on the maximum values of a and b and then in Eq. (14-68) allowed them to be small enough to replace $\delta_a(t)$ and $\delta_b(t)$ by $\delta(t)$. Assuming that this has been done, consistent notation would be to imply the same small values in the integral just treated, by leaving off the subscripts. Thus, when a and b are small enough for Eqs. (14-69) to be valid, we then also have

$$\int_0^t \delta(\tau - T_1)\delta(t - \tau - T_2)\,d\tau = \delta(t - T_1 - T_2) \tag{14-73}$$

When the values of the four integrals in Eq. (14-68) are combined, the result is in agreement with Eq. (14-66). This fact establishes that the integration principles developed in Sec. 14-10 can be used in the convolution integral to give results consistent with the definition of the symbolic inverse transforms of functions of the form

$$G(s) + A_0 + A_1 s + \cdots + A_n s^n$$

The proof given applies only for the first two terms of the above, but by following similar arguments the general case can be confirmed.

14-12. Superposition. The convolution formula of Theorem 11-3, when applied to a linear system, can be interpreted as an expression of the superposition principle. In the notation of Sec. 14-7, the response of an

Initially relaxed linear system is given, in transform functions, by Eq. (14-49), which we repeat:

$$R(s) = H(s)F(s)$$

If $\mathcal{L}^{-1}[H(s)]$ exists, from Theorem 11-3 it is known that the response $r(t)$ is

$$r(t) = \int_0^t h(t - \tau)f(\tau)\, d\tau \qquad (14\text{-}74)$$

To any desired degree of approximation, the above integral can be written

$$r(t) = \sum_{i=0}^N h(t - \tau_i)f(\tau_i)\, \Delta\tau_i \qquad (14\text{-}75)$$

In Eq. (14-75), the factor $h(t - \tau_i)$ is the response of the system due to a unit impulse occurring at $t = \tau_i$. Therefore,

$$h(t - \tau_i)f(\tau_i)\, \Delta\tau_i$$

is the response due to an impulse of strength $f(\tau_i)\, \Delta\tau_i$, and the total response, given by the summation in Eq. (14-75), is the superposition of the responses of a sequence of impulses whose strengths are proportional to $f(\tau)$ at all values of τ less than t. Figure 14-7 illustrates how $f(\tau)$ can be thought of as consisting of a sequence of impulses.

In view of the extension of the convolution theorem, as presented in Sec. 14-11, it is possible to extend the present concept to include impulse excitations and systems having impulsive responses to a discontinuous driving function. In other words, it is permissible for $F(s)$ and $H(s)$, or both, to have various-order

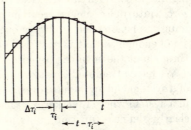

FIG. 14-7. Approximation of a function by a sequence of pulses. The strength of each pulse is equal to the area of its rectangular representation.

impulse functions in their inverses, so long as the convolution integral is interpreted in the manner described in Sec. 14-11.

14-13. Summary. In this chapter we have attempted to provide a rational development of a topic which has sometimes been considered controversial. The controversy has arisen because, as we have seen, discontinuous functions do not have ordinary derivatives at the points of discontinuity and an impulse function does not approach a function in the limit as the width goes to zero. It is not surprising, then, when we formally get the transform of the derivative of a discontinuous function, that we obtain a function which does not have an inverse, in the ordinary sense. In spite of these anomalies, it has long been known that correct answers can be obtained by a purely formal process, by regarding certain functions of s as Laplace transforms, even though they are not.

No attempt to develop a theory to cover these exceptional cases, based on the usual calculus, can succeed. Two basic reasons can be given for this statement: The basic definition of the derivative cannot be applied at a point of discontinuity, and an integral is a continuous function of a variable limit of integration. Thus, no true function $\delta(t - T_1)$ can exist for which

$$\int_0^t \delta(t - T_1) = \begin{cases} 0 & t < T_1 \\ 1 & t > T_1 \end{cases}$$

Another way to put it is to say that the operations of integration and differentiation are noncommutative. In the treatment presented here this difficulty has been overcome by defining $\delta(t)$ as always being of finite width.

By adopting this policy, it is possible to proceed without violating any mathematical or physical principles, although some symbolic notation like $\mathcal{L}[\delta(t)] = 1$ is used. This notation is not precise, to the extent that here the symbol does not mean a Laplace transform. The idea of keeping the pulse width finite is satisfying from a physical standpoint, because zero-width pulses never occur in physical phenomena. In practice, an impulse phenomenon is always one whose duration (in time or space) is small enough so that variation of other function components is negligible throughout the span of the pulse.

We have attempted to make a distinction between the *impulse response* of a system and an *impulsive response*. Under certain conditions, namely, in a system whose function $H(s)$ has an inverse, the impulse response is a clearly defined function, obtained when the width of a driving pulse goes to zero. However, when $H(s)$ does not meet this condition, or when the driving function is a higher-order impulse, there will be impulse components in the response, giving what we have called an impulsive response. An impulsive response can be obtained even when there is no impulse in the driving function. An example of this occurs when a step function of voltage is applied to a capacitor, current being regarded as the response. In the analysis present here, the impulse components of an impulsive response are always of finite width.

It may be disturbing to you that we use the unique symbol $\delta(t)$ to imply a pulse whose width and shape are left unspecified. Of course, as has been pointed out, the area under the pulse is the unique feature in determining the impulse response; and when a response includes an impulse term, we understand this to be a condition in which the pulse is so short in duration that its detailed shape is not observable and is of no importance. If we are interested in pulse details, *the pulse cannot be called an impulse.*

Mention should certainly be made of the work of L. Schwartz and others* in which *distributions* are used instead of functions. It is then possible to define an operation which is analogous to differentiation but which has meaning for situations which bear a similarity to the discontinuous functions. However, in order to apply this theory, it is necessary to change the conceptual models used to represent physical devices, in such a way that their characteristics are described by distributions rather than functions. Therefore, there does not seem to be any way to append the theory of distributions to the present theory, which is built on the concept of a function.

PROBLEMS

14-1. In Fig. 14-1 let the excitation be a unit voltage pulse

$$v_0(t) = \begin{cases} \dfrac{1}{2} \sin \dfrac{\pi t}{T} & 0 < t < T \\ 0 & T < t \end{cases}$$

Obtain the response to this pulse, and the limit of this response, as T goes to zero. Compare your result with the limit obtained in the text.

14-2. Show that a pulse can be negative during part of its duration, and still be considered a unit pulse, so long as its area is unity.

14-3. For each of the circuits of Fig. P 14-3 what duration of excitation pulse $v_1(t)$ would be adequately short, so that the pulse could be considered an impulse?

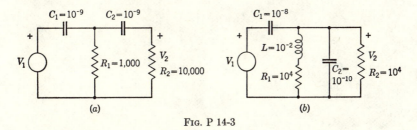

(a)   (b)

FIG. P 14-3

14-4. Consider a system whose response function is

$$H(s) = \frac{s}{s+1}$$

* L. Schwartz, "Théorie des distributions," vols. I, II, Hermann & Cie, Paris, 1957; I. Halperin, "Introduction to the Theory of Distributions," University of Toronto Press, Toronto, 1952; M. J. Lighthill, "An Introduction to Fourier Analysis and Generalized Functions," Cambridge University Press, New York, 1958; Sir George Temple, *J. London Math. Soc.*, **28**: 134–138 (1953).

Design a unit excitation pulse of finite duration which can be used in an exact application of the Laplace transform theory, but such that the response differs negligibly from the impulse response. If the system function were

$$H(s) = \frac{1}{1+s}$$

how could the pulse specification be changed?

14-5. Show that the driving-point response of a passive network cannot have higher than a second-order impulse function (doublet) in its impulse response.

14-6. Prove that all orders of singularity functions, higher than the first, have zero area.

14-7. What is the response, to a second-order impulse of unit strength, of a system whose function is

(a) $H(s) = \dfrac{(s+0.5)(s+1.5)}{(s+1)(s+2)}$ (b) $H(s) = \dfrac{s+0.5}{(s+1)(s+2)}$

14-8. Discuss the difference between the concepts of an *impulse response* and an *impulse function*.

14-9. Refer to the functions defined in (a) and (b) in Prob. 12-12. In each case specify $df(t)/dt$, using symbolic impulse functions. Also, obtain $\mathcal{L}[df(t)/dt]$, and obtain the inverse of this transform divided by s. It should be the same as the original function.

14-10. For the following examples, obtain the response to a unit impulse and to a unit step, and observe that the former is the derivative of the latter:

(a) $H(s) = \dfrac{1}{(s+1)(s+2)}$ (b) $H(s) = \dfrac{s+1}{s^2+s+1}$

(c) $H(s) = \dfrac{s+1}{s+2}$ (d) $H(s) = \dfrac{s-1}{s+2}$

CHAPTER 15

PERIODIC FUNCTIONS

15-1. Introduction. In many important physical systems the driving function is periodic. In the classical development of the theory of the behavior of such systems, the sinusoidal excitation occurred first, and in many respects, it is the simplest. It was recognized early that the solution could be broken into two parts, usually called the transient response and the steady-state response, as we have pointed out in some detail in Chap. 1. In that chapter we also showed that for a non-sinusoidal periodic driving function the Fourier series provided a solution.

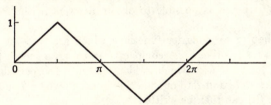

Fig. 15-1. Example of a periodic function.

The Fourier-series solution for the nonsinusoidal case is certainly of great theoretical importance, providing an important link in the theory of linear systems. However, in many cases the Fourier series is not convenient for computation, owing to slowness of convergence. In the present chapter we shall be concerned with a method of obtaining the response of a system to a periodic driving function, which is preferable when specific numerical answers are required.

Our attention will be confined to those cases where the driving and response functions are at least piecewise continuous. The gist of the method is very simple and can be described in terms of the triangular function shown in Fig. 15-1. This function can be described by the Fourier series

$$f(t) = \frac{8}{\pi^2}\left(\sin \omega t - \frac{1}{9} \sin 3\omega t + \frac{1}{25} \sin 5\omega t \cdots\right) \qquad (15\text{-}1)$$

and this series could be used to calculate values of the function, but with some effort if a high degree of accuracy is required. Computation of such a series is more difficult if it represents a discontinuous function, because convergence is then slower. Now we observe the simple idea that this same function can be specified by the pair of formulas

$$f(t) = \begin{cases} \dfrac{2}{\pi}\,\omega t & 0 \leqq \omega t \leqq \dfrac{\pi}{2} \\[2mm] 2 - \dfrac{2}{\pi}\,\omega t & \dfrac{\pi}{2} \leqq \omega t \leqq \dfrac{3\pi}{2} \\[2mm] 4 + \dfrac{2}{\pi}\,\omega t & \dfrac{3\pi}{2} \leqq \omega t \leqq 2\pi \end{cases} \tag{15-2}$$

and
$$f(t) = f\left(t - \frac{2\pi}{\omega}\right) \tag{15-3}$$

The first formula establishes the function throughout the first period, and the second equation tells us that the function is periodic. In other words, we recognize that it is necessary to know the function only for the first period in order to know it for all values of t. It is obvious that Eq. (15-2) is preferable to Eq. (15-1) for purposes of numerical evaluation. We shall find that it is possible to obtain answers in the form of Eq. (15-2).

In this chapter we consider functions which are zero for negative t and periodic for positive t. A compact way to write such a function is to define a function $f_0(t)$ which is identical with the periodic function $f(t)$, for $0 < t < T$, and zero outside this interval, where T is the period. We see that $f(t)$ can then be written

$$f(t) = f_0(t) + f_0(t - T) + f_0(t - 2T) + \cdots$$
$$= \sum_{n=0}^{\infty} f_0(t - nT) \tag{15-4}$$

where $T = 2\pi/\omega$. Thus we see that most of our attention can be directed toward finding the function $f_0(t)$.

15-2. Laplace Transform of a Periodic Function. We shall now consider the basic properties of the transform of a piecewise continuous periodic function $f(t)$, of period T, for which we also define the function

$$f_0(t) = \begin{cases} f(t) & 0 \leqq t \leqq T \\ 0 & T < t \text{ and } t < 0 \end{cases} \tag{15-5}$$

It is possible to write the complete function

$$f(t) = f_0(t) + f(t - T)u(t - T) \tag{15-6}$$

in view of the fact that $f(t - T) = f(t)$. Theorem 10-8 can be used on the second term of the right. Therefore, if $F(s) = \mathcal{L}[f(t)]$ and

$$F_0(s) = \mathcal{L}[f_0(t)]$$

we have

$$F(s) = F_0(s) + e^{-sT}F(s)$$

or

$$F(s) = \frac{F_0(s)}{1 - e^{-sT}} \tag{15-7}$$

where

$$F_0(s) = \int_0^\infty f_0(t)e^{-st}\,dt$$

$$= \int_0^T f(t)e^{-st}\,dt \tag{15-8}$$

The last equation shows that $F_0(s)$ can be obtained by an integration between finite limits. Accordingly, there is no difficulty in differentiation under the integral sign, giving

$$\frac{dF_0(s)}{ds} = \int_0^T -tf(t)e^{-st}\,dt \tag{15-9}$$

which exists if $f(t)$ is integrable, as we assume. Equation (15-9) is valid for all finite values of s, and so it is concluded that $F_0(s)$ is an entire function. This same property of $F_0(s)$ is available from Theorem 12-1, from which we also learn that $F_0(s)$ approaches zero uniformly as $|s|$ goes to infinity in any *closed* right half plane.* Later on we shall find this to be a useful property. We collect these ideas and state them formally as a theorem:

Theorem 15-1. If $f(t)$ is a piecewise continuous periodic function, of period T, its Laplace transform exists and is given by

$$F(s) = \frac{F_0(s)}{1 - e^{-sT}}$$

where

$$F_0(s) = \int_0^T f(t)e^{-st}\,dt$$

is an entire function which approaches zero uniformly as $|s|$ goes to infinity, for Re $s \geqq \sigma_0$, where σ_0 is any real number.

We shall now prove the converse of this theorem, starting with a function $F_0(s)$ which is known to be an entire function having an inverse transform

$$f_0(t) = \mathcal{L}^{-1}[F_0(s)] \tag{15-10}$$

which is identically zero for $T < t$. It will now be shown that

$$F(s) = \frac{F_0(s)}{1 - e^{-sT}} \tag{15-11}$$

* The closed half plane includes points on the vertical dividing line.

is a transform which has a periodic inverse. It being recognized that $F_0(s)$ is the given function, $f_0(t)$ is determined by taking its inverse, and from this the periodic function

$$f(t) = \sum_{n=0}^{\infty} f_0(t - nT) \tag{15-12}$$

can be formed. From Theorem 15-1 it is known that Eq. (15-11) is the transform of $f(t)$. In view of the uniqueness of the relationship between a transform and its inverse (except for the possibility of the function of t being arbitrarily defined over a set of measure 0) we conclude that Eq. (15-12) is the inverse of $F(s)$ as given by Eq. (15-11). Thus we have proved the following theorem:

Theorem 15-2. Let an entire function $F_0(s)$ be given such that it has an inverse transform

$$f_0(t) = \mathcal{L}^{-1}[F_0(s)]$$

which is identically zero for $T < t$ and for $t < 0$. Then

$$\mathcal{L}^{-1}\left[\frac{F_0(s)}{1 - e^{-sT}}\right]$$

is a periodic function, of period T, and is identical with $f_0(t)$ for $0 \leqq t \leqq T$.

15-3. Application to the Response of a Physical Lumped-parameter System. Let $H(s)$ represent the system function for a linear lumped-parameter system, which is without stored energy at $t = 0$. A periodic driving function $x(t)$, of period T, begins to act on the system at $t = 0$. The response will be a function $y(t)$ which, from intuitive reasoning, we can expect to be the sum of a periodic function $p(t)$ and an aperiodic function $a(t)$. Thus we shall tentatively write

$$y(t) = p(t) + a(t) \tag{15-13}$$

with the expectation of obtaining independent solutions for $p(t)$ and $a(t)$. Tentatively it is assumed that $H(s)$ is regular at infinity, but this restriction will be removed later.

From Theorem 15-1 it is known that the transform of the driving function is

$$X(s) = \frac{X_0(s)}{1 - e^{-sT}} \tag{15-14}$$

where

$$X_0(s) = \mathcal{L}[x_0(t)] = \int_0^T x(t)e^{-st} \, dt \tag{15-15}$$

and where

$$x_0(t) = \begin{cases} x(t) & 0 \leqq t \leqq T \\ 0 & T < t \text{ and } t < 0 \end{cases} \tag{15-16}$$

The transform of the output function is

$$Y(s) = \frac{H(s)X_0(s)}{1 - e^{-sT}} \tag{15-17}$$

This function has simple poles at the roots of the equation

$$1 - e^{-sT} = 0 \tag{15-18}$$

and also is singular at the poles of the rational function $H(s)$. We exclude from consideration the degenerate case (resonance) where poles of $H(s)$ coincide with roots of Eq. (15-18). Now let $A(s)$ designate the sum of the principal parts of $Y(s)$ at the poles s_1, s_2, etc., of $H(s)$, and form the function

$$P(s) = Y(s) - A(s) \tag{15-19}$$

which will exhibit poles only at the roots of Eq. (15-18). These poles can be placed in evidence by defining

$$P_0(s) = H(s)X_0(s) - (1 - e^{-sT})A(s) \tag{15-20}$$

in terms of which Eq. (15-19) becomes

$$P(s) = \frac{P_0(s)}{1 - e^{-sT}} \tag{15-21}$$

In the following section it is shown that $P_0(s)$ satisfies the conditions of Theorem 15-2, and so from that theorem the inverse of $P(s)$ is known to be periodic. The inverse of $A(s)$ certainly exists and is not periodic, since $A(s)$ is a finite sum of terms like

$$\frac{a_k}{(s - s_k)^n}$$

for which the inverse is known, from Eq. (10-72), to be of the form

$$\frac{a_k}{(n - 1)!} t^{n-1}e^{s_k t}$$

Accordingly, it is evident that the functions $p(t)$ and $a(t)$ of Eq. (15-13) will be given by

$$a(t) = \mathcal{L}^{-1}[A(s)] \tag{15-22}$$
$$p(t) = \mathcal{L}^{-1}[P(s)] \tag{15-23}$$

In fact, referring to Theorem 15-2, we have an explicit formula for $p(t)$, namely,

$$p(t) = \sum_{n=0}^{\infty} p_0(t - nT) \tag{15-24}$$

where

$$p_0(t) = \mathcal{L}^{-1}[P_0(s)] \tag{15-25}$$

is identical with $p(t)$ during the first period. Thus, it is emphasized that the inversion indicated in Eq. (15-23) need never be carried out: Eq. (15-25) is sufficient.

15-4. Proof That $\mathcal{L}^{-1}[P_0(s)]$ Is Periodic. The crucial item omitted from the previous section is a proof that $P_0(s)$, as defined by Eq. (15-20),

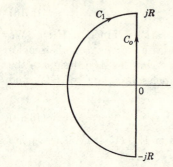

FIG. 15-2. Two ways to approach the Br contour when the integrand is regular in the right half plane.

has an inverse transform which is identically zero for $T < t$ and for $t < 0$. First we show that $P_0(s)$ is an entire function. From Eq. (15-19) it is known that $P(s)$ is regular at the poles of $H(s)$, and so it follows from Eq. (15-21) that $P_0(s)$ is regular at these points. Furthermore, $A(s)$ and $H(s)$ are regular at the points where $e^{-sT} = 1$; and with this information Eq. (15-20) can be used to show that $P_0(s)$ is regular at these points. Finally, $X_0(s)$ is known to be an entire function, by virtue of Theorem 15-1. Thus, all possible singular points have been accounted for, and it is concluded that $P_0(s)$ is an entire function.

Each term in Eq. (15-20) has an inverse which is zero for $t < 0$, by virtue of being a Laplace transform. We now consider $p_0(t)$, for $T < t$. The Br contour for the inversion formula is defined in terms of the geometry shown in Fig. 15-2. Accordingly, we have

$$p_0(t) = \frac{1}{2\pi j} \int_{\mathrm{Br}} P_0(s) e^{st} \, ds = \frac{1}{2\pi j} \lim_{R \to \infty} \int_{C_0} P_0(s) e^{st} \, ds \qquad (15\text{-}26)$$

Since the integrand is an entire function, the function is regular in the region between curves C_0 and C_1, as well as on C_1. Therefore, the contour can be distorted from C_0 to C_1 without changing the value of the integral, giving

$$\frac{1}{2\pi j} \int_{\mathrm{Br}} P_0(s) e^{st} \, ds = \frac{1}{2\pi j} \lim_{R \to \infty} \int_{C_1} P_0(s) e^{st} \, ds \qquad (15\text{-}27)$$

The form of this last integral immediately suggests Jordan's lemma. However, since we wish to show that the integral is identically zero for $T < t$, it is convenient to rewrite it as follows:

$$\int_{C_1} P_0(s) e^{st} \, ds = \int_{C_1} P_0(s) e^{sT} e^{s(t-T)} \, ds \qquad (15\text{-}28)$$

From Eq. (15-20),

$$P_0(s) e^{sT} = H(s) X_0(s) e^{sT} - A(s) e^{sT} + A(s) \qquad (15\text{-}29)$$

and reference to Theorem 8-14 shows that the integral on the right of Eq. (15-28) approaches zero, as R becomes infinite, when $b = t - T > 0$, if $P_0(s)$ approaches zero uniformly, as $|s|$ becomes infinite, for Re $s \leqq 0$. We shall show that each of the three terms of Eq. (15-29) satisfies this condition. Upon recalling Eq. (15-15), it is evident that the first term can be written

$$H(s)X_0(s)e^{sT} = H(s) \int_0^T x(t)e^{s(T-t)} \, dt = H(s) \int_0^T x(T - u)e^{su} \, du \quad (15\text{-}30)$$

Theorem 12-1 is available to show that the last integral approaches zero uniformly, as $|s|$ becomes infinite, for Re $s \leqq 0$. Since $H(s)$ is regular at infinity, multiplication by this function does not disturb the behavior at infinity. The function $A(s)$ goes to zero at least as fast as $1/s$, because it is a sum of principal parts, and hence it goes to zero uniformly with respect to ang s, for all angles, as $|s|$ goes to infinity. Thus, the last term of Eq. (5-29) satisfies the required condition. Furthermore, multiplication by e^{sT} does not disturb this property, for Re $s \leqq 0$, and so the second term also goes to zero uniformly, for Re $s \leqq 0$. It having thus been proved that $P_0(s)$ goes to zero uniformly, for Re $s \leqq 0$, it follows from Jordan's lemma that the integral on the right of Eq. (15-27) goes to zero as R becomes infinite. Thus,

$$\mathcal{L}^{-1}[P_0(s)] = 0 \qquad T < t \text{ and } t < 0 \quad (15\text{-}31)$$

and from Theorem 15-2 it follows that $p(t)$, as defined by Eq. (15-24), is periodic.

Referring to Eq. (15-20), and considering its inverse, we see that $\mathcal{L}^{-1}[H(s)X_0(s)]$ is the response of the system to a single pulse which is identical to the first cycle of $x(t)$ and which certainly is nonzero for $0 \leqq t \leqq T$. Likewise, $\mathcal{L}^{-1}[A(s)]$ is nonzero for the same interval. However,

$$\mathcal{L}^{-1}[e^{-sT}A(s)]$$

is zero for $t < T$. Accordingly,

$$p_0(t) = \mathcal{L}^{-1}[H(s)X_0(s) - A(s)] \qquad 0 \leqq t \leqq T \quad (15\text{-}32)$$

and since we know that $p_0(t)$ is identically zero when $T < t$, the inverse of $e^{-sT}A(s)$ is not needed in a numerical problem. It is this term that cancels the other two, when $t > T$, making $p_0(t)$ identically zero.

15-5. The Case Where $H(s)$ Has a Pole at Infinity. Most practical situations yield $H(s)$ functions which are regular at infinity. However, systems do occur for which a discontinuous driving function yields an impulsive response, at least for certain idealized systems. We shall recognize two cases for which $H(s)$ has a pole of order N at infinity. The first case is the more practical one, in which $x(t)$ and the first

$N - 1$ derivatives are zero at $t = 0$ and continuous for $t \geqq 0$, a condition which will ensure absence of impulse components in the response.* During the first period following $t = 0$, the derivatives of $x_0(t)$ will be identical with the derivatives of $x(t)$, and hence they exist; but these derivatives [up to and including the $(N - 1)$st] are zero at $t = 0$, and hence from Theorem 10-9 we can say that the Nth derivative $x_0^{(N)}(t)$ has a transform

$$X_N(s) = \int_0^T x^{(N)}(t)e^{-st} \, dt \tag{15-33}$$

and

$$X_0(s) = \frac{X_N(s)}{s^N} \tag{15-34}$$

The transform of the response is then

$$Y(s) = \frac{H(s)}{s^N} \frac{X_N(s)}{1 - e^{-sT}} \tag{15-35}$$

$H(s)/s^N$ is regular at infinity, and $X_N(s)$ has the same properties at infinity as $X_0(s)$. Therefore, Eq. (15-35) can be treated in the same manner as Eq. (15-17), with $H(s)/s^N$ used in place of $H(s)$ and $X_N(s)$ in place of $X_0(s)$. $A(s)$ continues to be the sum of the principal parts of $Y(s)$, and

$$P_0(s) = \frac{H(s)}{s^N} X_N(s) - (1 - e^{-sT})A(s) \tag{15-36}$$

replaces Eq. (15-20). The response is still given by Eqs. (15-24) and (15-25), and the proof of periodicity is still valid, subject to the interchange of functions mentioned above.

In the case just discussed, the effect of the pole of $H(s)$ at infinity is nullified by the existence of a sufficient number of derivatives of $x(t)$. The second case occurs when less than $N - 1$ derivatives are continuous. If only $N - 2$ derivatives are continuous, the first term in Eq. (15-36) is

$$\frac{H(s)}{s^{N-1}} X_{N-1}(s)$$

which has a symbolic inverse in the sense defined in Chap. 14. A first-order impulse will appear, due to the first-order pole of $H(s)/s^{N-1}$ at infinity. Thus, by following the concepts presented in Chap. 14, the present theory is applicable to the general case, for all $x(t)$ functions which are piecewise continuous.

15-6. Illustrative Example. The periodic response will be found for the equation

$$y + \int_0^t y(\tau) \, d\tau = x(t)$$

---

* See Theorem 12-11.

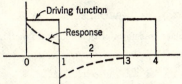

FIG. 15-3. Periodic driving function and response during first period, for illustrative example.

where $x(t)$ is the periodic function shown in Fig. 15-3. Following the plan outlined, in transform functions,

$$Y(s) = \frac{s}{1 + s} \frac{X_0(s)}{1 - e^{-3s}}$$

and

$$X_0(s) = \int_0^1 e^{-st} \, dt = \frac{1 - e^{-s}}{s}$$

giving

$$Y(s) = \frac{1}{1 + s} \frac{1 - e^{-s}}{1 - e^{-3s}}$$

The above has a first-order pole at $s = -1$, for which the principal part is

$$A(s) = \frac{1 - e}{1 - e^3} \frac{1}{1 + s}$$

giving, from Eq. (15-20),

$$P_0(s) = \frac{1}{1 + s} (1 - e^{-s}) - \frac{1 - e}{1 - e^3} \frac{1}{1 + s} (1 - e^{-3s})$$

$$= \frac{e^3 - e}{e^3 - 1} \frac{1}{1 + s} - \frac{1}{1 + s} e^{-s} + \frac{e - 1}{e^3 - 1} \frac{1}{1 + s} e^{-3s}$$

The inverse of this is

$$p_0(t) = \frac{e^3 - e}{e^3 - 1} e^{-t} - e^{-(t-1)} u(t - 1) + \frac{e - 1}{e^3 - 1} e^{-(t-3)} u(t - 3)$$

$$= \begin{cases} \dfrac{e^3 - e}{e^3 - 1} e^{-t} & 0 \leqq t \leqq 1 \\[2mm] -\dfrac{e^4 - e^3}{e^3 - 1} e^{-t} & 1 \leqq t \leqq 3 \\[2mm] 0 & 3 \leqq t \end{cases}$$

We observe that the third term in the first line is not needed, since it merely cancels the other two terms when $t > 3$. This fact was pointed out in Sec. 15-4, and the term was included here merely to confirm Eq. (15-32) by a numerical example.

PROBLEMS

15-1. Obtain the Laplace transform of the following functions:

(a) $f(t) = |\sin at|$ (b) $f(t) = |\cos at|$

(c) $f(t) = \begin{cases} \sin at & \dfrac{2n\pi}{a} < t < \dfrac{(2n+1)\pi}{a} \\ 0 & \dfrac{(2n+1)\pi}{a} < t < \dfrac{(2n+2)\pi}{a} \end{cases}$

(d) $f(t) = \begin{cases} 1 & nT < t < (n+\alpha)T \\ 0 & (n+\alpha)T < t < (n+1)T \end{cases}$

In (d), α is a number between 0 and 1.

15-2. Since the Laplace transform of a sine function is rational in s, it follows that the Laplace transform of a periodic function which can be expressed as a finite number of harmonic terms will also be *rational*. How can this fact be reconciled with the form of Eq. (15-7)?

15-3. Determine which of the following functions has an inverse and whether or not the inverse is zero for $t > T$. [That is, determine whether or not they satisfy the conditions on $F_0(s)$ in Theorem 15-2.]

(a) $F_0(s) = \dfrac{e^{-Ts/4} \sinh Ts/4}{s}$ (b) $F_0(s) = \dfrac{1 - e^{-2Ts}}{s^2}$

(c) $F_0(s) = \dfrac{1 - e^{-Ts}}{s^2}$ (d) $F_0(s) = \left(\dfrac{1 - e^{-Ts/2}}{s}\right)^2$

(e) $F_0(s) = \dfrac{(1 - e^{-Ts/2})^2}{s^3}$ (f) $F_0(s) = \dfrac{1 - e^{-Ts}}{s + 1}$

(g) $F_0(s) = \dfrac{1 - e^{-T(s+1)}}{(s + 1)^2}$ (h) $F_0(s) = \dfrac{1 - e^{-T(s+1)/2}}{s + 1}$

15-4. For each of the functions specified in Prob. 15-1, obtain the steady-state response for the system function

$$H(s) = \frac{1}{1 + s}$$

15-5. For each of the functions specified in Prob. 15-1, obtain the steady-state response for the system function

$$H(s) = \frac{s}{(s + 1)^2}$$

15-6. For each of the functions specified in Prob. 15-1, obtain the steady-state response for the system function

$$H(s) = \frac{s}{(s + 1)(s + 2)}$$

CHAPTER 16

THE Z TRANSFORM

16-1. Introduction. The Laplace transform is particularly well adapted to the solution of lumped linear systems because the system function is a comparatively simple rational function of s. When the driving function is a sum of exponentials multiplied by powers of t, its transform is also a rational function (see Chap. 12).

Of course, the Laplace transform is applicable to any linear equation, with any transformable driving function. There are cases, however, in which the transform functions are transcendental and not particularly

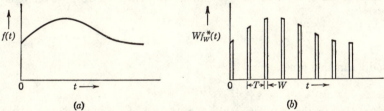

Fig. 16-1. A continuous function sampled at uniform intervals. (a) The continuous function; (b) the sampled function.

simple to handle. In this chapter we shall consider a particular class of problems in which the transform functions are simplified through the transformation

$$z = e^{sT} \qquad (16\text{-}1)$$

where T is a constant. A Laplace transform, when converted to a function of z, is called a Z transform. Although all the Laplace theory could be presented in terms of the z variable, it is particularly advantageous to do so when $F(s)$ is a rational function in the variable e^{sT}. This situation occurs when $f(t)$ is a sequence of equally spaced pulses of variable strength.

Signals such as those described above can occur in practice as inherent inputs to a system, as in pulse radar, or by virtue of a *sampling process* whereby a pulse signal is created from a continuous signal by a periodic switching operation. In either case we have two related signals, as shown in Fig. 16-1. In a radarlike system the train of pulses can be used to define a continuous envelope function; and in a sampled-data system the system creates the pulses from a continuous function. We shall use the symbols $f(t)$ and $f_W^*(t)$ to designate the continuous and pulse signals, respectively. The subscript W implies the pulse width. The signal $f_W^*(t)$, in which the pulse period is T, can be related to $f(t)$ by

$$f_W^*(t) = \sum_{n=0}^{\infty} f(t)\delta_W(t - nT) \tag{16-2}$$

where

$$\delta_W(t) = \begin{cases} \dfrac{1}{W} & 0 \leq t \leq W \\ 0 & W < t \end{cases} \tag{16-3}$$

is a pulse of finite width. Since the height of $\delta_W(t)$ is $1/W$, rather than unity, there is a scale change between $f(t)$ and $f_W^*(t)$, as indicated in Fig. 16-1.

We shall immediately dispense with the subscript W, with the understanding that the symbol $\delta(t)$ implies a finite-width pulse, of negligible width in the sense described in Chap. 14. Then we define a *sampling function*

$$p(t) = \sum_{n=0}^{\infty} \delta(t - nT) \tag{16-4}$$

and a sampled function

$$f^*(t) = f(t)p(t)$$

where $f^*(t)$ implies a function like Fig. 16-1a with small but finite pulses.

16-2. The Laplace Transform of $f^*(t)$. Let $f(t)$ be continuous for $t \geq 0$ and of exponential order α_0, and consider the Laplace integral

$$\int_0^{\infty} f^*(t)e^{-st}\,dt = \int_0^{\infty} f(t)p(t)e^{-st}\,dt$$

$$= \sum_{n=0}^{\infty} \int_{nT}^{(n+1)T} f(t)\delta(t - nT)e^{-sT}\,dt \tag{16-5}$$

From Eq. (14-62) it is recalled that $\delta(t)$ is defined in such a way that we can write the formal equation

$$\int_{nT}^{(n+1)T} f(t)\delta(t - nT)e^{-st}\,dt = f(nT)e^{-snT} \tag{16-6}$$

if $f(t)$ is continuous at $t = nT$. Thus, we have

$$\mathcal{L}[f^*(t)] = \sum_{n=0}^{\infty} f(nT)e^{-snT} \qquad \operatorname{Re} s > \sigma_c \tag{16-7}$$

where σ_c is the abscissa of convergence. We stipulated that $f(t)$ should be of exponential order. Therefore, corresponding to an arbitrary small number $\epsilon > 0$, there is a number N such that, if $\alpha > \alpha_0$,

$$|f(nT)|e^{-\alpha nT} < \epsilon$$

and
$$|f(nT)|e^{-\sigma nT} = |f(nT)|e^{-\alpha nT}e^{-(\sigma-\alpha)nT} < \epsilon e^{-(\sigma-\alpha)nT} \qquad (16\text{-}8)$$

when
$$n > N$$

The series

$$\sum_{n=N}^{\infty} e^{-(\sigma-\alpha)nT}$$

is known to converge if $\sigma > \alpha$. Therefore, the series in Eq. (16-7) converges absolutely, and therefore also converges, for $\sigma > \alpha_0$. It can be shown to diverge for $\sigma < \alpha_0$. Thus, the abscissa of convergence of the Laplace integral, which is also the abscissa of convergence of the series, is

$$\sigma_c = \alpha_0$$

and can be determined from the behavior of $f(t)$ at infinity, since it is this behavior that determines α_0.

It is now convenient to make the variable change $z = e^{sT}$, as defined in Eq. (16-1), which transforms the axis of convergence $\operatorname{Re} s = \sigma_c$ into the circle

$$z = e^{\sigma_c T} e^{j\omega T}$$
$$|z| = e^{\sigma_c T} \qquad (16\text{-}9)$$

In the z plane we now have a new function

$$F_z(z) = F^*(s)$$

and, from Eq. (16-7),

$$F_z(z) = \sum_{n=0}^{\infty} f(nT)\left(\frac{1}{z}\right)^n \qquad |z| > e^{\sigma_c T} \qquad (16\text{-}10)^*$$

By the principle of analytic continuation, it is known that $F_z(z)$ is an analytic function; and the series on the right of Eq. (16-10) is the Laurent expansion about the origin. $F_z(z)$ is called the Z transform of $f(t)$. A shorthand notation is sometimes useful, corresponding to $F(s) = \mathcal{L}[f(t)]$. Accordingly, we define

$$F_z(z) = \mathcal{Z}[f(t)] \qquad (16\text{-}11)$$

to mean the function defined by Eq. (16-10).

* In this equation it is understood that $f(0)$ means $f(0+)$.

16-3. *Z* **Transform of Powers of** *t*. The *Z* transform of t^k, as a series, is found from Eq. (16-10) to be

$$Z(t^k) = \sum_{n=0}^{\infty} n^k T^k z^{-n} \qquad |z| > 1$$

$$= Tz \sum_{n=0}^{\infty} n^{k-1} T^{k-1} n z^{-(n+1)} \qquad (16\text{-}12)$$

and also

$$Z(t^{k-1}) = \sum_{n=0}^{\infty} n^{k-1} T^{k-1} z^{-n} \qquad |z| > 1 \qquad (16\text{-}13)$$

Observe that the series can be differentiated term by term, and so

$$\frac{d}{dz} Z(t^{k-1}) = -\sum_{n=0}^{\infty} n^{k-1} T^{k-1} n z^{-(n+1)} \qquad |z| > 1$$

Comparing with Eq. (16-12), we get the recurrence formula

$$Z(t^k) = -Tz \frac{d}{dz} Z(t^{k-1}) \qquad (16\text{-}14)$$

For $k = 0$, we have

$$Z(1) = \sum_{n=0}^{\infty} z^{-n} \qquad |z| > 1$$

which is recognized as the Laurent expansion of

$$\frac{z}{z-1}$$

Thus, a sequence of formulas is obtained from Eq. (16-14), as follows:

$$Z(1) = \frac{z}{z-1}$$

$$Z(t) = \frac{Tz}{(z-1)^2}$$

$$Z(t^2) = \frac{T^2 z (z+1)}{(z-1)^3} \qquad (16\text{-}15)$$

. . . .

Since the *Z* transforms are obtained from the Laplace transforms merely by a change of variable, we conclude that the *Z* transform of a sum is the sum of the transforms. Also, the *Z* transform of a constant times a function of *t* is equal to that constant times the *Z* transform of the function. Accordingly, it is seen that the Z transform of a polynomial in t is a rational function of z.

16-4. Z Transform of a Function Multiplied by e^{-at}. From Eq. (16-10), we have

$$Z[e^{-at}f(t)] = \sum_{n=0}^{\infty} f(nT)e^{-anT}z^{-n}$$

$$= \sum_{n=0}^{\infty} f(nT)(e^{aT}z)^{-n} \qquad |z| > e^{\sigma_c T} \qquad (16\text{-}16)$$

Thus, if

$$F_z(z) = Z[f(t)]$$

it follows from Eq. (16-16) that

$$Z[e^{-at}f(t)] = F_z(e^{aT}z) \qquad (16\text{-}17)$$

Referring to Eqs. (16-15) and taking $f(t) = t^k$, we have the additional formulas

$$Z(e^{-at}) = \frac{z}{z - e^{-aT}}$$

$$Z(te^{-at}) = \frac{Te^{-aT}z}{(z - e^{-aT})^2} \qquad (16\text{-}18)$$

$$Z(t^2 e^{-at}) = \frac{T^2 e^{-aT} z(z + e^{-aT})}{(z - e^{-aT})^3}$$

$$. \quad . \quad . \quad . \quad . \quad . \quad . \quad . \quad . \quad . \quad . \quad . \quad .$$

These are rational functions of z, and a sum of a finite number of them will be rational. From the region of convergence of Eq. (16-10) we also know that the poles of this function will lie inside a circle of radius $e^{\alpha_0 T}$, where $f(t)$ is of exponential order α_0.

Since functions of this type are used so frequently in analysis, it is useful to have the following theorem, for which the above development constitutes a proof:

Theorem 16-1. If $f(t)$ is a sum of a finite number of terms of the form t^k and $t^k e^{-at}$, the sum being of exponential order α_0, the Z transform is a rational function of z, having poles inside a circle of radius $e^{\alpha_0 T}$.

A useful corollary is obtained by referring to Theorem 10-11, which tells us that, if $F(s)$ is a rational function with a zero at infinity, then its inverse will be a function of t such as is described in Theorem 16-1. Thus, we have the following corollary:

Corollary. If $F(s)$ is a rational function with a zero at infinity, and if α_0 is the largest real part of the poles of $F(s)$, then $Z\{\mathcal{L}^{-1}[F(s)]\}$ is a rational function of z, with poles lying inside a circle of radius $e^{\alpha_0 T}$.

16-5. The Shifting Theorem. Consider a function $f(t)$ for which the Z transform $F_z(z)$ is given by

$$F_z(z) = f(0) + f(T)z^{-1} + f(2T)z^{-2} + \cdots + f(NT)z^{-N} + \cdots$$
$$|z| > e^{\sigma_c T} \quad (16\text{-}19)$$

Suppose that we now find the Z transform of the shifted function

$$f(t - NT)u(t - NT)$$

where the shift is an integral number of sampling periods NT. In view of the unit step $u(t - NT)$, this function is zero for $t < NT$, and therefore the Z transform is

$$0 + 0 + \cdots + f(0)z^{-N} + f(T)z^{-(n+1)} + \cdots \quad |z| > e^{\sigma_c T}$$

and, comparing with Eq. (16-19), gives

$$Z[f(t - NT)u(t - NT)] = z^{-N}Z[f(t)] \quad (16\text{-}20)$$

as the Z-transform counterpart of Theorem 10-10. We observe that the proof is simpler with Z transforms than with Laplace transforms.

16-6. Initial- and Final-value Theorems. The values $f(0)$ and $f(\infty)$, when the latter exists, are related to properties of the Z transform of $f(t)$. The initial-value theorem will be considered first. From Eq. (16-10)

$$F_z(z) = Z[f(t)] = f(0) + f(T)z^{-1} + f(2T)z^{-2} + \cdots \quad |z| > e^{\sigma_c T}$$

The series converges uniformly, for $|z| \geq R' > e^{\sigma_c T}$, and therefore the limit of the series as $|z| \to \infty$ can be obtained by taking the limit of the individual terms, giving

$$\lim_{|z| \to \infty} F_z(z) = f(0) \quad (16\text{-}21)$$

For consideration of the final-value theorem, we are interested in the behavior of $F_z(z)$ as $z \to 1$, as we might suspect by recalling the final-value theorem of the Laplace transform, which involves the point $s = 0$. If $f(t)$ is of exponential order α_0, where $\alpha_0 < 0$, the Z-transform series

$$Z[f(t)] = \sum_{n=0}^{\infty} f(nT)z^{-N} \quad (16\text{-}22)$$

converges for $|z| > e^{\alpha_0 T}$. Since $\alpha_0 < 0$, it is evident that the series converges and the function is regular, for $z = 1$. The result is a useful theorem, which we state as follows:

Theorem 16-2. If $f(t)$ is of exponential order $\alpha_0 < 0$, then the Z transform of $f(t)$ is finite, and is also regular, at $z = 1$.

This theorem can be used to get a final-value theorem for Z transforms. Let the general function be written

$$f(t) = f_0(t) + K \qquad (16\text{-}23)$$

where $f_0(t)$ is of exponential order $\alpha_0 < 0$. It follows that $f(\infty) = K$. Referring to Eqs. (16-15), it is evident that the Z transformation of Eq. (16-23) is

$$F_z(z) = Z[f_0(t)] + K \frac{z}{z-1}$$

and therefore $\qquad K = \lim_{z \to 1} \frac{z-1}{z} \{F_z(z) - Z[f_0(t)]\}$

However, by Theorem 16-2, $Z[f_0(t)]$ exists at $z = 1$, and so

$$\lim_{z \to 1} \frac{z-1}{z} Z[f_0(t)] = 0$$

giving $\qquad K = \lim_{z \to 1} \frac{z-1}{z} F_z(z) = \lim_{z \to 1} (z-1)F_z(z)$

But $K = f(\infty)$, and so, under the conditions stated,

$$\lim_{z \to 1} (z-1)F_z(z) = f(\infty) \qquad (16\text{-}24)$$

The two limit-value theorems derived in this section are now stated formally, as follows:

Theorem 16-3. If $f(t)$ has a Z transform $F_z(z)$, the initial value of $f(t)$ is equal to the limit approached by $F_z(z)$ as $|z|$ approaches infinity.

Theorem 16-4. If $f(\infty)$ exists, and if $f(t) - f(\infty)$ is of exponential order $\alpha_0 < 0$, then

$$f(\infty) = \lim_{z \to 1} (z-1)F_z(z)$$

where $F_z(z)$ is the Z transform of $f(t)$.

16-7. The Inversion Formula. The inversion formula for $F_z(z)$ can be obtained by substituting $z = Re^{j\theta}$ in Eq. (16-10), to give

$$F_z(Re^{j\theta}) = \sum_{n=0}^{\infty} f(nT) \frac{1}{R^n} e^{-jn\theta} \qquad R > e^{\sigma_c T} \qquad (16\text{-}25)$$

which can also be written

$$f_1(\theta) + jf_2(\theta) = \sum_{n=0}^{\infty} f(nT) \frac{1}{R^n} \cos n\theta - j \sum_{n=1}^{\infty} f(nT) \frac{1}{R^n} \sin n\theta$$

where $\qquad f_1(\theta) = \text{Re}\,[F_z(Re^{j\theta})] \qquad f_2(\theta) = \text{Im}\,[F_z(Re^{j\theta})]$

Using the formulas for coefficients of a Fourier series,

$$f(0) = \frac{1}{2\pi} \int_0^{2\pi} f_1(\theta)\, d\theta$$

$$f(nT) = \frac{R^n}{\pi} \int_0^{2\pi} f_1(\theta)\, \cos n\theta\, d\theta \qquad (16\text{-}26)$$

$$f(nT) = -\frac{R^n}{\pi} \int_0^{2\pi} f_2(\theta)\, \sin n\theta\, d\theta$$

Adding the last two, and dividing by 2, gives

$$f(nT) = \frac{R^n}{2\pi} \int_0^{2\pi} [f_1(\theta)\, \cos n\theta - f_2(\theta)\, \sin n\theta]\, d\theta$$

$$= \frac{1}{2\pi} \int_0^{2\pi} \operatorname{Re} [F_z(Re^{j\theta}) R^n e^{jn\theta}]\, d\theta \qquad (16\text{-}27)$$

Note that $F_z(z)$ is a real function of z, and therefore $\operatorname{Re}[F_z(Re^{j\theta})]$ and $\operatorname{Im}[F_z(Re^{j\theta})]$ are, respectively, even and odd functions of θ. Thus,

$$\operatorname{Im}[F_z(Re^{j\theta}) R^n e^{jn\theta}] = \operatorname{Re}[F_z(Re^{j\theta})] \operatorname{Im}(R^n e^{jn\theta}) - \operatorname{Im}[F_z(Re^{j\theta})] \operatorname{Re}(R^n e^{jn\theta})$$

is an odd function of θ, showing that

$$\int_0^{2\pi} \operatorname{Im}[F_z(Re^{j\theta}) R^n e^{nj\theta}]\, d\theta = 0$$

It follows that Eq. (16-27) can be written

$$f(nT) = \frac{1}{2\pi} \int_0^{2\pi} F_z(Re^{j\theta}) R^n e^{jn\theta}\, d\theta$$

Let C_0 designate a counterclockwise circular integration contour of radius $R > e^{\sigma_c T}$, as shown in Fig. 16-2, and convert the above integral to a contour integral for which

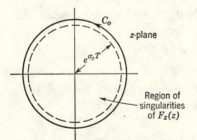

$$Re^{j\theta} = z \qquad d\theta = \frac{1}{jz}\, dz$$

giving

$$f(nT) = \frac{1}{2\pi j} \int_{C_0} F_z(z) z^{n-1}\, dz \quad (16\text{-}28)$$

FIG. 16-2. The integration contour used for the Z-transform inversion integral.

The above integral is the inversion formula for the Z transform. We observe particularly that it gives $f(nT)$ only for integral values of n; it does not uniquely specify $f(t)$. If $F_z(z)$ is single-valued, the calculus of residues is readily available to evaluate this integral. C_0 encloses all the singular points of $F_z(z)$.

It is interesting to observe a difference between the Laplace inversion

$$\mathcal{L}^{-1}[F^*(s)] = f(t)p(t) = f^*(t)$$

and the result just obtained, whereby $f(nT)$ is obtained from $F_z(z)$. We recognize $f^*(t)$ as a symbolic function which has an impulse of strength $f(nT)$ at $t = nT$. Equation (16-28) gives the more useful $f(nT)$, which will be regarded as the inverse of $F_z(z)$, and accordingly we write

$$f(nT) = Z^{-1}[F_z(z)] \qquad (16\text{-}29)$$

Although there is an exact functional equivalence

$$H^*(s) = F_z(z)$$

the inverses $\mathcal{L}^{-1}[H^*(s)]$ and $Z^{-1}[F_z(z)]$ are different.

As an illustration of the inversion formula, consider

$$F_z(z) = \frac{Tz}{(z - 1)^2}$$

which is known to be the Z transform of $f(t) = t$. The inversion integral gives

$$f(nT) = \frac{1}{2\pi j} \int_{C_0} \frac{Tz}{(z - 1)^2} z^{n-1}\, dz = \frac{1}{2\pi j} \int_{C_0} \frac{Tz^n}{(z - 1)^2}\, dz$$

with the understanding that C_0 has a radius greater than 1. The second-order pole at 1 is therefore enclosed, and the integral is equal to $2\pi j$ times the residue, where

$$\text{Residue} = \frac{d}{dz} Tz^n \Big|_{z=1} = nTz^{n-1} \Big|_{z=1} = nT$$

Thus, the expected result

$$f(nT) = nT$$

is obtained.

We conclude these remarks about the inversion integral with the observation that the inversion integral is not normally needed. For a given $F_z(z)$ it is necessary only to expand in a Laurent series about the origin, say by a division algorithm. The successive values of $f(nT)$ are the coefficients in this series.

16-8. Periodic Properties of $F^*(s)$, and Relationship to $F(s)$. In this section we return to a consideration of $F^*(s)$, recalling that it is related to the Z transform by

$$z = e^{sT} \qquad F^*(s) = F_z(z)$$

By virtue of the Laurent-series representation for $F_z(z)$, we are assured that $F_z(z)$ is an analytic function of z. Since

$$e^{sT} = e^{(s + j2\pi n/T)T}$$

it follows that the Z transform is a periodic function of s, which fact is put into evidence by writing

$$F^*(s) = F^*\left(s + \frac{j2n\pi}{T}\right) \tag{16-30}$$

Corresponding to each singular point of $F_s(z)$, there will be an infinite set of singularities of $F^*(s)$. Specifically, if $z_1 = x_1 + jy_1$ is a singularity of $F_s(z)$, $F^*(s)$ is singular at

$$s_k = \frac{1}{T} \log z_1 = \frac{1}{2T} \log (x_1{}^2 + y_1{}^2) + j\frac{1}{T}\left(\tan^{-1}\frac{y_1}{x_1} + 2\pi n\right)$$

where n takes on all integral values. If $F(s)$ is rational, implying that $F_s(z)$ is also rational, we see that $F^*(s)$ is meromorphic.

In view of the above comments, when $F(s)$ is rational, we suspect that $F^*(s)$ can be expressed in a Mittag-Leffler (partial-fraction) expansion, in the manner described in Sec. 5-16. The objective of this section is to show that this is true and to obtain the expansion. $F(s)$ is assumed to be a given rational function, with a zero at infinity. Assume that it is written in partial-fraction form, for which we write three specific terms as follows:

$$F(s) = \cdots + \frac{a}{s - \alpha} + \cdots + \frac{b}{(s - \beta)^2} + \cdots + \frac{c}{(s - \gamma)^3} + \cdots \tag{16-31}$$

The total number of terms is finite. The corresponding $f(t)$ function is

$$f(t) = \cdots + ae^{\alpha t} + \cdots + bte^{\beta t} + \cdots + \frac{c}{2} t^2 e^{\gamma t} + \cdots \tag{16-32}$$

and the Z transform, according to Eqs. (16-18), is

$$F_s(z) = \cdots + \frac{az}{z - e^{\alpha T}} + \cdots + \frac{bTe^{\beta T}z}{(z - e^{\beta T})^2} + \cdots$$
$$+ \frac{cT^2e^{\gamma T}z(z + e^{\gamma T})}{2(z - e^{\gamma T})^3} + \cdots \tag{16-33}$$

Since the factors $e^{\beta T}$ and $e^{\gamma T}$ cause only a scale change in the z variable, Eq. (16-14) is applicable, giving

$$\frac{bTe^{\beta T}z}{(z - e^{\beta T})^2} = -Tz\frac{d}{dz}\left(\frac{bz}{z - e^{\beta T}}\right)$$
$$\frac{cT^2e^{\gamma T}z(z + e^{\gamma T})}{2(z - e^{\gamma T})^3} = -Tz\frac{d}{dz}\left[\frac{cTe^{\gamma T}z}{2(z - e^{\gamma T})^2}\right] \tag{16-34}$$

with similar relations for higher-order terms. Noting from Eq. (16-1) that

$$\frac{ds}{dz} = \frac{1}{Tz}$$

we see that in the variable s the above becomes somewhat simpler, as follows:

$$\frac{bTe^{\beta T}e^{sT}}{(e^{sT} - e^{\beta T})^2} = -\frac{d}{ds}\left(\frac{be^{sT}}{e^{sT} - e^{\beta T}}\right)$$
$$\frac{cT^2e^{\gamma T}e^{sT}(e^{sT} + e^{\gamma T})}{2(e^{sT} - e^{\gamma T})^3} = \frac{1}{2}\frac{d^2}{ds^2}\left(\frac{ce^{sT}}{e^{sT} - e^{\gamma T}}\right) \qquad (16\text{-}35)$$

Thus, by using Eqs. (16-33) and (16-35), a formula for $F^*(s)$ is

$$F^*(s) = \cdots \frac{ae^{sT}}{e^{sT} - e^{\alpha T}} + \cdots - \frac{d}{ds}\left(\frac{be^{sT}}{e^{sT} - e^{\beta T}}\right) + \frac{1}{2}\frac{d^2}{ds^2}\left(\frac{ce^{sT}}{e^{sT} - e^{\gamma T}}\right)$$
$$+ \cdots \quad (16\text{-}36)$$

This form is useful because it shows the importance of the first term, the other terms being derivatives of a similar function of s. Accordingly, the function

$$\frac{ae^{sT}}{e^{sT} + e^{\alpha T}} = \frac{a}{1 - e^{-(s-\alpha)T}} \qquad (16\text{-}37)$$

will be studied in detail. By the principle established in Sec. 5-16, this can be written as an infinite sum of the principal parts at each of the poles, which occur at

$$s_n = \alpha + \frac{j2\pi n}{T}$$

Each pole is simple, with residue

$$\lim_{s \to s_n} \frac{a(s - \alpha - j2\pi n/T)}{1 - e^{-(s-\alpha)T}} = \frac{a}{T}$$

Thus, the Mittag-Leffler expansion of Eq. (16-37) is

$$\frac{ae^{sT}}{e^{sT} - e^{\alpha T}} = \frac{a}{T}\sum_{n=-\infty}^{\infty}\frac{1}{s - \alpha - j2\pi n/T} \qquad (16\text{-}38)$$

This expansion can be differentiated term by term, and so, in view of Eqs. (16-35), from the above we get the additional expansions

$$\frac{bTe^{\beta T}e^{sT}}{(e^{sT} - e^{\beta T})^2} = \frac{b}{T}\sum_{n=-\infty}^{\infty}\frac{1}{(s - \alpha - j2\pi n/T)^2} \qquad (16\text{-}39a)$$

and

$$\frac{cT^2e^{\gamma T}e^{sT}(e^{sT} + e^{\gamma T})}{2(e^{sT} - e^{\gamma T})^3} = \frac{c}{T}\sum_{n=-\infty}^{\infty}\frac{1}{(s - \alpha - j2\pi n/T)^3} \qquad (16\text{-}39b)$$

with similar results for higher-order terms.   Equation (16-36) can now be written

$$F^*(s) = \frac{1}{T}\left[\cdots + a \sum_{n=-\infty}^{\infty} \frac{1}{s - \alpha - j2\pi n/T} + \cdots \right.$$

$$+ b \sum_{n=-\infty}^{\infty} \frac{1}{(s - \alpha - j2\pi n/T)^2} + \cdots$$

$$\left. + c \sum_{n=-\infty}^{\infty} \frac{1}{(s - \alpha - j2\pi n/T)^3} + \cdots \right]$$

$$= \frac{1}{T}\left[\sum_{n=-\infty}^{\infty} \cdots + \frac{a}{s - \alpha - j2\pi n/T} + \cdots \right.$$

$$\left. + \frac{b}{(s - \alpha - j2\pi n/T)^2} + \cdots + \frac{c}{(s - \alpha - j2\pi n/T)^3} + \cdots \right]$$

Reference to Eq. (16-31) shows that this can be given by the simpler expression

$$F^*(s) = \frac{1}{T} \sum_{n=-\infty}^{\infty} F\left(\frac{s - j2\pi n}{T}\right) \tag{16-40}$$

This result can be obtained formally in a much simpler way by using s-plane convolution of the transforms of $f(t)$ and $p(t)$.   However, the result is an integral which is not known to converge, and so the above more extended derivation is presented.   The following theorem has been proved:

Theorem 16-5.   If $F(s)$ is a rational function, with a zero at infinity, the Z transform, as a function of s, is given by

$$F^*(s) = \frac{1}{T} \sum_{n=-\infty}^{\infty} F\left(\frac{s - j2\pi n}{T}\right)$$

16-9. Transmission of a System with Synchronized Sampling of Input and Output.   As a practical application of the Z transform, consider a system function $H(s)$ which is rational and has a zero at infinity.   If $f(t)$ is sampled by multiplying by $p(t)$, the transform of the output is

$$R(s) = H(s)F^*(s) \tag{16-41}$$

The output $\mathcal{L}^{-1}[R(s)]$ is continuous.   Now suppose that the output is also sampled, using the *same* sampling function $p(t)$.   Its transform $R^*(s)$

is to be found. From Theorem 16-5, Eq. (16-41) becomes

$$R(s) = \frac{H(s)}{T} \sum_{n=-\infty}^{\infty} F\left(\frac{s - j2\pi n}{T}\right)$$

and so

$$R^*(s) = \frac{1}{T^2} \sum_{k=-\infty}^{\infty} H\left(\frac{s - j2\pi k}{T}\right) \sum_{n=-\infty}^{\infty} F\left[\frac{s - j2\pi(n+k)}{T}\right] \quad (16\text{-}42)$$

In the second summation, for fixed k, we can write $v = n + k$ and sum v from $-\infty$ to ∞, with no change in the sum. The result is

$$R^*(s) = \frac{1}{T^2} \sum_{k=-\infty}^{\infty} H\left(\frac{s - j2\pi k}{T}\right) \sum_{v=-\infty}^{\infty} F\left(\frac{s - j2\pi v}{T}\right)$$

or
$$R^*(s) = H^*(s)F^*(s) \quad (16\text{-}43)$$

This result can also be expressed in terms of Z transforms:

$$R_z(z) = H_z(z)F_z(z) \quad (16\text{-}44)$$

The function $H^*(s)$ is called the *sampled transfer function* of the system. Upon recalling that

$$h(t) = \mathcal{L}^{-1}[H(s)]$$

is the impulse response, it is evident that

$$h(nT) = Z^{-1}[H_z(z)] \quad (16\text{-}45)$$

In view of the comments at the end of Sec. 16-7, it is observed that Eqs. (16-43) and (16-44) are not exactly equivalent. Inversion of $R^*(s)$ by the Laplace inversion integral gives

$$r^*(t) = \mathcal{L}^{-1}[R^*(s)]$$

whereas the Z inversion of $R_z(z)$ gives

$$r(nT) = Z^{-1}[R_z(z)]$$

which are related by

$$r^*(t) = \sum_{n=0}^{\infty} r(nT)\delta(t - nT)$$

Neither inversion defines $r(t)$ for values of t other than integral multiples of T.

16-10. Convolution. The Z transform provides a useful convolution formula. In the notation of Sec. 16-9, we write

$$R_z(z) = H_z(z)F_z(z)$$

and direct our attention toward relating $Z^{-1}[R_z(z)]$ to the inverse transforms of $H_z(z)$ and $F_z(z)$. Equation (16-10) gives

$$F_z(z) = \sum_{u=0}^{\infty} f(uT)z^{-u} \qquad |z| > e^{\sigma_f T}$$

$$H_z(z) = \sum_{v=0}^{\infty} h(vT)z^{-v} \qquad |z| > e^{\sigma_h T}$$

and for their product we have

$$H_z(z)F_z(z) = \sum_{v=0}^{\infty} h(vT)z^{-v} \sum_{u=0}^{\infty} f(uT)z^{-u}$$

$$= \sum_{n=0}^{\infty} A_n z^{-n} \qquad (16\text{-}46)$$

where A_n is the sum of all products $h(vt)f(ut)$ for which $v + u = n$. This sum is given by either of the following:

$$A_n = \sum_{k=0}^{n} h(kT)f[(n-k)T] \qquad (16\text{-}47a)$$

$$A_n = \sum_{k=0}^{n} h[(n-k)T]f(kT) \qquad (16\text{-}47b)$$

Comparing Eq. (16-46) with Eq. (16-10) shows that A_n is $Z^{-1}[H_z(z)F_z(z)]$ at $t = nT$. But $r(nT)$ is our designation for the above inverse, and therefore we can write the following convolution theorem:

Theorem 16-6. If $f(t)$ and $h(t)$ have respective Z transforms $F_z(z)$ and $H_z(z)$, then the inverse Z transform of the product $F_z(z)H_z(z)$ is a function $r(nT)$ which can be expressed as

$$r(nt) = \sum_{k=0}^{n} h(kT)f[(n-k)T] \quad \text{or} \quad r(nt) = \sum_{k=0}^{n} h[(n-k)T]f(kT) \quad (16\text{-}48)$$

16-11. The Two-sided Z Transform. The Z transform can be extended to those situations for which the Fourier transform and two-sided Laplace transforms are used, by defining a two-sided Z transform

$$F_{2z}(z) = \sum_{n=-\infty}^{\infty} f(nT)z^{-n} \qquad e^{\sigma_1 T} < |z| < e^{\sigma_2 T} \qquad (16\text{-}49)$$

The above can be written as the sum of two ordinary Z transforms:

$$F_{zs}(z) = F_{s1}(z) + F_{s2}\left(\frac{1}{z}\right) \tag{16-50}$$

where

$$F_{s1}(z) = \sum_{n=0}^{\infty} f(nT)z^{-n} \qquad |z| > e^{\sigma_1 T}$$

$$\tag{16-51}$$

$$F_{s2}(z) = \sum_{n=1}^{\infty} f(-nT)z^{-n} \qquad |z| > e^{\sigma_2 T}$$

The series for $F_{s2}(1/z)$ converges inside a circle, whereas the first series converges outside a circle. If $\sigma_1 < \sigma_2$, there is a ring of convergence.

In regard to the inversion integral and uniqueness, the situation is very similar to the two-sided Laplace transform. The inversion formula is simpler to derive than for the single-sided case. Referring to Eq. (16-49), if

$$z = Re^{j\theta} \qquad e^{\sigma_1 T} < R < e^{\sigma_2 T}$$

we have

$$F_{2s}(Re^{j\theta}) = \sum_{n=-\infty}^{\infty} f(nT)R^{-n}e^{-jn\theta} \tag{16-52}$$

which shows that $f(nT)R^{-n}$ is the general Fourier coefficient of the expansion of $F_{2s}(Re^{j\theta})$. Thus, from the formula for Fourier coefficients, we immediately get

$$f(nT) = \frac{1}{2\pi} \int_0^{2\pi} F_{2s}(Re^{j\theta}) R^n e^{jn\theta}\, d\theta$$

which converts to the contour integral

$$f(nT) = \frac{1}{2\pi j} \int_{C_0} F_{2s}(z) z^{n-1}\, dz \tag{16-53}$$

This formula is identical in form with Eq. (16-28). However, we note that now the radius of C_0 is restricted to being in the annular region between the two circles of convergence. If a radius is used which carries C_0 outside this region, the inversion formula will give a different $f(t)$, but one which has the same two-sided Z transform. In other words, in similarity with the two-sided Laplace transform, the two-sided Z transform is not uniquely related to the $f(t)$ function, unless the contour of integration of the inversion integral is specified. The corresponding situation for the Laplace case is discussed in Secs. 10-17 and 10-18.

16-12. Systems with Sampled Input and Continuous Output. If a system has a sampled input but a nonsampled (continuous) output, the Laplace transform of the output is given by Eq. (16-41) rather than Eq. (16-43). Z transforms are particularly suited to Eq. (16-43) because $H^*(s) = H_s(z)$ is rational in z if $H(s)$ is rational in s. However, $H(s)$

is not rational in z, and for that reason Eq. (16-41) is not simplified by introduction of the Z transform. It is best to remain with the s variable, Laplace theory being used to obtain $\mathcal{L}^{-1}[R(s)]$. Of course, $F^*(s)$ is transcendental in s, but this is less disturbing than the multivaluedness that occurs if we try to use

$$s = \frac{1}{T} \log z$$

in the rational function $H(s)$. Further thought on this question will show that, if we attempt to get an inversion formula like Eq. (16-28) for continuous functions, the integrand is multivalued except at $t = nT$.

It is possible, however, to adapt the Z-transform theory to obtain information about a continuous output, by periodically sampling the output at values of $t = \tau, \tau + T, \tau + 2T$, etc., where the variable parameter τ is in the range $0 \leqq \tau \leqq T$. In this way, the output can be determined at any sequence of points, arbitrarily located between the original sampling points.

The Laplace transform of the continuous output is

$$R(s) = H(s)F^*(s) \tag{16-54}$$

where we are assuming that the input is sampled at $0, T, 2T$, etc. We recall Eq. (16-10) and write

$$H(s)F^*(s) = \sum_{k=0}^{\infty} H(s)f(kT)e^{-skT} \tag{16-55}$$

Since $h(t) = \mathcal{L}^{-1}[H(s)]$, we have, according to Theorem 10-8,

$$\mathcal{L}^{-1}[H(s)e^{-skT}] = h(t - kT)u(t - kT) \tag{16-56}$$

Thus, assuming that we can take inverse transforms inside the summation of Eq. (16-55), the output $r(t) = \mathcal{L}^{-1}[R(s)]$ is

$$r(t) = \sum_{k=0}^{\infty} f(kT)h(t - kT)u(t - kT) \tag{16-57}$$

Although this is the function we want, it is not in a convenient form for computation. It is nothing more than the summation of responses due to impulses of strength $f(kT)$ occurring at $t = kT$, a result which could have been derived without the aid of Z transforms.

Now let us sample $r(t)$ at the points $\tau, \tau + T$, etc., by multiplying $r(t)$ by a sampling function like $p(t)$, but for which the impulses are shifted an amount τ. Thus, in similarity with Eq. (16-4), we define

$$p(t,\tau) = \sum_{n=0}^{\infty} \delta(t - \tau - nT) \tag{16-58}$$

which yields the sampled output

$$r^*(t,\tau) = r(t)p(t,\tau) \tag{16-59}$$

In comparison with Eq. (16-7) the Laplace transform of this is

$$\mathcal{L}[r^*(t,\tau)] = \sum_{n=0}^{\infty} r(\tau + nT)e^{-s(\tau+nT)} \tag{16-60}$$

The function $\mathcal{L}[r^*(t,\tau)]$ is really superfluous, as we now show by defining

$$R_z(z,\eta) = \sum_{n=0}^{\infty} r(\tau + nT)z^{-n} \tag{16-61}$$

where $\eta = \tau/T$. $R_z(z,\eta)$ is called the *modified Z transform* of $r(t)$. Thus,

$$\mathcal{L}[r^*(t,\tau)] = e^{-s\tau}R_z(z,\eta) \tag{16-62}$$

The factor $e^{-s\tau}$ shows that $R_z(z,\eta)$ is the Laplace transform of the output function shifted an amount τ to the left and then sampled at $t = 0$, T, $2T$, etc. Equation (16-61) yields the same conclusion by emphasizing that $R_z(z,\eta)$ is the ordinary Z transform of $r(t + \tau)$.

Equation (16-57) can be used in Eq. (16-61). This would introduce a unit-step factor $u[\tau + (n - k)T]$ which is zero for $k > n$. Therefore, the upper summation limit in Eq. (16-57) can be changed to n, giving

$$R_z(z,\eta) = \sum_{n=0}^{\infty} \sum_{k=0}^{n} f(kT)h[\tau + (n - k)T]z^{-n} \tag{16-63}$$

This result is similar in form to Eq. (16-46) when Eq. (16-47b) is used for A_n. This fact can be used to establish that Eq. (16-63) can be put in a form similar to the first form of Eq. (16-46), namely,

$$R_z(z,\eta) = \sum_{v=0}^{\infty} h(\tau + vT)z^{-v} \sum_{u=0}^{\infty} f(uT)z^{-u} \tag{16-64}$$

The second summation on the right is $F_z(z)$. The first summation on the right is similar to Eq. (16-61), and so we define the following modified Z transform of $h(t)$:

$$H_z(z,\eta) = \sum_{v=0}^{\infty} h(\tau + vT)z^{-v} \tag{16-65}$$

The final result can thus be written

$$R_z(z,\eta) = H_z(z,\eta)F_z(z) \tag{16-66}$$

The function $H_z(z,\eta)$ is readily found from Eq. (16-65), being the normal Z transform of $h(t + \tau)$.

In concept, this result is very simple. To obtain a solution, we proceed as with ordinary Z transforms, but using an ordinary Z transform of $h(t + \tau)$, namely, $H_z(z,\eta) = Z[h(t + \eta T)]$, in place of $H_z(z)$. The inverse of the resulting Z-transform function then gives values of $r(\eta T + nT)$ at sampling points displaced an amount $\eta T = \tau$ from the original sampling points.

It is interesting to observe that Eq. (16-44) can be obtained by the above proof. This is evident when we note that, if η goes to zero, the modified transform reduces to an ordinary Z transform and Eq. (16-44) becomes a special case of Eq. (16-66). In fact, to prove Eq. (16-44) by the present method, although perhaps more complicated, is preferable because it does not require $F_z(z)$ and $H_z(z)$ to be rational functions.

16-13. Discontinuous Functions. Throughout the discussion of the Laplace transform, much attention is given to the possibility of $f(t)$ having points of discontinuity. In contrast, the discussion of the Z transform has been based on the assumption that $f(t)$ is continuous. This is not a significant omission, because $f^*(t)$ is insensitive to discontinuities occurring between the sampling points. Furthermore, if $f(t)$ should be discontinuous at a sampling instant, the theory presented is still applicable. In that case, if there is a discontinuity at $n_1 T$, it is necessary only to replace $f(\eta_1 T)$ by $f(n_1 T+)$, wherever it occurs.

PROBLEMS

16-1. Obtain the Z transforms of

(a) $\sin bt$

(b) $e^{at} \sin bt$

(c) $\cos bt$

(d) $\cos at \sin bt$

and locate their singular points in the z plane.

16-2. Show that

$$Z^{-1}[F_z(z)] = \int_0^{nT+} f^*(t)\, dt$$

where $nT+$ signifies a value slightly greater than nT. Both sides of this equation are functions of n.

16-3. Let the input function of a system be

$$f(t) = e^{-at}$$

and suppose that the system function is

$$H(s) = \frac{1}{s + b}$$

The system is initially relaxed.

(a) Assuming sampling intervals of duration T, and using series expressions for the appropriate Z transforms, find the first three terms of the synchronously sampled output function (corresponding to $t = 0$, T, $2T$).

(b) Using the inversion integral, obtain a general expression for the output function required in (a), and compare with the three values obtained in (a).

16-4. Let a system be described by a function

$$H(s) = \frac{s}{s + b}$$

and assume a driving function

$$f(t) = \sin bt$$

sampled at integral multiples of T. Using the inversion integral, obtain an expression for the synchronously sampled response, assuming that the system is initially relaxed.

16-5. Obtain the solution to the system specified in Prob. 16-3, by using the convolution theorem for Z transforms.

16-6. Show that at time $nT < t < (n + 1)T$ the nonsampled response of the system defined in Prob. 16-3 is

$$e^{-bt} \frac{e^{-(a-b)(n+1)T} - 1}{e^{-(a-b)T} - 1}$$

16-7. Obtain the solution for the nonsampled response, for the situation described in Prob. 16-4.

16-8. A system with input function $f(t)$ which has been sampled at $0, T, 2T, \ldots$ is to operate on this sampled input in such a way as to give an output of the form

$$r(t) = \begin{cases} f(0) & 0 < t < T \\ f(T) & T < t < 2T \\ f(2T) & 2T < t < 3T \end{cases}$$

$\cdots\cdots\cdots\cdots \qquad \cdots\cdots\cdots\cdots$

Show that a system function

$$H(s) = \frac{1 - e^{-sT}}{s}$$

will meet these requirements.

16-9. A system is described by the function

$$H(s) = \frac{1}{s + 1}$$

Assume that the input $f(t)$ is sampled at integral multiples of T. The output is unsampled. If $r(nT)$ is the corresponding sampled output, show that

$$r(t) = e^{-(t-nT)}f(nT)$$

where $nT < t < (n + 1)T$.

16-10. For the general function

$$f(t) = \begin{cases} e^{at} & t > 0 \\ e^{bt} & t < 0 \end{cases}$$

obtain a formula for the two-sided Z transform, giving conditions on a and b for the existence of this transform.

16-11. Obtain the two-sided Z transform of

$$f(t) = e^{-|t|}$$

and then use the inversion integral to recover $f(nT)$. (HINT: A change in variable of the inversion integral will simplify the case for negative n.) Also, determine what function would have a single-sided Z transform identical with the two-sided transform obtained above.

16-12. In the case of the two-sided Z transform, let $f(t)$ be written

$$f(t) = f_e(t) + f_o(t)$$

where $f_e(t)$ and $f_o(t)$ are, respectively, even and odd functions of t. Let $F_{2ze}(z)$ and $F_{2zo}(z)$ be the respective Z transforms of $f_e(t)$ and $f_o(t)$.

(a) Show that

$$F_{2ze}(z) = F_{2ze}\left(\frac{1}{z}\right)$$

$$F_{2zo}(z) = -F_{2zo}\left(\frac{1}{z}\right)$$

(b) Show that, for all n, the inversion integral can be written

$$f(nT) = \frac{1}{2\pi j} \int_{C_0} \left[F_{2ze}(z) + \frac{n}{|n|} F_{2zo}(z) \right] z^{|n|-1}\, dz$$

16-13. In Eqs. (16-15) we find

$$Z(t) = \frac{Tz}{(z-1)^2}$$

and in Eq. (16-40) we have

$$F^*(s) = \frac{1}{T} \sum_{n=-\infty}^{\infty} F\left(s - j\frac{2\pi n}{T}\right)$$

Show that the condition

$$F^*(s) = F_z(z)$$

is satisfied for this case.

16-14. The inversion integral for the Z transform [Eq. (16-28)] yields $f(nT)$, whereas the Laplace inversion integral applied to $F^*(s)$ yields $f_*^*(t)$.

(a) Discuss how $f^*(t)$ differs from $f(nT)$.

(b) Observe that a formal change of variable to z in

$$\frac{1}{2\pi j} \int_{Br} F^*(s) e^{snT}\, ds$$

yields a formula identical to Eq. (16-28), but that according to (a) this is $f^*(t)$. Explain the apparent paradox.

16-15. Carry out the details of the derivation of Eq. (16-60).

16-16. Observe that $f^*(t)$ is given as a product $f(t)p(t)$. Although the symbolic Laplace transform does not meet the conditions for s-plane convolution, carry out a formal derivation of Eq. (16-40), using s-plane convolution. Explain how it can be determined that the convolution theorem is not applicable in this case.

16-17. Give a derivation of Eq. (16-44) based on the method suggested in the last paragraph of Sec. 16-12.

16-18. Use the inversion formula to obtain $f(nT)$ for the one-sided Z transform

$$F_z(z) = \frac{z^2 + az + b}{(z-\alpha)(z-\beta)}$$

where $\alpha \neq \beta$. Determine by trial whether or not the initial- and final-value theorems are valid in this case, and appraise your findings.

16-19. Prove that $Z^{-1}[F_z(z)]$ is of exponential order < 0 if all poles of $F_z(z)$ lie inside the unit circle.

APPENDIX A

With a working knowledge of the basic concepts and theorems presented in this book, you are equipped with techniques for finding the Laplace transform, or the inverse, of almost any admissible function. However, when doing applied work, a person does not always remember all the theoretical background. This being the case, it is useful, for practical work, to have a tabulation of transform pairs. Reasonably complete tables are found in many of the books on the Laplace transform listed in the Bibliography, and very extensive tables have been compiled.* Since this text is primarily designed to present concepts, rather than procedures, an extensive tabulation is not given. Table 10-1 takes care of a large number of cases, and a few more of the more common functions are presented in the following table. It should be recognized, of course, that the two shifting theorems and the theorems on the derivative and integral of a function, in both t and s variables, are available to extend these tables to many other cases, almost on sight.

* A. Erdelyi (ed.), "Tables of Integral Transforms," vol. I, McGraw-Hill Book Company, Inc., 1954.

TABLE OF FUNCTIONS AND THEIR LAPLACE TRANSFORMS

The unit step $u(t)$ is implied as a multiplier of $f(t)$ in each case. Defining formulas for many of the functions are given in Appendix B.

$f(t)$	$F(s)$
$\dfrac{\sin at}{t}$	$\tan^{-1}\dfrac{a}{s}$
$\dfrac{1}{t}(1 - \cos at)$	$\frac{1}{2}\log\dfrac{s^2 + a^2}{s^2}$
$\dfrac{1}{t}(e^{at} - e^{bt})$	$\log\dfrac{s - b}{s - a}$
$\dfrac{\cos\sqrt{at}}{\sqrt{t}}$	$\sqrt{\dfrac{\pi}{s}}\,e^{-a/4s} \qquad a \geqq 0$
$\sin\sqrt{at}$	$\dfrac{1}{2}\sqrt{\dfrac{a\pi}{s^3}}\,e^{-a/4s} \qquad a > 0$
$\dfrac{1}{\sqrt{t}}e^{-a^2/t}$	$\sqrt{\dfrac{\pi}{s}}\,e^{-2a\sqrt{s}} \qquad a \geqq 0$
$\dfrac{1}{\sqrt{t^3}}(e^{at} - e^{bt})$	$2\sqrt{\pi}\,(\sqrt{s - b} - \sqrt{s - a})$
$\dfrac{e^{-\sqrt{at}}}{\sqrt{t}}$	$\sqrt{\dfrac{\pi}{s}}\,e^{a/4s}\left(1 - \operatorname{erf}\sqrt{\dfrac{a}{4s}}\right) \qquad a \geqq 0$
$e^{-\sqrt{at}}$	$\sqrt{\dfrac{a\pi}{4s^3}}\,e^{a/4s}\left(\operatorname{erf}\sqrt{\dfrac{a}{4s}} - 1\right) + \dfrac{1}{s} \qquad a \geqq 0$
$\log t$	$\dfrac{\Gamma'(1) - \log s}{s}$
$t^{n+1}\left[\dfrac{\Gamma'(n)}{\Gamma(n)} - \log t\right]$	$\dfrac{\Gamma(n)}{s^n}\log s \qquad n > 0$
$P_n(at)$	$\dfrac{1}{2^n}\displaystyle\sum_{k=\left[\frac{n+1}{2}\right]^*}^{n}\dfrac{(-1)^{k+n}(2k)!\,a^{2k-n}}{k!\,(n - k)!\,s^{2k-n+1}} \qquad n > 0$
$J_n(at)$	$\dfrac{(\sqrt{s^2 + a^2} - s)^n}{a^n\sqrt{a^2 + s^2}} \qquad n > -1$
$\sqrt{t^n}\,J_n\sqrt{at}$	$\left(\dfrac{\sqrt{a}}{2}\right)^n\dfrac{e^{-a/4s}}{s^{n+1}} \qquad a > 0,\, n > -1$
$\dfrac{1}{t}e^{-at}I_1(at)$	$\dfrac{\sqrt{s + 2a} - \sqrt{s}}{\sqrt{s + 2a} + \sqrt{s}}$
$L_n(at)$	$\dfrac{n!}{s}\left(\dfrac{s - a}{s}\right)^n$

TABLE OF FUNCTIONS AND THEIR LAPLACE TRANSFORMS (*Continued*)

$f(t)$	$F(s)$
$\dfrac{1}{\sqrt{t}\,(t+a)}$	$\dfrac{\pi}{\sqrt{a}}\,e^{as}(1-\operatorname{erf}\sqrt{as})\qquad a>0$
$\dfrac{1}{\sqrt{t+a}}$	$\sqrt{\dfrac{\pi}{s}}\,e^{as}(1-\operatorname{erf}\sqrt{as})\qquad a>0$
$\dfrac{1}{t+a}$	$-e^{as}\operatorname{Ei}(-as)\qquad a>0$
e^{-at^2}	$\dfrac{1}{2}\sqrt{\dfrac{\pi}{a}}\,e^{s^2/4a}\left(1-\operatorname{erf}\dfrac{s}{2\sqrt{a}}\right)\qquad a>0$
$\operatorname{Si}(at)$	$\dfrac{1}{s}\tan^{-1}\dfrac{a}{s}$
$\operatorname{erf}\sqrt{at}$	$\dfrac{\sqrt{a}}{s\sqrt{s+a}}\qquad a>0$
$\operatorname{erf}(at)$	$\dfrac{1}{s}\,e^{s^2/4a^2}\left(1-\operatorname{erf}\dfrac{s}{2a}\right)\qquad a>0$
$1-\operatorname{erf}\dfrac{a}{\sqrt{t}}$	$\dfrac{1}{s}\,e^{-2a\sqrt{s}}\qquad a\geqq 0$
$\operatorname{Ei}(-at)$	$-\dfrac{1}{s}\log\dfrac{s+a}{a}\qquad a>0$
$\operatorname{Ci}(at)$	$-\dfrac{1}{2s}\log\dfrac{a^2+s^2}{a^2}$
$e^{-at}\displaystyle\int_0^{\sqrt{at}}e^{x^2}\,dx$	$\dfrac{\sqrt{a\pi}}{2\sqrt{s}\,(s+a)}\qquad a>0$
$\sqrt{at}-e^{-at}\displaystyle\int_0^{\sqrt{at}}e^{x^2}\,dx$	$\dfrac{\sqrt{a^3\pi}}{2\sqrt{s^3}\,(s+a)}\qquad a>0$

* Brackets denote the greatest integer.

APPENDIX B

$$P_n(x) = \frac{1}{2^n n!} \frac{d^n}{dx^n} (x^2 - 1)^n \qquad \text{Legendre polynomial}$$

$$L_n(x) = e^x \frac{d^n}{dx^n} (x^n e^{-x}) \qquad \text{Laguerre polynomial}$$

$$J_n(x) = \frac{(-j)^n}{\pi} \int_0^\pi e^{jx \cos u} \cos nu \, du \qquad \text{Bessel function}$$

$$I_n(x) = (-j)^n J_n(jx) \qquad \text{Modified Bessel function}$$

$$\operatorname{erf} x = \frac{2}{\sqrt{\pi}} \int_0^x e^{-u^2} \, du \qquad \text{Error function}$$

$$\operatorname{Ei} x = -\int_1^\infty \frac{e^{xu}}{u} \, du \qquad \text{Exponential integral function}$$

$$\operatorname{Si} x = \int_0^1 \frac{\sin xu}{u} \, du \qquad \text{Sine integral function}$$

$$\operatorname{Ci} x = -\int_1^\infty \frac{\cos xu}{u} \, du \qquad \text{Cosine integral function}$$

BIBLIOGRAPHY

Works dealing mainly with functions of a complex variable:

Churchill, R. V.: "Introduction to Complex Variables and Applications," 2d ed., McGraw-Hill Book Company, Inc., New York, 1960.

Copson, E. T.: "Theory of Functions of a Complex Variable," Oxford University Press, New York, 1935.

Franklin, Philip: "A Treatise on Advanced Calculus," John Wiley & Sons, Inc., New York, 1940.

———: "Functions of Complex Variables," Prentice-Hall, Inc., Englewood Cliffs, N.J., 1958.

Guillemin, E. A.: "Mathematics of Circuit Analysis," John Wiley & Sons, Inc., New York, 1949.

Kaplan, W.: "Lectures on Functions of a Complex Variable," University of Michigan Press, Ann Arbor, Mich., 1955.

Knopp, K.: "Theory of Functions," Dover Publications, New York, 1945–1947.

Macrobert, T. M.: "Functions of a Complex Variable," St. Martin's Press, Inc., New York, 1950.

Nehari, Z.: "Conformal Mapping," McGraw-Hill Book Company, Inc., New York, 1952 (Dover Publications, Inc., New York, 1975).

Works dealing mainly with the Laplace and related transforms:

Aseltine, J. A.: "Transform Method in Linear System Analysis," McGraw-Hill Book Company, Inc., New York, 1958.

Bochner, S.: "Vorlesungen über Fouriersche Integrale," Chelsea Publishing Company, New York, 1948.

——— and K. Chandrasekharan: "Fourier Transforms," Princeton University Press, Princeton, N.J., 1949.

Campbell, G. A., and R. M. Foster: "Fourier Integrals for Practical Applications," D. Van Nostrand Company, Inc., Princeton, N.J., 1948.

Carslaw, H. S.: "Theory of Fourier Series and Integrals," 2d ed., St. Martin's Press, Inc., New York, 1921 (3d ed., Dover Publications, Inc., New York, 1950).

——— and J. C. Jaeger: "Operational Methods in Applied Mathematics," Oxford University Press, New York, 1941.

Cheng, D. K.: "Analysis of Linear Systems," Addison-Wesley Publishing Company, Reading, Mass., 1959.

Churchill, R. V.: "Fourier Series and Boundary Value Problems," McGraw-Hill Book Company, Inc., New York, 1941.

———: "Operational Mathematics," 2d ed., McGraw-Hill Book Company, Inc., New York, 1958.

Doetsch, G.: "Theorie und Anwendung der Laplace Transformation," Springer-Verlag, Berlin, 1937.

————: "Handbuch der Laplace Transformation," vols. I–III, Birkhauser, Stuttgart, 1950–1956.

————: "Einfuhrung in Theorie und Anwendung der Laplace Transformation," Birkhauser, Stuttgart, 1958.

Gardner, M. F., and J. L. Barnes: "Transients in Linear Systems," vol. I, John Wiley & Sons, Inc., New York, 1942.

Goldman, S.: "Transformation Calculus," Prentice-Hall, Inc., Englewood Cliffs, N.J., 1949.

Halperin, I.: "Introduction to the Theory of Distributions," University of Toronto Press, Toronto, 1952.

Holl, D. L., C. G. Maple, and B. Vinograde: "Introduction to the Laplace Transform," Appleton-Century-Crofts, Inc., New York, 1959.

Jury, E. I.: "Sampled Data Control Systems," John Wiley & Sons, Inc., New York, 1958.

Lago, G. V., and D. L. Waidelich: "Transients in Electrical Circuits," The Ronald Press Company, New York, 1958.

Lighthill, M. J.: "An Introduction to Fourier Analysis and Generalized Functions," Cambridge University Press, New York, 1958.

McLachlan, N. W.: "Modern Operational Calculus, with Applications in Technical Mathematics," St. Martin's Press, Inc., New York, 1948.

Paley, R., and N. Wiener: "Fourier Transforms in the Complex Domain," American Mathematical Society, Providence, R.I., 1934.

Ragazzini, J. R., and G. F. Franklin: "Sampled-data Control Systems," McGraw-Hill Book Company, Inc., New York, 1958.

Schwartz, L.: "Theorie des distributions," vol. I, Herman & Cie, Paris, 1950.

Sneddon, I. N.: "Fourier Transforms," McGraw-Hill Book Company, Inc., New York, 1951.

Titchmarsh, E. C.: "Theory of Fourier Integrals," Oxford University Press, New York, 1937.

van der Pol, B., and H. Bremmer: "Operational Calculus Based on the Two-sided Laplace Integral," Cambridge University Press, New York, 1950.

Weber, E.: "Linear Transient Analysis," vols. I, II, John Wiley & Sons, Inc., New York, 1954–1956.

Widder, D. V.: "The Laplace Transform," Princeton University Press, Princeton, N.J., 1941.

Wiener, N.: "The Fourier Integral," Cambridge University Press, New York, 1933 (Dover Publications, Inc., New York, 1958).

INDEX

Abscissa of convergence, one-sided
    Laplace integral, 292, 293
    two-sided Laplace integral, 295
Algebraic singularity, 185
Almost piecewise continuity, 240
Amplitude modulation, application of
    convolution integral to, 347
Analytic continuation, 147
    significance of, in definition of Laplace
        transform, 298
Analytic function, 32, 152
Analytic geometry plane, 24
Angle, of complex number, 21
    initial, for sinusoidal wave, 2
    preservation of, by conformal
        mapping, 75
Angle function, 229
Arc, differentiable, 85
    simple, 85
Axis of convergence of Laplace integral,
    293

Bessel function, 468
    Laplace transform of, 466
Bilinear transformation, properties of, 70
Branch cut, 171, 185
Branch point, 177
    integration around, 180
    methods of locating, 186
    order of, 173
Bromwich contour, for one-sided trans-
        form, 320
    for two-sided transform, 319

Cauchy integral formulas, 106
Cauchy integral theorem, 94–98
Cauchy principle of convergence, for
    improper integrals, 240

Cauchy principle of convergence, for
    infinite series, 117
Cauchy-Riemann differential equations,
    34
Characteristic equation, 389
Characteristic values, 326, 389
Complex number, angle of, 21
    exponential form, 37
    imaginary component of, 20
    imaginary part of, 20
    magnitude of, 21
    as an ordered pair, 19
    polar form, 21
    real component of, 20
    real part of, 20
    rectangular form, 21
Complex plane, 24
Conformal mapping, by analytic func-
        tion, 73
    preservation, of angles by, 75
    of shapes by, 76
Conformal maps, bilinear function, 70
    exponential function, 61
    hyperbolic cosine, 62
    reciprocal function, 56, 66
    trigonometric sine and cosine, 63
Conjugate, complex, 4, 23
Connected set, 86
Connectivity, order of, 87
    of a region, 87
    significance of, for integrals, 99
Continuity, definition of, 28
Contour integral, 89, 92
Convergence (see Improper real inte-
        grals; Infinite series)
Convolution in s plane, for one-sided
        transform, 343
    for two-sided transform, 349
Convolution integral, for Fourier trans-
        form, 338

Convolution integral, for impulse function, 428
  for one-sided Laplace transform, 343
  for two-sided Laplace transform, 340
Convolution theorem for Z transform, 458
Cosine integral function, 468
  Laplace transform of, 467
Current source, equivalent, 405
Curve, simple closed, 85

Deleted neighborhood, 28
Derivative, definition for function of a complex variable, 29
  for discontinuous function of a real variable, 367
  of impulse function, 417
  of multivalued function, 177
Distribution, 433

Element of an analytic function, 148
Entire function, 153
Error function, 468
  Laplace transform of, 467
Essential singularity, first kind, 142
  second kind, 143
Exponential form, 37
Exponential function, 36
Exponential integral function, 468
  Laplace transform of, 467
Exponential order, function of, 287
Exponential type, function of, 357
Extended plane, 88

Field problems, solution in two dimensions, 77–80
Final-value theorem, for Laplace transform, 315
  for Z transform, 451
Finite plane, 88
Forced response, 390
Fourier integral, 6, 268
Fourier integral theorem, 268–272
Fourier series, 5
Fourier transform, 274
  derivative of, 275
  symmetry of, 276
Frequency, 10

Frequency variable, 10
  generalized, 12
Function, of a complex variable, 24
  analytic, 32, 152
  entire, 152
  meromorphic, 152
  rational, 152
  transcendental, 35
  global definition of, 151
  of a real variable, almost piecewise continuity of, 240
  piecewise continuity of, 235

Gain function, 229
Gauss mean-value theorem, 205
Global definition of a function, 151

Half plane of convergence for Laplace integral, 293
Harmonic function, 41
Heaviside expansion theorem, 325
Helmholtz theorem, 403
  Norton's theorem as special case, 405
  Thévenin's theorem as special case, 404
Hilbert transform, 225
Hyperbolic functions, 38

Imaginary component of complex number, 20
Imaginary number, 20
Imaginary part of complex number, 20
Immittance function, 398
  combination of, 400
  self-immittance, 400
  transfer immittance, 400
Improper real integrals, 237, 238
  convergence of, 237, 239, 240
    absolute, 239
    uniform, 248
Impulse function, 414
  higher-order, 416, 418
  symbolic transform of, 417, 419
Impulse response, 414
Impulsive response, 415
Index principle, 211
Infinite series, 116–122
  convergence of, 116, 117

Infinite series, convergence, absolute, 118
    uniform, 120
    power (*see* Power series)
    ratio test, 119
    root test, 119
Infinity, point at, 64
Initial-value theorem, for Laplace
        transform, 315
    for Z transform, 451
Integral, contour, 89, 92
    upper bound, 94
    improper (*see* Improper real integrals)
    line, 90
    real, theorems for, 236
Integration, around branch points, 180
    over large circular arcs, 254
    by parts, 236
    by primitive functions, 109
    by residues, 145
Inversion formula, for Fourier trans-
        form, 268, 273
    nonuniqueness of, for two-sided
        Laplace transform, 315, 321
    for one-sided Laplace transform, 320
    for one-sided Z transform, 452
    for two-sided Laplace transform, 319
    for two-sided Z transform, 459
    uniqueness of, for Fourier transform
        and one-sided Laplace transform,
        276, 320
Iterated integrals, finite limits, 242
    infinite limits, 247

Jordan curve, closed, 86
Jordan curve theorem, 88
Jordan's lemma, 259

Laguerre polynomial, 113, 468
    Laplace transform of, 466
Laplace integral, convergence of, 288,
        289
    one-sided, 8, 287
    related to Fourier integral, 286
    two-sided, 286
Laplace transform, as an analytic func-
        tion, 298
    behavior at infinity, 306
    conditions to make an entire function,
        309

Laplace transform, derivative of, 308
    of derivative, 311
    of integral, 312
    inverse of (*see* Inversion formula)
    linear combination of, 299
    one-sided, 9, 287
    sufficient conditions for, 374
    uniqueness of inverse, 320
Laplace's equation, solution of, by func-
        tions of a complex variable, 77–80
    in two dimension, 41, 77
Laurent series, 134
    expanded about a singular point, 139
    properties of coefficients for real
        function, 203
    uniqueness of, 136
Legendre polynomial, 113, 468
    Laplace transform of, 466
Limit, definition of, 27
Line integral, 90
Linear function, mapping properties of,
        46
Lipshitz condition, 271
Logarithm, definition, 102
    mapping properties of, 176
    Riemann surface, 102
Logarithmic singularity, 185

M test (*see* Weierstrass M test)
Magnitude of complex number, 21
Maximum, principle of, 207
Meromorphic function, 152
Minimum, principle of, 207
Mittag-Leffler theorem, 157
Modified Z transform, 461
Morera's theorem, 109
Multivalued function, expansion in
        series, 188

Natural boundary, 150
Natural mode, 326, 389
Natural response, 389
Neighborhood, deleted, 28
Nyquist criterion, 213

Open region, 86
Order, of branch point, 142
    of connectivity, 87

Order, of a zero, 212
Ordered pair, 19

Parseval's theorem for Fourier transform, 279
Partial fraction expansion, 153
  of meromorphic function, 157
Partial sum of series, 116
Periodic function, Laplace transform of, 436–438
  response of system to, 438–442
Piecewise continuity, 235
Point set, 86
Poisson's integrals, 215
  transformed to imaginary axis, 220
Pole, definition of, 142
  order of, 142
  of system function related to natural response, 389
Positive real function, 208
Power series, 125–129
  circle of convergence, 125
  term-by-term differentiation, 129
  term-by-term integration, 128
Primitive function, 100
  used in evaluating integrals, 109
Principal part of Laurent expansion, 142

Radius of convergence for power series, 126
Real component of complex number, 20
Real function, 201
Real number, 20
Real part of complex number, 20
Region, 86
  bounded, 88
  closed, 87
  connectivity of, 87
  open, 86
Regular point, 31
Residue, definition, 144
  formulas for, 145
  at infinity, 146
Residue theorem, 145
Resonance, 391
Riemann sphere, 64
Riemann surface, 103, 171

Riemann's theorem for trigonometric integrals, 252–254
  generalization of, 354–355
Root locus, 190–197
Roots, of complex numbers, 22, 37
  of equations, 212

Saddle point, 186
Sampled transfer function, 457
Sampling function, 446
Series (see Infinite series)
Set, connected, 86
  of measure zero, integration over, 237
  point, 86
Shapes, preservation of, 76
Shifting theorems, for Laplace transform, 309, 310
  for Z transform, 450
Sine integral function, 468
  Laplace transform of, 467
Singular point, 31
  classification (see Algebraic singularity; Branch point; Essential singularity; Logarithmic singularity; Pole)
  at infinity, 143
Singularity functions, 424
Sinusoidal pulse, Laplace transform of, 304
Sinusoidal steady state, 1–5, 397
Stability, 12, 390
Strip of convergence for two-sided Laplace integral, 296
Superposition, 430
Symmetry of Fourier transforms, 276
System function, 2, 406
  angle function, 229
  gain function, 229
  relationships between real and imaginary parts, 223–228

Taylor series, 129–133
  uniqueness of, 132
Tchebysheff polynomial, 199, 468
Thévenin's theorem as special case of Helmholtz theorem, 404
Transcendental functions, 35
Transformation point due to a function, 48

Transient response, 389
Triangular pulse, Laplace transform of, 304
Trigonometric functions, 38
Two-sided Laplace transform (*see* Laplace transform)
Two-sided Z transform, 458

Uniform convergence, improper integrals, 248
    infinite series, 121
    Laplace integrals, 289, 291, 294
    M test, for integrals, 251
        for series, 127
    power series, 127
Unit doublet, 417
Unit impulse, 414
Unit pulse, 410

Unit ramp, 424
Unit step, 310

Voltage source, equivalent, 405

Weierstrass M test, for integrals, 251
    for series, 127

Z transform, 447
    convolution theorem for, 457
    of function multiplied by e^{-at}, 449
    inversion formula for, 451, 459
    modified, 461
    periodic properties of, 453
    of powers of t, 448
    two-sided, 458
Zeros of a function, 212

A CATALOGUE OF SELECTED DOVER BOOKS
IN ALL FIELDS OF INTEREST

A CATALOGUE OF SELECTED DOVER
BOOKS IN ALL FIELDS OF INTEREST

CELESTIAL OBJECTS FOR COMMON TELESCOPES, T. W. Webb. The most used book in amateur astronomy: inestimable aid for locating and identifying nearly 4,000 celestial objects. Edited, updated by Margaret W. Mayall. 77 illustrations. Total of 645pp. 5⅜ x 8½.
20917-2, 20918-0 Pa., Two-vol. set $9.00

HISTORICAL STUDIES IN THE LANGUAGE OF CHEMISTRY, M. P. Crosland. The important part language has played in the development of chemistry from the symbolism of alchemy to the adoption of systematic nomenclature in 1892. ". . . wholeheartedly recommended,"—Science. 15 illustrations. 416pp. of text. 5⅝ x 8¼.
63702-6 Pa. $6.00

BURNHAM'S CELESTIAL HANDBOOK, Robert Burnham, Jr. Thorough, readable guide to the stars beyond our solar system. Exhaustive treatment, fully illustrated. Breakdown is alphabetical by constellation: Andromeda to Cetus in Vol. 1; Chamaeleon to Orion in Vol. 2; and Pavo to Vulpecula in Vol. 3. Hundreds of illustrations. Total of about 2000pp. 6⅛ x 9¼.
23567-X, 23568-8, 23673-0 Pa., Three-vol. set $26.85

THEORY OF WING SECTIONS: INCLUDING A SUMMARY OF AIR-FOIL DATA, Ira H. Abbott and A. E. von Doenhoff. Concise compilation of subatomic aerodynamic characteristics of modern NASA wing sections, plus description of theory. 350pp. of tables. 693pp. 5⅜ x 8½.
60586-8 Pa. $7.00

DE RE METALLICA, Georgius Agricola. Translated by Herbert C. Hoover and Lou H. Hoover. The famous Hoover translation of greatest treatise on technological chemistry, engineering, geology, mining of early modern times (1556). All 289 original woodcuts. 638pp. 6¾ x 11.
60006-8 Clothbd. $17.95

THE ORIGIN OF CONTINENTS AND OCEANS, Alfred Wegener. One of the most influential, most controversial books in science, the classic statement for continental drift. Full 1966 translation of Wegener's final (1929) version. 64 illustrations. 246pp. 5⅜ x 8½.   61708-4 Pa. $4.50

THE PRINCIPLES OF PSYCHOLOGY, William James. Famous long course complete, unabridged. Stream of thought, time perception, memory, experimental methods; great work decades ahead of its time. Still valid, useful; read in many classes. 94 figures. Total of 1391pp. 5⅜ x 8½.
20381-6, 20382-4 Pa., Two-vol. set $13.00

THE CURVES OF LIFE, Theodore A. Cook. Examination of shells, leaves, horns, human body, art, etc., in "*the* classic reference on how the golden ratio applies to spirals and helices in nature . . . ."—Martin Gardner. 426 illustrations. Total of 512pp. 5⅜ x 8½.   23701-X Pa. $5.95

AN ILLUSTRATED FLORA OF THE NORTHERN UNITED STATES AND CANADA, Nathaniel L. Britton, Addison Brown. Encyclopedic work covers 4666 species, ferns on up. Everything. Full botanical information, illustration for each. This earlier edition is preferred · by many to more recent revisions. 1913 edition. Over 4000 illustrations, total of 2087pp. 6⅛ x 9¼.   22642-5, 22643-3, 22644-1 Pa., Three-vol. set $24.00

MANUAL OF THE GRASSES OF THE UNITED STATES, A. S. Hitchcock, U.S. Dept. of Agriculture. The basic study of American grasses, both indigenous and escapes, cultivated and wild. Over 1400 species. Full descriptions, information. Over 1100 maps, illustrations. Total of 1051pp. 5⅜ x 8½.   22717-0, 22718-9 Pa., Two-vol. set $15.00

THE CACTACEAE,, Nathaniel L. Britton, John N. Rose. Exhaustive, definitive. Every cactus in the world. Full botanical descriptions. Thorough statement of nomenclatures, habitat, detailed finding keys. The one book needed by every cactus enthusiast. Over 1275 illustrations. Total of 1080pp. 8 x 10¼.   21191-6, 21192-4 Clothbd., Two-vol. set $35.00

AMERICAN MEDICINAL PLANTS, Charles F. Millspaugh. Full descriptions, 180 plants covered: history; physical description; methods of preparation with all chemical constituents extracted; all claimed curative or adverse effects. 180 full-page plates. Classification table. 804pp. 6½ x 9¼.
23034-1 Pa. $10.00

A MODERN HERBAL, Margaret Grieve. Much the fullest, most exact, most useful compilation of herbal material. Gigantic alphabetical encyclopedia, from aconite to zedoary, gives botanical information, medical properties, folklore, economic uses, and much else. Indispensable to serious reader. 161 illustrations. 888pp. 6½ x 9¼. (Available in U.S. only)
22798-7, 22799-5 Pa., Two-vol. set $12.00

THE HERBAL or GENERAL HISTORY OF PLANTS, John Gerard. The 1633 edition revised and enlarged by Thomas Johnson. Containing almost 2850 plant descriptions and 2705 superb illustrations, Gerard's *Herbal* is a monumental work, the book all modern English herbals are derived from, the one herbal every serious enthusiast should have in its entirety. Original editions are worth perhaps $750. 1678pp. 8½ x 12¼.
23147-X Clothbd. $50.00

MANUAL OF THE TREES OF NORTH AMERICA, Charles S. Sargent. The basic survey of every native tree and tree-like shrub, 717 species in all. Extremely full descriptions, information on habitat, growth, locales, economics, etc. Necessary to every serious tree lover. Over 100 finding keys. 783 illustrations. Total of 986pp. 5⅜ x 8½.
20277-1, 20278-X Pa., Two-vol. set $10.00

THE DEPRESSION YEARS AS PHOTOGRAPHED BY ARTHUR ROTH-STEIN, Arthur Rothstein. First collection devoted entirely to the work of outstanding 1930s photographer: famous dust storm photo, ragged children, unemployed, etc. 120 photographs. Captions. 119pp. 9¼ x 10¾.
23590-4 Pa. $5.00

CAMERA WORK: A PICTORIAL GUIDE, Alfred Stieglitz. All 559 illustrations and plates from the most important periodical in the history of art photography, *Camera Work* (1903-17). Presented four to a page, reduced in size but still clear, in strict chronological order, with complete captions. Three indexes. Glossary. Bibliography. 176pp. 8⅜ x 11¼.
23591-2 Pa. $6.95

ALVIN LANGDON COBURN, PHOTOGRAPHER, Alvin L. Coburn. Revealing autobiography by one of greatest photographers of 20th century gives insider's version of Photo-Secession, plus comments on his own work. 77 photographs by Coburn. Edited by Helmut and Alison Gernsheim. 160pp. 8⅛ x 11.
23685-4 Pa. $6.00

NEW YORK IN THE FORTIES, Andreas Feininger. 162 brilliant photographs by the well-known photographer, formerly with *Life* magazine, show commuters, shoppers, Times Square at night, Harlem nightclub, Lower East Side, etc. Introduction and full captions by John von Hartz. 181pp. 9¼ x 10¾.
23585-8 Pa. $6.00

GREAT NEWS PHOTOS AND THE STORIES BEHIND THEM, John Faber. Dramatic volume of 140 great news photos, 1855 through 1976, and revealing stories behind them, with both historical and technical information. Hindenburg disaster, shooting of Oswald, nomination of Jimmy Carter, etc. 160pp. 8¼ x 11.
23667-6 Pa. $5.00

THE ART OF THE CINEMATOGRAPHER, Leonard Maltin. Survey of American cinematography history and anecdotal interviews with 5 masters—Arthur Miller, Hal Mohr, Hal Rosson, Lucien Ballard, and Conrad Hall. Very large selection of behind-the-scenes production photos. 105 photographs. Filmographies. Index. Originally *Behind the Camera.* 144pp. 8¼ x 11.
23686-2 Pa. $5.00

DESIGNS FOR THE THREE-CORNERED HAT (LE TRICORNE), Pablo Picasso. 32 fabulously rare drawings—including 31 color illustrations of costumes and accessories—for 1919 production of famous ballet. Edited by Parmenia Migel, who has written new introduction. 48pp. 9⅜ x 12¼. (Available in U.S. only)
23709-5 Pa. $5.00

NOTES OF A FILM DIRECTOR, Sergei Eisenstein. Greatest Russian filmmaker explains montage, making of *Alexander Nevsky,* aesthetics; comments on self, associates, great rivals (Chaplin), similar material. 78 illustrations. 240pp. 5⅜ x 8½.
22392-2 Pa. $4.50

HOLLYWOOD GLAMOUR PORTRAITS, edited by John Kobal. 145 photos capture the stars from 1926-49, the high point in portrait photography. Gable, Harlow, Bogart, Bacall, Hedy Lamarr, Marlene Dietrich, Robert Montgomery, Marlon Brando, Veronica Lake; 94 stars in all. Full background on photographers, technical aspects, much more. Total of 160pp. 8⅜ x 11¼.   23352-9 Pa. $6.00

THE NEW YORK STAGE: FAMOUS PRODUCTIONS IN PHOTO-GRAPHS, edited by Stanley Appelbaum. 148 photographs from Museum of City of New York show 142 plays, 1883-1939. *Peter Pan, The Front Page, Dead End, Our Town*, O'Neill, hundreds of actors and actresses, etc. Full indexes. 154pp. 9½ x 10.   23241-7 Pa. $6.00

MASTERS OF THE DRAMA, John Gassner. Most comprehensive history of the drama, every tradition from Greeks to modern Europe and America, including Orient. Covers 800 dramatists, 2000 plays; biography, plot summaries, criticism, theatre history, etc. 77 illustrations. 890pp. 5⅜ x 8½.
20100-7 Clothbd. $10.00

THE GREAT OPERA STARS IN HISTORIC PHOTOGRAPHS, edited by James Camner. 343 portraits from the 1850s to the 1940s: Tamburini, Mario, Caliapin, Jeritza, Melchior, Melba, Patti, Pinza, Schipa, Caruso, Farrar, Steber, Gobbi, and many more—270 performers in all. Index. 199pp. 8⅜ x 11¼.   23575-0 Pa. $6.50

J. S. BACH, Albert Schweitzer. Great full-length study of Bach, life, background to music, music, by foremost modern scholar. Ernest Newman translation. 650 musical examples. Total of 928pp. 5⅜ x 8½. (Available in U.S. only)   21631-4, 21632-2 Pa., Two-vol. set $10.00

COMPLETE PIANO SONATAS, Ludwig van Beethoven. All sonatas in the fine Schenker edition, with fingering, analytical material. One of best modern editions. Total of 615pp. 9 x 12. (Available in U.S. only)
23134-8, 23135-6 Pa., Two-vol. set $15.00

KEYBOARD MUSIC, J. S. Bach. Bach-Gesellschaft edition. For harpsichord, piano, other keyboard instruments. English Suites, French Suites, Six Partitas, Goldberg Variations, Two-Part Inventions, Three-Part Sinfonias. 312pp. 8⅛ x 11. (Available in U.S. only)   22360-4 Pa. $6.95

FOUR SYMPHONIES IN FULL SCORE, Franz Schubert. Schubert's four most popular symphonies: No. 4 in C Minor ("Tragic"); No. 5 in B-flat Major; No. 8 in B Minor ("Unfinished"); No. 9 in C Major ("Great"). Breitkopf & Hartel edition. Study score. 261pp. 9⅜ x 12¼.
23681-1 Pa. $6.50

THE AUTHENTIC GILBERT & SULLIVAN SONGBOOK, W. S. Gilbert, A. S. Sullivan. Largest selection available; 92 songs, uncut, original keys, in piano rendering approved by Sullivan. Favorites and lesser-known fine numbers. Edited with plot synopses by James Spero. 3 illustrations. 399pp. 9 x 12.   23482-7 Pa. $7.95

AN AUTOBIOGRAPHY, Margaret Sanger. Exciting personal account of hard-fought battle for woman's right to birth control, against prejudice, church, law. Foremost feminist document. 504pp. 5⅜ x 8½.
20470-7 Pa. $5.50

MY BONDAGE AND MY FREEDOM, Frederick Douglass. Born as a slave, Douglass became outspoken force in antislavery movement. The best of Douglass's autobiographies. Graphic description of slave life. Introduction by P. Foner. 464pp. 5⅜ x 8½.
22457-0 Pa. $5.50

LIVING MY LIFE, Emma Goldman. Candid, no holds barred account by foremost American anarchist: her own life, anarchist movement, famous contemporaries, ideas and their impact. Struggles and confrontations in America, plus deportation to U.S.S.R. Shocking inside account of persecution of anarchists under Lenin. 13 plates. Total of 944pp. 5⅜ x 8½.
22543-7, 22544-5 Pa., Two-vol. set $11.00

LETTERS AND NOTES ON THE MANNERS, CUSTOMS AND CONDITIONS OF THE NORTH AMERICAN INDIANS, George Catlin. Classic account of life among Plains Indians: ceremonies, hunt, warfare, etc. Dover edition reproduces for first time all original paintings. 312 plates. 572pp. of text. 6⅛ x 9¼.
22118-0, 22119-9 Pa.. Two-vol. set $11.50

THE MAYA AND THEIR NEIGHBORS, edited by Clarence L. Hay, others. Synoptic view of Maya civilization in broadest sense, together with Northern, Southern neighbors. Integrates much background, valuable detail not elsewhere. Prepared by greatest scholars: Kroeber, Morley, Thompson, Spinden, Vaillant, many others. Sometimes called Tozzer Memorial Volume. 60 illustrations, linguistic map. 634pp. 5⅜ x 8½.
23510-6 Pa. $7.50

HANDBOOK OF THE INDIANS OF CALIFORNIA, A. L. Kroeber. Foremost American anthropologist offers complete ethnographic study of each group. Monumental classic. 459 illustrations, maps. 995pp. 5⅜ x 8½.
23368-5 Pa. $10.00

SHAKTI AND SHAKTA, Arthur Avalon. First book to give clear, cohesive analysis of Shakta doctrine, Shakta ritual and Kundalini Shakti (yoga). Important work by one of world's foremost students of Shaktic and Tantric thought. 732pp. 5⅜ x 8½. (Available in U.S. only)
23645-5 Pa. $7.95

AN INTRODUCTION TO THE STUDY OF THE MAYA HIEROGLYPHS, Syvanus Griswold Morley. Classic study by one of the truly great figures in hieroglyph research. Still the best introduction for the student for reading Maya hieroglyphs. New introduction by J. Eric S. Thompson. 117 illustrations. 284pp. 5⅜ x 8½.
23108-9 Pa. $4.00

A STUDY OF MAYA ART, Herbert J. Spinden. Landmark classic interprets Maya symbolism, estimates styles, covers ceramics, architecture, murals, stone carvings as artforms. Still a basic book in area. New introduction by J. Eric Thompson. Over 750 illustrations. 341pp. 8⅜ x 11¼.
21235-1 Pa. $6.95

A MAYA GRAMMAR, Alfred M. Tozzer. Practical, useful English-language grammar by the Harvard anthropologist who was one of the three greatest American scholars in the area of Maya culture. Phonetics, grammatical processes, syntax, more. 301pp. 5⅜ x 8½. 23465-7 Pa. $4.00

THE JOURNAL OF HENRY D. THOREAU, edited by Bradford Torrey, F. H. Allen. Complete reprinting of 14 volumes, 1837-61, over two million words; the sourcebooks for *Walden*, etc. Definitive. All original sketches, plus 75 photographs. Introduction by Walter Harding. Total of 1804pp. 8½ x 12¼. 20312-3, 20313-1 Clothbd., Two-vol. set $50.00

CLASSIC GHOST STORIES, Charles Dickens and others. 18 wonderful stories you've wanted to reread: "The Monkey's Paw," "The House and the Brain," "The Upper Berth," "The Signalman," "Dracula's Guest," "The Tapestried Chamber," etc. Dickens, Scott, Mary Shelley, Stoker, etc. 330pp. 5⅜ x 8½. 20735-8 Pa. $3.50

SEVEN SCIENCE FICTION NOVELS, H. G. Wells. Full novels. *First Men in the Moon, Island of Dr. Moreau, War of the Worlds, Food of the Gods, Invisible Man, Time Machine, In the Days of the Comet*. A basic science-fiction library. 1015pp. 5⅜ x 8½. (Available in U.S. only) 20264-X Clothbd. $8.95

ARMADALE, Wilkie Collins. Third great mystery novel by the author of *The Woman in White* and *The Moonstone*. Ingeniously plotted narrative shows an exceptional command of character, incident and mood. Original magazine version with 40 illustrations. 597pp. 5⅜ x 8½. 23429-0 Pa. $5.00

MASTERS OF MYSTERY, H. Douglas Thomson. The first book in English (1931) devoted to history and aesthetics of detective story. Poe, Doyle, LeFanu, Dickens, many others, up to 1930. New introduction and notes by E. F. Bleiler. 288pp. 5⅜ x 8½. (Available in U.S. only) 23606-4 Pa. $4.00

FLATLAND, E. A. Abbott. Science-fiction classic explores life of 2-D being in 3-D world. Read also as introduction to thought about hyperspace. Introduction by Banesh Hoffmann. 16 illustrations. 103pp. 5⅜ x 8½. 20001-9 Pa. $1.75

THREE SUPERNATURAL NOVELS OF THE VICTORIAN PERIOD, edited, with an introduction, by E. F. Bleiler. Reprinted complete and unabridged, three great classics of the supernatural: *The Haunted Hotel* by Wilkie Collins, *The Haunted House at Latchford* by Mrs. J. H. Riddell, and *The Lost Stradivarius* by J. Meade Falkner. 325pp. 5⅜ x 8½. 22571-2 Pa. $4.00

AYESHA: THE RETURN OF "SHE," H. Rider Haggard. Virtuoso sequel featuring the great mythic creation, Ayesha, in an adventure that is fully as good as the first book, *She*. Original magazine version, with 47 original illustrations by Maurice Greiffenhagen. 189pp. 6½ x 9¼. 23649-8 Pa. $3.50

ART FORMS IN NATURE, Ernst Haeckel. Multitude of strangely beautiful natural forms: Radiolaria, Foraminifera, jellyfishes, fungi, turtles, bats, etc. All 100 plates of the 19th-century evolutionist's *Kunstformen der Natur* (1904). 100pp. 9⅜ x 12¼.   22987-4 Pa. $4.50

CHILDREN: A PICTORIAL ARCHIVE FROM NINETEENTH-CENTURY SOURCES, edited by Carol Belanger Grafton. 242 rare, copyright-free wood engravings for artists and designers. Widest such selection available. All illustrations in line. 119pp. 8⅜ x 11¼.
23694-3 Pa. $3.50

WOMEN: A PICTORIAL ARCHIVE FROM NINETEENTH-CENTURY SOURCES, edited by Jim Harter. 391 copyright-free wood engravings for artists and designers selected from rare periodicals. Most extensive such collection available. All illustrations in line. 128pp. 9 x 12.
23703-6 Pa. $4.50

ARABIC ART IN COLOR, Prisse d'Avennes. From the greatest ornamentalists of all time—50 plates in color, rarely seen outside the Near East, rich in suggestion and stimulus. Includes 4 plates on covers. 46pp. 9⅜ x 12¼.   23658-7 Pa. $6.00

AUTHENTIC ALGERIAN CARPET DESIGNS AND MOTIFS, edited by June Beveridge. Algerian carpets are world famous. Dozens of geometrical motifs are charted on grids, color-coded, for weavers, needleworkers, craftsmen, designers. 53 illustrations plus 4 in color. 48pp. 8¼ x 11. (Available in U.S. only)   23650-1 Pa. $1.75

DICTIONARY OF AMERICAN PORTRAITS, edited by Hayward and Blanche Cirker. 4000 important Americans, earliest times to 1905, mostly in clear line. Politicians, writers, soldiers, scientists, inventors, industrialists, Indians, Blacks, women, outlaws, etc. Identificatory information. 756pp. 9¼ x 12¾.   21823-6 Clothbd. $40.00

HOW THE OTHER HALF LIVES, Jacob A. Riis. Journalistic record of filth, degradation, upward drive in New York immigrant slums, shops, around 1900. New edition includes 100 original Riis photos, monuments of early photography. 233pp. 10 x 7⅞.   22012-5 Pa. $6.00

NEW YORK IN THE THIRTIES, Berenice Abbott. Noted photographer's fascinating study of city shows new buildings that have become famous and old sights that have disappeared forever. Insightful commentary. 97 photographs. 97pp. 11⅜ x 10.   22967-X Pa. $5.00

MEN AT WORK, Lewis W. Hine. Famous photographic studies of construction workers, railroad men, factory workers and coal miners. New supplement of 18 photos on Empire State building construction. New introduction by Jonathan L. Doherty. Total of 69 photos. 63pp. 8 x 10¾.
23475-4 Pa. $3.00

THE ANATOMY OF THE HORSE, George Stubbs. Often considered the great masterpiece of animal anatomy. Full reproduction of 1766 edition, plus prospectus; original text and modernized text. 36 plates. Introduction by Eleanor Garvey. 121pp. 11 x 14¾.   23402-9 Pa. $6.00

BRIDGMAN'S LIFE DRAWING, George B. Bridgman. More than 500 illustrative drawings and text teach you to abstract the body into its major masses, use light and shade, proportion; as well as specific areas of anatomy, of which Bridgman is master. 192pp. 6½ x 9¼. (Available in U.S. only)
22710-3 Pa. $3.00

ART NOUVEAU DESIGNS IN COLOR, Alphonse Mucha, Maurice Verneuil, Georges Auriol. Full-color reproduction of *Combinaisons orne-mentales* (c. 1900) by Art Nouveau masters. Floral, animal, geometric, interlacings, swashes—borders, frames, spots—all incredibly beautiful. 60 plates, hundreds of designs. 9⅜ x 8-1/16.   22885-1 Pa. $4.00

FULL-COLOR FLORAL DESIGNS IN THE ART NOUVEAU STYLE, E. A. Seguy. 166 motifs, on 40 plates, from *Les fleurs et leurs applications decoratives* (1902): borders, circular designs, repeats, allovers, "spots." All in authentic Art Nouveau colors. 48pp. 9⅜ x 12¼.
23439-8 Pa. $5.00

A DIDEROT PICTORIAL ENCYCLOPEDIA OF TRADES AND IN-DUSTRY, edited by Charles C. Gillispie. 485 most interesting plates from the great French Encyclopedia of the 18th century show hundreds of working figures, artifacts, process, land and cityscapes; glassmaking, paper-making, metal extraction, construction, weaving, making furniture, clothing, wigs, dozens of other activities. Plates fully explained. 920pp. 9 x 12.
22284-5, 22285-3 Clothbd., Two-vol. set $40.00

HANDBOOK OF EARLY ADVERTISING ART, Clarence P. Hornung. Largest collection of copyright-free early and antique advertising art ever compiled. Over 6,000 illustrations, from Franklin's time to the 1890's for special effects, novelty. Valuable source, almost inexhaustible.
Pictorial Volume. Agriculture, the zodiac, animals, autos, birds, Christmas, fire engines, flowers, trees, musical instruments, ships, games and sports, much more. Arranged by subject matter and use. 237 plates. 288pp. 9 x 12.
20122-8 Clothbd. $13.50

Typographical Volume. Roman and Gothic faces ranging from 10 point to 300 point, "Barnum," German and Old English faces, script, logotypes, scrolls and flourishes, 1115 ornamental initials, 67 complete alphabets, more. 310 plates. 320pp. 9 x 12.   20123-6 Clothbd. $15.00

CALLIGRAPHY (CALLIGRAPHIA LATINA), J. G. Schwandner. High point of 18th-century ornamental calligraphy. Very ornate initials, scrolls, borders, cherubs, birds, lettered examples. 172pp. 9 x 13.
20475-8 Pa. $6.00

DRAWINGS OF WILLIAM BLAKE, William Blake. 92 plates from Book of Job, *Divine Comedy, Paradise Lost,* visionary heads, mythological figures, Laocoon, etc. Selection, introduction, commentary by Sir Geoffrey Keynes. 178pp. 8⅛ x 11. 22303-5 Pa. $4.00

ENGRAVINGS OF HOGARTH, William Hogarth. 101 of Hogarth's greatest works: *Rake's Progress, Harlot's Progress, Illustrations for Hudibras, Before and After, Beer Street and Gin Lane,* many more. Full commentary. 256pp. 11 x 13¾. 22479-1 Pa. $7.95

DAUMIER: 120 GREAT LITHOGRAPHS, Honore Daumier. Wide-ranging collection of lithographs by the greatest caricaturist of the 19th century. Concentrates on eternally popular series on lawyers, on married life, on liberated women, etc. Selection, introduction, and notes on plates by Charles F. Ramus. Total of 158pp. 9⅜ x 12¼. 23512-2 Pa. $5.50

DRAWINGS OF MUCHA, Alphonse Maria Mucha. Work reveals drafts-man of highest caliber: studies for famous posters and paintings, render-ings for book illustrations and ads, etc. 70 works, 9 in color; including 6 items not drawings. Introduction. List of illustrations. 72pp. 9⅜ x 12¼. (Available in U.S. only) 23672-2 Pa. $4.00

GIOVANNI BATTISTA PIRANESI: DRAWINGS IN THE PIERPONT MORGAN LIBRARY, Giovanni Battista Piranesi. For first time ever all of Morgan Library's collection, world's largest. 167 illustrations of rare Piranesi drawings—archeological, architectural, decorative and visionary. Essay, detailed list of drawings, chronology, captions. Edited by Felice Stampfle. 144pp. 9⅜ x 12¼. 23714-1 Pa. $7.50

NEW YORK ETCHINGS (1905-1949), John Sloan. All of important American artist's N.Y. life etchings. 67 works include some of his best art; also lively historical record—Greenwich Village, tenement scenes. Edited by Sloan's widow. Introduction and captions. 79pp. 8⅜ x 11¼.
23651-X Pa. $4.00

CHINESE PAINTING AND CALLIGRAPHY: A PICTORIAL SURVEY, Wan-go Weng. 69 fine examples from John M. Crawford's matchless private collection: landscapes, birds, flowers, human figures, etc., plus calligraphy. Every basic form included: hanging scrolls, handscrolls, album leaves, fans, etc. 109 illustrations. Introduction. Captions. 192pp. 8⅞ x 11¾.
23707-9 Pa. $7.95

DRAWINGS OF REMBRANDT, edited by Seymour Slive. Updated Lipp-mann, Hofstede de Groot edition, with definitive scholarly apparatus. All portraits, biblical sketches, landscapes, nudes, Oriental figures, classical studies, together with selection of work by followers. 550 illustrations. Total of 630pp. 9⅛ x 12¼. 21485-0, 21486-9 Pa., Two-vol. set $15.00

THE DISASTERS OF WAR, Francisco Goya. 83 etchings record horrors of Napoleonic wars in Spain and war in general. Reprint of 1st edition, plus 3 additional plates. Introduction by Philip Hofer. 97pp. 9⅜ x 8¼.
21872-4 Pa. $3.75

THE COMPLETE WOODCUTS OF ALBRECHT DURER, edited by Dr. W. Kurth. 346 in all: "Old Testament," "St. Jerome," "Passion," "Life of Virgin," Apocalypse," many others. Introduction by Campbell Dodgson. 285pp. 8½ x 12¼. 21097-9 Pa. $7.50

DRAWINGS OF ALBRECHT DURER, edited by Heinrich Wolfflin. 81 plates show development from youth to full style. Many favorites; many new. Introduction by Alfred Werner. 96pp. 8⅛ x 11. 22352-3 Pa. $5.00

THE HUMAN FIGURE, Albrecht Dürer. Experiments in various techniques—stereometric, progressive proportional, and others. Also life studies that rank among finest ever done. Complete reprinting of *Dresden Sketchbook*. 170 plates. 355pp. 8⅜ x 11¼. 21042-1 Pa. $7.95

OF THE JUST SHAPING OF LETTERS, Albrecht Dürer. Renaissance artist explains design of Roman majuscules by geometry, also Gothic lower and capitals. Grolier Club edition. 43pp. 7⅞ x 10¾ 21306-4 Pa. $8.00

TEN BOOKS ON ARCHITECTURE, Vitruvius. The most important book ever written on architecture. Early Roman aesthetics, technology, classical orders, site selection, all other aspects. Stands behind everything since. Morgan translation. 331pp. 5⅜ x 8½. 20645-9 Pa. $4.00

THE FOUR BOOKS OF ARCHITECTURE, Andrea Palladio. 16th-century classic responsible for Palladian movement and style. Covers classical architectural remains, Renaissance revivals, classical orders, etc. 1738 Ware English edition. Introduction by A. Placzek. 216 plates. 110pp. of text. 9½ x 12¾. 21308-0 Pa. $8.95

HORIZONS, Norman Bel Geddes. Great industrialist stage designer, "father of streamlining," on application of aesthetics to transportation, amusement, architecture, etc. 1932 prophetic account; function, theory, specific projects. 222 illustrations. 312pp. 7⅞ x 10¾. 23514-9 Pa. $6.95

FRANK LLOYD WRIGHT'S FALLINGWATER, Donald Hoffmann. Full, illustrated story of conception and building of Wright's masterwork at Bear Run, Pa. 100 photographs of site, construction, and details of completed structure. 112pp. 9¼ x 10. 23671-4 Pa. $5.50

THE ELEMENTS OF DRAWING, John Ruskin. Timeless classic by great Viltorian; starts with basic ideas, works through more difficult. Many practical exercises. 48 illustrations. Introduction by Lawrence Campbell. 228pp. 5⅜ x 8½. 22730-8 Pa. $2.75

GIST OF ART, John Sloan. Greatest modern American teacher, Art Students League, offers innumerable hints, instructions, guided comments to help you in painting. Not a formal course. 46 illustrations. Introduction by Helen Sloan. 200pp. 5⅜ x 8½. 23435-5 Pa. $4.00

THE EARLY WORK OF AUBREY BEARDSLEY, Aubrey Beardsley. 157 plates, 2 in color: *Manon Lescaut, Madame Bovary, Morte Darthur, Salome,* other. Introduction by H. Marillier. 182pp. 8⅛ x 11. 21816-3 Pa. $4.50

THE LATER WORK OF AUBREY BEARDSLEY, Aubrey Beardsley. Exotic masterpieces of full maturity: *Venus and Tannhauser, Lysistrata, Rape of the Lock, Volpone,* Savoy material, etc. 174 plates, 2 in color. 186pp. 8⅛ x 11. 21817-1 Pa. $4.50

THOMAS NAST'S CHRISTMAS DRAWINGS, Thomas Nast. Almost all Christmas drawings by creator of image of Santa Claus as we know it, and one of America's foremost illustrators and political cartoonists. 66 illustrations. 3 illustrations in color on covers. 96pp. 8⅜ x 11¼. 23660-9 Pa. $3.50

THE DORÉ ILLUSTRATIONS FOR DANTE'S DIVINE COMEDY, Gustave Doré. All 135 plates from Inferno, Purgatory, Paradise; fantastic tortures, infernal landscapes, celestial wonders. Each plate with appropriate (translated) verses. 141pp. 9 x 12. 23231-X Pa. $4.50

DORÉ'S ILLUSTRATIONS FOR RABELAIS, Gustave Doré. 252 striking illustrations of *Gargantua and Pantagruel* books by foremost 19th-century illustrator. Including 60 plates, 192 delightful smaller illustrations. 153pp. 9 x 12. 23656-0 Pa. $5.00

LONDON: A PILGRIMAGE, Gustave Doré, Blanchard Jerrold. Squalor, riches, misery, beauty of mid-Victorian metropolis; 55 wonderful plates, 125 other illustrations, full social, cultural text by Jerrold. 191pp. of text. 9⅜ x 12¼. 22306-X Pa. $6.00

THE RIME OF THE ANCIENT MARINER, Gustave Doré, S. T. Coleridge. Dore's finest work, 34 plates capture moods, subtleties of poem. Full text. Introduction by Millicent Rose. 77pp. 9¼ x 12. 22305-1 Pa. $3.50

THE DORE BIBLE ILLUSTRATIONS, Gustave Doré. All wonderful, detailed plates: Adam and Eve, Flood, Babylon, Life of Jesus, etc. Brief King James text with each plate. Introduction by Millicent Rose. 241 plates. 241pp. 9 x 12. 23004-X Pa. $6.00

THE COMPLETE ENGRAVINGS, ETCHINGS AND DRYPOINTS OF ALBRECHT DURER. "Knight, Death and Devil"; "Melencolia," and more—all Dürer's known works in all three media, including 6 works formerly attributed to him. 120 plates. 235pp. 8⅜ x 11¼. 22851-7 Pa. $6.50

MAXIMILIAN'S TRIUMPHAL ARCH, Albrecht Dürer and others. Incredible monument of woodcut art: 8 foot high elaborate arch—heraldic figures, humans, battle scenes, fantastic elements—that you can assemble yourself. Printed on one side, layout for assembly. 143pp. 11 x 16. 21451-6 Pa. $5.00

UNCLE SILAS, J. Sheridan LeFanu. Victorian Gothic mystery novel, considered by many best of period, even better than Collins or Dickens. Wonderful psychological terror. Introduction by Frederick Shroyer. 436pp. 5⅜ x 8½.   21715-9 Pa. $6.00

JURGEN, James Branch Cabell. The great erotic fantasy of the 1920's that delighted thousands, shocked thousands more. Full final text, Lane edition with 13 plates by Frank Pape. 346pp. 5⅜ x 8½.
                                                                    23507-6 Pa. $4.50

THE CLAVERINGS, Anthony Trollope. Major novel, chronicling aspects of British Victorian society, personalities. Reprint of Cornhill serialization, 16 plates by M. Edwards; first reprint of full text. Introduction by Norman Donaldson. 412pp. 5⅜ x 8½.   23464-9 Pa. $5.00

KEPT IN THE DARK, Anthony Trollope. Unusual short novel about Victorian morality and abnormal psychology by the great English author. Probably the first American publication. Frontispiece by Sir John Millais. 92pp. 6½ x 9¼.   23609-9 Pa. $2.50

RALPH THE HEIR, Anthony Trollope. Forgotten tale of illegitimacy, inheritance. Master novel of Trollope's later years. Victorian country estates, clubs, Parliament, fox hunting, world of fully realized characters. Reprint of 1871 edition. 12 illustrations by F. A. Faser. 434pp. of text. 5⅜ x 8½.   23642-0 Pa. $5.00

YEKL and THE IMPORTED BRIDEGROOM AND OTHER STORIES OF THE NEW YORK GHETTO, Abraham Cahan. Film *Hester Street* based on *Yekl* (1896). Novel, other stories among first about Jewish immigrants of N.Y.'s East Side. Highly praised by W. D. Howells—Cahan "a new star of realism." New introduction by Bernard G. Richards. 240pp. 5⅜ x 8½.   22427-9 Pa. $3.50

THE HIGH PLACE, James Branch Cabell. Great fantasy writer's enchanting comedy of disenchantment set in 18th-century France. Considered by some critics to be even better than his famous *Jurgen*. 10 illustrations and numerous vignettes by noted fantasy artist Frank C. Pape. 320pp. 5⅜ x 8½.   23670-6 Pa. $4.00

ALICE'S ADVENTURES UNDER GROUND, Lewis Carroll. Facsimile of ms. Carroll gave Alice Liddell in 1864. Different in many ways from final Alice. Handlettered, illustrated by Carroll. Introduction by Martin Gardner. 128pp. 5⅜ x 8½.   21482-6 Pa. $2.00

FAVORITE ANDREW LANG FAIRY TALE BOOKS IN MANY COLORS, Andrew Lang. The four Lang favorites in a boxed set—the complete *Red, Green, Yellow* and *Blue* Fairy Books. 164 stories; 439 illustrations by Lancelot Speed, Henry Ford and G. P. Jacomb Hood. Total of about 1500pp. 5⅜ x 8½.   23407-X Boxed set, Pa. $14.95

HOUSEHOLD STORIES BY THE BROTHERS GRIMM. All the great Grimm stories: "Rumpelstiltskin," "Snow White," "Hansel and Gretel," etc., with 114 illustrations by Walter Crane. 269pp. 5⅜ x 8½.
21080-4 Pa. $3.00

SLEEPING BEAUTY, illustrated by Arthur Rackham. Perhaps the fullest, most delightful version ever, told by C. S. Evans. Rackham's best work. 49 illustrations. 110pp. 7⅞ x 10¾.
22756-1 Pa. $2.50

AMERICAN FAIRY TALES, L. Frank Baum. Young cowboy lassoes Father Time; dummy in Mr. Floman's department store window comes to life; and 10 other fairy tales. 41 illustrations by N. P. Hall, Harry Kennedy, Ike Morgan, and Ralph Gardner. 209pp. 5⅜ x 8½.
23643-9 Pa. $3.00

THE WONDERFUL WIZARD OF OZ, L. Frank Baum. Facsimile in full color of America's finest children's classic. Introduction by Martin Gardner. 143 illustrations by W. W. Denslow. 267pp. 5⅜ x 8½.
20691-2 Pa. $3.50

THE TALE OF PETER RABBIT, Beatrix Potter. The inimitable Peter's terrifying adventure in Mr. McGregor's garden, with all 27 wonderful, full-color Potter illustrations. 55pp. 4¼ x 5½. (Available in U.S. only)
22827-4 Pa. $1.25

THE STORY OF KING ARTHUR AND HIS KNIGHTS, Howard Pyle. Finest children's version of life of King Arthur. 48 illustrations by Pyle. 131pp. 6⅛ x 9¼.
21445-1 Pa. $4.95

CARUSO'S CARICATURES, Enrico Caruso. Great tenor's remarkable caricatures of self, fellow musicians, composers, others. Toscanini, Puccini, Farrar, etc. Impish, cutting, insightful. 473 illustrations. Preface by M. Sisca. 217pp. 8⅜ x 11¼.
23528-9 Pa. $6.95

PERSONAL NARRATIVE OF A PILGRIMAGE TO ALMADINAH AND MECCAH, Richard Burton. Great travel classic by remarkably colorful personality. Burton, disguised as a Moroccan, visited sacred shrines of Islam, narrowly escaping death. Wonderful observations of Islamic life, customs, personalities. 47 illustrations. Total of 959pp. 5⅜ x 8½.
21217-3, 21218-1 Pa., Two-vol. set $12.00

INCIDENTS OF TRAVEL IN YUCATAN, John L. Stephens. Classic (1843) exploration of jungles of Yucatan, looking for evidences of Maya civilization. Travel adventures, Mexican and Indian culture, etc. Total of 669pp. 5⅜ x 8½.
20926-1, 20927-X Pa., Two-vol. set $7.90

AMERICAN LITERARY AUTOGRAPHS FROM WASHINGTON IRVING TO HENRY JAMES, Herbert Cahoon, et al. Letters, poems, manuscripts of Hawthorne, Thoreau, Twain, Alcott, Whitman, 67 other prominent American authors. Reproductions, full transcripts and commentary. Plus checklist of all American Literary Autographs in The Pierpont Morgan Library. Printed on exceptionally high-quality paper. 136 illustrations. 212pp. 9⅛ x 12¼.
23548-3 Pa. $7.95

AMERICAN ANTIQUE FURNITURE, Edgar G. Miller, Jr. The basic coverage of all American furniture before 1840: chapters per item chronologically cover all types of furniture, with more than 2100 photos. Total of 1106pp. 7⅞ x 10¾.   21599-7, 21600-4 Pa., Two-vol. set $17.90

ILLUSTRATED GUIDE TO SHAKER FURNITURE, Robert Meader. Director, Shaker Museum, Old Chatham, presents up-to-date coverage of all furniture and appurtenances, with much on local styles not available elsewhere. 235 photos. 146pp. 9 x 12.   22819-3 Pa. $5.00

ORIENTAL RUGS, ANTIQUE AND MODERN, Walter A. Hawley. Persia, Turkey, Caucasus, Central Asia, China, other traditions. Best general survey of all aspects: styles and periods, manufacture, uses, symbols and their interpretation, and identification. 96 illustrations, 11 in color. 320pp. 6⅛ x 9¼.   22366-3 Pa. $6.95

CHINESE POTTERY AND PORCELAIN, R. L. Hobson. Detailed descriptions and analyses by former Keeper of the Department of Oriental Antiquities and Ethnography at the British Museum. Covers hundreds of pieces from primitive times to 1915. Still the standard text for most periods. 136 plates, 40 in full color. Total of 750pp. 5⅜ x 8½.
23253-0 Pa. $10.00

THE WARES OF THE MING DYNASTY, R. L. Hobson. Foremost scholar examines and illustrates many varieties of Ming (1368-1644). Famous blue and white, polychrome, lesser-known styles and shapes. 117 illustrations, 9 full color, of outstanding pieces. Total of 263pp. 6⅛ x 9¼. (Available in U.S. only)   23652-8 Pa. $6.00

Prices subject to change without notice.

Available at your book dealer or write for free catalogue to Dept. GI, Dover Publications, Inc., 180 Varick St., N.Y., N.Y. 10014. Dover publishes more than 175 books each year on science, elementary and advanced mathematics, biology, music, art, literary history, social sciences and other areas.